JN180561

数学 英和·和英辞典

増補版

MATHEMATICS ENGLISH-JAPANESE & JAPANESE-ENGLISH DICTIONARY

小松勇作 編

東京理科大学 数学教育研究所 増補版編集

共立出版

編 著 者

小 松 勇 作　東京工業大学名誉教授・理学博士
浅 野 重 初　拓 殖 大 学 教 授・理学博士
窪 田 佳 尚　東京学芸大学教授・理学博士
奥 村 正 文　埼 玉 大 学 教 授・理学博士
酒 井 　 良　東京都立大学教授・理学博士

増補版編集

東京理科大学数学教育研究所
増補版編集委員会

まえがき

本書は数学用語に関する辞典であり，前半は英和に，後半は和英にあてられている．

数学用語については，これまで簡便な用語集のほか，かなりの数の辞典ないしは事典が刊行され，それぞれに役割を果してきている．しかし，これらのほとんどは用語の解説を主眼とし，副次的に英和・和英の対訳を兼ねたものである．

このところ，数学はそれ自体として進歩をつづけるとともに，その応用を通して広く社会に浸透しつつある．いろいろな分野の著書や論文に，数多くの数学用語が現れる時代である．数学の全域にわたる用語を便利な形に収録したものがあれば，多くの人たちにとって有益であろうと考えられる．本書はこのような事情に鑑みて企画されたものである．

本書は，ふつうの英和辞典・和英辞典に準じた形態で，数学に関連する用語を網羅的に収録したものである．すなわち，数学における術語を核とし，類縁分野の用語を含め，なるべく広くなるべく多くの語彙を集めて，その対訳を示している．収録した範囲は，初等数学の基本的な用語から研究段階の術語にまでわたり，統計学やコンピュータ部門はもちろん，ひろく数理的な科学に及んでいる．また，事項のほかかなりの数の数学者名も載せてある．

内容を豊富にするために，各項目ごとに関連する用例を配した．それによって，熟語辞典としての性格も具えている．他方で，使用上の便宜をはかるために，合成語や用例については，重複をいとわずそれぞれの該当箇所にあげてある．また，関連項目間の引用の便をはかり，相互参照を示してある．巻末には，有用と思われる付録を副えてある．

以上の主旨で編まれた本書が，多くの人たちに活用されることを期待したい．もちろん，日進月歩の科学界にあって，今後とも内容の改善充実に絶えず努力すべきことはいうまでもない．必要に応じて，増補ないしは改訂を重ね，良い辞典に育てあげるつもりである．

本書を編むにあたって，項目選出の段階で隅田信子さんに，企画から校正までの全期間にわたって共立出版株式会社の小山　透氏ほか多くの方々に，多大の御協力をいただいた．ここに記して感謝の意を表したい．

1979 年 6 月

小 松 勇 作

増補にあたって

英語で書かれた数学書を読む際に，数学用語の標準的な日本語訳は何なのか知りたくなることがよくあります．そんなとき，数学の専門語はふつうの辞書にはほとんど載っていませんが，本書のような数学用語の英和辞典は大いに助けになると思います．一方で，将来的に英語の数学書を読みこなすために，和書を読みながら日本語の数学用語の標準的な英訳に徐々に親しむことや，英文で論文を執筆するときの参考として，本書の和英の部は役に立つと思います．読者の実際の利用方法はさまざまであると思いますが，刊行されてから 40 年近くの間に非常に多くの版を重ねていることは，本書のような辞典を求める方が数多いことを示しています．しかし，さすがに数学用語といえども，最近では使われなくなった術語や標準的な訳が変化したものなどもあり，数学の新理論で登場してきた用語も多くあります．また，コンピュータ関係の発展に伴い，グラフ理論をはじめ情報理論に関連する数学の用語も変化してきました．そこで，共立出版の提案で本書を改訂増補することになりました。編集に当たっては必要な新語を十分取り入れるとともに，既存掲載項目を全面的にチェックして訳や用例をアップデートしました．また，論文や学術書に特有の言い回しに現れる単語なども多少補いましたが，もはや不要となった項目は削除しました．

編集委員一同は，この増補版が旧版の改善充実となっていることを信じ，いっそう読者に活用していただけることを願っています．

2016 年 1 月

東京理科大学数学教育研究所
増補版編集委員
池 田 文 男
眞 田 克 典
新 妻 弘
宮 岡 悦 良
宮 島 静 雄
矢 部 博
（五十音順）

凡　　例

英和の部

項目の配列や体裁など，一般的にはふつうの英和辞典に準じる．

見出し語

1. 見出し語は太字の立体フォントで示し，アルファベット順に配列する．

2. 接頭辞および接尾辞を見出し語とする．例: **anti-, bi-; -able, -ply.**

3. 略字も見出し語とする．例: **a.e., i.e.; OR; fn., func.**

4. 人名は姓だけを大(太)文字とし，第二文字以下をスモール・キャピタルとする．

品詞と語形変化

5. 品詞は見出し語ごとに，略字や人名を除き，ふつうの辞典にみられる省略形で斜体活字で付記してある．(次ページの略語表参照)

6. 同一語がいくつかの品詞として用いられる場合には，それぞれの品詞ごとに併記する．

7. 名詞の不規則な複数形については，その見出し語の直後に示してある．例: **index** (*pl.* -dexes, -dices); **hypothesis** (*pl.* -ses).

語義

8. 語義を列挙するさいに，同義のいいかえ程度のものに限りコンマで区切り，一般にはセミコロンで区切る．例: **basic** *a.* 基礎的な，基本的な; 基底の.

9. 同一語の異なる品詞の間はピリオドで区切る．

10. 略字ないしは略記号の見出し語には，由来する本来の綴りを示す．例: **pt.** point; **sup** ＝supremum; **COBOL** COmmon Business Oriented Language.

11. 外来の見出し語に対しては，***F.***（フランス語），***G.***（ドイツ語），***It.***（イタリア語），***L.***（ラテン語）を括弧をつけて示し，同義の英語があるときはそれを付記する．例: **faisceau** *n.* (***F.*** ＝sheaf.); **Schmiegungsfunktion** *n.* (***G.*** ＝proximity function); **id est** (***L.*** ＝that is.)

12. 英語と米語の相違の著しいものは，区別を付記する．

13. 人名の見出し語には生歿年を付記し，ほぼ慣用のカタカナ読みを添える．

14. 同義語または反義語の参照は，矢印⇨をもって示す．例: **flow** ⇨ stream; **increase** ⇨ decrease; **hyper-** ⇨ hypo.

用例と記法

15. 見出し語の大部分には，引き続き用例としてそれを含む熟語を添えてある．

16. 各用例に含まれる見出し語の部分は記号～で代用し，用例は全体としてアルファベット順に配列する．例: 見出し語 **acceleration** の用例　～ vector は acceleration vector，angular ～ は angular acceleration を表す．

17. 用例の多くは，それを構成する単語を見出し語とする箇所に重複してあげてある．例: 見出し語 **almost** の用例として　～ everywhere，見出し語 **everywhere** の用例として　almost ～．

18. 用例(ならびに見出し語)において，その語義で省略してもよい文字または語句は記号(　)で示す．例: 見出し語 **algebraic** の用例　～ complex の語義が '代数(的)複体' とあるのは，'代数的複体' あるいは '代数複体' のいずれでもよい意．また，見出し語 **star(-)like** は star-like または starlike の意; 見出し語 **simply** の用例にある ～(-)connected は simply-connected または simply connected の意．

19. 用例ならびに見出し語の語義において，直前の字句と可換な字句は記号 [　] で示す．例: **factorization** の語義が '因数[因子]分解' とあるのは，'因数分解' あるいは '因子分解' の意．

20. 用例において，スペースを節約するために，同類のものを併記するさいの「それぞれ」の意味にも記号 [　] を用いる．例: 見出し語 **differential** の用例のうちに 'holomorphic [meromorphic] ～ 正則[有理型]微分' とあるのは，'holomorphic ～ 正則微分' と 'meromorphic ～ 有理型微分' を併記したもの．

21. 見出し語にはなっていないが本辞典記載の熟語(フレーズ)による同義の言い換えを，等号 = をもって表す．

略　語　表

a.	adjective（形容詞）	*prep.*	preposition（前置詞）
ad.	adverb（副詞）	*pron.*	pronoun（代名詞）
conj.	conjunction（接続詞）	*rel. pron.*	relative pronoun（関係代名詞）
def. art.	definite article（定冠詞）	*sing.*	singular（単数形）
indef. art.	indefinite article（不定冠詞）	*suf.*	suffix（接尾辞）
n.	noun（名詞）	*v.*	verb（動詞）
pl.	plural（複数形）	*vi.*	intransitive verb（自動詞）
pref.	prefix（接頭辞）	*vt.*	transitive verb（他動詞）

注: 分類 "*a*" は現在分詞の形容詞的用法も含む．

和英の部

項目の配列や体裁など，一般的にはふつうの和英辞典に準じる.

見出し語

1. 見出し語(読み)はゴシック体フォントで示し，原則として五十音順に配列し，見出し語が該当する術語表記や日本語表記を添える.

2. 外来語が仮名書きのまま見出し語となる場合には，片仮名をもって五十音順の該当箇所に配置する．ただし，長音記号 ‘ー’ は無視して配置するものとする.

3. 同名異義の見出し語は別項目として併記する．例: **たい**　体　body; solid / **たい**　体　field; (*G.*) Körper.「**しゃこう**」の見出しには三項目あり，斜高／斜交／斜航.

4. 接頭辞および接尾辞も見出し語とする．例: **へん-**　偏，**-ばい**　倍.

5. 人名はその形容詞が慣用されるものを見出し語とする．広範囲の人名については，英和の部を参照されたい.

訳語

6. 見出し語はなるべく名詞の形であげ，その形容詞，副詞ないしは動詞の形は引き続く用例で示す．訳語を列挙するさいには，原則としてセミコロンで区切る.

7. 名詞の不規則な複数形については，該当する訳語の直後に示す．例: **えんかん**　円環　annulus (*pl.* -li, -luses); circular ring.

8. 外来語の訳語に対しては，*F.*, *G.*, *L.* などを括弧をつけて示す．例: **アデール**　(*F.*) adèle / **たとえば**　例えば　for instance; for example; (*L.*) exempli gratia; e.g.

9. 同義語または反義語の参照は，矢印⇨をもって示す．例: **ない-**　内　internal; interior; intra-; inner.　⇨内(うち); 外(がい-).

用例と記法

10. 見出し語の大部分には，引き続き用例としてそれを含む熟語を添えてある.

11. 各用例に含まれる見出し語の部分（括弧内を除く）は記号～で代用して配列する．例: 見出し語 ‘**ぼすう**　母数　modulus (*pl.* -li)’ の用例 ‘～の’ は ‘母数の’, ‘周期～’ は ‘周期母数’ を表す．また，見出し語 ‘**せいじゅん**　正準(の)　canonical’ の用例 ‘～方程式’ は ‘正準方程式’ を表す.

12. 用例の多くはそれを構成する単語を見出し語とする箇所に重複してあげてある．例: 見出し語 ‘**きゅうめん**　球面’ の用例として ‘～三角形’, 見出し語 ‘**さんかくけい**　三角形’ の用例として ‘球面～’ がある．別に独立にも ‘**きゅうめんさんかくけい**　球面三角形’ がある.

13. 行末のハイフンが単なる切れ目を示す場合は，ふつうのルールにしたがう．例: 行末が func- となり，次行が tion と続く場合は，行末で切れなければ function となる．

14. 本来ハイフンをもつ合成語のハイフンが行末にある場合には，次行のはじめにハイフンを反復する．例: 行末が skew- となり，次行が -symmetric と続く場合は，行末で切れなければ skew-symmetric となる．

15. 見出し語に添えた術語表記や日本語表記が用例に直結するように，その格を表す語尾を括弧（ ）に入れたものがある．引き続く訳語は括弧内を含めている．例: ‘**だいいち** 第一(の) first; primary’ における訳語は ‘第一の’ に対するものであり，その用例 ‘～象限’ は ‘第一象限’ を表す．

16. 省略してもよい文字または語句は記号（ ）で示す．例: ‘**ごうけい** 合計’ の訳語のうち ‘total (amount)’ とあるのは，‘total amount’ または ‘total’ を表す．‘**ごかん** 互換(性)’ とあるのは，‘互換’ または ‘互換性’ を表し，その用例 ‘～的な’ は ‘互換的な’ を表す．また，‘**ごさ** 誤差’ の用例で ‘～(の)伝播’ とあるのは，‘誤差伝播’ または ‘誤差の伝播’ を表す．‘**ほしがた-** 星型’ の訳語で ‘star(-)like’ とあるのは，‘star-like’ または ‘starlike’ を表し，‘**たんれんけつ** 単連結(性)’ の用例の訳語で ‘simply(-)connected’ とあるのは，‘simply-connected’ または ‘simply connected’ を表す．

17. 直前の字句と可換な字句は記号[ ]で示す．例: ‘**せつぞく** 接続’ の用例のうち ‘解析～’ の訳語に ‘analytic continuation [prolongation]’ とあるのは，‘analytic continuation’ または ‘analytic prolongation’ を表す．‘**どくりつ** 独立(性)’ の用例のうち ‘一次[線形]～な’ とあるのは，‘一次独立な’ または ‘線形独立な’ を表す．

18. 同類の用例を併記するさいの「それぞれ」の意味にも記号 [ ] を用いる．例: ‘**ベクトル** vector’ の用例のうち ‘接[法]～ tangent [normal] vector’ とあるのは，‘接ベクトル tangent vector’ と ‘法ベクトル normal vector’ を表す．

I

英和の部

ENGLISH-JAPANESE

A

a *indef. art.* ある一つの; 一つの.【発音が母音で始まる語の前では an】.

a-, an- *pref.*「非, 無 (non-, without)」の意: acyclic, anharmonic, asymmetry.

a. a. almost all. 殆んどすべての.

ab- *pref.*「離脱」の意: abstract, abuse.

abacus (*pl.* -cuses, -ci) *n.* そろばん(算盤).
~ school 算盤派.

abbreviate *vt.* 簡約する; 約する.
~d notation 簡略化(した)記法[号].

abbreviation *n.* 簡約; 約分; 省略形.

ABEL, Niels Henrik (1802-1829) アーベル.

Abelian, abelian *a.* アーベル的な, (群が)可換な.
~ category アーベル圏. ~ differential アーベル微分. ~ extension アーベル拡大. ~ function field アーベル関数体. ~group アーベル群, 可換群. ~ integral アーベル積分. ~ variety アーベル多様体.

aberration *n.* 収差, 光行差.
diurnal ~ 日周光行差.

-able *suf.*「…できる」,「…に適する」の意: countable. ⇨ -ible.

abnormal *a.* 異常な, 変則の; 不正規-.
~ series 不正規級数.

about *prep.* について; のまわりに. *ad.* およそ.
~ ten meters 約十メートル. circle ~ C C のまわりの[を中心とする]円. circle circumscribed ~ triangle 三角形に外接する円. textbook ~ calculus 微積分に関するテキスト.

above *prep., ad.* ~より上に; 上に, 以上【≧】. *n.* 上方. ⇨ below.
bounded (from) ~ 上に有界な.

abridge *vt.* 省略する; 要約する.
~d notation 略記法.

abscissa (*pl.* -sas, -sae) [æbsísə] *n.* 横座標; 横軸. ⇨ ordinate.
~ of boundedness 有界(横)座標. ~ of convergence 収束(横)座標. ~ of regularity 正則(横)座標.

abridg(e)ment *n.* 省略, 摘要.

absolute *a.* 絶対の, 絶対-. *n.* 絶対形.
~ class field 絶対類体. ~ constant 絶対定数. ~ convergence 絶対収束. ~ curvature 絶対曲率. ~ differential (calculus) 絶対微分(学). ~ error 絶対誤差. ~ geometry 絶対幾何学. ~ inequality 絶対不等式. ~ invariant 絶対不変式. ~ moment 絶対積率[能率]. ~ norm 絶対ノルム. ~ number 無名数. ~ parallelism 遠隔平行性. ~ summability 絶対総和可能性. ~ value 絶対値. ~ velocity 絶対速度.

absolutely *ad.* 絶対的に, 無条件に.
~ continuous 絶対連続な. ~ convergent 絶対収束の. ~ convex 円形凸な. ~ integrable 絶対可積な, 絶対積分可能な. ~ irreducible character 絶対既約指標. ~ stable 絶対安定な.

absoluteness *n.* 絶対(性); 完全(性).

absorb *vt.* 吸収する.

absorbant *a.* (*F.*) 吸収的な.

absorbing *a.* 吸収的な. ⇨ reflecting.
~ barrier 吸収壁. ~ chain 吸収連鎖. ~ state 吸収状態.

absorption *n.* 吸収.
~ cross section 吸収断面(積). ~ law 吸収法則[律].

absorptive *a.* 吸収する, 吸収的な.

〜 barrier 吸収壁.

abstract *a.* 抽象的な. *n.* 抜粋, 摘要, 要約; 抽象. *vt.* 抽象する; 要約する.
〜 algebra 抽象代数(学). 〜 algebraic variety 抽象的代数多様体. 〜 group 抽象群. 〜 integral 抽象積分. 〜 number 無名数, 不名数. 〜 space 抽象空間.

abstraction *n.* 抽象, 抽象化.

abstractly *ad.* 抽象的に.

abstractness *n.* 抽象的なこと, 抽象性.

absurd *a.* 不合理な, 不条理な, 道理に合わない.

absurdity *n.* 不合理, 不条理.
reduction to 〜 背理法.

Abul Wefa, Abû 'l-Wafâ' (940–998) アブル・ウェファ.

abundance *n.* 豊富.

abundant *a.* 豊富な.
〜 number 過剰数.

a.c.c. ascending chain condition. 昇鎖条件.

accelerate *vt., vi.* 加速する.
〜d failure model 加速モデル, 加速故障モデル.

acceleration *n.* 加速度.
〜 vector 加速度ベクトル. angular 〜 角加速度. constant [uniform] 〜 等加速度.

accept *vt.* (仮説を)採択する.

acceptable *a.* 受け入れ可能な, 容認できる; 採択できる.
〜 quality level 合格品質水準[AQL].

acceptance *n.* 受入, 採択, 合格.
〜 inspection 受入検査. 〜 line 合格線. 〜 number 採択(個)数, 合格判定個数. 〜 region 採択域.

accessibility *n.* 到達可能性.

accessible *a.* 到達可能な.
〜 boundary point 到達可能境界点. 〜 space 追接空間.

accessory *a.* 付随の, 副次的な.
〜 condition 付帯条件. 〜 point 副点.

accidental *a.* 偶然な.
〜 error 偶然誤差.

accomplish *vt.* (約束, 義務などを)果たす; 成し遂げる.

accordingly *adv.* よって, それゆえ; それに応じて.

account *n.* 会計, 計算; 理由.
〜 machine 会計機. cast 〜s 計算する. on 〜 of のために, という理由で. take into 〜 を考慮する.

accounting *n.* 決算. cost 〜 原価計算.

accumulate *vt., vi.* 集積する, 累積する.
〜d error 累積誤差.

accumulation *n.* 集積, 累積.
〜 by annual instal(l)ments 年金積立. 〜 point 集積点. 〜 value 集積値.

accuracy *n.* 精密さ, 精度, 正確さ.

accurate *a.* 正確な, 精密な.

accurately *ad.* 正確に, 精密に.

achieve *vt.* (仕事, 目的を)達成する, 実現する; 勝ち取る.

Achilles アキレス【Homer 作 Iliad に現れるギリシアの勇士】.
〜 and tortoise アキレスと亀.

achromatic *a.* 色消しの; 無色の.
〜 number 消色数.

acnode *n.* (グラフの)孤立点.

acoustics *n.* 音響学.

across *prep., ad.* 横切って, 向う側に; 交差して; 差し渡しで.
circle 1cm 〜 直径1センチメートルの円.

act *n.* 行動. *vi.* 作用する.
〜 effectively 効果的に作用する.

action *n.* 作用, 処置.
〜 integral 作用積分. 〜 of group 群の作用. 〜 space 行動空間. least 〜 最小作用. reductive 〜 簡約可能(な)作用.

active *a.* 能動的な; 活動的な.
〜 constraint 有効制約.

activity *n.* 活動; 作業; 活動力, 積極性.
〜 analysis 活動分析.

actual *a.* 現実の, 実際の.
〜 line 実線. 〜 measurement 実測.

actuary *n.* 保険数理士【保険統計専門家】.

acute *a.* 鋭角の, 鋭い.
〜 angle 鋭角. 〜-angled 鋭角の. 〜 triangle 鋭角三角形.

acyclic *a.* 非輪状の, 無輪状の, 非循環の.
〜 complex 非輪状複体.

A.D. Anno Domini. (*L.* =in the year of our Lord.) 西暦年数.
~1979, 1979 ~ 西暦 1979 年.

ad- *pref.*「方向, 変化, 添加, 増加」などの意: adhere, advance.

ad absurdum (*L.* =to absurdity.) reductio ~, ~ 背理法.

adapt *vt.* 適合[応]させる, 応用する.

adaptive *a.* 適合する, 適応的な.
~ control 適応制御. ~ design 適応的デザイン.

add *vt.* 加える, 加算する.
~ up to 総計…になる.

addend *n.* 加数[量].

addendum (*pl.* -da) *n.* 補遺; 付録.

addition *n.* 寄せ算, 足し算, 加え算, 加法.
~ formula 加法公式. ~ table 加法九々. ~ theorem 加法定理. approximate ~ 省略加法.

additional *a.* 追加の, 付加の.
~ condition 追加された条件. ~ sample 追加の資料.

additive *a.* 加法の; 加法的な.
~ category 加法圏, 加法的カテゴリ. ~ class 加法族. ~ constant 付加定数. ~ family 加法族. ~ function 加法的関数. ~ functional 加法的汎関数. ~ functor 加法的関手. ~ group 加法群, 加群. ~ model 加法モデル. ~ number theory 加法的整数論. ~ operator 加法的作用素. ~ process 加法過程. ~ valuation 加法付値. almost ~ 概加法的な. completely ~ 完全加法的な. countably ~ 可算加法的な. finitely ~ 有限加法的な. totally ~ 完全加法的な.

additivity *n.* 加法性.
~ of area [probability] 面積[確率]の加法性. complete ~ 完全加法性.

address *n.* アドレス; 講演. *vt.* 取り組む; 扱う.

adèle *n.* (*F.*) アデール.
~ group アデール群. ~ ring アデール環. principal ~ 主アデール.

adhérence *n.* (*F.*) 触集合.

ad hoc (*L.*) for this (special purpose).

ADI Alternating Direction Iteration. 交互方向法.

-adic *suf.*「-進の, -進法の, …を法とする」の意.
$\mathfrak{a}$~ $\mathfrak{a}$ 進の. $\mathfrak{a}$~ completion $\mathfrak{a}$ 進完備化. $\mathfrak{a}$~ topology $\mathfrak{a}$ 進位相. p~ integer p 進整数. p~ number field p 進(数)体.

ad infinitum (*L.* =to infinity.) 無限に.

adjacency *n.* 隣接, 近隣, 接近.
~ matrix 隣接行列.

adjacent *a.* 接近した, 隣接した.
~ angle 接角, 隣接角. ~ part 隣接部分. ~ side 隣辺.

adjoin *vt., vi.* 付加する, 添加する.
~ an element to a field 体にある元を添加する. ~ed element 添加された元.

adjoining *a.* 隣接の. ⇨ adjacent.
~ side 隣辺.

adjoint *n.* 随伴式, 同伴式. *a.* 随伴の, 同伴の; 補助の.
~ determinant 随伴行列式. ~ differential equation 随伴微分方程式. ~ equation 随伴方程式, 補助方程式. ~ functor 随伴関手. ~ group 随伴群. ~ kernel 随伴核. ~ line 補助線. ~ matrix 随伴行列, 共役転置行列. ~ operator 随伴作用素, 共役作用素. ~ representation 随伴表現. ~ transformation 随伴変換. left [right] ~ 左[右]随伴の.

adjunction *n.* 添加, 付加.
algebraic ~ 代数的(な元の)添加.

adjust *vt., vi.* 調整する.
~ed annual worth 年金換算値.

adjustment *n.* 調整.
fine ~ 微調整. sampling with ~ 調整型抜取(り).

admissibility *n.* 許容性.

admissible *a.* 許容される, 許容の.
~ control 可容制御. ~ function 許容関数. ~ ideal 許容イデアル. ~ isomorphism 許容同形[型]. ~ sequence 許容列. ~ series 許容列. ~ subgroup 許容部分群.

admit *vt., vi.* 許容する, 容認する.

admittance *n.* アドミッタンス.
~ matrix アドミッタンス行列.

advance *n.* 前払; 進行.
~ estimate 事前推定値. in ~ 前もって; 前払で.

advanced *a.* 前進した, 進んだ; 高等な.
~ algebra [calculus] 高等代数[微積分]学. ~ type 進み型.

adverse *n.* 副逆元.

a. e. almost everywhere. 殆んどいたるところ.

affect *vt.* 影響する, 影響を及ぼす, 作用する.

affine *a.* アフィン的な, アファイン的な.
~ algebraic group アフィン代数群. ~ arc length アフィン(的)弧長. ~ bundle アフィン束. ~ collineation アフィン共線変換. ~ connection アフィン接続. ~ coordinates アフィン座標. ~ correspondence アフィン対応. ~ frame アフィン枠, アフィン標構. ~ geometry アフィン幾何学. ~ group アフィン群. ~ isomorphism アフィン同形[型]. ~ mapping アフィン写像. ~ normal アフィン法線. ~ property アフィン的性質. ~ ring アフィン環. ~ scheme アフィンスキーム. ~ space アフィン空間. ~ transformation アフィン変換. ~ (algebraic) variety アフィン(代数)多様体.

affinely *ad.* アフィン的に, アファイン的に.
~ congruent アフィン合同な.

affinity *n.* 親近感; 親近性; アフィン変換. =affine transformation.

affinor *n.* アフィノール【アフィン幾何学でベクトル, テンソルの総称】.

affirm *vt.* 肯定する.

affirmation *n.* 肯定.

affirmative *a.* 肯定的な. *n.* 肯定(命題).
~ proposition [sentence] 肯定命題[文].

affirmatively *ad.* 肯定的に.

affix *n.* 付随値.

afflux *n.* 流入. ⇨ efflux.

a fortiori *ad.* (*L.*) いっそう強い理由で, なおさら.

against *prep.* に向けて, に対して; に対比して.
concave [convex] ~ pole 極に(向けて)凹[凸]な. two ~ three 二対三.

aggregate *vt., vi.* 集合する. *a.* 集合した; 合計の, 合併の. *n.* 集合.

aggregation *n.* 集合; 集約単位; 括弧の総称.

Agnesi, Maria Gaetana (1718-1799) アニェージ, アーネシ, アグネシ.
witch of ~, ~'s witch アニェージの魔女.

Ahlfors, Lars Valerian (1907-1996) アールフォルス.

AIC エイ・アイ・シー.
Akaike Information Criterion 赤池情報量基準.

aim *vi., vt.* ねらう; 目ざす. *n.* 目標, 目的.
~ed at precision 目標精度.

Airy, Sir George Bidell (1801-1892) エアリ.

Akaike, Hirotugu (1927-2009) 赤池弘次.
Akaike Information Criterion; AIC 赤池情報量基準.

al Battani (850?-929) アル・バッタニ.

aleph *n.* アレフ【ヘブライ語アルファベットの第1字 $\aleph$: 連続体濃度の記号】.
~ zero [null] アレフ・ゼロ【$\aleph_0$: 可算濃度の記号】.

Alexandrov, Pavel Sergeevič (1896-1982) アレクサンドロフ.

algebra *n.* 代数学, 代数; 多元環.
abstract ~ 抽象代数(学). advanced ~ 高等代数(学). basic ~ 基本多元環. ~ class 多元環類. ~ extension 多元環拡大. ~ homomorphism 多元環準同型[形]. ~ isomorphism 多元環同型[形]. ~ of logic 論理代数. Banach ~ バナッハ環. Banach *~ バナッハ*環. Boolean ~ ブール代数. C^*- ~ C^*環. computer ~ 計算機代数. division ~ 多元体. elementary ~ 初等代数(学). exterior ~ 外積多元環. group ~ 群(多元)環. Lie ~ リー代数[環]. linear ~ 線形代数(学). multiplier ~ 乗法子環. operator ~ 作用素代数[環]. path ~ 道多元環. slice ~ スライス環. W^*- ~ W^*環.

algebraic *a.* 代数学の, 代数的な, 代数-.

～ algebra 代数的多元環. ～ analysis 代数解析. ～ branch point 代数分岐点. ～ closure 代数的閉包. ～ complex 代数(的)複体. ～ correspondence 代数的対応. ～ curve 代数曲線, ～ element 代数的元. ～ equation 代数方程式. ～ expression 代数式. ～ extension 代数(的)拡大. ～ function 代数関数. ～ function field 代数関数体. ～ geometry 代数幾何学. ～ group 代数群. ～ integer 代数的整数. ～ Lie algebra 代数的リー代数. ～ method 代数的方法. ～ multiplicity 代数的重複度. ～ number 代数的数. ～ number field 代数的数体, 代数体. ～ scheme 代数的概型. ～ singularity 代数(的)特異点. ～solution 代数的解(法). ～ spiral 代数螺[渦]線. ～ structure 代数的構造. ～ subgroup 代数的部分群. ～ sum 代数和. ～ surface 代数曲面. ～ symbol 代数記号. ～ system 代数系. ～ theory of numbers 代数的整数論. ～ topology 代数的位相幾何学. ～ variety 代数(的)多様体.

algebraical ＝algebraic.

algebraically *ad.* 代数的に.
～ closed field 代数的に閉じた体, 代数的閉体. ～ dependent [independent] 代数的(に)従属[独立]な. ～ equivalent 代数的(に)同値な.

algebroid *a.* 代数型の.
～ function 代数型関数. integral ～ 整代数型の.

algebroidal ＝algebroid.

algorism ＝algorithm.

algoristic *a.* アルゴリズムの.
～ school 筆算派.

algorithm *n.* アラビア記数法; アルゴリズム【計算手続き, 算法】, 互除法.
composite ～ 複合(算)法. division ～ 連除法. Euclid's [Euclidean] ～ ユークリッドの算法[互除法]. polynomial time ～ 多項式時間アルゴリズム.

algorithmic *a.* アルゴリズムの.
～ language 算法言語.

AL-HAYYAM, Umar (1044-1123) アル-ハイヤーミー

AL-HWÂRAZMI ＝AL-KHWARIZMI.

alienation *n.* 離間.

align *vt.* 直線に並べる, 一列にそろえる. *vi.* 一列をなす.

alignment *n.* 一直線にする[なる]こと.
～ chart 共線図表. ＝collinear nomograph.

aline ＝align.

alinement ＝alignment.

aliquant *a.* 割切れない, 整除できない. *n.* 割切れない数, 非整除数.

aliquot *a.* 割切れる, 整除できる. *n.* 割切れる数, 約数
～ part 約数.

AL-KHWARIZMI, Muhammad ibn Mūsā (780?-850?) アル-クワリズミ, アル-フワリズミー.

all *a.* すべての, 全部の. *n.* 全部.
～ permutation 全順列. almost ～ 殆んどすべての.

allied *a.* 関連のある.
～ series 共役級数.

alligation *n.* 混合法.
～ alternate 和較法. ～ medial 混和法.

allocation *n.* 割当, 配当, 配分.
～ model 配分模型. multistage ～ 多段配分. optimal ～ 最適配当[割当; 配分].

allow *vt.* 許す, 許容する.
～ed homomorphism 許容準同型[形].

allowance *n.* 余裕; 許可.
safety ～ 安全余裕.

almost *ad.* 殆んど, 概-.
～ additive 概加法的な. ～ all 殆んどすべての[a.a.]. ～ certainly 殆んど確実に. ～ complex structure 概複素構造. ～ convergence 概収束. ～ effective 殆んど効果的な. ～ everywhere 殆んどいたるところ[a.e.]. ～ everywhere convergent 概収束の. ～ metric structure 概計量構造. ～ periodic function 概周期関数. ～ surely 殆んど確実に[a.s.]

alpha *n.* アルファ. ギリシア語アルファベットの1番目の文字【A, α】.

alphabet *n.* アルファベット, 字母.

alphabetical *a.* アルファベットの, アル

ファベット順の.
～ order アルファベット順.
alphabetically *ad.* アルファベット順に.
alter *vt.* 変える. *vi.* 変わる.
alternant *n.* 交代式; 交代行列.
alternate *a.* 交互の, たがいちがいの. *vi., vt.* 交互に並ぶ[現れる].
alligation ～ 和較法. ～ (interior) angles 錯角, ～ expression [function] 交代式. ～ matrix 交代行列.
alternately *ad.* 交互に.
alternating *a.* 交互の, 交代-.
～ derivation 交代微分. ～ direction iteration 交互方向法[ADI]. ～ expression 交代式. ～ form 交代形式. ～ function 交代式. ～ group 交代群. ～ knot 交代結び目. ～ matrix 交代行列. ～ operation 交代演算. ～ polynomial 交代(多項)式. ～ series 交代級数. ～ tensor 交代テンソル. elementary ～ expression 基本交代式. elementary ～ polynomial 基本交代(多項)式. simplest ～ polynomial [expression] 最簡交代式, 差積.
alternation *n.* 交替.
alternative *a.* どれか一つの; いずれか一方の. *n.* 代替, 代案; 選択肢; 二者択一.
～ algebra 交代代数. ～ approximation 交互近似. ～ field 交代体. ～ hypothesis 対立仮説. ～ proof 別証(明). theorem of the ～ 二者択一定理.
alternatively *ad.* 二者択一的に, 代用として.
alternendo 更迭の理.
alternizer *n.* 交代化作用素.
altitude *n.* (三角形などの)高さ, 高度; 標高.
Krull's ～ [height] theorem クルルの標高定理.
amalgamate *vt., vi.* 融合する. 混合する.
～d product 融合積.
amalgamation *n.* 融合.
ambig *a.* アンビグ, 特異-.
～ class 特異類, アンビグ類. ～ ideal 特異イデアル.
ambiguous *a.* あいまいな; 不確かな.
～ case 不確定な場合. ～ grammar あいまいな文法. ～ point 不確定点.
amicable *a.* 親和的な.
～ number 友数, 親和数.
among *prep.* の間に, の中に.
～-class 級間. ～-class variation 級間の変動.
amount *n.* 量, 総計; 元利合計. *vi.* (全体で)…になる【to】.
～ of annual instal(l)ments 年賦金. ～ of annuity 年金終価. ～ of discount 割引高. ～ of information 情報量. in ～ 総計して. small ～ 少量.
AMPÈRE, André-Marie (1775-1836) アンペール.
amphicheiral *a.* 両手型の.
ample *a.* 豊かな, 豊富な
～ divisor 豊富な因子. ～ sheaf 豊富層.
amplitude *n.* (図形の)幅. (複素数の)偏角; (振動の)振幅.
～ function 振幅関数. probability ～ 確率振幅.
an- ⇨ a-.
anagram *n.* 字謎.
anagrammatic *a.* 字謎(的)の; 自反転の.
anallagmatic *a.* 自反転の.
～ surface 自反転曲面.
analog *n.* アナログ, 相似.
～ data アナログ・データ.
analogous *a.* 類似の.
analogue *n.* 類似物. *a.* 類似の.
analogy *n.* 類比, 等比, 類似.
analysis (*pl.* -ses) *n.* 解析; 分析; 解析学.
activity ～ 活動分析. algebraic(al) ～ 代数解析. ～ of covariance 共分散分析. ～ of variance 分散分析. ～ situs 位相幾何. complex ～ 複素解析. convex ～ 凸解析. interval ～ 区間解析. mathematical ～ 数学解析(学). numerical ～ 数値解析. real ～ 実解析. sensitivity ～ 感度解析. stability ～ 安定性解析.
analytic *a.* 解析の, 解析的な. ⇨ analytical.
～ bundle 解析的ファイバー束. ～ capacity 解析(的)容量. ～ center 解析的中心.

~ completion 解析的完備化. ~ continuation 解析接続. ~ curve 解析曲線. ~ differential 解析的微分. ~ dynamics 解析力学. ~ function 解析関数. ~ geometry 解析幾何学. ~ manifold 解析多様体. ~ mapping 解析写像. ~ neighbo(u)rhood 解析的近傍. ~ polyhedron 解析的多面体. ~ prolongation 解析接続. ~ set 解析(的)集合. ~ space 解析空間. ~ structure 解析構造. ~ subspace 解析的部分空間. ~ theory of numbers 解析(的)整数論. complex [real] ~ 複素[実]解析的な. real ~ function 実解析関数.

analytical ＝analytic.
~ dynamics 解析力学. ~ geometry 解析幾何学. ~ mechanics 解析力学. ~ theory 解析的理論.

analytically *ad.* 解析的に.
~ complete 解析的完備な. ~ continuable 解析接続可能な. ~ dependent [independent] 解析的(に)従属[独立]な. ~ normal 解析的正規な. ~ thin 解析的に細い. ~ unramified 解析的非分岐の.

analyticity *n.* 解析性.
real ~ 実解析性.

analyze *vt.* 解析する.

analyzer *n.* 解析機.
differential ~ 微分解析機. harmonic ~ 調和解析機.

Anaxagoras of Klazomenai (500–428 B.C.) クラゾメナイのアナクサゴラス.

Anaximandros (610–546 B.C.) アナクシマンドロス.

anchor *n.* 錨.
~ ring 輪環面, 円環面, トーラス; ドーナツ形. ＝torus.

ancient *a.* 古代の.
~ mathematics 古代(の)数学.

ancillary *a.* 補助の.
~ statistic 補助統計量.

and *conj.* および. *n.* 論理積【記号で P and Q は, $P \cdot Q$, PQ, $P \wedge Q$ などと表される】.

and/or …および…ないしはそれらの一方【非排他的論理和】.

angle *n.* 角; 角度.
acute ~ 鋭角. alternate (interior) ~ 錯角. ~ of circumference 円周角. ~ of depression 俯[伏]角. ~ of elevation 仰角. ~ of incidence 入射角. ~ of intersection 交角. ~ of parallelism (非ユークリッド幾何の)平行角. ~ of segment 弓形の角. ~-side table 角辺表. base [basic] ~ 底角. central ~ 中心角. complementary ~ 余角. corresponding ~ 同位角. direction ~ 方向角. eccentric ~ 離心角. external ~ 外角. included ~ 夾角. obtuse ~ 鈍角. phase ~ 位相角. right ~ 直角. round ~ 周角. side-~ table 辺角表. solid ~ 立体角. straight ~ 平角. supplementary ~ 補角. vertical ~ 頂角. vertical ~(s) 対頂角. vertically opposite ~(s) 対頂角. visual ~ 視角.

angular *a.* 角の, 角-.
~ derivative 角微係数. ~ distance 角距離. ~ domain 角領域. ~ frequency 角振動数. ~ limit 角(内からの)極限. ~ measure 角測度. ~ momentum 角運動量. ~ perspective 有角透視図. ~ transformation 角変換. ~ velocity 角速度.

anharmonic *a.* 非調和的な.
~ ratio 非調和比, 複比, 十字比.

anisotropic *a.* 非等方的な.

anisotropy *n.* 非等方性.

annals (*pl.*) *n.* 年報.

annihilate *vt.* 零にする, 零化する, 消す.

annihilation *n.* 零化.
~ operator 消滅作用素, 零化子.

annihilator [ənáiəlèitər] *n.* 零化元, 零化群, 零化イデアル; 零化域.
~ ideal 零化イデアル. ~ set 零化集合.

annual *a.* 一年の; 毎年の.
~ interest 年利. ~ payment 年賦償還.

annuity *n.* 年金.
amount of ~ 年金終価. ~ certain 確定年金. life ~ 終身年金. perpetual ~ 永続年金. terminable ~ 定期年金.

annular *a.* 環状の; 輪状の; 円環の.
~ domain 円環領域. ~ surface 管状曲面.

annulus (*pl.* -li, -luses) *n.* 環形; (同心)円環; 球殻.

concentric ～ 同心円環.

anode *n.* 陽極.

anomalous *a.* 異常な，不規則な.
～ threshold 異常閾(しきい).

anomaly *n.* 近点離角; 例外; 特異性.
eccentric ～ 離心近点離角. true ～ 真近点離角.

another *a.* 別の，もう一つの.
～ proof 別証明.

ANSCII American National Standard Code for Information Interchange. 情報変換用標準コード.

answer *n.* 答，解答. *vt., vi.* 答える.

antecedent *a.* 前の; 先行の. *n.* 前項，前率; 先行(者)，先行元; 仮定，前件.

anterior *a.* 前の; 事前の. ⇨ posterior.

anti- *pref.* 「反対」の意. 逆-，反-.

anti-automorphism *n.* 反[逆]自己同型[形].

anti-clockwise *a., ad.* 左回りの[に]，時計の針と反対方向の[に].

anti-commute *vt.* 反可換である.

anti-conformal *a.* 逆等角な.
～ mapping 逆等角写像.

anti-derivative *n.* 不定積分.

anti-endomorpism *n.* 反[逆]自己準同型[形].

anti-equivalence *n.* 反[逆]同値.

anti-Hermitian *a.* 反[逆]エルミート的な.
～ matrix 反[逆]エルミート行列.

anti-homomorphism *n.* 反[逆]準同型[形].

anti-isomorphism *n.* 反[逆]同型[形].

anti-logarithm *n.* 真数.

antinomy *n.* 不合理，矛盾; 二律背反.

anti-parallel *a.* 逆平行な.

anti-parallelism *n.* 逆平行性.

antipodal *a.* 対蹠的な，対心-.
～ point 対心点，対蹠点.

antipode *n.* 対蹠的なもの; 対心点，対蹠点.

anti-regular *a.* 反[逆]正則な.
～ transformation 反正則変換.

anti-symmetric *a.* 反[逆]対称な，歪対称な.
～ matrix 反対称行列. ～ tensor 反対称テンソル.

anti-symmetry *n.* 反[逆]対称.

any *a.* 任意な; どの…でも.

Anzahlfunktion *n.* (*G.* ＝counting function.) 個数関数.

A.P. Arithmetical Progression. 等差数列.

aperiodic *a.* 非周期的な.

apex (*pl.* -xes, apices) *n.* 頂点，最高点; 向点.

apolar *a.* 無極の.

apolarity *n.* 無極性.
～ theorem 無極定理.

Apollonian *a.* アポロニウスの.
～ [Apollonius'] circle アポロニウスの円.

Apollonius of Perga (262?-190? B.C.) ペルガのアポロニウス.

a posteriori *a., ad.* (*L.*) 後の[に]，後天的な[に]，帰納的な[に]. ⇨ a priori.
～ probability 事後確率. ～ risk 事後危険.

apothem *n.* 辺心距離【正多角形の中心から辺までの距離】.

apparent *a.* 見掛けの，明白な.
～ degree 見掛けの次数. ～ singular point 見掛けの特異点.

appear *vi.* 現れる.

Appell, Paul Emile (1855-1930) アッペル.

appendix (*pl.* -dixes, -dices) *n.* 付録，補遺.

applicable *a.* 適用[応用]できる; 展開可能な.
～ surface 可展面，展開可能曲面.

application *n.* 適用，応用; 展開; 写像.
～ problem [question] 応用問題.

applied *a.* 応用の，実用の.
～ function(s) 実用関数. ～ mathematics 応用数学.

apply *vt., vi.* 適用する，応用する; 適合する.

approach *vt., vi.* 近づく. *n.* 接近; 処理法; 取り組み方，研究方法.

approximate *vt., vi.* 近似する. *a.* 近似的な，省略の.
～ addition 省略加法. ～ algebra 省略代数. ～ construction 近似作図. ～ deriva-

tive 概微分係数. ~ expression 近似式. ~ functional equation 近似関数等式. ~ polynomial 近似多項式. ~ solution 近似解. ~ value 近似値. lower [upper] ~ value 過小[大]近似値.

approximately *ad.* 近似的に; お(お)よそ, ほぼ; 概-.
~ analytic 概解析的な. ~ continuous 概連続な. ~ derivable 概微分可能な. ~ equal ほぼ等しい.

approximation *n.* 近似, 近似値, 近似法.
~ in deficiency 不足近似. ~ in excess 過剰近似. ~ theorem 近似定理. best (-fitted) ~ 最良近似. cellular ~ 胞体近似. difference ~ 差分近似. first ~ 第一近似. least square ~ 最小二[自]乗近似. linear ~ 線形近似. polynomial ~ 多項式近似. successive ~ 逐次近似.

a priori *a., ad.* (*L.*) 先天[験]的な[に]; 演繹的な[に]. ⇨ a posteriori.
~ estimate アプリオリ評価(式). ~ probability 事前[先験]確率.

AQL Acceptable Quality Level. 合格品質水準.

Arabian *a.* アラビア(人)の.
~ cypher [cipher] アラビア数字, 算用数字. ~ mathematics アラビア(の)数学.

Arabic =Arabian.
~ numeral アラビア数字; 算用数字.

arbitrarily *ad.* 任意に, かってに.

arbitrary *a.* 任意な, かってな; 不定の.
~ constant 任意定数. ~ function 任意関数.

arborescence *n.* 樹木状.

arboricity *n.* 樹相度.

arc *n.* 弧.
analytic ~ 解析(的)弧. ~ element 弧(長)要素. ~ length 弧長. ~ of segment 弓形の弧. closed ~ 閉(じた)弧. conjugate ~ 共役弧. continuous ~ 連続弧. Jordan ~ ジョルダン弧. major ~ 優弧. minor ~ 劣弧. open ~ 開(いた)弧. regular ~ 正則弧. simple ~ 単純弧. smooth ~ 滑らかな弧.

arccosine *n.* 逆余弦, アークコサイン.

arch *n.* アーチ.

Archimedean *a.* アルキメデスの, アルキメデス的な.
~ ordered field アルキメデス順序体. ~ spiral アルキメデス螺線. ~ valuation アルキデス(的)付値.

ARCHIMEDES of Syracus (287-212 B.C.) シラキュスのアルキメデス.

ARCHITAS of Tarent (428?-365? B.C.) タレントのアルキタス.

ARCH model アーチモデル.
autoregressive conditional heteroskedasticity model 自己回帰条件付き異分散モデル.

arcograph *n.* 円弧規.

arcsine *n.* 逆正弦, アークサイン.
~ law 逆正弦法則. ~ transformation 逆正弦変換.

arcwise *ad.* 弧状に.
~ connected 弧状連結の.

area *n.* 面積; 面分.
~ coordinate(s) 面積座標. ~ element 面積要素, 面素. ~ graph 面積グラフ. ~ of base 底面積. ~ theorem 面積定理. inner [outer] ~ 内[外]面積. lateral ~ 側面積. mixed ~ 混合面積. plane ~ 平面積. surface ~ 曲[表]面積.

areal *a.* 面積の, 範囲の.
~ element 面積要素. ~ sampling 地域抽出(法).

area-preserving *a.* 面積を保存する.
~ map 保積写像.

ARGAND, Jean Robert (1768-1822) アルガン.

argument *n.* (複素数の)偏角; (独立変数の)引数, 変数; 推論, 論法, 議論, 主張, 論証, 要旨.
~ function 変関数. ~ principle 偏角(の)原理.

ARIMA model アリマモデル.
autoregressive integrated moving average model 自己回帰和分移動平均モデル.

ARISTARCHOS of Samos (310?-230? B.C.) サモスのアリスタルコス.

ARISTOTELES of Stagira (384-322 B.C.) スタギラのアリストテレス.

arithmetic *n.* 算数，算術; 算法; 算術者，計算学習者. *a.* 算術的な.
~ function (整)数論的関数. ~ genus 算術種数. ~ mean 相加平的，算術平均; 等差中項. ~ of algebraic number fields 代数体の整数論. ~ progression 等差[算術]数列[A.P.]. ~ series 等差[算術]級数. interval ~ 区間演算.

arithmetical *a.* ＝arithmetic.

arithmetically *ad.* 算術的に.

arithmetician *n.* 算術家.

arithmetico- 「算術」の意の結合辞.
~geometric mean 算術幾何平均. ~harmonic mean 算術調和平均.

arithmetization *n.* 算術化.

arm *n.* 腕.
~ of angle 角の辺.

ARONSZAJN, Nathan (1907-1980) アロンシャイン.

arrangement *n.* 順列; 配列; 並べ方.
random ~ 無作為配置.

array *vt.* 配列する. *n.* 配列.
orthogonal ~ 直交配列.

arrival *n.* 到着.
batch ~ 集団到着.

arrow *n.* 矢，射.
~ diagram 矢線図.

artful *a.* 技巧的な，巧妙な.

article *n.* 論文，論説; 条項.

artificial *a.* 人工の，人為的な.
~ intelligence 人工知能. ~ variable 人為変数. ~ vector 補助ベクトル.

ARTIN, Emil (1898-1962) アルティン.
~-Schreier extension アルティン・シュライアー拡大. ~-Rees lemma アルティン・リースの補題.

Artinian *a.* アルティン的な.
~ module アルティン(的)加群. ~ ring アルティン環.

-ary *suf.* 「…する所，者，項」の意.
n~ relation *n* 項関係.

ARYABHATA (476-550?) アリアバータ.

ARZELA, Cesare (1847-1912) アルツェラ.

ascend *vi., vt.* 昇る; さかのぼる. ⇨ descend.

ascending *a.* 上昇の，昇-.
~ central series 昇中心列，~ chain 昇鎖. ~ chain condition [a.c.c.] 昇鎖条件. ~ order of powers 昇ベキの順. ~ powers 昇ベキ.

ASCOLI, Giulio (1843-1896) アスコリ.

assemble *vt.* アセンブルする【記号言語から機械言語に変換する】.

assembler *n.* アセンブラ.

assembly *n.* 組立(部品).
~ line 組立ライン; 流れ作業工程.

assert *vt.* 主張する; 言明する.

assertion *n.* 主張; 言明; 所説.

assignment *n.* 割当; 指定.
~ problem 割当の問題.

associate *vt., vi.* 連合させる[する]，結合させる[する]. *a.* 同伴の，随伴の.
~ matrix 随伴行列.

associated *a.* 連合した，結合した; 同伴-，随伴-; 陪-; 付随する.
~ bilinear form 同伴双線形形式. ~ convergence radii 同伴[関連]収束半径. ~ differential equation 陪微分方程式. ~ director circle [line] 連合準円[線]. ~ equation 同伴方程式. ~ form 同伴形式. ~ function 陪関数. ~ linear mapping 同伴線形写像. ~ matrix 随伴行列. ~ prime ideal 随伴素イデアル. ~ surface 随伴曲面.

association *n.* 結合; 連関.
~ algebra 結合環. degree of ~ 関連度.

associative *a.* 連合する，結合-.
~ algebra 結合的代数[多元環]. ~ law 結合法則，結合律. ~ ring 結合環.

associativity *n.* 結合性.

assume *vt., vi.* 仮定する; (値を)とる.

assumption *n.* 仮定，仮設; 条件; (値を)とること.
consistent ~(s) 無矛盾の仮定. inductive ~ 帰納法の仮定.

asterisk *n.* 星印【*】.

asteroid *n.* アステロイド，星芒形.

astrodynamics *n.* 天体力学.

astroid ＝asteroid.

astronomical *a.* 天文学(上)の.

~ observation 天文観測. ~ triangle 天文三角形.

astronomy *n.* 天文学.
spherical ~ 球面天文学.

asymmetric *a.* 非対称な, 反対称な.
~ distribution 非対称分布. ~ law 反対称法則[律].

asymmetrical =asymmetric.
~ curve 非対称曲線. ~ distribution 非対称分布.

asymmetry *n.* 非対称(性), 反対称(性).

asymptote *n.* 漸近線.

asymptotic *a.* 漸近的な.
~ analysis 漸近解析. ~ behavio(u)r 漸近(的)性状. ~ circle 漸近円. ~ cone 漸近錐面. ~ convergence 漸近収束. ~ curve 漸近曲線, 主接線曲線. ~ direction 漸近方向. ~ distribution 漸近分布. ~ expansion 漸近展開. ~ formula 漸近公式. ~ line 漸近線. ~ method 漸近法. ~ normality 漸近正規性. ~ parabola 漸近放物線. ~ path 漸近路. ~ series 漸近級数. ~ stability 漸近安定性. ~ tangent 漸近接線. ~ value 漸近値.

asymptotical =asymptotic.

asymptotically *ad.* 漸近的に.
~ stable 漸近安定な. ~ unbiased estimator 漸近不偏推定量.

asynchronous *a.* 非同期(式)の.

at *prep.* …に, …で.
~ least 少なくとも, 以上. ~ most 高々, せいぜい, 以下. ~ random 無作為に, 確率的に, ランダムに. point [line] ~ infinity 無限遠点[直線].

Atiyah, Michael Francis (1929-) アティヤー.

ATLAS ATLAS of Finite Groups. 有限群のアトラス.

atlas *n.* 地図帳, 地図; アトラス.

atled *n.* アトレッド【記号 ∇, delta の逆読み】. ⇨ nabla.

atom *n.* 原子.

atomic *a.* 原子の, 極微の; 純不連続な.
~ act (チューリング機械の)動作. ~ element (束の)原子元.

attach *vi., vt.* 付属[随]する[させる]. *vt.* 結びつける, 対応させる.
~ed figure 付図. ~ing space 接着空間.

attaching *n.* 貼り合わせ.

attain *vt.* (値を)とる.

attempt *vt.* 試みる. *n.* 試み.

attenuation *n.* 減衰.
~ distortion 減衰ひずみ.

attraction *n.* 引力.
gravitational ~ 重力.

attractive *a.* 引きつける.
~ force 引力.

attractor *n.* アトラクター.
stable ~ 安定アトラクター.

attribute *n.* 属性.
~ statistics 属性統計.

augend *n.* 被加(算)数[量].

augment *vt., vi.* 増す; 付加する, 添加する.

augmentation *n.* 付加, 添加; 拡大; 基本写像.

augmented *a.* 付加された, 添加された; 拡大された.
~ algebra 係数添加環. ~ matrix 拡大行列.

Auslander, Maurice (1926-1994) アウスランダー.

autbor *n.* 著者.
co~; joint ~ 共著者.

auto- *pref.* 「自らの, 自己…」の意: automaton, automorphism.

autocorrelation *n.* 自己相関.
~ function 自己相関関数.

autocovariance *n.* 自己共分散.
~ function 自己共分散関数.

automatic *a.* 自動の.
~ control 自動制御. ~ control engineering 自動制御工学. ~ differentiation 自動微分.

automation *n.* オートメーション.

automaton (*pl.* -ta) *n.* オートマトン.
cellular ~ セルオートマトン. finite ~ 有限オートマトン. pushdown ~ プッシュダウン・オートマトン.

automorphic *a.* 保形-, 保型-, 形を保つ.
~ form 保型[形]形式. ~ function 保型

[形]関数.

automorphism *n.* 自己同型[形](写像).
~ group 自己同型[形]群. inner ~ 内部自己同型[形]. involutive ~ 対合的自己同型[形].

autonomic =autonomous.

autonomous *a.* 自律[励]的な.
~ oscillation 自励振動. ~ system 自励系.

autoregression *n.* 自己回帰.

autoregressive *a.* 自己回帰の.
~ conditional heteroskedasticity model (ARCH model) 自己回帰条件付き異分散モデル. ~ integrated moving average model (ARIMA model) 自己回帰和分移動平均モデル. ~ process 自己回帰過程.

auxiliary *a.* 補助の, 副-.
~ axis 副軸. ~ circle 補助円. ~ diagonal 副対角線. ~ equation 補助方程式. ~ ground line 副基線. ~ line 補助線.

available *a.* 役に立つ, 利用可能な.
~ time 使用可能時間.

average *n.* 平均, 平均値. *a.* 平均の. *vt.* 平均する.
~ error 平均誤差. ~ mark 平均点. ~ number of defects 平均欠点数. ~ quality 平均品質. ~ rate of change 平均変化率. ~ sample number 平均抜取り個数. ~ speed 平均の速さ. ~ velocity 平均速度. moving ~ 移動平均. on an ~ 平均して. phase ~ 相平均. time ~ 時間平均.

axial *a.* 軸の; 軸性の.
~ symmetry 軸対称, 線対称. ~ vector 軸性ベクトル.

axiom *n.* 公理.
~ of choice 選択[出]公理. ~(s) of congruence 合同(の)公理. ~(s) of continuity 連続性(の)公理. ~ of parallels 平行線公理. common ~ 普通[普遍]公理. countability ~ 可算公理. system of ~s 公理系.

axiomatic *a.* 公理の, 公理的な.
~ method 公理的方法. ~ set theory 公理的集合論.

axiomatism *n.* 公理主義.

axiomatization *n.* 公理化.

axiomatize *vt.* 公理化する.

axis (*pl.* axes) *n.* 軸.
auxiliary ~ 副軸. ~ of abscissa 横(座標)軸. ~ of convergence 収束軸. ~ of ordinate 縦(座標)軸. ~ of revolution 回転軸. ~ of symmetry 対称軸. conjugate ~ 共役軸. coordinate ~ 座標軸. imaginary ~ 虚軸. major ~ 長軸. minor ~ 短軸. oblique ~ 斜交軸. orthogonal ~ 直交軸. polar ~ (極座標系の)始線. principal ~ 主軸. real ~ 実軸. transverse ~ 横軸.

axisymmetric *a.* 軸対称な.

axonometric *a.* 軸測の.
~ line 軸測線. ~ projection 軸測投影[象](法).

axonometry *n.* 軸測投象(法).
isometric ~ 等軸測投象(法). oblique ~ 斜軸測投象(法).

azimuth *n.* 方位, 方位角.
~ angle 方位角.

azimuthal *a.* 方位(角)の.
~ angle 方位角.

AZUMAYA, Goro (1920-2010) 東屋五郎.

B

Bachmann, Paul (1837-1920) バッハマン.

back *n.* 背, うしろ; 裏. *a.* うしろの, 背後の.
~ and face 裏(と)表. call-~ 再調査. pull ~ ひきもどす. pull-~ 引き戻し, プルバック.

background *n.* 背景; 基礎知識.

backward *ad.* 後方に[へ]. *a.* 後退の; 後方の. ⇨ forward.
~ difference 後進差分. ~ equation 後進方程式. ~ moving average 後退移動平均. ~ step 後退[逆進]段階. ~ substitution 後退代入.

Baire, René Louis (1874-1932) ベール.

balance *n.* 平衡, 釣合い; 秤; 残高. *vt., vi.* 釣り合わす; 釣り合う.
~ due 不足額. ~ sheet 貸借対照表.

balanced *a.* 釣合いのとれた, 平衡した.
A-~ mapping *A* 平衡写像. ~ block design 釣合いブロック計画. ~ convex set 均衡凸集合. ~ incomplete block design 釣合い不完備ブロック計画[BIBD]. ~ mapping 平衡写像. ~ sample 釣合いのとれた標本.

balancing *n.* 釣合いをとること.
line ~ 生産ラインの最適編成.

balayage *n.* (*F.*) 掃くこと, 掃き出し; 掃散. *vt.* 掃散する. ⇨ sweeping.
~d measure 掃散測度[分布]. ~ principle 掃散原理.

ball *n.* 球(体).
~ pair 胞体対. open ~ 開球. unit ~ 単位球(体).

ballistic *a.* 弾道の.
~ curve 弾道曲線.

ballistics *n.* 弾道学.

BAN Best Asymptotically Normal. 最良漸近正規な.

Banach, Stefan (1892-1945) バナッハ.

band *n.* 帯; バンド.
~ graph 帯グラフ. ~ matrix 帯行列. confidence ~ 信頼帯. Möbius(') ~ メビウスの帯.

bandwidth *n.* 帯域幅, バンド幅.

bang *n.* 強打; ばたん.
~-~ type 衝撃形.

bang-bang ⇨ bang.
~ control 衝撃制御. ~ type 衝撃型.

bankruptcy *n.* 破産.

bar *n.* 棒; 横線【—】.
~ chart 横線工程表. ~-construction 棒構成. ~ graph 棒グラフ.

Barnes, Ernest William (1874-1953) バーンズ.

barrel *n.* 樽, 樽集合, バレル.
~led set 樽集合. ~led space 樽型空間. ~(-)shape 樽型.

barrier *n.* バーリア, 障壁; 柵関数.
absorbing [absorptive] ~ 吸収壁. reflecting [reflective] ~ 反射壁.

barycentre, -ter *n.* 重心.

barycentric *a.* 重心の.
~ coordinates 重心座標. ~ mapping 重心写像. ~ subdivision 重心細分.

base *n.* 基, 基底; 基数; 基線; 底, 底辺, 底面; (点列の)台. ⇨ basis.
area of ~ 底面積. ~ angle 底角. ~-centred lattice 底心格子. ~ change matrix 基(底)変換行列. ~ line 底線. ~ of logarithm 対数の底. ~ point 基点. ~ space 底空間, 基礎空間. canonical ~ 標

準基(底). free ~ 自由基. lower [upper] ~ 下[上]底.

basic *a.* 基礎的な, 基本的な; 基底の.
~ algebra 基本多元環. ~ angle 底角. ~ component 基底部分, 基礎成分. ~ concept 基礎概念. ~ equation 基底方程式. ~ field 基礎体. ~ solution 基底解. ~ surface 基礎面. ~ variable 基底変数. ~ vector 基底ベクトル.

basis (*pl.* -ses) *n.* 基, 基底. ⇨ base.
~ vector 基底ベクトル. canonical ~ 標準基. change of ~ 基底の変換. countable ~ 可算基. dual ~ 双対基. Gröbner [Groebner] ~ グレブナー基底. integral ~ 整数基. ordered ~ 順序基底. orthogonal ~ 直交基. orthonormal ~ 正規直交基. separable transcendence ~ 分離超越基底. transcendence ~ 超越基底.

batch *n.* バッチ【一括処理されるデータの一回分】.
~ arrival 集団到着. ~ processing 一括処理, バッチ処理.

BAYES, Thomas (1702–1761) ベイズ.

Bayesian *a.* ベイジアン, ベイズ流の.
~Information Criterion (BIC) ベイズ情報量基準.

BCD Binary(-)Coded Decimal notation. 二進化十進法.

BCH Bose-Chaudhuri-Hocquenghem. ビー・シー・エイチ.
~ code BCH 符号.

become *vi.* となる, になる.

begin *vi., vt.* 始まる, …し始める.

behave *vt.* 振る舞う, 挙動する.

behavior,【英】-iour *n.* 性状; ふるまい, 挙動.
asymptotic ~ 漸近(的)性状. boundary ~ 境界性状. output ~ 出力のふるまい.

behavio(u)ral *a.* 挙動の.
~ science 行動科学.

BEHNKE, Heinrich (1898–1971) ベーンケ.

bell *n.* ベル, 鐘.
~ (-shaped) curve 鐘形曲線.

BELLMAN, Richard Ernest (1920–1984) ベルマン.

belong *vi.* (元が集合に)属する【to】, 所属する.

below *prep., ad.* ~より下に; 下に, 以下【≦】, 未満【<】. ⇨ above.
bounded (from) ~ 下に有界な.

BELTRAMI, Eugenio (1835–1900) ベルトラミ.

bend *vt., vi.* 曲げる(がる). *n.* 曲り, 屈曲.

BERGMAN(N), Stefan (1895–1977) ベルクマン, バーグマン.

BERLEKAMP, Elwyn Ralph (1940–) バーレカンプ.

BERNAYS, Paul Isaak (1888–1977) ベルナイス.

BERNOULLI, Daniel (1700–1782), Jakob [James, Jacques] (1654–1705), Johann [John, Jean] (1667–1748) ベルヌーイ.

Bernoullis ベルヌーイ家.

BERNSTEIN, Sergei Natanovič (1880–1968) ベルンシュタイン.

BERS, Lipman (1914–1993) ベルス.

BERTRAND, Joseph Louis François (1822–1900) ベルトラン.

Berührungsmenge *n.* (*G.*) 触集合.

beschränktartig *a.* (*G.*) 有界型の.

BESOV, Oleg Vladimirovich (1933–) ベゾフ.

BESSEL, Friedrich Wilhelm (1784–1846) ベッセル.

best *a.* 最良の【good, well の最上級】.
~ approximation 最良近似. ~ critical region 最良棄却域. ~ invariant estimator 最良不変推定量. ~ linear unbiased estimator (BLUE) 最良線形不偏推定量. ~ osculation 最大接触. ~ possible 最良の. ~ predictor 最良予報値.

beta *n.* ベータ. ギリシア語アルファベットの 2 番目の文字【B, β】.
~ function ベータ[β]関数.

BETTI, Enrico (1823–1892) ベッチ.

between *prep.* (二つの)間に[の, を, で]. *n.* 間.
~-class 級間. ~-class variation 級間変動. ~-groups 群間.

BEURLING, Arne Karl-August (1905–1986)

ボイルリンク.

beyond *prep.* を越[超]えて; より以上に.

Bézout, Etienne (1730-1783) ベズー.

BG Bernays-Gödel set theory.

Bhaskara (1114-1185?) バスカラ.

bi- *pref.*「二, 両, 双, 重, 複, 陪の」などの意.

biadditive *a.* 双加法的な.

biadditivity *n.* 双加法性.

bianalytic *a.* 双解析的な, 両側解析的な.
~ mapping 双解析写像.

bianalyticity *n.* 双解析性.

Bianchi, Luigi (1856-1928) ビアンキ.

biannual *a.* 半年ごとの.

bias *n.* 偏り. *vt.* 偏らせる.
~ed estimate 偏った推定量.

biaxial *a.* 二軸の.
~ spherical surface function 二軸球面関数.

BIBD Balanced Incomplete Block Design. 釣合不完備ブロック計画.

bibliography *n.* 参考書目(録), 文献目録.

BIC ビー・アイ・シー.
Bayesian Information Criterion ベイズ情報量基準.

bicharacteristic *a.* 陪[従]特性の. *n.* 陪特性.
~ curve 陪[従]特性曲線. ~ strip 陪特性帯.

bicompact *a.* ビコンパクトな.【現在では=compact.】

bicomplex *n.* 二重複体.

biconditional *a.* 双条件的な.
~ proposition [statement] 同値命題.

bicontinuous *a.* 両連続な.
~ mapping 両連続写像.

bidual *a.* 重双対な, 重共役な. *n.* 重双対.
~ space 第二共役空間.

Bieberbach, Ludwig (1886-1982) ビーベルバッハ.

Bienaymé, Jules (1796-1878) ビアネメ.

biennial *a.* 隔年の.

bifocal *a.* 双焦の.
~ chord 双焦弦.

bifurcation *n.* 分岐.
~ point 分岐点.

biharmonic *a.* 重調和な.
~ function 重調和関数.

biholomorphic *a.* 両解析的な.
~ mapping 両解析的写像.

bijection *n.* 全単射; 双射; 1 対 1 対応.

bijective *a.* 全単射的な; 双射的な.

bilateral *a.* 両側の, 二面のある.
~ network 相反回路. ~ surface 両側曲面.

bilinear *a.* 双一次の, 双[二重]線形の.
~ form 双一次[線形]形式. ~ function 双一次[線形]関数. ~ functional 双線形汎関数. ~ mapping 双[二重]線形写像.

bilinearity *n.* 双[二重]線形性.

billion *n.*【米・仏】十億,【英・独】兆.

bimatrix *n.* 双行列.
~ game 双行列ゲーム.

bimodal *a.* 双峰の.
~ distribution 双峰分布. ~ frequency curve 双峰頻度曲線.

bimodule *n.* 二重加群, 両側加群.

binal *a.* 二倍の, 二重の.

binary *a.* 二元の, 双体の; 二進(法)の, 二分的な; 二値の; 二項-.
~ cell 二値素子. ~ code 二進コード. ~ digit 二進数字. ~ logic 二値論理. ~ notation 二進法. ~ operation 二項演算, 二項算法. ~ relation 二項関係. ~ scale 二進法. ~ search 二分探索. ~ symmetric channel 二進対称通信路. ~ tree 二分木.

binary-coded *a.* 二進符号化した.
~ decimal notation 二進化十進法.

bind *vt.* 結ぶ; 束縛する.

binomial *a.* 二項の, 二項式の. *n.* 二項式.
~ coefficient 二項係数. ~ distribution 二項分布. ~ equation 二項方程式. ~ expansion 二項展開. ~ series 二項級数. ~ theorem 二項定理.

binormal *n.* 従法線, 陪法線. *a.* 双正規な.
~ indicatrix 従[陪]法線標形. ~ vector 従法線ベクトル.

binormality *n.* 双正規性.

bioassay *n.* 生物検定法.

biomathematics *n.* 生物数学.

biometric *a.* 計量生物学の, 生物測定学の.

biometrics, biometry *n.* 計量生物学; 生物測定学.

bionics *n.* バイオニクス.

biophysics *n.* 生物物理学.

biorthogonal *a.* 二重直交の.
～ system 二重直交系.

bipartite *a.* 二部から成る; 共同の.
～ graph 二部グラフ.

bipoint *n.* 複点.

bipolar *a.* 双極の. *n.* 双極集合.
～ coordinates 双極座標. ～ potential 双極ポテンシャル.

biprojective *a.* 双射影的な.
～ space 双射影空間.

biquadrate *n.* 四乗したもの, 四乗ベキ.

biquadratic *a.* 四乗ベキの; 四次の. *n.* 四乗ベキ, 四次(方程)式.
～ equation 四次方程式.

biquinary *a.* 2-5 進(法)の.
～ notation 2-5 進法.

birational *a.* 双有理的な.
～ correspondence 双有理対応. ～ equivalence 双有理同値. ～ isomorphism 双有理同形[型]. ～ mapping [transformation] 双有理写像[変換].

birectangular *a.* 両直角の.
～ triangle 両直角三角形.

biregular *a.* 双正則な.
～ mapping 双正則写像.

Birkhoff, Garrett (1911–1996), George David (1884–1944) バーコフ.

birth *n.* 出生.
～ and death chain 出生死亡連鎖. ～ function 出生関数. ～ process 出生過程. ～ ratio [rate] 出生率.

bisect *vt.* 二等分する.
～ing normal [perpendicular] 垂直二等分線.

bisectability *n.* 二分(可能)性.

bisection *n.* 二等分; 二等分された一方.
～ method 二分法.

bisector *n.* 二等分線; 二等分面.
～ of angle 角の二等分線. perpendicular ～ 垂直二等分線[面].

bispinor *n.* バイスピノル.

bit *n.* ビット【二進法の 1 桁】.
check ～ 検査ビット.

bivariate *a.* 二変量の.
～ correlation 二変量相関. ～ distribution 二変量分布.

bivector *n.* 二重ベクトル.

Black-Scholes ブラック・ショールズ.
～model ブラック・ショールズモデル.

blank *a.* 空白の; 白紙の. *n.* 空所; 空白.

Blaschke, Wilhelm (1885–1962) ブラシュケ.

blend *vt., vi.* 混合する.
～ing problem 混合問題.

Bliss, Gilbert Ames (1876–1951) ブリッス.

Bloch, André (1893–1948) ブロック.

block *n.* ブロック; 区画.
～ design ブロック計画. ～ diagram ブロック・ダイアグラム. ～ effect ブロック効果. ～ records ブロック・レコード. initial ～ 初ブロック. Jordan ～ ジョルダン細胞. randomized ～ method 乱塊法.

blocking *n.* ブロック化, ブロッキング.

blow-down *n.* ブローダウン.

blow-up *n.* (解の)爆発, ブローアップ.

BLUE ブルー.
best linear unbiased estimator 最良線形不偏推定量.

Blumenthal, Otto (1876–1944) ブルーメンタール.

body *n.* 物体; 立体, 体.
～-centred lattice 体心格子. ～ of rotation 回転体. convex ～ 凸体, 卵形体. elastic ～ 弾性体. falling ～ 落体. individual ～ 個品. *n*-～ problem *n* 体問題.

Bohr, Harald (1887–1951) ボーア.

Bolyai, János [Johann] (1802–1860), Wolfgang (1775–1856) ボヤイ, ボリアイ.

Bolzano, Bernard (1781–1848) ボルツァノ.

Bonferroni, Carlo Emilio (1892–1960) ボンフェロニ.

Bonnet, Pierre Ossian (1819–1892) ボンネ.

Boole, George (1815–1864) ブール.

Boolean *a.* ブールの; ブール的な.
~ algebra ブール代数. ~ lattice ブール束. ~ operations ブール演算. ~ ring ブール環.

bootstrap *n.* ブートストラップ, 靴ひも.
~-argument ブートストラップ論法.

border *n.* 縁; 境界. *vt.* へりを付ける, 限定する.
~ed matrix 縁取った行列. ~ set 縁集合.

bordering *n.* 境界を作ること; 縁取り.

Borel, Armand (1923-2003) ボレル.

Borel, Émile Félix Edouard Justin (1871-1956) ボレル.

bornological =bornologique.

bornologique *a.* (*F.*) ボルノロジック, 有界形の.

bornology *n.* ボルノロジー.

borrow *vt., vi.* (引算で一けた上から)借りる, 降ろす.

borrowing *n.* 繰下り.

both *a.* 両方の. *prop.* 両者. *ad.* 両方とも.
~ sides 両辺.

Bott, Raoul (1923-2005) ボット.

bottle *n.* びん[瓶], 徳利.
~-neck problem 隘路の問題. Klein's ~ クラインの管[壷].

bound *n.* 境界(線); 限界. *vt.* 制限する, 限界を設ける; 境界をつける.
~s of integral 積分(の)限界. branch and ~ 分枝限定. greatest lower ~ 下限. least upper ~ 上限. lower [upper] ~ 下[上]界.

bound *a.* 束縛された. ⇨ bind.
~ variable 束縛変数.

boundary *n.* 境界, 周; 境界輪体.
~ behavio(u)r 境界性状. ~ cluster set 境界集積値集合. ~ condition 境界条件. ~ cycle 境界輪体. ~ element 境界要素. ~ layer 境界層. ~ operator 境界作用素. ~ point 境界点. ~ simplex 辺単体. ~ space 境界値の空間. ~ strip 境界帯. ~ value 境界値. ~ value problem 境界値問題. harmonic ~ 調和境界. ideal ~ 理想境界. natural ~ 自然境界. Šilov ~ シロフ境界.

bounded *a.* 限られた, 有界な.
~ convergence 有界収束. ~ downwards [upwards] 下に[上に]有界な. ~ from above [below] 上に[下に]有界な. ~ function 有界関数. ~ linear functional 有界線形汎関数. ~ matrix 有界行列. ~ measure 有限測度. ~ operator 有界作用素. ~ set 有界集合. of ~ variation 有界変動の. totally ~ 全有界な. vaguely ~ 漠有界な.

boundedly *ad.* 有界的に.
~ complete 有界的完備な. ~ convergent 有界収束の.

boundedness *n.* 有界性.
uniform ~ 一様有界性.

Bouquet, Jean Claude (1819-1885) ブーケ.

Bourbaki, Nicolas (1938-) ブールバキ【フランスの数学者集団名】.

box *n.* 箱, ボックス; 直方体.
black ~ ブラック・ボックス, 暗箱. ~ topology 箱位相.

box-whisker
~ plot 箱ひげ図.

brace (*pl.*) *n.* 大括弧【{ }】.

brachistochrone *n.* 最短降下線. (=line of swiftest descent 最速降下線)

bracket (*pl.*) *n.* 角(型)括弧【おもに[ ], 〔 〕, 【 】】. ⇨ parenthesis, brace.
~ operation 括弧演算. ~ product 括弧積【$[x, y]$】.

Brahmagupta (598-630) ブラマグプタ.

braid *n.* 組みひも, 組み糸.
~ group 組み糸群.

branch *n.* 枝, 分枝; 分科, 部門. *vi.* 分岐する.
~ and bound 分枝限定. ~ and cut 分枝カット. ~ divisor 分岐因子. ~ (-) point 分岐点. principal ~ 主枝.

branching *a.* 分岐の. *n.* 分岐.
~ process 分岐[分枝]過程.

Brauer, Richard Dagobert (1901-1977) ブラウアー.

breadth *n.* 幅.
~ first search 幅優先探索. curve of constant ~ 定幅曲線. of constant ~ 定幅の.

shortest ~ 短幅.

break *vt.* こわす，割る; 破る; 中断させる. *vi.* こわれる．割れる; 中絶する．*n.* 破れ，裂け目，断絶; 区切り.
~-even 損益分岐点．~ in 割込む．~-point 区切り点．even ~ 五分五分，同点.

Brelot, Marcel (1903-1987) ブルロ.

Brezis, Haim (仏 Brézis, Haïm) (1944-) ブレジス.

Brianchon, Charles-Julian (1785-1864) ブリアンション.

brief *a.* 簡単な，簡潔な.
~ method 簡便法．~ proof 簡単な証明.

Briggs, Henry (1561-1631) ブリッグス.

Briot, Saint Hyppolyte (1817-1882) ブリオ.

broken *a.* 破れた; 断線的な.
~ line 破線【------; ⇨ dotted line, solid line】; 折線，屈折線．~ line of correlation 相関折線．~ number(s) 端数，分数.

Bromwich, Thomas John I' Anson (1875-1929) ブロムウィッチ.

Brouwer, Luitzen Egbertus Jan (1881-1966) ブロウエル，ブラウアー.

Buchberger, Bruno (1942-) ブッフバーガー.

buffer *n.* 緩衝; 緩衝器，緩衝回路.
~ effect 緩衝効果．~ inventory 緩衝用在庫.

Buffon, George Louis Leclerc Compte de (1707-1788) ビュッフォン.

built-in *a.* (ハードウェアに)組み込みの.
~ check 自動検査.

bulk *n.* 大きさ，容積; 集団.
~ queue 集団待ち行列．~ service 集団サービス.

bulletin *n.* 報告(誌).

bunch *n.* 束.

bundle *n.* 束，バンドル; 空間束.
affine ~ アフィン束．analytic ~ 解析的ファイバー束．~ mapping 束写像．~ space 束空間．cotangent ~ 余接束．exterior power ~ 外積バンドル．fibre ~ ファイバー束，ファイバー・バンドル．frame ~ フレーム・バンドル，枠束．line ~ 直線束．normal ~ 法束．principal ~ 主束．sphere ~ 球面束．tangent ~ 接束．tensor ~ テンソル束．vector ~ ベクトル束.

Bunyakovskii, Viktor Jakovlevič (1804-1889) ブニャコフスキ.

Bürgi, Jobst (1552-1632) ビュルギ.

Burnside, William Snow (1852-1927) バーンサイド.

bus *n.* 母線.

busy *a.* 稼働中の.
~ period (全)稼働期間.

but *conj.* しかし．*prep.* を除いて.
the last ~ one 最後から二番目.

C

C, c （ローマ数字の）100.

ca. ＝circa.

calculate *vt., vi.* 計算する，算出する.

calculation *n.* 計算，演算，算法. ⇨ computation.
～ by writing 筆算. ～ on abacus 珠算. contracted ～ 省略算. mental ～ 暗算. numerical ～ 数値計算.

calculator *n.* 計算器[機].

calculus （*pl.* -li, -luses） *n.* 計算法; 微(分)積分法[学].
advanced ～ 高等微積分学. ～ of contingency 保険数学. ～ of finite differences 差分[定差]法[算]. ～ of logic 論理演算. ～ of variations 変分法[学]. class ～ （論理)集合算. difference ～ 差分法. differential ～ 微分法[学]. infinitesimal ～ 無限小算法; 微(分)積分(法). integral ～ 積分法[学]. predicate ～ 述語計算. tensor ～ テンソル計算(法).

Calderón, Alberto Pedro（1920–1998） カルデロン.

calendar *n.* 暦，カレンダー.
～ year 暦年.

calipers （*pl.*） *n.* キャリパー.
slide ～ ノギス.

call *vi., vt.* 呼ぶ，名づける，称する. *n.* 請求.
～-back 再調査.

canal *n.* 溝; 管.
～ surface 管状曲面.

cancel *vt., vi.* 約する，簡約する，消す. *n.* 取消.

cancellation *n.* 簡約.
～ law 簡約法則.

cancelling *n.* 簡約; 桁落ち.

canonical *a.* 標準的な，規準の; 正準の.
～ base [basis] 標準基. ～ bundle 標準束. ～ coordinates 標準座標. ～ correlation analysis 正準相関分析. ～ decomposition 標準[規準]分解. ～ divisor 標準因子. ～ domain 規準領域. ～ equation 正規[準]方程式，規[標]準方程式. ～ form 標準形; 基底形式. ～ homomorphism 標準的準同型(写像). ～ injection 標準的単射. ～ isomorphism 標準的同形[型]. ～ map 標準(的)写像. ～ parameter 標準助変数[パラメータ]. ～ product 規[標]準(乗)積. ～ region 規準領域. ～ scale 標準尺度. ～ sheaf 標準層. ～ surjection 標準的全射. ～ variable 正準変数.

Cantor, Georg Ferdinand Ludwig Philipp（1845–1918） カントール.

cap *n.* 縁なし帽子; キャップ【記号 ∩】，交わり. ⇨ cup.
～ product キャップ積.

capacitability *n.* 可容性.

capacitable *a.* 可容な.
～ set 可容集合.

capacitary *a.* 容量の.
～ measure 容量測度.

capacity *n.* 容量.
analytic ～ 解析(的)容量. ～ function 容量関数. ～ of memory 記憶容量. channel ～ 通信路容量. inner ～ 内容量. logarithmic ～ 対数容量. Newtonian ～ ニュートン容量. outer ～ 外容量. transmission ～ 伝送容量.

capital *a., n.* かしら文字(の)，大文字(の); 元金(の)，資本(の).

~ coefficient 資本係数. ~ letter 大文字. ~ redeeming coefficient 資本回収[償却]係数. ~ turn 資本回転率.

capture-recapture *n.* 捕獲・再捕獲.
~ method 捕獲・再捕獲法.

Caratheodory, Constantin (1873-1950) カラテオドリ.

Cardano, Hieronymo [Girolamo, Gironimo] (1501-1576) カルダノ.

cardinal *n.* 基数. *a.* 基本的な; 計数論的な.
~ number 基数, 濃度, カージナル数; 計量数. inaccessible ~ 到達不能基数.

cardinality ＝cardinal number.

cardioid *n.* 心臓形, カージオイド.

Carleman, Torsten (1892-1949) カーレマン.

carrier *n.* 台; キャリア. ⇨ support.
~ of function 関数の台. ~ space 台空間. compact ~ コンパクトな台.

carry *vt.* 運ぶ, 移す, 動かす; 繰上[下]げる; 帯びる, 所有する. *n.* 繰上[下]げ.
~ down 繰下げ(る). ~ out 果たす(約束, 義務などを). ~ up 繰上げ(る). partial ~ 部分桁上げ.

carrying *n.* 繰上り.

Cartan, Elie Joseph (1869-1951), Henri Paul (1904-2008) カルタン.

Cartesian *a.* デカルト[Descartes]の, カルテシアン.
~ coordinates デカルト座標. ~ product (集合の)直積; カルテシアン積. ~ space カルテシアン空間.

cartography *n.* 地図製作(法); 製図.

cascade *n.* 小滝, 段瀑.
~ method カスケード法.

case *n.* 場合; 事実, 例.
exceptional ~ 例外の場合. in any ~ どの場合にも, いずれにせよ. in ~ of の場合には. in this ~ この場合には.

Casorati, Felice (1835-1890) カゾラチ.

Cassini, Jean Dominique (1625-1712) カッシーニ.

Cassinian *a.* カッシーニの. *n.* カッシーニ曲線.

cast *vt., vi.* 計算する, 足し算する, 合計する. *n.* 計算, 勘定.
~ accounts 計算する. ~ing out (合同式を利用する)検査法. ~ing out nines 九去法.

Castelnuovo, Guido (1865-1952) カステルヌオボ.

catacaustic *a.* 反射火[焦]線の. *n.* 反射火[焦]線.
~ curve 反射火[焦]線.

catalog, catalogue *n.* 目録, カタログ; 要覧, 便覧.

catastrophe *n.* カタストロフィー; 破局.
~ map カタストロフィー写像. ~ theory カタストロフィー理論.

categorical *a.* 圏の, カテゴリーの; 範疇的な; 定言の.
~ algebra 圏[カテゴリー]代数 ~ proposition 範疇[定言]命題. ~ syllogism 定言三段論法.

category *n.* 圏, カテゴリー; 類, 範疇.
additive ~ 加法圏, 加法的カテゴリー. ~ of sets 集合(の)圏. derived ~ 導来圏. dual ~ 双対圏. quotient ~ 商圏. set of the first [second] ~ 第一[二]類集合.

catenary *n.* 懸垂線, カテナリー. *a.* 懸垂線の.
~ ring カテナリー環.

catenoid *n.* 懸垂面, カテノイド.

Cauchy, Augustin Louis (1789-1857) コーシー.
~-Riemann equation コーシー・リーマンの方程式. ~ sequence コーシー列, 基本列.

causality *n.* 因果律, 因果関係.
law of ~ 因果律.

cause *n.* 原因; 動機; 理由, 根拠.
law of ~ and effect 因果律. probability of ~(s) 原因の確率.

caustic *a.* 火[焦]線の, 火[焦]面の. *n.* 火[焦]線, 火[焦]面.
~ curve 火線, 焦線. ~ surface 火面, 焦面.

caution *vt.* 注意する.

Cavalieri, Franesco Bonaventura (1598-1647) カヴァリエリ.

CAYLEY, Arthur (1821-1895) ケーリー.

Cayley-Hamilton ケーリー・ハミルトン.
~ formula ケーリー・ハミルトンの公式.

cc, c.c. cubic centimeter(s). 立方センチ.

C.D.F. Cumulative Distribution Function. 累積分布関数.

celestial *a.* 天の; 天体の.
~ axis 天軸. ~ body 天体. ~ coordinates 天球座標. ~ equator 天球赤道. ~ globe 天球儀. ~ mechanics 天体力学. ~ navigation 天測航法. ~ pole 天極. ~ sphere 天球. ~ surveying 天文測量.

cell *n.* 細胞; (位相幾何学の)胞体; 枡目.
binary ~ 二値素子. ~ complex 胞複体. convex ~ 凸胞体. dual ~ 双対胞体. Jordan ~ ジョルダン細胞. unit ~ 単位胞体.

cellular *a.* 細胞の; 胞体の; セルの.
~ approximation 胞体近似. ~ automaton セルオートマトン. ~ decomposition 胞体分割. ~ mapping 胞体写像.

censoring *n.* 打ち切り.

cent *n.* セント; 百.
per ~ [per~] 百分率, パーセント【%】.

center *n.* 中心. *vi., vt.* 中心とする; 中心におく. (英:centre).
analytic ~ 解析的中心. ~ circle 中心円. ~ of circumcircle 外心. ~ of curvature 曲率中心. ~ of figure 図心. ~ of gravity 重心. ~ of group 群の中心. ~ of incircle 内心. ~ of inversion 反転の中心. ~ of mass 質量中心, 重心. ~ of perspectivity 配景の中心. ~ of rotation 回転の中心. ~ of similitude 相似の中心. ~ surface 中心曲面. circle ~ed at C C を中心とする円. visual ~ 視心.

centering *n.* 中心化.

centesimal *a.* 百分法の, 百進法の.
~ measure 百分法.

centi- *pref.* 「百; 百分の一」の意.
~ grade 百分度の. ~ meter センチ[百分の一]メートル.

centile *n.* 百分位数.

centillion *n., a.* 【米】千の 101 乗(の); 【英】百万の百乗(の).

central *a.* 中心の, 中心的な; 有心の; 中央の.
~ angle 中心角. ~ conics 有心二次曲線. ~ difference 中心差分. ~ homothety 中心相似. ~ limit theorem 中心極限定理. ~ line 中心線. ~ perspective drawing 中心投象画法. ~ projection 中心射影. ~ quadrics 有心二次曲面. ~ series 中心列. ~ simple algebra 中心的単純多元環. ~ symmetry 中心対称(変換). ~ tendency 中心傾向. ~ value 代表値. 中心値.

centralization *n.* 中心化.

centralizer *n.* 中心化群[環].

centre =center.

centred *a.* 中心化された.
~ process 中心化された過程.

centring (=**centering**) *n.* 中心化.

centroid *n.* 図心, 重心, 中心軌跡: (環の) セントロイド.
five ~s 五心. five ~s of triangle 三角形の五心.

centuple *n., a.* 百倍(の). *vt.* 百倍する.

centuplicate *a.* 百倍の. *vt.* 百倍する.

certain *a.* 確かである, 確定した,

certainly *ad.* 確かに, 間違いなく.

certification *n.* 証明, 保証; 認証.

CESÀRO, Ernesto (1859-1906) チェザロ.

CEVA, Giovanni (1648-1734) チェバ.

cf. =confer. *vt.* 参照せよ【命令形】.

C.G.S. Centimeter-Gram-Second.

chain *n.* 鎖; 連鎖. *vt.* 鎖でつなぐ.
absorbing ~ 吸収連鎖. ascending ~ 昇鎖. ~ complex 鎖複体. ~ condition 連鎖条件. ~ed line 鎖線. ~ homomorphism 鎖準同型[形]. ~ homotopy 鎖ホモトピー. ~ mapping 鎖写像. ~ ratio 連鎖比. ~ rule 連鎖律. descending ~ 降鎖. finite [infinite] ~ 有限[無限]鎖. Markov ~ マルコフ連鎖. zero ~ 零鎖.

chance *n.* 偶然; 機会. *a.* 偶然な.
~ hit まぐれ当り. ~ variable 確率変数.

change *vt.* 変える. *vi.* 変わる. *n.* 変化; 変更, 変換; 順列.
base ~ 基底変換. ~ of basis 基底の変換. ~ of coordinate(s) 座標変換. ~ of

variable(s) 変数変換. ~ sign 符号を変える. rate of ~ 変化率.

channel *n.* 情報路, 通信路, チャンネル; 窓口.
binary symmetric ~ 二進対称通信路. ~ capacity 通信路容量.

chaos [kéɪɑs/-ɔs] *n.* カオス; 混沌.

chaotic [keɪ ɑ́ t̬ɪk/-ɔ́t-] *a.* カオス的な.

chapter *n.* (書籍・論文の)章[ch., chap.].

character *n.* 指標; 特徴, 性質; 字, キャラクタ, 文字.
~ group 指標群. ~ of group 群の指標. ~ recognition 文字認識. ~ system 指標系. Chinese ~ 漢字. group ~ 群指標. induced ~ 誘導指標. inherited ~ 遺伝形質. integral ~ 積分指標. irreducible ~ 既約指標.

characteristic *a.* 特性的な, 固有な. *n.* (対数の)指標, (体, 整域の)標数, 特性類.
~ class 特性類. ~ cone 特性錐. ~ curve 特性曲線. ~ direction 特性方向. ~ equation 特性[固有]方程式. ~ function 特性[特徴]関数; 固有関数. ~ functional 特性汎関数. ~ hypersurface 特性超曲面. ~ mapping 特性写像. ~ multiplier 特性乗数. ~ number 特性数. ~ polynomial 特性[固有]多項式. ~ root 特性根. ~ subgroup 特性部分群. ~ value 固有値. ~ vector 特性ベクトル. ~ zero 標数零. Euler ~ オイラー指標[特性類].

characterization *n.* 特徴づけ.

characterize *vt.* 特徴づける.

charge *vt.* 充電する; 帯電させる. *n.* 電荷, 荷電; 充電.
~ independence 荷電不変性, 荷電独立. ~ simulation method 代用電荷法. point ~ 点電荷.

CHARLIER, Carl Ludwig (1861-1934) シャリエ.

chart *n.* 図表, チャート, グラフ.
alignment ~ 共線図表. bar ~ 横線工程表. control ~ 管理図. flow ~ 流れ図, フローチャート. intersection ~ 共点図表. local ~ 局所座標系. projection ~ 投影図.

CHASLES, Michel (1793-1880) シャール.

CHEBYCHEV, Pafnuty Lwowich (1821-1894) チェビシェフ.

check *n.* 検算; 検査, 照合. *vt., vi.* 検査する, チェックする.
automatic [built-in] ~ 自動検査. ~ bit 検査ビット. ~ formula 検算公式. ~ing routine 検査ルーチン. ~ inspection 監査検査. ~ limit 監査限界. ~ up 照合する; 検査する.

chemotaxis *n.* 走化性.

CHERN Shing-Shen (1911-2004) 陳省身.

CHEVALLEY, Claude (1909-1984) シュバレー.

chi *n.* カイ. ギリシア語アルファベット22番目の文字【X, χ】.
~-square distribution カイ二[自]乗分布. ~-square test カイ二[自]乗判定法.

choice *n.* 選択, 選出.
axiom of ~ 選択[出]公理. ~ function 選択関数. ~ set 選択[出]集合. multistage ~ 多段選択.

choose *vi., vt.* 選択する, 選ぶ.

CHOQUET, Gustave (1915-2006) ショケ.

chord *n.* 弦.
bifocal ~ 双焦弦. common ~ 共通弦. complementary ~ 補弦. focal ~ 焦弦.

chordal *a.* 弦の.
~ distance 弦距離.

CHRISTOFFEL, Erwin Bruno (1829-1900) クリストッフェル.

chromatic *a.* 色の; 染色性の.
~ index 染色指数. ~ number 染色数, 彩色数.

chronologic *a.* 経時的な.

chronological *a.* 経時的な.

cipher *n.* 零, 無; アラビア数字の各記号; 暗号(文). *vi., vt.* 計算する.
number of five ~s 五桁の数.

circa *prep.* (*L.*) 約, およそ [ca.].

circle *n.* 円; 円周; 循環. *vt., vi.* 回る.
asymptotic ~ 漸近円. auxiliary ~ 補助円. center ~ 中心円. ~ geometry 円幾何学. ~ graph 円グラフ. ~ method 円周法. ~ of convergence 収束円. ~ of curvature 曲

率円. ~ of inversion 反転円. ~ of latitude [longitude] 緯[経]線. ~ ratio 円周率. ~-to-~ correspondence 円々対応. coaxial ~s 共軸円. concentric ~s 同心円. convergence ~ 収束円. Euler('s) ~ オイラー円. full ~ 周角. great ~ 大円. imaginary ~ 虚円. small ~ 小円. unit ~ 単位円. vicious ~ 循環論法.

circled *a.* 円形の, 丸い.

circling *n.* 循環.

circuit *n.* 回路.
electric ~ 電気回路. logical ~ 論理回路.

circulant *n.* 循環行列(式).

circular *a.* 円の, 円-; 輪状の; 循環の; 巡回する.
~ arc 円弧. ~ argument [reasoning] 循環論法. ~ cone 円錐. ~ congruence 円叢. ~ constant 円周率. ~ cylinder 円柱. ~ cylindrical coordinates 円柱座標. ~ disc [disk] 円板. ~ domain 円領域. ~ frequency 角振動数. ~ function 円関数. ~ measure 弧度法. ~ motion 円運動. ~ parts 円部分. ~ permutation 円順列. ~ ring 円環. ~ slit 円弧截線, 波紋(截線).

circulate *vt.* 循環させる. *vi.* 循環する.

circulating *a.* 循環する. ⇨ recurring.
~ decimal 循環小数.

circulation *n.* 循環; 流通.

circulator *n.* 循環小数; 循環器.

circum- *pref.*「周, 外接, 境界」などの意.

circumcenter *n.* 外心.

circumcircle *n.* 外接円.
~ center; center of ~ 外心.

circumference *n.* 円周; 周, 周線, 境界線. ⇨ periphery.
angle at ~ 円周角.

circumferential *n.* 周の. ⇨ peripheral.
~ angle 円周角.

circumscribe *vt.* 外接させる.
~d circle 外接円. ~d polygon 外接多角形. ~d sphere 外接球.

circumscriptible *a.* (円に)外接できる.
~ polygon 円に外接する多角形.

circumscription *n.* 外接; 外接させる[する]こと.

cissoid *n.* シッソイド, 疾走線.
generalized ~ 一般シッソイド. skew ~ 歪疾走線.

cissoidal *a.* シッソイドの, シッソイド的な.
~ curve (一般)シッソイド曲線.

claim *vt.* 主張する, (事実として)申し立てる; 要求する. *n.* 主張, 申し立て; 要求.

Clairaut, Alexis Claude (1713-1765) クレーロー.

class *n.* 類, 族; 級; 階級; 組. ⇨ family.
additive ~ 加法族. ambig ~ 特異類. Baire ~ ベール族. characteristic ~ 特性類. ~ calculus 集合論理; (論理)集合算. ~ equation 類等式. ~ field 類体. ~ field theory 類体論. ~ frequency 級度数. ~ function 類関数. ~-group of paths 道類群. ~ interval 級の幅. ~ mark 階級値. ~ number 類数. ~ of curves 曲線族. ~ of functions 関数族. ~ value 級代表値. cohomology ~ コホモロジー類. conjugate ~ 共役類. curve of the second ~ 二級曲線. differential ~ 微分類. equivalence ~ 同値類. Hardy ~ ハーディ族. upper [lower] ~ 上[下]組.

classic *a.* 古典的な. *n.* (古典)文献.

classical *a.* 古典的な.
~ analysis 古典解析学. ~ group 古典群. ~ Lie algebra 古典リー代数. ~ logic 古典論理. ~ theory 古典(理)論.

classification *n.* 分類, 類別.
~ theorem 分類定理.

classify *vt.* 分類する, 類別する.
~ing space 分類空間.

clear *a.* 明らかな; 分かり易い. *vt., vi.* 明らかにする; 払う, 御破算にする.

clearly *ad.* 明らかに, 明瞭に, 分かり易く.

Clifford, William Kingdon (1845-1879) クリフォード.

clique *n.* クリーク.
~ number クリーク数.

clockwise *a., ad.* 右回りの[に], 時計の針の方向の[に]. ⇨ counterclockwise.

close *vt., vi.* 閉じる. *a.* 密接な; 狭い, 近い.

~st packing 最密充填.

closed *a.* 閉じた，閉-. ⇨ open.
algebraically ~ 代数的に閉じた. ~ arc 閉(じた)弧. ~ convex hull 閉凸包. ~ curve 閉曲線. ~ disc [disk] 閉円板. ~ domain 閉領域. ~ game 閉鎖ゲーム. ~ geodesic 閉測地線. ~ interval 閉区間. ~ mapping 閉写像. ~ path 閉道. ~ region 閉領域. ~ set 閉集合. ~ subroutine 閉じたサブルーチン. ~ surface 閉曲面. locally ~ set 局所閉集合. multiplicatively ~ set 乗法的閉集合. semi-~ 半閉の.

closure *n.* 閉包; 包; 触集合.
algebraic ~ 代数的閉包. ~ finite 閉包有限な. ~ problem 閉形問題. convex ~ 凸包. integral ~ 整閉包. projective ~ 射影閉包.

clothoid *n.* クロソイド，コルニュ[Cornu]の螺線.

clover-leaf *n.* クローバー[三葉]型のもの; 四つ葉クローバー型のもの.
~ knot 三葉結び目.

cluster *n.* 集積，束; 集落. *vi.* 集積する.
~ sampling 集落抜取り. ~ set 集積値集合. ~ value 集積値. inner [boundary] ~ set 内[境界]集積値集合. ~ point 集積点.

co- *pref.*「双対-，余-; 共-」の意.

coalgebra *n.* 双対代数.

coalition *n.* 結託; 連携.

coaltitude *n.* 天頂距離 (=zenith distance).

coarse *a.* あら[粗]い，粗な.

coarseness *n.* 粗さ.

coauthor *n.* 共著者.

coaxial *a.* 同軸の，共軸の.
~ circles 共軸円. ~ conics 共軸円錐曲線.

cobordant *a.* 同境の，コボルダントな.
~ mod 2 2を法として同境の.

cobordism *n.* 同境，コボルディズム.
~ class 同境類.

coboundary *n.* 双対境界輪体.
~ operator 双対境界作用素.

cochain *n.* 双対鎖(体).
~ complex 双対鎖複体. ~-equivalent 双対鎖同値な. ~ subcomplex 部分双対鎖複体.

cocycle *n.* 双対輪体.
continuous ~ 連続双対輪体. group of ~s 双対輪体群.

Coddington, Earl Alexander (1925-) コディントン.

code *n.* 符号，コード. *vt.* 符号化する.
binary ~ 二進コード. ${}_nC_r$ ~ ${}_nC_r$定マーク符号. ~d character 符号化文字. ~d decimal notation 符号化十進法. ~ length 符号長. ~ rate 符号化率. ~ word 符号語. cyclic ~ 巡回符号. error correcting ~ 誤り訂正符号. Golay ~ ゴレイ符号. Huffman ~ ハフマン符号. linear ~ 線形符号. Morse ~ モールス符号. q-ary ~ q-進符号.

coderivative *n.* 余微分.

co-differential *n.* 双対微分. *a.* 双対微分の.

codimension *n.* 余次元，補次元.

coding *n.* 符号化.
~ problem 符号化の問題. ~ rate 符号化レート，符号化率. ~ theory 符号理論.

codomain *n.* 終域，余域.

coefficient *n.* 係数.
capital ~ 資本係数. ~ field 係数体. ~ matrix 係数行列. ~ of concordance 一致係数. ~ of determination 寄与率; 決定係数. ~ of displacement 変位係数. ~ of variation 変動係数. ~ problem 係数問題. confidence ~ 信頼係数. connection ~ 接続係数. correlation ~ 相関係数. differential ~ 微分係数. direction ~ 方向係数. leading ~ 主係数，最高次の係数. maximal [maximum] ~ 最大係数. regression ~ 回帰係数. undetermined ~(s) 未定係数. with constant ~(s) 定(数)係数の.

coefficientwise *a., ad.* 係数ごとの[に].

coercive *a.* 強圧的な.
~ boundary condition 強圧的境界条件.

cofactor *n.* 余因数，余因子.
~ expansion 余因子展開. ~ matrix 余因子行列.

cofinal *a.* 共終の.
~ object 始対象. ~ subset 共終部分集合.

cofinality *n.* 共終度.
cogenarator *n.* 余生成素, 余生成対象.
cogradient *n.* 共傾. *a.* 共傾の.
COHEN, Irvin Sol (1917-1955) コーエン.
Cohen-Macaulay ring コーエン・マコーレー環.
COHEN, Paul Joseph (1934-2007) コーエン.
coherence *n.* 同調, 連接.
coherency =coherence.
coherent *a.* 同調した, 連接した; 首尾一貫した.
~ sheaf 連接層.
coherently *ad.* 同調的に.
~ oriented 同調的に向きづけられた.
cohesion *n.* 粘着; 結合.
COHN-VOSSEN, Stefan (1902-1936) コーンフォッセン.
cohomological *a.* コホモロジーの.
~ dimension コホモロジー次元. ~ functor コホモロジー関手.
cohomologous *a.* コホモローグな.
cohomology *n.* コホモロジー.
Čech ~ チェックのコホモロジー. ~ class コホモロジー類. ~ group コホモロジー群. ~ ring コホモロジー環. étale ~ エタール・コホモロジー. exact ~ sequence コホモロジー完全系列.
cohomotopy *n.* コホモトピー.
~ group コホモトピー群.
coimage *n.* 余像.
coincide *vi.* 一致する【with】.
coincidence *n.* 一致.
coincident *a.* 一致した.
cokernel *n.* 余核.
col *n.* とうげ[峠]点, 鞍点.
method of ~ 鞍点法.
collaboration *n.* 協力, 共同研究; 共著.
collapse [kəlǽps] *vt.,vi.* 消滅する, つぶれる, 崩壊する.
collar *n.* カラー; 襟. *vt.* カラー[襟]をつける.
~ed neighbo(u)rhood 襟(つき)近傍.
collared *a.* カラー[襟]をもつ.
collating 照合.
collect *vt.* 集める, 収集する.
~ed works [papers] 著作[論文]集.
collection *n.* 全体, 合併, 集合; 収集.
collective *a.* 集合的な. *n.* 集団.
~ risk theory 集合的危険論.
collinear *a.* 同一直線上の, 共線(的)の.
~ diagram 共線図表. ~ planes 共線平面.
~ points 共線点.
collinearity *n.* 共線性.
collinearly *ad.* 共線的に.
collineation *n.* 共線変換.
affine ~ アフィン共線変換. ~ in the wider sense 共線写像. projective ~ 射影的共線変換.
collineatory *a.* 共線の.
collocation *n.* 選点; 配置.
~ method 選点法.
colloquium (*pl.* -quia, -ums) *n.* 談話会; コロキウム; 共同討議; セミナー.
cologarithm *n.* 余対数.
colonial *a.* 集落の.
colonization *n.* 集落化.
colony *n.* 集落.
colo(u)r *n.* 色. *vt., vi.* 着色する.
~ed tree 色つき樹木. ~ graph 色グラフ.
four ~ problem 四色問題.
colo(u)rable *a.* 色分けできる.
column *n.* (行列の)列; 欄; コラム.
~ -finite matrix 列有限行列. ~ vector 列[縦]ベクトル. elementary ~ operation 列基本変形.
co-lunar *a.* 共月の.
~ triangle 共月三角形.
com- =con-.
comatrix (*pl.* -trices, -trixes) *n.* 余行列.
combination *n.* 組合せ; 結合.
convex ~ 凸結合. linear ~ 一次[線形]結合. repeated ~ 重複組合せ.
combinatorial *a.* 組合せ的, 組合せ論の.
~ analysis 組合せ論. ~ manifold 組合せ多様体. ~ optimization 組合せ最適化. ~ topology 組合せ(論的)位相数[幾何]学.
combinatorics *n.* 組合せ論.
combinatory =combinatorial.
combine *vt., vi.* 結合する; 組合わせる.
command *n.* 命令, 指令.

commensurability *n.* 通約可能性，可約性.

commensurable *n.* 通約可能な.
~ number 尽数. ~ quantity 尽量.

commensuration *n.* 通約.

common *a.* 共通な，共有の，公-; 普通の.
~ axiom 普通[普遍; 共通]公理. ~ chord 共通弦. ~ denominator 公分母. ~ difference 公差. ~ divisor 公約数[元]. ~ factor 共通因数[因子，元]. ~ logarithm 常用対数. ~ measure 公約数. ~ multiple 公倍数[元]. ~ part 共通部分. ~ perpendicular 共通垂線. ~ point 共有点. ~ ratio 公比. ~ root 共通根. ~ set 共通集合. ~ tangent 共通接線.

communicate *vt.* 伝達する. *vi.* 通信する.

communication *n.* 伝達，通信.
~ matrix 通信行列.

commutable *a.* 可換な; 交換可能な.

commutation *n.* 交換.
~ relation 交換関係.

commutative *a.* 交換的な，可換な.
~ algebra 可換多元環. ~ convergence 可換収束. ~ diagram 可換な図式. ~ field 可換体. ~ group 可換群，アーベル群. ~ law 可換律[法則]; 交換法則. ~ ring 可換環.

commutatively *ad.* 可換的に.
~ converge 可換収束する.

commutativity *n.* 可換性.

commutator *n.* 交換子; 交換子積.
~ subgroup 交換子群.

commute *vt., vi.* 交換する，可換である.

commutor *n.* 可換子; 可換子環.

compact *a.* コンパクトな.
~-open topology コンパクト開位相. ~ set コンパクト集合. ~ space コンパクト空間. ~ type コンパクト型. locally ~ 局所コンパクトな.

compactification *n.* コンパクト化.
one-point ~ 一点コンパクト化. resolutive ~ 可解コンパクト化. X ~ X(氏)コンパクト化.

compactum (*pl.* -ta) *n.* コンパクト集合.

comparability *n.* 比較可能性.
~ theorem 比較定理.

comparable *a.* 比較可能な【with】; 匹敵する【to】.

compare *vt.* 比較する【with; to】; 参照する.

comparison *n.* 比較.
~ test 比較判定(法). ~ theorem 比較定理. in ~ with と比較して.

compass *n.* (通例 *pl.*) コンパス，両脚器.

compatibility *n.* 適合性，両立性; 互換性.
~ condition 両立条件.

compatible *a.* 両立する; 適合する; 矛盾のない.
~ with と両立する.

competing *a.* 競争する.
~ function 競争関数.

competition *n.* 競争，競合.

competitive *a.* 競争の.
~ economy 競争経済. ~ equilibrium (完全)競争均衡.

complement *n.* 補元; 余集合，補集合; 補空間; 補群; 補数，余数.
~ graph 補グラフ. normal ~ 正規補群. orthogonal ~ 直交補空間.

complementarity *n.* 相補性.
~ problem 相補性問題.

complementary *a.* 補足の，補充の; 補-，余-. *n.* 余形.
~ angle 余角. ~ chord 補弦. ~ degree 補充次数. ~ event 余事象. ~ function 余関数. ~ law 補充法則. ~ module 補加群，相補加群. ~ modulus 補母数. ~ series 補系列. ~ set 補[余]集合. ~ submatrix 余[補]小行列.

complemented *a.* 可補的な.
~ lattice 可補束，相補束.

complete *a.* 完全な，十分の，全くの; 完備な. *vt.* 完成する.
analytically ~ 解析的完備な. ~ additivity 完全加法性. ~ block 完備計画. ~ census 完全センサス. ~ condition 完全条件. ~ count 全部調査. ~ elliptic integral 完全楕円積分. ~ field 完備体. ~ induction 完全[数学的]帰納法. ~ inverse image 全逆像. ~ lattice 完備束. ~ linear system 完備線形系. ~ number 完全数. ~ ortho-

normal system 完全正規直交系. ~ quadrangle [quadrilateral] 完全四角[辺]形. ~ set of representative 完全代表系. ~ solution 完全解. ~ space 完備空間. ~ square 平方を完成する. ~ system 完全系. ~ works 全集. conditionally ~ 条件[制限]完備な. geodesically ~ 測地的完備な.

completely *ad.* 完全に, 完全-.
~ additive 完全加法的な. ~ integrable 完全積分可能な. ~ reducible 完全可約な. ~ regular 完全正則な.

completeness *n.* 完全; 完全性; 完備性.
~ of system 系の完全性. ~ theorem 完全[備]性定理.

completion *n.* 完備化.
analytic ~ 解析的完備化.

complex *a.* 複素数の, 複素-; 複雑な. *n.* 複体; 複素数 (*pl.* -lexes, -lices).
acyclic ~ 非輪状複体. algebraic ~ 代数(的)複体. almost ~ structure 概複素構造. cell ~ 胞複体. chain ~ 鎖複体. cochain ~ 双対鎖複体. ~ analysis 複素解析. ~ conjugate 共役複素数; 複素共役な. ~ conjugate number 共役複素数. ~ differentiation 複素微分. ~ dimension 複素次元. ~ form 複素形; 複素形式. ~ fraction 繁分数. ~ function 複素関数. ~ integration 複素積分. ~ linear space 複素線形空間. ~ manifold 複素多様体. ~ multiplication 虚数乗法. ~ number 複素数. ~ of lines 線網. ~ plane 複素平面. ~ potential 複素ポテンシャル. ~ projective space 複素射影空間. ~ quadratic form 複素二次形式. ~ representation 複素表現. ~ sequence [series] 複素数列[級数]. ~ sphere 複素球面. ~ structure 複素構造. ~ system 複雑系. ~-valued function 複素数値関数. ~ variable 複素変数. ~ vector space 複素ベクトル空間. conjugate ~ number 共役複素数. double ~ 二重複体. geometric cell ~ 幾何学的胞複体. Koszul ~ コスツル[コジュール, コスール, コシュール]複体. ordered ~ 順序複体. simplicial ~ 単体(的)複体. standard ~ 標準複体. tilting ~ 傾斜複体.

complexification *n.* 複素化.

complexity *n.* 複雑さ; 複雑度.
average-case ~ 平均計算量. computational ~ 計算量, 計算複雑性. worst-case ~ 最悪計算量.

componendo *n.* 加比[合比]の理.
~ and dividendo 合除比の理.

component *n.* 成分, 構成要素.
~ function 成分関数. ~ vector 成分ベクトル. ~ velocity 分速度. connected ~ 連結成分. diagonal ~ 対角成分. irreducible ~ 既約成分. normal ~ 法線成分. orthogonal ~ 直交成分. simple ~ 単純成分. tangential ~ 接線成分.

componentwise *a.*, *ad.* 成分ごとの[に].

compose *vt.* 組織する, 構成する, 結合する; 構図する.
be ~d of から成り立つ. ~d function [mapping] 合成関数[写像].

composite *a.* 合成の; 複合の. *n.* 合成物.
~ algorithm 複合法. ~ function 合成関数. ~ hypothesis 複合仮説. ~ mapping 合成写像. ~ number 合成数. ~ proposition 複合命題. ~ simplex algorithm 複合単体法.

composition *n.* 合成; 結合; 組成.
~ factor 組成因子. ~ product 合成積. ~ series 組成列. ~ theorem 結合定理. law of ~ 算法. secondary ~ 二次合成.

compound *a.* 複合の; 合成の. *n.* 混合物.
~ division 複名数[諸等数]の除法. ~ event 複(合)事象. ~ fraction 繁分数. ~ interest 複利. ~ number 複名数, 諸等数. ~ probability 複確率. ~ process 複合過程. ~ ratio [proportion] 複比[比例]. ~ statement 合成命題; 複合文. ~ subtraction 複名数[諸等数]の減法.

comprehension *n.* 理解; 内包; 包含.

compress *vt.* 圧縮する.
~ed limit 圧縮限界.

compressible *a.* 縮む, 圧縮性の.
~ fluid 圧縮(性)流体.

computable *a.* 計算可能な.
~ function 計算可能な関数.

computation *n.* 計算, 計算過程. ⇨ calculation.

~ theory 計算論. ~ with [by] logarithm 対数計算. non-deterministic ~ 非決定性計算. numerical ~ 数値計算. parallel ~ 並列計算.

computational *a.* 計算的な, 計算上の.

~ complexity 計算量, 計算複雑性. ~ error 計算誤差. ~ geometry 計算幾何学. ~ linguistics 計算言語学.

compute *vt., vi.* 計算する.

computer *n.* 計算機, コンピュータ; 計算者.

~ algebra 計算機代数. ~ science コンピュータ科学. digital ~ デジタル[計数型]計算機.

con- *pref.* 「共-, 共に」の意.

concave *a.* 凹の. ⇨ convex.

~ function 凹関数. ~ polygon 凹多角形. ~ programming problem 凹計画問題. ~ regular polygon 星正多角形. strictly ~ 狭義(に)凹の.

concavity *n.* 凹; 凹であること.

concentration *n.* 濃度; 集中.

~ function 濃度関数. ~ point 濃度点.

concentric *a.* 同心の.

(~) annulus 同心円環. ~ circle(s) 同心円. ~ circular ring 同心円環.

concept *n.* 概念.

basic ~ 基礎概念. fundamental ~(s) 基礎[基本]概念. undefined ~ 無定義概念.

conceptual *a.* 概念的な, 概念上の.

concern *vt.* 関与する, 影響する. *n.* 関心, 関心事; 関係.

be ~ed with に関係する, に関心がある. region [domain] of ~ 影響領域.

conchoid *n.* コンコイド, 螺獅線.

conchoidal *a.* コンコイドの[的な].

~ curve (一般)コンコイド曲線.

concircular *a.* 共円の; 共円的な.

~ curvature tensor 共円曲率テンソル. ~ points 共円点. ~ transformation 共円変換.

concircularly *ad.* 共円的に.

~ flat 共円的に平坦な.

conclude *vt.* 終える, 終結する; (…と)結論する.

This ~s the proof. (これで)証明終り.

conclusion *n.* 終結, 結論.

concomitant *a.* 随伴の, 付値の. *n.* 随伴式.

bilinear ~ 双一次随伴式. ~ variable 随伴変数.

concordance *n.* 同調(性); 一致.

coefficient of ~ 一致係数.

concordant *a.* 同調した; 一致する.

concrete *a.* 有形の. 具体的な. ⇨ discrete.

~ number 名数【three men, five days など】.

concur *vi.* 一点に集まる; 一致する【with】.

concurrence *n.* (線・面の)集合; 集合点; 共点性.

concurrent *a.* 一点に集まる, 共点の. *n.* 共点.

concyclic ＝concircular.

condensation *n.* 凝集.

~ point, point of ~ 凝集点. ~ test 凝集判定法.

condition *n.* 条件.

accessory ~ 付帯条件. boundary ~ 境界条件. chain ~ 連鎖条件. compatibility ~ 両立条件. complete ~ 完全条件. ~ number 条件数. ~ of convergence, convergence ~ 収束条件. ~ of integrability, integrability ~ 積分可能性(の)条件. finiteness ~ 有限(性)条件. initial ~ 初期条件. Lipschitz ~ リプシッツ(の)条件. necessary and sufficient ~ 必要十分条件. necessary ~ 必要条件. optimality ~ 最適性条件. regularity ~ 正則性(の)条件. side ~, subsidiary ~ 付帯条件. sufficient ~ 十分条件. transversality ~ 横断(性)条件.

conditional *a.* 条件つきの. *n.* 仮言命題; 条件文.

~ convergence 条件収束. ~ expectation 条件つき期待値. ~ extremum 条件つき極値. ~ inequality 条件つき不等式. ~ measurement 条件つき測定. ~ probability 条件つき確率. ~ proposition 条件命題.

〜 syllogism 仮言三段論法.

conditionally *ad.* 条件つきで.
〜 complete 条件[制限]完備な. converge 〜 条件収束する.

conductance *n.* コンダクタンス.

conduction *n.* 伝導.
heat 〜 熱伝導.

conductor *n.* 導手; 導体.
〜 potential 導体ポテンシャル. 〜-ramification theorem 導手分岐定理.

cone *n.* 錐, 錐体, 錐面, 円錐.
asymptotic 〜 漸近錐面. characteristic 〜 特性錐. circular 〜 円錐. 〜 condition 円錐条件. elliptic 〜 楕円錐. enveloping 〜 包絡錐(面). frustum of 〜 錐台. mapping 〜 写像錐. normal 〜 法錐. oblique circular 〜 斜円錐. polar 〜 極錐. quadratic 〜 二次錐(面). reduced 〜 約錐, 被約錐. reduced mapping 〜 約写像錐, 被約写像錐. regular 〜 正則錐. right circular 〜 直円錐. spherical 〜 球錐. tangent 〜 接錐.

confer *vt.* (*L.*) 参照せよ [cf.].

conference *n.* 会議.

confidence *n.* 信頼.
〜 band, 〜 belt 信頼帯. 〜 coefficient 信頼係数[度]. 〜 interval 信頼区間. 〜 level 信頼水準. 〜 limit 信頼限界. 〜 region 信頼域. unbiased 〜 region 不偏信頼域.

configuration *n.* 配置; 配列.
Pascal's 〜 パスカルの配置. symmetric 〜 対称配置.

confluence *n.* 合流.
〜 of singularities 特異点の合流.

confluent *a.* 合流した.
〜 hypergeometric function 合流(型)超幾何関数. of 〜 type 合流型の.

confocal *a.* 共焦の, 同焦点の.
〜 conicoids 共焦二次曲面. 〜 conics 共焦円錐曲線. 〜 cyclides 共焦サイクライド. 〜 ellipses 共焦楕円. 〜 hyperbolas 共焦双曲線. 〜 parabolas 共焦放物線. 〜 quadrics 共焦二次曲面.

conformable *a.* 適合した.

conformal *a.* 等角な, 共形的な.
〜 connection 共形接続. 〜 correspondence 共形(的)対応. 〜 curvature 共形曲率. 〜 geometry 共形幾何学. 〜 invariant 等角不変量. 〜 mapping 等角写像. 〜 projection 等角投影(法). 〜 representation 等角写像. 〜 structure 等角構造. 〜 transformation 等角写像; 共形変換. directly [indirectly] 〜 順[逆]等角な.

conformality *n.* 等角性, 共形性.

conformally *ad.* 等角に, 共形的に.
〜 equivalent 等角同値な. 〜 flat 共形的に平坦な. 〜 invariant 等角不変な.

confound *vt.* 混同する; 混乱させる.

confounding *n.* 交絡.
partial 〜 部分交絡(法).

confusion *n.* 混乱; 混同【with】.

congress *n.* 会議.
International 〜 of mathematicians 国際数学者会議.

congruence *n.* 合同, 相合; 合同式; 叢; 適合.
axiom(s) of 〜 合同(の)公理. circular 〜 円叢. 〜 equation 合同方程式. 〜 expression 合同式. 〜 subgroup 合同部分群. 〜 zeta function 合同式ゼータ関数. line 〜 (直)線叢. simultaneous 〜s 連立合同式.

congruent *a.* 合同な.
affinely 〜 アフィン合同な. 〜 expression 合同式. 〜 figures 合同(な)図形. 〜 transformation 合同変換.

congruity *n.* 合同性.

congruous ＝congruent.

conic *a.* (円)錐の, 円錐形の. *n.* 円錐曲線 (*pl.*). ⇨ conics.
〜 pencil 円錐曲線束. 〜 section 円錐曲線; 円錐の断面.

conical ＝conic.
〜 surface 錐面.

conicoid *n.* 二次曲面.
confocal 〜s 共焦二次曲面.

conics (*pl.*) *n.* 円錐曲線; 円錐曲線論.
central 〜 有心二次曲線. confocal 〜 共焦円錐曲線. 〜 with center 有心二次曲線. 〜 without center 無心二次曲線. degenerate 〜 退化円錐曲線. noncentral 〜 無心

二次曲線. polar ~ 極円錐曲線. proper ~ 固有な円錐曲線.

conjecture *n.* 予想; 推測. *vt.* 予想する; 推測する.

conjugate *a.* 共役[共軛]な. *n.* 共役.
complex ~ 共役複素数; 複素共役な. ~ arc 共役弧. ~ axis (双曲線の)副[共役]軸. ~ class 共役類. ~ complex (number) 共役複素数. ~ diameter 共役(直)径. ~ direction 共役方向. ~ element 共役元. ~ gradient method 共役勾配法. ~ harmonic function 共役調和関数. ~ hyperbola 共役双曲線. ~ kernel 共役核. ~ number 共役数. ~ operator 共役作用素. ~ point 共役点. ~ pole [polar] 共役な極[極線]. ~ root 共役根. ~ space 共役空間. ~ vector 共役ベクトル. harmonic ~ 共役調和関数; 共役調和な.

conjugation *n.* 共役写像; 共役化.
complex ~ 複素共役化. ~ mapping 共役(化)写像.

conjunction *n.* 合接; 連合.

conjunctive *a.* 合接の; 連合の.
~ proposition 合接[連合]命題.

connect *vt.* 結合する, 連結する.

connected *a.* 連結した, 連結な.
arcwise ~ 弧状連結の. ~ component 連結成分. ~ graph 連結グラフ. ~ set 連結集合. locally ~ 局所連結の. multiply-~ (重)複連結の. *n*-ply ~ *n* 重連結の. simply-~ 単連結の.

connectedness *n.* 連結(性).
~ of real number system 実数系の連結性.

connecting *a.* 連結する; 連絡する.
~ homomorphism 連結準同型[形](写像). ~ mapping 連結写像. ~ tensor 連絡テンソル.

connection *n.* 連結; 関係; 接続.
affine ~ アフィン接続. conformal ~ 共形接続. ~ coefficient 接続係数. ~ form 接続形式. ~ formula 接続公式. linear ~ 線形接続. metric ~ 計量接続. projective ~ 射影接続.

connective *n.* 結合記号. *a.* 連結な.

connectivity *n.* 連結度; 連結性.
local ~ 局所連結度. of finite [infinite] ~ 有限[無限]連結の.

connexion =connection.

connotation *n.* 内包.

conoid *a.* 擬円錐(形)の, 尖円の. *n.* 擬円錐体, コノイド.
right ~ 正コノイド.

conormal *n.* 余法線.
~ sheaf 余法線層.

CONS Complete OrthoNormal System. 完全正規直交系.

consecutive *a.* 連続な, 引き続く.
~ integers 引き続く[継続する]整数.

consequence *n.* 結果; 帰結, 結論.

consequent *a.* 結果として起こる; 必然の. *n.* 後項, 後率; 結論.

consequently *ad.* したがって; その結果として; それゆえに.

conservation *n.* 保存, 保持, 不変.
~ law 保存則. principle of ~ of form [calculation] 形式[算法]不易の原理.

conservative *a.* 保存的な.
~ force 保存力. ~ system 保存系.

consider *vt.* 考察する, 熟考する, 検討する.

consideration *n.* 考察, 考慮, 熟慮, 検討.

consist *vi.* …から成る【of】; …に在る【in】.

consistence =consistency.

consistency *n.* 一致性, 適合性; 無矛盾性; 共存性.
~ requirement 無矛盾の要請.

consistent *a.* 一致した; 無矛盾な, 両立する.
~ assumption(s) 無矛盾な仮定. ~ estimator 一致推定量. ~ statistic 一致統計量. ~ test 一致検定.

const, const. constant.

constancy *n.* 不変性.

constant *n.* 定数, 常数; 恒数. *a.* 一定な; 定まった, 定数の.
absolute ~ 絶対定数. additive ~ 付加定数. arbitrary ~ 任意定数. ~ curvature 定曲率. ~ factor 定数因子. ~ function 定数関数. ~ mapping 定値写像. ~ of

proportion (ality) 比例定数. ～ polynomial 定数多項式. ～ sheaf 定数層. ～ sum 定和. ～ term 定数項, ～ value 定値. ～ vector 定ベクトル. integration ～ 積分定数. proportional ～ 比例定数. structure ～(s) 構造定数. universal ～ 宇宙定数, 普遍定数. with ～ coefficient(s) 定(数)係数の.

constituent *n.* 要素, 成分.

constitute *vt.* 構成する, 構成要素をなす.

constrain *vt.* 束縛する.
～ed extremum 条件つき極値. ～ed motion 束縛[拘束]運動. ～ed optimization 制約付き最適化. ～ed vector 束縛ベクトル.

constraint *n.* 拘束; 制約; 制約条件.
active ～ 有効制約. ～ qualification 制約想定. force of ～ 束縛力.

construct *vt.* 作図する; 構成する.

constructible *a.* 作図可能な; 構成可能な.

construction *n.* 作図; 構成, 構造.
approximate ～ 近似作図. ～ problem 作図題. of impossible [possible] ～ 作図不能[可能]な. ruler and compass ～ 定規とコンパスによる作図.

constructive *a.* 作図の; 構成的な.
～ method 構成的方法. ～ ordinal 構成的順序数.

consume *vt.* 消費する.

consumer *n.* 消費者, 消費財. ⇨ producer.

contact *n.* 接触. *vt., vi.* 接触させる[する].
～ angle 接触角. ～ element 接触要素. ～ network 接点回路. ～ structure 接触構造. ～ transformation 接触変換. order of ～ 接触位数. point of ～ 接点.

contain *vt.* 含む, (辺が角を)はさむ; 包む, 囲む; …で割り切れる, …を因子として含む.

contemporary *a.* 現代の, 当今の.

contend *vt.* 主張する.

content *n.* 容積, 面積; 拡度; 内容.
～ of polynomial 多項式の内容. inner [outer] ～ 内[外]拡度. linear ～ 長さ, 延長. solid [cubic(al)] ～ 容積, 体積. superficial ～ 面積.

contention *n.* コンテンション; 競争; 論点, 主張.

context *n.* 文脈.
～-free language 文脈自由言語. ～-sensitive language 文脈依存言語.

contiguous *a.* 隣接する【to】.
～ function 隣接関数.

contingence *n.* ＝contingency.

contingency *n.* 偶然性.
～ table 分割表; 偶然表. two-by-two ～ table 2 × 2 分割表.

contingent *a.* 偶然的な, 経験的な; あり得る.

continuable *a.* 接続可能な.
analytically ～ 解析接続可能な.

continuation *n.* 継続; 接続, 延長. ⇨ prolongation.
analytic ～ 解析接続. ～ of sign 符号の継続. direct [indirect] ～ 直接[間接]接続. harmonic ～ 調和接続.

continue *vi., vt.* 続く; 続ける; 連続する[させる]; 接続する[させる].
～ analytically 解析接続する. ～d fraction 連分数. ～d multiplication 連乗(積). ～d product 連乗積. ～d proportion 連比例. ～d ratio 連比.

continuity *n.* 連続(性).
axiom(s) of ～ 連続性(の)公理. ～ correction 連続修正, 連続補正. ～ modulus 連続度[率]. ～ principle 連続性原理. ～ theorem 連続(性)定理. equi～ 同程度連続性. uniform ～ 一様連続性.

continuous *a.* 連続な, 連続的な, 連続-. ⇨ discontinuous; discrete.
absolutely ～ 絶対連続な. ～ arc 連続弧. ～ curve 連続曲線. ～ data 連続型データ. ～ deformation 連続(的)変形. ～ distribution 連続分布. ～ function 連続関数. ～ geometry 連続幾何学. ～ group 連続群. ～ mapping 連続写像. ～ on the left [right] 左[右]側連続な. ～ random variable 連続(型)確率変数. ～ series 連続系列. ～ simplex 曲単体. ～ spectrum 連続スペクトル. uniformly ～ 一様連続な.

vaguely ～ 漠連続な.

continuously *ad.* 連続して，連続的に. *n*-times ～ differentiable *n* 回連続(的)微分可能な.

continuum (*pl.* -nua) *n.* 連続体. ～ hypothesis 連続体仮説.

contour *n.* 輪郭; 等高[深]線; 積分路. closed ～ 閉じた(積分)路. ～ integral 路に沿う積分. ～ line, ～(-)line 等高線.

contra- *pref.*「逆，反，対」などの意.

contract *vt.* 縮小する; 縮約する.

contracted *a.* 縮小した; 縮小-，縮約-. ～ calculation 省略算. ～ ideal 縮約イデアル. ～ tensor 縮約テンソル.

contractible *a.* 縮小可能な，可縮な. locally ～ 局所可縮な.

contraction *n.* 縮小; 縮約; 縮合. ～ mapping 縮小写像. ～ operator 縮小作用素.

contradict *vt.* 矛盾する.

contradiction *n.* 矛盾，不合理; 矛盾命題. in ～ to に反して. lead to ～ 矛盾を生じる. proof by ～ 背理法.

contradictory *a.* 矛盾した. ～ proposition 矛盾命題.

contragradient *n.* 反傾. *a.* 反傾の. ～ mapping 反傾写像. ～ representation 反傾表現.

contraposition *n.* 対偶.

contrapositive *n.* 対偶; 対偶命題. converse of ～ (含意命題の)裏.

contrary *a.* 矛盾した【to】; *ad.* 反対に.

contrast *n.* 対比，対照. normalized ～ 標準化対比. treatment ～ 処理対比.

contravariant *a.* 反変的な. *n.* 反変. ～ component 反変成分. ～ differentiation 反変微分. ～ functor 反変関手. ～ index 反変指標. ～ tensor 反変テンソル. ～ vector 反変ベクトル.

contribution *n.* 寄与，貢献.

control *n.* 制御; 管理. *vt.* 管理[制御]する. adaptive ～ 適応制御. admissible ～ 可容制御. automatic ～ 自動制御. ～ chart 管理図. ～ function 制御関数. ～ inspection 管理検査. ～ level 管理水準. ～ limit 管理限界. ～ theory 制御理論. ～ variable 制御変数. optimal ～ 最適制御. quality ～ 品質管理. statistical ～ 統計的管理.

convention *n.* 規約，とりきめ; 条件. summation ～ 総和規約.

converge *vi., vt.* 収束[収斂]する. ⇨ diverge.

convergence *n.* 収束[収斂]. abscissa of ～ 収束(横)座標. absolute ～ 絶対収束. almost (everywhere) ～ 概収束. asymptotic ～ 漸近収束. bounded ～ 有界収束. circle of ～ 収束円. conditional ～ 条件収束. ～ circle 収束円. ～ criterion 収束判定法[条件]. ～ domain 収束領域. ～ in law 法則収束. ～ in the mean 平均収束. ～ in measure 測度収束. ～ in probability 確率収束. ～ radius 収束半径. global ～ 大域的収束性. local ～ 局所的収束性. mean ～ 平均収束. pointwise ～ 点別収束. radius of ～ 収束半径. rate of ～ 収束率，収束速度. simple ～ 単純収束. stochastic ～ 確率収束. strong ～ 強収束. superlinear ～ 超一次収束. uniform ～ 一様収束. vague ～ 漠収束. weak ～ 弱収束.

convergency ＝convergence.

convergent *a.* 収束[斂]の. *n.* 近似分数. absolutely ～ 絶対収束の. boundedly ～ 有界収束の. commutatively ～ 可換収束の. conditionally ～ 条件収束の. ～ sequence 収束列. ～ series 収束級数. *n* th ～ 第 *n* 近似分数. uniformly ～ 一様収束の.

conversation *n.* 会話，談話.

conversational *a.* 会話の.

converse *a.* 逆の. *n.* 逆; 転換，換位. ～ of contrapositive (含意命題の)裏. ～ of theorem 定理の逆. ～ proposition 逆[転換]命題.

conversely *ad.* 逆に.

conversion *n.* 換算; (比例の)転換. ～ table 換算表.

convex *a.* 凸の. ⇨ concave. absolutely ～ 円形凸な. closed ～ hull 閉凸包. ～ analysis 凸解析. ～ body 凸体，卵

形体. ~ closure 凸包. ~ combination 凸結合. ~ cone 凸錐. ~ curve 凸曲線, 卵形線 ~ domain 凸領域. ~ envelope 凸包. ~ function 凸関数. ~ hull 凸包. ~ mapping 凸写像. ~ neighbo(u)rhood 凸近傍. ~ polygon 凸多角形. ~ polyhedron 凸多面体. ~ programming 凸(形)計画法. ~ sequence 凸数列. ~ set 凸集合. ~ surface 凸曲面. ~ toward pole 極に(向けて)凸な. holomorphically ~ 正則凸な. locally ~ 局所凸な. logarithmically ~ 対数的凸な. proper ~ function 真凸関数, 適正(な)凸関数. proper ~ set 真凸集合. strictly ~ 狭義(に)凸な. uniformly ~ 一様凸な.

convexity *n.* 凸, 凸性.
~ theorem 凸(性)定理. holomorphical ~ 正則凸性. logarithmic ~ 対数的凸性. radius of ~ 凸性[凸型]半径. uniform ~ 一様凸性.

convolution *n.* たたみ込み; 合成(変換); 合成積.
~ kernel 合成核. ~ transform (ation) 合成変換.

cooperative *a.* 協力の.
~ game 協力ゲーム.

cooperator *n.* 協力者.

coordinate *n.* 座標.
barycentric ~s 重心座標. Cartesian ~s デカルト座標. ~ axis 座標軸. ~ curve 座標曲線. ~-free 座標によらない. ~ geometry 座標幾何学. ~ plane 座標平面. ~ surface 座標曲面. ~ system 座標系. ~ transformation 座標変換. current ~s 流通座標. curvilinear ~s 曲線座標. cyclic ~s 円点座標; 循環座標. cylindrical ~s (円)柱[筒]座標. generalized ~s 一般座標. geodesic ~s 測地座標. local ~ 局所座標. parallel ~s 平行座標. orthogonal ~s 直交座標. polar ~s 極座標. rectangular ~s 直角[交]座標. rectilinear ~s 直角[直線]座標. spherical ~s 球座標.

coplanar *a.* 同平面の, 共面の, 共角の.
~ points 共面点.

coplanarity *n.* 共面性.

coprime *a.* 互いに素な.

coproduct *n.* 双対積.
direct ~ 双対直積.

copula *n.* コピュラ.

copunctal *a.* 共点の.
~ planes 共点平面.

copyright *n.* 版権, 著作権【©】.

cor. corollary.

coregular *a.* 双対正則な.
~ representation 双対正則表現.

corner *n.* 角, かど.
~ condition 角点条件, ~ point 角点.

corollary *n.* (定理の)系.

corona *n.* 冠, コロナ.
~ problem コロナ問題

correct *a.* 正しい, 正確な. *vt.* 訂正する; 校正する.

correction *n.* 補正, 修正; 校正; 訂正.

corrective *a.* 修正の.

corrector *n.* 修正子; 校正者.

correlation *n.* 相反[相関]変換; 相関.
bivariate ~ 二変量相関. ~ coefficient 相関係数. ~ diagram 相関図. ~ ellipse 相関楕円. ~ ratio 相関比. ~ table 相関表. ~ tensor 相関テンソル. cross ~ 相互相関. curved ~ 曲(線)相関. involutive ~ 対合的相反変換. linear ~ 線形相関. multiple ~ 重相関. partial ~ 偏相関. perfect ~ 完全相関. population [sample] ~ 母(集団)[標本]相関. total ~ 単相関, 全相関.

correlative *a.* 相関(的)の.

correlogram *n.* コレログラム, 相関図(表).

correspond *vi.* 対応する; 通信する【to, with】.

correspondence *n.* 対応; 通信.
affine ~ アフィン対応. algebraic ~ 代数的対応. birational ~ 双有理対応. circle-to-circle ~ 円々対応. conformal ~ 共形(的)対応. ~ principle 対応原理. inverse ~ 逆対応. one-to-one ~ 一対一対応. projective ~ 射影(的)対応. similar ~ 相似(的)対応.

corresponding *a.* 対応する, 相当する.

~ angles 同位角. ~ points 対応点. ~ sides 対応辺. ~ vertices 対応頂点.

corrigendum (*pl.* -da) *n.* (*pl.*) 正誤表.

cosecant *n.* コセカント, 余割.

coset *n.* 剰余類.
left [right] ~ 左[右]剰余類.

cosine *n.* コサイン, 余弦.
~ circle 余弦円. ~ formula 余弦公式. ~ law [rule] 余弦法則. ~ transformation 余弦変換. direction ~(s) 方向余弦. first [second] ~ formula 第一[第二]余弦公式. hyperbolic ~ 双曲(線)余弦. integral ~ 積分余弦.

cost *n.* 費用; 原価.
~ accounting 原価計算. ~ function 費用関数. ~ price, prime ~ 原価. ~ slope 費用増加率.

cotangent *n.* コタンジェント, 余接.
~ bundle 共変接ベクトル束, 余接束.

coterminal *a.* 境界共有の.
~ angles 両辺共有の角.

Cotes, Roger (1682-1716) コーツ.

co-tree *n.* 補木.

count *vt.* 数える, 計算する. *n.* 計算; 総数.
complete ~ 全部調査. ~ing function 個数関数. ~ing measure 個数測度. ~ing process 計数過程.

countability *n.* 可算性, 可付番(性).
~ axiom 可算公理.

countable *a.* 可算の, 可付番の. ⇨ denumerable, enumerable.
~ basis 可算基. ~ covering 可算被覆. ~ ordinal number 可算順序数. ~ set 可算集合.

countably *ad.* 可算(的)に.
~ additive 可算加法的な. ~ compact 可算コンパクトな. ~ infinite 可算無限の. ~ normed space 可算ノルム空間.

counter- *pref.* 「反対, 逆, 対」の意.
~ example 反例.

counterclockwise *a., ad.* 左回りの[に], 時計の針と反対の方向の[に].

counter(-)example *n.* 反例.

couple *n.* 対; 偶力.
~ of forces 偶力.

coupling *n.* 連関; 対結合, 相引.

courant *n.* (*F.* =current.) カレント.

Courant, Richard (1888-1972) クーラント.

Cournot, Antoine Augustin (1801-1877) クールノー.

covariance *n.* 共分散; 共変性.
analysis of ~ 共分散分析. ~ matrix (共)分散行列.

covariant *a.* 共変的な, 共変の.
~ component 共変成分. ~ differentiation 共変微分. ~ functor 共変関手. ~ index 共変指標. ~ tensor 共変テンソル. ~ vector 共変ベクトル.

covariate *n.* 共変量.

covector *n.* コベクトル.
~ field コベクトル場.

cover *vt.* おおう, 被覆する. *n.* おおい; 包.
measurable ~ 等測包. node ~ 頂点被覆. open ~ 開被覆. projective ~ 射影被覆.

covering *n.* 被覆. *a.* おおっている.
countable ~ 可算被覆. ~ group 被覆群. ~ manifold 被覆多様体. ~ space 被覆空間. ~ surface 被覆面. ~ theorem 被覆定理. ~ transformation 被覆変換. finite ~ 有限被覆. open ~ 開被覆.

coversine *n.* coversed sine. 余矢【covers】.

Coxeter, Harold Scott (1907-2003) コクセター.

CPM Critical Path Method. 臨界路法.

Cramer, Gabriel (1704-1752) クラーメル.

Cramer, Harald (1893-1985) クラメル.

creation *n.* 生成.
~ operator 生成作用素.

Crelle, August Leopold (1780-1855) クレレ.

Cremona, Luigi (1830-1903) クレモナ.

crescent *n.* 弓形, 月形. ⇨ lune.

criterion (*pl.* -ria, -ons) *n.* 基準; 判定条件, 判定法. ⇨ test.
convergence ~, ~ of convergence 収束判定法[条件]. ~ function 判定関数. Eisenstein ~ アイゼンシュタイン既約判定法. Jacobian ~ (of regularity) (正則性の)ヤコビ判定法. simplex ~ 単体規準.

critical *a.* 臨界の.
～ angle 臨界角. ～ determinant 臨界行列式. ～ lattice 臨界格子. ～ path 臨界路. ～ point 臨界点, 危点; 棄却点, 棄却限界値. ～ region 臨界域, 棄却域. ～ value 臨界値.

cross *n.* 十字形; 交差[交叉]. *a.* 交差した, 相互の. *vt., vi.* 交差する, 横切る.
～ at right angles 直交する. ～ correlation 相互相関. ～ cut 横断線. ～ figure 断面図. ～ product クロス乗積, 交差積. ～ ratio 十字比, 非調和比, 複比, 交差比. ～ reference 相互参照. ～ section 断面.

crossed *a.* 交差[交叉]した.
～ homomophism 一双対鎖; 微分作用素. ～ product 接合積.

crossover *n.* 雑種; 交配.
～ design 交配法.

cross(-)section *n.* 断面.
～ paper 方眼紙.

crosswise *a.* 交差[交叉]した. *ad.* 交差[交叉]させて.

crunode *n.* 二重点.

cryptography *n.* 暗号(法).

cryptosystem *n.* 暗号系.

crystal *n.* 結晶.
～ class 結晶類. ～ system 結晶系.

crystalline *a.* 結晶性の.
～ cohomology クリスタルコホモロジー.

crystallographic *a.* 結晶の, 結晶学上の.
～ group 結晶群.

cubage *n.* 体積, 容積.

cubature *n.* 立体求積(法), 体積計算.

cube *n.* 立方体; 立方, 三乗. *vt.* 三乗する. ⇨ square.
～ root 立方根. perfect ～ 完全立方. unit ～ 単位立方体.

cubic *a.* 立方体の, 正六面体の; 立方の, 三次の, 三乗の; 等軸晶系の. *n.* 三次式, 三次方程式, 三次曲線, 三次関数.
～ curve 三次曲線. ～ equation 三次方程式. ～ field 三次体. ～ function 三次関数. ～ lattice 立方格子(形). ～ number 立方数. ～ remainder 開立剰余. ～ resolvent 三次分解(方程)式. ～ root 立方根. ～ system 等軸晶系, 立方晶系.

cubical *a.* 立方体の, 正六面体の; 三次の.

cuboid *n.* 直平行六面体. *a.* 立方(体)形の.

cuboidal *a.* 立方(体)形の.

cumulant *n.* キュムラント.

cumulation *n.* 累加.

cumulative *a.* 累積的な, 累積-.
～ distribution 累積分布. ～ distribution function 累積分布関数[C.D.F.]. ～ error 累積誤差. ～ frequency 累積度数.

cuneiform 楔形文字.

cup *n.* コップ, カップ【記号 ∪】, 結び. ⇨ cap.
～ product カップ積.

curl *n.* 回転; 回転(演算)子. ⇨ rotation.

current *n.* 流れ; カレント.
～ coordinate(s) 流通座標. random ～ 彷徨カレント.

curriculum (*pl.* lums, -la) *n.* 教科課程, カリキュラム.
～ vitae 履歴(書).

curse *n.* のろい, 災い.
～ of dimensionality 次元の呪い.

cursor *n.* カーソル.
～ line カーソル線.

curtail *vt.* 短縮する; 切り詰める.

curvature *n.* 曲率.
absolute ～ 絶対曲率. center of ～ 曲率中心. circle of ～ 曲率円. ～ form 曲率形式. ～ tensor 曲率テンソル. Gauss(ian) ～ ガウス(の)曲率. line of ～ 曲率線. mean ～ 平均曲率. normal ～ 法曲率. principal ～ 主曲率. radius of ～ 曲率半径. Ricci ～ リッチ曲率. scalar ～ スカラー曲率. sectional ～ 断面曲率. total ～ 全曲率.

curve *n.* 曲線.
algebraic ～ 代数曲線. analytic ～ 解析曲線. coordinate ～ 座標曲線. ～ length 曲線の長さ. ～-fitting 曲線のあてはめ. ～ of constant breadth 定幅曲線. ～-plotting 曲線を画くこと. ～ theorem 曲線定理. ～ tracing 曲線の追跡. elliptic ～ 楕円曲線. frequency ～ 度数[頻度]曲線. Jordan ～ ジョルダン曲線. meromorphic ～ 有理型[形]曲線. modular ～ モジュラー曲線.

Peano ~ ペアノ曲線. plane ~ 平面曲線. quadratic ~ 二次曲線. rational ~ 有理曲線. regular ~ 正則曲線. simple closed ~ 単純[単一]閉曲線. sine ~ 正弦曲線. space ~ 空間曲線. tooth ~ 歯形曲線. transcendental ~ 超越曲線. twisted ~ 空間曲線, ねじれた曲線.

curved *a.* 曲がった.
~ correlation 曲(線)相関. ~ line 曲線. ~ surface 曲面.

curvilinear *a.* 曲線の.
~ coordinates 曲線座標. ~ integral (曲)線積分.

cusp *n.* 尖点, カスプ.
~ form 尖点形式. ~-locus 尖点の軌跡. double ~ 二重尖点. parabolic ~ 放物的尖点.

cusped *a.* とがった; 尖点の.
three-~ hypocycloid 三尖点内擺線.

cuspidal *a.* とがった, 尖点の.
~ point 尖点.

customer *n.* (顧)客.

cut *vt., vi.* 切る, 切断する. *n.* 切断; カット; 切れ目.
cross ~ 横断線. ~ of rational numbers 有理数の切断. ~ orthogonally 直交する. ~ set 切断集合. ~ting plane 切除平面. Dedekind ~ デデキント(の)切断. golden ~ [section] 黄金分割. irrational [rational] ~ 無理[有理]切断. minimum ~ 最小カット.

cut-off *n.* 切り捨て. *vt.* 切り落とす.

cybernetics *n.* サイバネティックス.

cycle *n.* 輪体; 周期, サイクル; 巡回置換.
boundary ~ 境界輪体. ~ time サイクル・タイム. dividing ~ 分割輪体. Hamiltonian ~ ハミルトン閉路. limit ~ 極限周期軌道.

cyclic *a.* 循環する; 巡回的な; 輪環の.
~ algebra 巡回多元環. ~ code 巡回符号. ~ coordinates 円点座標; 循環座標. ~ determinant 巡回行列式. ~ extension 巡回拡大. ~ group 巡回群. ~ homology 巡回ホモロジー. ~ order 輪環の順. ~ permutation 巡回置換; 輪環; 円順列. ~ vector 巡回ベクトル.

cyclide *n.* サイクライド.
confocal ~s 共焦サイクライド.

cycling *n.* 循環; 巡回.

cycloid *n.* サイクロイド, 擺線.
~ pendulum サイクロイド振子.

cycloidal *a.* サイクロイドの.

cyclometry *n.* 円の求積.

cyclosymmetry *n.* 循環対称.

cyclotomic [sàiklətámik, sàiklətɔ́mik] *a.* 円分の, 円等分の.
~ (number) field 円分体. ~ polynomial 円分多項式.

cylinder *n.* 円柱, 柱, 筒[壔], 柱面[体].
circular ~ 円柱. ~ function (円)柱関数. elliptic ~ 楕円柱. ~ set 筒[柱]集合. hyperbolic [parabolic] ~ 双曲[放物]柱. quadratic ~ 二次柱(面). right ~ 直角柱.

cylindric =cylindrical.

cylindrical *a.* 円筒[壔](状)の, (円)柱(状)の.
~ coordinates (円)柱[筒]座標. ~ function 円柱[筒]関数. ~ surface 柱[筒]面.

cylindroid *n.* 曲線柱. *a.* 円柱形の.

cypher =cipher.

D

D, d (ローマ数字の) 500.

daily *a.* 毎日の.
~ interest 日歩.

D'ALEMBERT, Jean Baptiste le Rond (1717–1783) ダランベール.

d'Alembertian *a.* ダランベールの. *n.* ダランベール演算子.

damp *vt.* 減衰させる. *n.* 減衰.
~ed Newton method 減速ニュートン法. ~ed oscillation 減衰振動. ~ ratio, ~ing ratio 減衰比.

DANIELL, Percy John (1889–1946) ダニエル.

DARBOUX, Jean Gaston (1842–1917) ダルブー.

dash *n.* (句読法での)ダッシュ【—】.【本邦では prime【'】の俗用.】

data (*pl.*) *n.* (*sing.* -tum) データ, 資料.
analog(ue) ~ アナログ・データ. continuous ~ 連続型データ. digital ~ デジタル・データ. initial ~ 初期データ. multivariate ~ 多変量データ.

datum (*pl.* -ta) *n.* (一つの)事実, 情報, データ(の一項目); 既知数. ⇨ data.

d.c.c. descending chain condition. 降鎖条件.

dead *a.* 死んだ; 静かな.
~ water 静水; 航跡渦流.

deal *vt.* 分ける. *vi.* 分配する【out】; 扱う【with】.
~ with を処理する, に従事する.

death *n.* 死亡.
birth and ~ chain 出生死亡連鎖. ~ process 死亡過程. ~ rate 死亡率.

deca- *pref.*「十」の意.

decade *n.* 十の一組; 十年間; 十巻, 十編.

decagon *n.* 十角形, 十辺形.
regular ~ 正十角[辺]形.

decagonal *a.* 十角形の, 十辺形の.

decahedral *a.* 十面体の.

decahedron (*pl.* -rons, -dra) *n.* 十面体.

deceleration *n.* 減速; 負の加速度.

deci- *pref.*「十分の一」の意.
~meter デシ[十分の一]メートル.

decidable *a.* 決定可能な.

decide *vt.* 決定する.

decile *n.* 十分位数.

decillion *n.*【米】千の十一乗数;【英・仏・独】百万の十乗数.

decimal *a.* 十進法の; 小数の. *n.* 小数.
circulating ~ 循環小数. ~ fraction 十進小数. ~ notation, ~ numeration system 十進記数法. ~ part 小数部(分). ~ place 小数位. ~ point 小数点. ~ system 十進法. recurring [repeating] ~ 循環小数. terminating ~ 有限小数.

decimalization *n.* 十進法化[にすること].

decimalize *vt.* 十進法にする.

decimally *ad.* 十進法で; 小数で, 小数の形で.

decipher *vt.* 解読する.

decision *n.* 決定, 判定, 判断.
~ element 論理素子. ~ function 決定関数. ~ problem 決定問題. ~ process 決定過程. ~ space 決定空間. ~ table デシジョン・テーブル, 判断表.

declination *n.* (下への)傾斜.
angle of ~ 俯角, 伏角.

decode *vt.* デコード, 復号する, 解読する. ⇨ encode.

decoder *n.* 解読器，デコーダ，復号器.

decoding *n.* 符号復元，復号; 解読.
~ function 復号化関数. maximum likelihood ~ 最尤復号法.

decomposability *n.* 分解可能性.

decomposable *a.* 分解可能な，可約な.

decompose *vt., vi.* (成分に)分解させる[する].

decomposition *n.* 分解; 分割.
canonical ~ 標準分解. cellular ~ 胞体分割. ~ field 分解体. ~ group 分解群. ~ into partial fractions 部分分数分解. ~ theorem 分解定理. direct sum ~ 直和分解. irreducible ~ 既約分解. orthogonal ~ 直交分解. Schur ~ シューア分解. simplicial ~ 単体分割. spectral ~ スペクトル分解.

decrease *n.* 減少. *vi., vt.* 減少する. ⇨ increase.
~ by one 一だけ減らす. increase and ~ 増減.

decreasing *a.* 減少の. ⇨ increasing.
~ function 減少関数. ~ sequence 減少列. ~ state 減少の状態. monotone ~ 単調減少の. rapidly ~ 急減少の. strictly ~ 狭義に減少の.

decrement *n.* 減少; 減少量; 減分. ⇨ increment.
logarithmic ~ 対数減衰率.

decrypt *vt.* 解読する; 復号する.

decryption *n.* 復号化; 暗号解読.

Dedekind, Julius Wilhelm Richard (1831-1916) デデキント.

deduce *vt.* 推論する，演繹する.

deduction *n.* 推論; 演繹(法).

deductive *a.* 推論の，演繹的な.
~ method 演繹法.

defect *n.* 欠点; 不足，不足指数; 除外指数.
~ group 不足群. ~ of triangle (三角形の)角不足. major [minor] ~ 重[軽]欠点.

deficiency *n.* 不足; 欠除次数; 除外指数.
approximation in ~ 不足近似. ~ index 不足指数.

deficient *a.* 不足している【in】; 不足の.
~ number 不足数，輪数. ~ value 欠除値; 除外値.

definability *n.* 定義可能性.

definable *a.* 定義できる，定義可能な.

define *vt.* 定義する; 限定する; (形状を)定める.

defining *a.* 定義する.
~ equation 定義方程式. ~ function 定義関数. ~ relation 定義関係.

definite *a.* 一定の; 限定的な; 確定した，明確な; 定-.
~ form 定符号形式. ~ integral 定積分. ~ language 限定言語. negative-~ 負(定)値の. of ~ area 面積確定の. of ~ sign 定符号の. positive-~ 正(定)値の.

definitely *ad.* 定まって.
~ divergent 定発散の.

definition *n.* 定義.
~ by induction 帰納法による定義. inductive ~ 帰納的定義. recursive ~ 再帰[帰納]的定義.

deflation *n.* 収縮(写像); 減次.

deflection *n.* 偏差; たわみ.
~ tensor 偏差テンソル.

deform *vt.* 変形させる.

deformation *n.* 変形; ゆがみ; 変位.
continuous ~ 連続(的)変形. ~ of structures 構造の変形. ~ ratio 変形率. ~ retract 変位レトラクト. ~ theory 変形理論. infinitesimal ~ 無限小変形.

deg. =degree.

degeneracy *n.* 退化，縮退，縮重; 退化性.
~ operator 退化作用素.

degenerate *vi.* 退化する. *a.* 退化した; 退化の.
~ bilinear form 退化双線形形式. ~ conics 退化円錐曲線. ~ kernel 退化核. ~ series 退化系列.

degeneration *n.* 退化.

degree *n.* 次; 次数; 度.
apparent ~ 見掛けの次数. complementary ~ 補充次数. ~ of association 関連度. ~ of freedom 自由度. ~ of scattering 分散度. ~ of separation 分離度. equation of higher ~ 高次方程式. extension ~ 拡大次数. inseparable ~ 非分離次数. map-

ping ~ 写像度. polynomial of ~ n n次の多項式. power ~ ベキ次数. separable ~ 分離次数. total ~ 総次数. transcendence ~, ~ of transcendence 超越次数. virtual ~ 仮想次[指]数.

del *n.* 勾配・発散の演算【記号 ∇】.

DE LA VALLÉE-POUSSIN, Charles Jean Gustav Nicolas (1866-1962) ド・ラ・バレ・プッサン.

delay *n.* 遅延, 遅れ.
~ system 待時式.

delete *vt.* 削除する.
~d neighbo(u)rhood 除外近傍.

deletion *n.* 削除.

DELSARTE, Jean (1903-1968) デルサルト.

delta *n.* デルタ. ギリシア語アルファベットの 4 番目の文字【Δ, δ】.
~ function デルタ関数. Kronecker's ~ クロネッカーのデルタ.

delta amplitudinis (*L.*) dn 関数の旧名.

deltoid *n.* デルトイド, 三尖点内擺線.

demand *n.* 需要; 要求. *vt.* 要求する.
~ and supply 需要供給. ~ curve 需要曲線. ~ forecasting 需要予測. ~ function 需要関数.

demi- *pref.* 「半」の意. ⇨ hemi-, semi-.
~lune 半月. ~space 半空間.

demography *n.* 人口統計学.

DE MOIVRE, Abraham (1667-1754) ド・モアブル.

demonstrate *vt., vi.* 証明する. 論証する.

demonstration *n.* 証明.

DE MORGAN, Augustus (1806-1871) ド・モルガン.

denary *a.* 十の; 十進の.
~ scale 十進法.

denial *n.* 否定.

DENJOY, Arnaud (1884-1974) ダンジョア.

denominate *a.* 特定の名のある. *vt.* 命名する, 名づける.
~d number 名数.

denomination *n.* (度量衡の)単位; 命名.

denominator *n.* 分母. ⇨ numerator.
common ~ 公分母. partial ~ 部分分母. rationalization of ~ 分母の有理化.

denote *vt.* 表示する, 示す, 表す; 意味する.

dense *a.* 密な; 稠密な.
~ in itself 自己(稠)密な. ~ matrix 密行列. everywhere ~ いたるところ稠密な. nowhere ~ 全疎な. self-~ 自己稠密な.

denseness *n.* 稠密(性).

density *n.* 稠密性; 密度; 密集.
~ function 密度関数. ~ point 密集点. ~ theorem 密度定理; 密集点定理. kinetic ~ 位置密度. line(ar) ~ 線密度. mean ~ 平均密度. packing ~ 充填密度. point of ~ 密集点. probability ~ 確率密度. surface ~ 面密度. volume ~ 体密度.

denumerability *n.* 可付番性, 可算性.

denumerable *a.* 可付番の, 可算の. ⇨ countable, enumerable.
~ set 可算[可付番]集合.

deny *vt.* 否定する.

department *n.* 部; 学部; 科, 教室.
~ of mathematics, mathematics ~, mathematical ~ 数学科; 数学教室.

departmental *a.* 部の; 科の; 教室の.

departure *n.* 退去; 出発.

depend *vi.* 従属する; 依存する.
~ on [upon] に依存する.

dependence *n.* 従属; 依存.
domain [region] of ~ 依存領域. functional ~ 関数的従属性. linear ~ 一次[線形]従属性.

dependency =dependence.

dependent *a.* 従属する; 従属な.
algebraically ~ 代数的(に)従属な. ~ event 従属事象. ~ variable 従属変数. functionally ~ 関数的(に)従属な. linearly ~ 一次[線形]従属な.

depression *n.* 低下; 俯角, 伏角.
angle of ~ 俯角, 伏角. ~ of order 階数低下.

dept. =department.

depth *n.* 深さ, 深度; 奥行き.
~ first search 深さ優先探索. ~ of prime ideal 素イデアルの深さ.

DE RHAM, Georges-William (1901-1990) ド・ラム.

derivable *a.* 誘導される; 微分可能な; 推論可能な.

derivate *n.* (=derivative) 導関数; 微分係数.

derivation *n.* 誘導; 微分; 導分; 微分子, 微分作用素.
alternating ~ 交代微分. exterior ~ 外微分. inner ~ 内(部)微分. partial ~ 偏微分作用素. point ~ 点微分. rightmost ~ 右端導出.

derivative *n.* 導関数; 微分係数, 微分商.
angular ~ 角微係数. approximate ~ 概微分係数. directional ~ 方向導関数. exterior ~ 外微分. Fréchet ~ フレシェ微分. Lie ~ リー微分. logarithmic ~ 対数(的)導関数. normal ~ 法線導関数. partial ~ 偏導関数, 偏微分係数. upper [lower] ~ 上[下]微分係数.

derive *vt.* 引き出す, 導き出す, 推論する.

derived *a.* 誘導された, 導来-.
~ category 導来圏. ~ equation 誘導方程式. ~ function 導関数. ~ functor 導来関手. ~ group 導来群. ~ series 導来列. ~ set 導集合.

DESARGUES, Gérard [Girard] (1591-1661) デザルグ.

Desarguesian *a.* デザルグの[的な].
non-~ 非デザルグ的な.

DESCARTES, René (1596-1650) デカルト.

descend *vi.* 下降する. ⇨ ascend.

descending *a.* 下降の, 降-.
~ central series 降中心列. ~ chain 降鎖. ~ chain condition [d.c.c.] 降鎖条件. ~ order of powers 降べキの順.

descent *n.* 降下.
line [curve] of steepest ~ 最急降下線. line of swiftest ~ 最速降下線. method of ~ 降下法. steepest ~ 最急降下. swiftest ~ 最速降下.

describe *vt.* 述べる. 記述する; (図形を)描く; 運行する.

description *n.* 記述; 説明; 作図.

descriptive *a.* 記述の.
~ geometry 画法幾何学. ~ set theory 記述集合論. ~ statistics 記述統計学.

design *n.* 計画, 設計. *vt., vi.* 計画する.
block ~ ブロック計画. crossover ~ 交配法. ~ed distance 設計距離. ~ matrix 計画行列. ~ of experiment 実験計画. ~ requirement 設計の(要求)品質. experimental ~ 実験計画. lattice ~ 格子計画. logical ~ 論理設計. two element ~ 二元計画(法).

designate *vt.* 命名する; 指定する; 記号づける.

destinator *n.* 受報者. ⇨ receiver.

destructive *a.* 破壊的な.
~ inspection 破壊検査.

detail *n.* 詳細; 細目.
in ~ 詳細に.

detection *n.* 探知; 検出, 発見.

determinant *n.* 行列式.
adjoint ~ 随伴行列式. cyclic ~ 巡回行列式. ~ divisor 行列式因子. expansion of ~ 行列式の展開. functional ~ 関数行列式. infinite ~ 無限行列式. minor ~ 小行列式.

determinantal *a.* 行列式の.
~ equation 行列式方程式.

determinate *a.* 一定の; 確定数[量]の, 既知(数)の.
~ divergence 定発散.

determination *n.* 決定; 限定.
coefficient of ~ 寄与率, 決定係数.

determine *vt.* 決定する; 位置を定める; 確定する.

determined *a.* 決定された, 確定した.
~ system 決定系.

determining *a.* 決定する.
~ condition 決定条件. ~ equation 決定方程式. ~ set 決定集合.

deterministic *a.* 決定的な, 決定論的な; 確定的な. ⇨ stochastic.
~ model 決定論的モデル[模型]. ~ process 確定的過程. ~ Turing machine 決定性チューリング機械.

detour [dí:tʊə, -tʊə] *n.* 迂回.
~ matrix 迂回行列.

DEURING, Max (1907-1984) ドイリング.

develop *vt., vi.* 展開する. *vt.* 発展させる.

vi. 発展する.

developable *a.* 展開可能な.
~ surface 可展面, 展開可能曲面.

development *n.* 展開; 発展.
series ~ 級数展開.

deviance *n.* 逸脱度, デビアンス.

deviate *vi.* 外れる【from】. *n.* 偏倚.

deviation *n.* 偏差, 偏倚.
~ point 偏差点. mean ~ 平均偏差. quartile ~ 四分位偏差. standard ~ 標準偏差.

device *n.* 装置; 機構.

devidendo *n.* 分[除]比の理. ⇨ dividendo. componendo and ~ 合除比の理.

dextrorse *a.* 右巻きの. ⇨ sinistrorse.

di- *pref.*「二, 二重」の意.

diacaustic *a.* 屈折火[焦]線の. *n.* 屈折火[焦]線

diadic *a.* 二進の.
~ integer 二進整数. ~ system 二進法.

diagnosis (*pl.* -noses) *n.* 診断

diagnostic *a.* 診断の. *n.* 診断.

diagonal *a.* 対角線の. *n.* 対角線; 対角集合; 距線.
auxiliary ~ 副対角線. ~ component 対角成分. ~ dominance 対角優位. ~ element 対角元[成分]. ~ line 対角線. ~ matrix 対角行列. ~ method [process] 対角(線)論法. ~ morphism 対角射. ~ partial sum 対角部分和. ~ point 対角点. ~ sequence 対角列. ~ set 対角集合. ~ triangle 対角三角形.

diagonalizable *a.* 対角化可能な.
~ matrix 対角化可能(な)行列.

diagonalization *n.* 対角化.

diagram *n.* 図, 図表; 図式, ダイアグラム; 作図.
arrow ~ 矢線図. block ~ ブロック・ダイアグラム. collinear ~ 共線図表. commutative ~ 可換な図式. correlation ~ 相関図. dot ~ 点図表. Dynkin ~ ディンキン図形, ディンキン・ダイアグラム. proof ~, ~ of proof 証明図. Venn ~ ベン(の)図式. Young ~ ヤング図形, ヤング・ダイアグラム.

diameter *n.* 直径.
conjugate ~ 共役直径. ~ of circle [sphere] 円[球]の直径. ~ of set 集合の直径. trans-finite ~ 超越直径.

diametral *a.* 直径の.
~ end points 直径の両端点. ~ plane 径面.

diametric *a.* 直径の.

diametrical ＝diametric.

dice (*pl.*) *n.* (*sing.* die) さいころ. *vi.* さいころ遊びをする.

dichotomizing *a.* 二分的な.
~ search 二分探索.

dichotomy *n.* 二分法; 二項対立; 半月.
double ~ 二重二分法.

Dickson, Leonard Eugene (1874–1954) ディックソン.

dictionary *n.* 辞書, 辞典.

die (*pl.* dice) *n.* さいころ.
random ~ 乱数賽.

Dieudonné, Jean Alexandre Eugène (1906–1992) ディユドンネ.

diffeomorphic *a.* 微分同相[型, 形]な.
~ mapping 微分同相写像.

diffeomorphism *n.* 微分同相[型, 形](写像).

diffeotopic *a.* 微分同位(な).

difference *n.* 差; 差分; 階差.
backward ~ 後退差分. calculus of ~s 差分法. central ~ 中心差分. common ~ 公差. ~ approximation 差分近似. ~ calculus 差分法. ~-differential equation 差分微分方程式. ~ equation 差分方程式, 階差[定差]方程式. ~ group 差群. ~ method 差分(的方)法. ~ product 差積. ~ quotient 差商. ~ table 差分表. divided ~ 差分商. finite ~ 有限差分. forward ~ 前進差分. mixed ~ 混合差分. progression of ~s 階差数列. symmetric ~ 対称差. tabular ~ 表差.

different *a.* 異なった. *n.* 共役差積.
~ from zero ゼロとは異なる.

Differente *n.* (*G.*) 共役差積.

differentiability *n.* 微分可能性, 可微分性.

differentiable *a.* 微分可能な，可微分な．
~ function 微分可能な関数．~ manifold 可微分多様体．~ mapping 可微分写像．~ structure 可微分構造．left [right] ~ 左[右]微分可能な．*n* times continuously ~ *n* 回連続微分可能な．partially ~ 偏微分可能な．piecewise ~ 区分的に微分可能な．totally ~ 全微分可能な．

differential *a.* 微分の; 差の．*n.* 微分．
analytic ~ 解析微分．~ analyzer 微分解析機．~ calculus 微分学．~ coefficient 微分係数．~ cost 差額費用．~ equation 微分方程式．~ form 微分形式．~ geometry 微分幾何(学)．~ invariant 微分不変式．~ manifold 微分多様体．~ of higher order 高階[次]微分．~ operator 微分作用素．~ process 加法過程．~ quotient 微分商．~ topology 微分位相幾何学．exact ~ 完全微分．harmonic ~ 調和微分．holomorphic [meromorphic] ~ 正則[有理型]微分．*n* th ~ 第 *n* 階[次]微分．ordinary [partial] ~ equation 常[偏]微分方程式．quadratic ~ 二次微分．stochastic ~ equation 確率微分方程式．total ~ 全微分．

differentiate *vt.* 微分する．
~ exteriorly 外微分する．~ partially 偏微分する．

differentiation *n.* 微分; 微分すること; 微分法．
automatic ~ 自動微分．complex ~ 複素微分．covariant ~ 共変微分．~ operator 微分作用素．exterior ~ 外微分．graphical ~ 図式微分．logarithmic ~ 対数(的)微分．numerical ~ 数値微分．partial ~ 偏微分．repeated ~ 累次微分．successive ~ 逐次微分．termwise ~ 項別微分．total ~ 全微分．

diffraction *n.* 回折．

diffusion *n.* 拡散．
~ coefficient 拡散係数．~ equation 拡散方程式．~ process 拡散過程．

digamma *n.* ディガンマ．
~ function ディガンマ関数．

digit *n.* 指; 指幅; 数字; 桁，ディジット．
binary ~ 二進数字．

digital *a.* 数字の; 計数型の; デジタルな．*n.* 指．
~ computer デジタル[計数型]計算機．~ data デジタル・データ．

digraph *n.* 有向グラフ．

dihedral *a.* 二面の．
~ angle 二面角，稜角．~ group 二面体群．

dilatation *n.* 拡大; 膨張; 膨張変換．
~ ratio 拡大率．maximal ~ 最大変形率．

dilate *vt.* 膨張させる．*vi.* 膨張する．

dilation [dɑɪléɪʃən, dɪ-] ＝dilatation.

dilemma *n.* 両刀論法，ディレンマ．

dim. dimension, dimensional.

dimension *n.* 次元，ディメンション．
cohomological ~ コホモロジー次元．complex ~ 複素次元．~ function 次元関数．~ theory 次元論．geometric ~ 幾何的次元．homological ~ ホモロジー次元．inductive ~ 帰納的次元．Krull ~ クルル次元．real ~ 実次元．

dimensional *a.* 次元の．
~ analysis 次元解析．even [odd] (-) ~ 偶[奇]数次元の．finite (-) [infinite (-)] ~ 有限[無限]次元の．*n*-~ *n* 次元(的)の．two-~ 二次元(的)の．

dimensionality *n.* 次元数
curse of ~ 次元の呪い．

diminish *vt.* 減少させる，縮小させる．*vi.* 減少する，縮小する．

diminution *n.* 減少，縮小．

DINI Ulisse (1845–1918) ディニ．

Diophantine *a.* ディオファンタスの．
~ approximation ディオファンタス近似．~ equation ディォファンタス方程式．

DIOPHANTUS of Alexandria (264?–330?) アレクサンドリアのディオファンタス．

dipole *n.* 双極(子); 極．
~ coordinates 双[二]極座標．~ potential 双極ポテンシャル．

DIRAC, Paul Adrien Maurice (1902–1984) ディラック．

direct *vt.* 向ける．*a.* 直接の．⇨ directed.
~ continuation 直接接続．~ image 順像．~ integral 直積分．~ limit 直[順]極限，

帰納的極限(=inductive limit). ~ measurement 直接測定. ~ method 直接(的方)法. ~ product 直積. ~ proof 直接証明(法). ~ proportion 正比例. ~ sum 直和. ~ summand 直和因子. ~ system 直[順]系, 帰納系(=inductive system).

directed *a.* 有向の. ⇨ oriented.
~ angle 有向角. ~ curve 有向曲線. ~ distance 有向距離. ~ family 有向族. ~ graph 有向グラフ. ~ (line) segment 有向線分; 矢線. ~ set 有向集合. ~ tangent 有向接線. ~ tree 有向木.

direction *n.* 方向, 方位; 向き.
asymptotic ~ 漸近方向. Borel [Julia] ~ ボレル[ジュリア]方向. characteristic ~ 特性方向. conjugate ~ 共役方向. descent ~ 降下方向. ~ angle 方向角. ~ coefficient 方向係数. ~ cosine(s) 方向余弦. ~ factor 方向因子. ~ of principal curvature 主曲率方向. ~ ratio 方向比. negative [positive] ~ 負[正]の方向. principal ~ 主方向. search ~ 探索方向. stable in both ~s 両側安定な.

directional *a.* 方位の; 方向(上)の.

directly *ad.* 直接に.
~ conformal 順等角な. ~ indecomposable 直既約な(=indecomposable).

director *n.* 指導物; 指導者.
~ circle 準円. ~ curve 導線. ~ line 準線.

directrix (*pl.* -trixes, -trices) *n.* 準線, 稜, 導面.
joint ~ 連合準線.

DIRICHLET, Peter Gustav Lejeune (1805-1859) ディリクレ, ディリシュレ.

dis- *pref.* 不-, 非-;「分離, 除去」の意: discontinuous; disjoint.

disable *vt.* 使用禁止にする, 割込み禁止にする.

disc *n.* (円)板, ディスク. ⇨ disk.
circular ~ 円板. closed ~ 閉円板. ~ algebra 円板環. unit ~ 単位円板.

discard *vt., vi.* 捨てる.
~ing of fraction 端数の切捨て.

discharge *n.* 除去; 放電.
~ of double negation 二重否定の除去.

discipline *n.* 分野, 学科.

disconnected *a.* 非連結な, 不連結な.
totally ~ (完)全不連結な.

discontinuity *n.* 不連続(性)[(点)].
point of ~ 不連続点.

discontinuous *a.* 不連続な.
~ function 不連続関数. ~ group 不連続群. ~ part 不連続部分. ~ point 不連続点. properly ~ 真性不連続な. purely ~ 純不連続な.

discontinuum (*pl.* -nua) *n.* 不連続体.

discount *n.* 割引. *vt.* 割引く.

discrete *a.* 離散的な; 疎な, 分離した. ⇨ continuous; concrete.
~ data 離散データ. ~ distribution 離散分布. ~ logarithm 離散対数. ~ mathematics 離散数学. ~ measure 離散測度. ~ optimization 離散最適化. ~ quantity 離散(的)量. ~ random variable 離散(型)確率変数. ~ series 疎系列. ~ set 離散集合. ~ spectrum 離散スペクトル. ~ topology 離散位相. ~ valuation 離散付値. ~ value 計数値, 離散値. ~ variable method 差分[離散変数]近似法.

discretization *n.* 離散化.
~ error 離散化誤差.

discriminant *n.* 判別式. *a.* 判別する.
Dedekind's ~ theorem デデキントの判別定理. ~ analysis 判別分析. ~ function 判別関数. fundamental ~ 基本判別式.

discuss *vt.* 議論する; 考察する; 吟味する.

discussion *n.* 吟味; 論議.

disjoint *a.* 排反の, 互いに素な.
~ sets 互いに素な集合. ~ union 直和(集合の). linearly ~ 線形離別[無関連]の. mutually ~ 互いに素な. pairwise ~ 対ごとに素な.

disjunction *n.* 選言, 離接.

disjunctive *a.* 選言的な, 離接的な. *n.* 選言命題.
~ syllogism 選言三段論法.

disk =disc.

dispersion *n.* ばらつき, 散らばり, 分散.
~ index 散布指数. ~ relation 分散(関係)

式.

dispersive *a.* 分散的な; 散逸的な; 伝播性の.
~ operator 散逸作用素. ~ wave 分散波.

displace *vt.* 置き換える; 移動させる.

displacement *n.* 移動; 変位.
infinitesimal ~ 無限小変位. normal ~ 法線(方向の)変位. parallel ~ 平行移動[変位]. perpendicular ~ 垂直変位. virtual ~ 仮想変位.

display *n.* 展示; 表示. *vt.* 展示する; 表示する.

disposal *n.* 処分, 売却; 処理.
at [in] one's ~ 人の自由になる. ~ value 処分価値, 純残価.

disproof *n.* 反証.

disprove *vt.* 反証する, 反証をあげる; の誤りを立証する.

dissection *n.* 分割.

dissemination *n.* 散布度; まき散らすこと.

dissertation *n.* (学位)論文.

dissimilar *a.* 非同類の.

dissipation *n.* 消散.

dissipative *a.* 消散的な.
~ operator 消散作用素. ~ part 消散部分. ~ system 消散系.

dissymmetry *n.* 非対称.

distance *n.* 距離, 道程, 道のり.
angular ~ 角距離. chordal ~ 弦距離. ~ circle 距離円. ~ function 距離関数. ~ point 距離点. Hamming ~ ハミング距離. non-Euclidean ~ 非ユークリッド距離. pseudo-~ 擬距離. shortest ~ 最短距離. spherical ~ 球面距離.

distinct *a.* (相)異なる, 別個の; 卓越した.
~ roots (相)異なる根. ~ system of parameters 卓越した助変数系. pairwise ~ 相異なる.

distinction *n.* 差別, 区別; 差異; 特質.

distinguish *vt., vi.* 識別する, 区別する.

distinguishable *a.* 判別可能な, 分離可能な.

distinguished *a.* 区別された; 格別の.
~ pseudo-polynomial 特殊擬多項式. ~ subgroup 正規部分群.

distortion *n.* 歪曲, ゆがみ; ひずみ.
delay ~ 遅延ひずみ. ~ inequality 歪曲不等式. ~ theorem 歪曲定理.

distribution *n.* 分布; 超関数; ディストリビューション.
binomial ~ 二項分布. cumulative ~ 累積分布. ~ curve 分布曲線. ~-free 分布によらない. ~ function 分布関数 ~ of errors 誤差(の)分布. ~ of prime numbers 素数分布. frequency ~ 度数分布. Gauss-(ian) ~ ガウス分布. joint ~ 結合同[時]分布. leptokurtic ~ 狭幅[幅ほそ; 急尖]分布. normal ~ 正規分布. platykurtic ~ 広幅[幅ひろ; 緩尖]分布. Poisson ~ ポアソン分布. population ~ 母集団分布. posterior ~ 事後分布. prior ~ 事前分布. probability ~ 確率分布. sample ~ 標本分布. simultaneous ~ 同時分布. stable ~ 安定分布. Student ~ ステューデント分布. *t*-~ *t*分布. theoretical ~ 理論的分布. value ~ 値分布.

distributional *a.* 超関数の, ディストリビューションの.
~ solution 超関数(の意味での)解.

distributive *a.* 分配の, 分配-.
~ algebra 分配的代数, 分配多元環. ~ lattice 分配束. ~ law 分配法則[律].

distributivity *n.* 分配性.

diverge *vi.* 発散する. ⇨ converge.

divergence *n.* 発散; 湧出量.
definite ~ 定発散. ~ (differential) expression 発散(微分)式. ~ theorem 発散定理. indefinite ~ 不定発散.

divergency ＝divergence.

divergent *a.* 発散する, 発散の.
definitely ~ 定発散の. ~ sequence 発散(数)列. ~ series 発散級数. indefinitely ~ 不定発散の. oscillating(ly) ~ 不定発散の. properly ~ 定発散の.

diverging ＝divergent.

divide *vt.* 分割する; 割る, 除する, 割り切る.
~ and conquer method 分割統治法. ~ externally [internally] 外分[内分]する. ~

harmonically 調和に分かつ.

divided *a.* 分割された; 割られた.
~ difference 差分商.

dividend *n.* 被除数.

dividendo *n.* 除比の理. ⇨ devidendo.

dividing *a.* 分割する; 分ける; 割る.
~ cycle 分割輪体.

divisibility *n.* 割り切れること, 整除性.
~ relation 整除関係.

divisible *a.* 分割できる; 割り切れる, 整除できる, 可除の.
~ subgroups 可除部分群. infinitely ~ 無限分解可能な.

division *n.* 割り算, 除法; 分割.
~ algebra 多元体. ~ algorithm 連除法, 互除法, 除法定理. ~ chain condition 約鎖律. ~ into subintervals 部分区間への分割. ~ modulo p pを法とする除法. ~ ring 斜体. extemal ~ 外分. harmonic ~ 調和分割. internal ~ 内分. synthetic ~ 組立除法.

divisional *a.* 除法の.

divisor *n.* 除数; 法; 因子, 因数; 約数.
canonical ~ 標[規]準因子. common ~ 公約数. ~ class 因子類. ~ function 約数関数. ~ ideal 約イデアル. ~ of function 関数の因子. effective ~ 有効因子. elementary ~ 単因子. integral ~ 整因子. positive ~ 正因子. prime ~ 素因子. principal ~ 主因子. zero ~ 零因子.

dodeca- *pref.*「十二」の意.

dodecagon *n.* 十二角[辺]形.
regular ~ 正十二角[辺]形.

dodecagonal *a.* 十二角[辺]形の.

dodecahedral *a.* 十二面体の.
~ group (正)十二面体群.

dodecahedron (*pl.* -rons, -dra) *n.* 十二面体.
regular ~ 正十二面体.

Dodgson, Charles Lutwidge (1832-1898) ドッジソン.【ルイス・キャロルのこと】

Doetsch, Gustav (1892-1977) デッチュ.

domain *n.* 領域; 域, 定義域; 整域.
angular ~ 角領域. annular ~ 円環領域. closed ~ 閉領域. convex ~ 凸領域. Dedekind ~ デデキント整域. ~ of convergence 収束域. ~ of definition 定義域. ~ of dependence 依存領域. ~ of holomorphy 正則領域. ~ of influence 影響領域. ~ of integrity 整域. ~ of variability 変域. ~ with boundary 境界のある領域. Euclidean ~ ユークリッド整域. integral ~ 整域. Jordan ~ ジョルダン領域. multiply(-)connected ~ 複連結領域. negative ~ 負領域. operator ~ 作用域. plane ~, planar ~ 面分, 平面領域. positive ~ 正領域. principal ideal ~ 主[単項]イデアル整域. Reinhardt ~ ラインハルト領域. ring ~ 環状領域. schlicht ~ 単葉領域. simply(-)connected ~ 単連結領域. spatial [space] ~ 空間領域. starlike [star-shaped] ~ 星型領域. unique factorization ~ 一意分解整域.

dominance *n.* 優越.
diagonal ~ 対角優位. ~ principle 優越原理. ~ relation 優越関係.

dominancy =dominance.

dominant *a.* 支配的な; 優性の; 優(越)の. ⇨ recessive.
~ game 優ゲーム. ~ representing measure 優表現測度. ~ series 優級数.

dominate *vt.* 支配する; 優越する.

dominated *n.* 支配された.

dominating *a.* 支配的な, 主たる.

domination *n.* 支配; 優越.
~ principle 優越原理.

Doob, Joseph Leo (1910-2004) ドゥーブ.

dot *n.* 点, ポツ; ドット. *vt., vi.* 点を打つ.
~ cross-product 二重クロス乗積. ~ diagram 点図表. ~ product 点乗積. ~ted line 点線【…; ⇨ broken line, solid line】. ~ted spinor 付点スピノル.

double *a.* 二重の; 複の; 両側の.
~(-)angle formula 二倍角の公式. ~ chain complex 二重鎖複体. ~ cusp 二重尖点. ~ dichotomy 二重二分法. ~ distribution 二重分布. ~ element 二重要素. ~ exponential formula 二重指数関数型公式. ~ induction 二重帰納法. ~ integral 二重積分. ~ lattice 二重格子. ~ layer 二

重層. ~ minimal surface 複極小曲面. ~ negation [negative] 二重否定. ~ point 二重点. ~ pole 二位の極. ~ precision (二)倍精度. ~ radical (sign) 二重根号. ~ ratio 複比. ~ root 二重根. ~ sampling 二回抽出. ~ sequence 二重(数)列. ~ series 二重級数 ~ series theorem 二重級数定理. ~ sign 複号. ~ suffix 二重添数. ~ tangent 二重接線. ~ vector 二重ベクトル.

doublet =dipole.

doubly *ad.* 二重に.
~(-)connected 二重連結の. ~(-)periodic 二重周期の.

doubt *n.* 疑い, 疑念.
with no ~, without ~ 確かに.

Douglas, Jesse (1897-1965) ダグラス.

down *ad.* 下へ, 下方へ.
carry ~ 繰下げる. going ~ theorem 下降定理.

down-ladder *n.* 下降演算子.

downtime *n.* 故障時間.

downward *ad.* 下に. ⇨ upward.

downwards *ad.* 下に. =downward.
bounded ~ 下(方)に有界な. ~ concave [convex] 下に凹[凸]な.

doz. =dozen(s).

dozen (*pl.* -(s)) *n.* ダース.
baker's ~ 十三個. ~ th 第十二の. long ~ 十三個. twelve ~(s) 十二ダース, 一グロス.

DP, D.P. Dynamic Programming.

dpt. department.

Dr., Dr, dr. doctor.

draft *vt.* 起草する; 製図する. *n.* 図案; 草稿.

draw *vt.* 引く. *vi.* 引く; 描く; 製図する.

drawing *n.* 線画; 製図.
instrumental [mechanical] ~ 用器画. perspective ~ 中心投影法.

drift *n.* ずれ, 偏流.

drop *n.* 滴; 降下. *vt.* 落とす. *vi.* 落下する.

dual *a.* 双対の, 双対的な. *n.* 双対.
~ basis, ~ base 双対基. ~ curve 双対曲線. ~ mapping 双対写像. ~ operator 双対作用素. ~ problem 双対問題. ~ representation 双対表現. ~ space 双対空間, 共役空間. ~ theorem 双対(な)定理. ~ variable 双対変数.

duality *n.* 双対(性), 双対関数.
~ gap 双対ギャップ. ~ principle, principle of ~ 双対原理. ~ theorem 双対定理.

du Bois-Reymond, Paul (1831-1889) デュ・ボア・レイモン.

due *a.* 帰すべき.
~ to に帰すべき, のおかげで.

dummy *a.* 見かけの, 擬装の, 擬似の.
~ index 無効添数. ~ treatment なぞらえ, 指定措置.

Dunford, Nelson (1906-1986) ダンフォード.

duodecimal *a.* 十二を単位とする, 十二進法の. *n.* 十二進法; 十二分の一.

duodenary *a.* 十二進(法)の; 十二倍の.
~ scale 十二進法.

Dupin, Charles (1784-1873) デュパン.

duple *a.* 二倍の; 二重の.
~ ratio 二倍比【2:1 の比】.

duplex *a.* 二重の; 二倍の.
~ system 二重システム.

duplicate *a.* 二重の; 重複した; 複の. *vt.* 二重にする, 二倍にする; 複写[製]する.
~ ratio 二重[乗]比. ~ sampling 重複抜取り.

duplication *n.* 二重にすること, 二倍にすること; 重複; 複写.
~ check 二重検査. ~ of cube 立方倍積.

durable *a.* 耐久性のある. *n.* 耐久消費財.
~ year(s) 耐用年数.

duration *n.* 所要時間; 持続期間.

dyad *n.* ディヤード, ダイアド; 二数, 一対.

dyadic *a.* ダイヤディック, ダイアディック; 二数の, 二個数の. *n.* ダイアドの和.
~ integer 二進整数. ~ system 二進法.

dynamic *a.* 動(力学)的な; 力学(上)の.
~ programming 動的計画法. ~ statistics 動態統計.

dynamical =dynamic.
~ system 力学系.

dynamics *n.* 力学, 動力学.

analytical ～ 解析力学. astro ～ 天体力学. ～ of rigid bodies 剛体の力学. electro ～ 電気力学. fluid ～; hydro ～ 流体力学. thermo ～ 熱力学.

Dynkin, Eugene Borisovich (1924–2014) ディンキン.

dz. ＝dozen(s).

E

e ネイピアの対数[自然対数]の底.
each *a.* おのおのの, 各-. *pron.* おのおの.
at ~ point 各点で. ~ other 互いに.
écart *n.* (*F.*) 距離; 外れ.
eccentric *a.* 離心の. *n.* 離心円.
~ angle 離心角, 心差角. ~ anomaly 離心近点離角. ~ circular ring 離心円環.
eccentricity *n.* 離心率, 心差率.
echelon *n.* 梯形; 階段.
reduced row ~ form [matrix] 簡約行階段行列. (row) ~ matrix [form] (行)階段行列.
econometrics *n.* 計量経済(学).
economy *n.* 経済.
competitive ~ 競争経済.
ed. edited; edition; editor; educated; education.
edge *n.* 辺, 稜; 縁.
~ homomorphism 辺準同型[形]. ~ of regression 反帰曲線. lateral ~ 側稜.
Edgeworth, Francis Ysidro (1845-1926) エッジワース.
edit *vt.* (書物, 雑誌などを)編集する; (原稿を)校訂する.
edit. edited; edition; editor.
editing *n.* 編集.
edition *n.* (刊行書籍の)体裁, 版.
~ [édition] de luxe 豪華版, 特製本. popular ~ 普及版. revised [enlarged] ~ 改訂[増補]版. second ~ 再[第二]版.
editorial *a.* 編集上の.
educate *vt.* 教育する.
educated *a.* 教育[教養]のある.
education *n.* 教育.
educational *a.* 教育の; 教育的な.
~ system 教育制度.
effect *n.* 効果; 結果.
block ~ ブロック効果. buffer ~ 緩衝効果. multiplier ~ 乗数効果. random ~ 変量効果. treatment ~ 処理効果.
effective *a.* 有効な.
almost ~ 殆んど効果的な. ~ degree 有効次数. ~divisor 有効因子. ~ error 実効誤差. ~ genus 有効種数. ~ range 有効距離[射程].
effectively *ad.* 有効に, 効果的に; 実際に.
~ calculable function 実際に計算可能な関数.
efficiency *n.* 効率, 能率; 有効性.
~ factor 効率因子.
efficient *a.* 有効な; 能率的な.
~ estimator 有効推定量. ~ point 有効点. ~ statistic 有効統計量.
efflux *n.* 流出. ⇨ afflux; influx.
e.g. exempli gratia (*L.* =for example, for instance.) 例えば.
Egorov, Dimitrii Fyodorovič (1869-1931) エゴロフ.
Eichler, Martin M. E. (1912-1992) アイヒラー.
eigen- *pref.* 固有な; 固有-.
eigenelement *n.* 固有元.
eigenfunction *n.* 固有関数.
eigen-oscillation *n.* 固有振動.
eigenpolynomial *n.* 固有多項式.
eigenspace *n.* 固有空間.
eigenvalue *n.* 固有値.
~ problem 固有値問題. multiple ~ 重複固有値.
eigenvector *n.* 固有ベクトル.

Eikonal *n.* (*G.*) アイコナール.

EILENBERG, Samuel (1913-1998) アイレンバーグ.

EINSTEIN, Albert (1879-1955) アインシュタイン.

EISENSTEIN, Friedrich Gotthold Max (1823-1852) アイゼンシュタイン.

elastic *a.* 弾性の; 弾力のある.
～ body 弾性体. ～ membrane 弾性膜. ～ plate 弾性板. ～ scattering 弾性散乱.

elasticity *n.* 弾性.
～ modulus 弾性率.

electric *a.* 電気の.
～ charge 電荷. ～ circuit [network] 電気回路. ～ current [stream] 電流. ～ displacement 電気変位. ～ field 電界, 電場. ～ wave 電波.

electrical ＝electric.

electro- *pref.*「電気の」の意.

electrode *n.* 電極.

electrodynamics *n.* 電気力学.

electrokinetics *n.* 動電学.

electromagnetic *a.* 電磁の.
～ equation 電磁方程式. ～ wave 電磁波.

electromagnetics *n.* 電磁気学.

electromagnetism *n.* 電磁気, 電磁気学.

electron *n.* 電子, エレクトロン.

electronic *a.* 電子の.
～ computer 電子計算機.

electronics *n.* 電子工学.
～ calculator [computer] 電子計算機.

electrostatics *n.* 静電気学.

Elemants 原論【エウクレイデスが著した数学書】.

element *n.* 要素; 元; エレメント, 素子; 初歩.
algebraic ～ 代数的元. arc ～ 弧(長)要素. area ～ 面素. boundary ～ 境界要素. conjugate ～ 共役元. decision ～ 論理素子. diagonal ～ 対角成分. ～ of set 集合の要素[元]. ～s of triangle 三角形の要素. finite ～ method 有限要素法. function ～ 関数要素. identity ～ 単位元. inverse ～ 逆元. line ～ 線素. logic(al) ～ 論理素子. neutral ～ 単位元, 中立元. order of ～ 元の位数. polar ～ 極要素. probability ～ 確率素分. surface ～ 面積要素, 面素. torsion ～ ねじれ元. transcendental ～ 超越的元. unit ～ 単位元. volume ～ 体積要素. zero ～ 零元.

elementary *a.* 基本の, 基本的な; 初等[初歩]の.
～ algebra 初等代数(学). ～ alternating expression 基本交代式. ～ alternating polynomial 基本交代(多項)式. ～ column operation 列基本変形. ～ divisor 単因子. ～ event 根元[素]事象. ～ figure 基本図形. ～ function 初等関数. ～ geometry 初等幾何(学). ～ mathematics 初等数学. ～ matrix 基本行列. ～ operation(s) 基本演算. ～ row operation 行基本変形. ～ solution 基本解; 素解. ～ symmetric expression 基本対称式. ～ symmetric polynomial 基本対称多項式. ～ theory 初等理論. ～ transcendental function 初等超越関数. ～ transfomation 初等変換; 基本変形.

elementwise *a.*, *ad.* 要素[元]ごとの[に].

elevation *n.* 立面図, 正面図. 前面図.
angle of ～, ～ angle 仰角. side ～ 側面図.

eliminable *a.* 消去できる.

eliminant *n.* 消去式.

eliminate *vt.* 消去する, 追い出す.

elimination *n.* 消去; 消去法.
～ method 消去法. forward ～ 前進消去. Gaussian ～ ガウスの消去法.

ellipse *n.* 楕円, 長円; 扁円.
confocal ～s 共焦楕円. imaginary ～ 虚楕円.

ellipsis (*pl.* -ses) *n.* 省略; 省略符号【…, —, *** など】.

ellipsoid *n.* 楕[長]円体[面].
～ of inertia 慣性楕円体. ～ of revolution 回転楕円体[面](＝spheroid).

ellipsoidal *a.* 楕円体[面]の.
～ coordinates 楕円体[面]座標. ～ harmonics 楕円体調和関数. ～ surface of revolution 回転楕円面.

elliptic *a.* 楕円の; 楕円型の, 楕円的な; 楕円形の.
～ cone 楕円錐. ～ coordinates 楕円座標.

～ curve 楕円曲線. ～ cylinder 楕円柱(面). ～ function 楕円関数. ～ geometry 楕円(的)幾何学. ～ integral 楕円積分. ～ modular function 楕円母数関数. ～ motion 楕円運動. ～ operator 楕円型作用素. ～ paraboloid 楕円放物面. ～ point 楕円(的)点. ～ quadrics 楕円型二次曲面. ～ transformation 楕円(型)変換. ～ type 楕円型.

elliptical ＝elliptic.

ellipticity *n.* 楕円(型)性.

elongation *n.* 延長, 伸長; 離角.

embed *vt.* 埋める, 埋蔵する, 埋めこむ.
～ into Euclidean space ユークリッド空間に埋蔵する.

embedded *a.* 埋蔵された, 埋めこまれた; 非孤立の.
～ prime divisor 非孤立素因子.

embedding *n.* 埋めこみ, 埋蔵.
～ theorem 埋めこみ[埋蔵]定理. regular ～ 正則(な)埋めこみ[埋蔵].

emission *n.* 射出; 射線錐体.

empirical *a.* 実験による, 経験的な.
～ constant 実験定数. ～ distribution function 経験分布関数. ～ formula 実験式.

empty *a.* 空の. ⇨ void.
～ event 空事象. ～ set 空集合【∅, ϕ, 0】. ～ word 空語.

emulate *vt.* 見習う, 模倣する.

en *prep.* (*F.*) において; として.
～ bloc 一括して.

enable *vt.* 可能にする.

encircle *vt.* 囲む; 一周する.

enclose *vt.* 囲む; 包囲する.

enclosure *n.* 包囲.
problem of ～ 閉形問題.

encode *vt.* エンコード, 符号化する. ⇨ decode.

encounter *n.* 遭遇. *vt., vi.* 遭遇する. ⇨ rencounter.

encrypt *vt.* 暗号化する; 符号化する.

encryption *n.* 暗号化; 符号化.

encyclop(a)edia *n.* 百科事典.

end *n.* 終り; 末端, 末尾. *vt.* 終える. *vi.* 終わる.
～ around carry 循環桁上げ. ～ point 端点, 終点.

endless *a.* 際限のない, 無限の.

endogen(e)ous *a.* 内生の. ⇨ exogenous.
～ variable 内生変数.

endomorphism *n.* 自己準同型[形](写像).
～ ring 自己準同型[形]環.

energetics *n.* エネルギー論.

energy *n.* エネルギー.
～ equation エネルギー方程式. ～ inequality エネルギー不等式. ～ integral エネルギー積分. ～ principle エネルギー原理. kinetic ～ 運動エネルギー. mutual ～ 相互エネルギー. potential ～ 位置エネルギー.

ENGEL, Friedrich (1861-1941) エンゲル.

enigma *n.* なぞ.

enlarge *vt.* 拡大する.
～d coefficient matrix 拡大係数行列.

enneagon ＝nonagon.

enneahedron (*pl.* -rons, -dra) *n.* 九面体.

enough *a.* 十分な. *pron.* 十分. *ad.* 十分に.
large ～ 十分大きい.

ENRIQUES, Federigo (1871-1946) エンリケス.

ensemble *n.* (*F.* ＝set.) 集合.

entire *a.* 整の; 全体の. *n.* 全体.
～ function 整関数.

entirely *ad.* 全く; もっぱら.

entity *n.* 存在者, 実在, 本質.

entrance *n.* はいること, 流入, 入り口. ⇨ exit.
～ boundary 流入境界.

entropy *n.* エントロピー.
conditional ～ 条件つきエントロピー. mean ～ 平均エントロピー.

entry *n.* 入場; 登録; 参加; 成分.

Entscheidungsproblem *n.* (*G.* ＝decision problem.) 決定問題.

enumerability *n.* 可算性.

enumerable *a.* 可付番の, 可算の. ⇨ countable, denumerable.
～ model 可算モデル. ～ set 可算[可付番]

集合.

enumerate *vt.* 数える; 表にして示す; 列挙する, 枚挙する.
~d data 計数値.

enumeration *n.* 計数; 計算; 列挙, 枚挙; 目録, 表.
~ data 計数データ. ~ function 算出関数. ~ theorem 枚挙(可能)定理.

enumerative *a.* 列挙する.
~ solution 列挙解.

envelop *vt.* 包む; 包絡する.

envelope *n.* 包絡線, 包絡面; 包.
~ of holomorphy 正則包. ~ power function 包絡検出力関数. injective ~ 単射[移入]包絡.

enveloping *a.* 包絡する.
~ algebra 包絡多元環. ~ cone 包絡錐面. ~ cylinder 包絡柱面. ~ surface 包絡面.

epi- *pref.*「上, 外」の意. ⇨ hypo-.

epicycle *n.* 周転円.

epicycloid *n.* 外(転)サイクロイド, 外擺線.

epicycloidal *a.* 外サイクロイドの.

epidemic *a.* 伝染性の. *n.* 伝染病; 流行.
~ curve 伝染曲線.

epidemics *n.* 伝染理論.

epigraph *n.* エピグラフ.

epimorphism *n.* 全射; 全射準同型[形].

epitrochoid *n.* 外トロコイド, 外余擺線.

epitrochoidal *a.* 外トロコイドの.

epsilon *n.* エプシロン, イプシロン. ギリシア語アルファベットの5番目の文字【E, ε】.
~-delta technique イプシロン・デルタ論法.

epsilontics *n.* イプシロン・デルタ論法.

eq. equal, equality; equation; equivalence.

equal *a.* 等しい【to】. *vt.* に等しい. *n.* 同等の人[物]; 等しい数量.
~ root 重根, 等根. ~ sign 等号.

equality *n.* 相等; 等式; 平等.
~ sign 等号.

equally *ad.* 等しく; 同等に, 一様に.
~ likely [probable] 同様[同程度]に確からしい. ~ spaced 等間隔の. ~ strong 同じ強さの.

equate *vt.* 等式にする; 等置する. *vi.* と一致する【with】.

equating *n.* 等化.

equation *n.* 方程式, 等式.
adjoint ~ 随伴方程式. algebraic ~ 代数方程式. associated ~ 同伴方程式. binomial ~ 二項方程式. characteristic ~ 特性方程式. class ~ 類等式. cubic ~ 三次方程式. difference ~ 差分方程式. differential ~ 微分方程式. Diophantine [Diophatus] ~ ディオファンタス方程式. ~ of continuity 連続の方程式. ~ with n unknowns n元方程式. exponential ~ 指数方程式. fractional ~ 分数方程式. functional ~ 関数方程式. Hamilton ~ ハミルトン方程式. homogeneous ~ 同[斉]次方程式. indeterminate ~ 不定方程式. inhomogeneous ~ 非同[斉]次方程式. integral ~ 積分方程式. irrational ~ 無理方程式. linear ~ 一次[線形]方程式. logarithmic ~ 対数方程式. numerical ~ 数値[字]方程式. parametric ~ 媒介変数方程式. polar ~ 極方程式. predator-prey ~ 捕食者-被食者方程式. quadratic ~ 二次方程式. quartic ~ 四次方程式. radical ~ 無理方程式. reciprocal ~ 相反方程式. simultaneous ~s, system of ~s 連立方程式, 方程式系. transcendental ~ 超越方程式. transposed ~ 転置方程式. trigonometric(al) ~ 三角方程式. variational ~ 変分方程式.

equational *a.* 方程式(上)の.
~ characterization 方程式的特徴づけ.

equationally *ad.* 方程式的に.

equator *n.* 赤道.
celestial ~ 天球赤道.

equi- *pref.*「等しい」の意.

equiangular *a.* 等角の.
~ hyperbola 等角[直角]双曲線. ~ polygon 等角多角形. ~ spiral 等角螺線.

equianharmonic *a.* 等非調和な.
~ range of points 等非調和列点.

equiareal =area-preserving.

equiaxial *a.* 等軸の.
~ system. 等軸晶系.

equicontinuity *n.* 同程度連続性.
equicontinuous *a.* 同程度連続な.
equidistance *n.* 等距離.
equidistant *a.* 等距離の.
~ curve 等距離(曲)線. ~ hypersurface 等距離(超曲)面.
equilateral *a.* 等辺の. *n.* 等辺形, 等辺.
~ hyperbola 等辺[直角]双曲線. ~ polygon 等辺多角形. ~ triangle 等辺[正]三角形.
equilibrium (*pl.* -riums, -ria) *n.* 平衡; 均衡.
complete competitive ~ 完全競争均衡. ~ distribution 平衡分布. ~ point 均衡点, 平衡点. ~ principle 平衡原理. stable ~ 安定平衡.
equimeasure *n.* 等測.
~ transformation 保測[等測]変換.
equipment *n.* 装置.
equipotence ＝equipotency.
equipotency *n.* 等濃度, 対等.
equipotent *a.* 対等な; 全単射同型[形]な.
equipotential *a.* 等位の, 等ポテンシャルの.
~ line [surface] 等ポテンシャル線[面].
equiprobable *a.* 同様に確からしい.
equiscalar *a.* 等スカラーの.
~ surface 等スカラー面.
equivalence *n.* 同値; 同等, 等価; 等積.
birational ~ 双有理同値. ~ class 同値類. ~ law 同値法則[律]. linear ~ 線形同値. ~ relation 同値関係. homotopy ~ ホモトピー同値. Morita ~ 森田同値.
equivalency ＝equivalence.
equivalent *a.* 同値な; 等価な; 等積の; 対等な. *n.* 同等なもの, 同値.
chain ~ 鎖同値な. conformally ~ 等角(的)同値な. ~ affinity 等積アフィン変換. ~ condition(s) 同値な条件. ~ map 等積地図. ~ projection 等積投影(法). ~ sets 対等な集合. ~ transformation 等積変換. homotopy ~ ホモトピー同値な. Morita ~ 森田同値な. stably ~ 安定同値な. topologically ~ 位相(的に)同値な. unitary ~ ユニタリ同値な.

equivocation *n.* あいまい度; 二[多]義性.
ERATOSTHENES of Kyrene (276?-194? B.C.) キレネのエラトステネス.
erg *n.* エルグ【エネルギーの単位(CGS 単位系)】.
ergodic *a.* エルゴード的な; エルゴードの.
~ hypothesis エルゴード仮説. ~ surface エルゴード面. ~ theorem エルゴード定理. ~ theory エルゴード理論. individual ~ theorem 個別エルゴード定理.
ergodicity *n.* エルゴード性.
ergonomics *n.* 人間工学.
erratum (*pl.* -ta) *n.* 誤り; 誤字, 誤植; (*pl.*) 正誤表.
error *n.* 誤差; 誤り.
absolute ~ 絶対誤差. accidental ~ 偶然誤差. accumulated ~ 累積誤差. computational ~ 計算誤差. discretization ~ 離散化誤差. ~ constant 誤差定数 ~ correcting code 誤り訂正符号. ~ curve 誤差曲線. ~ function 誤差関数. ~ of input data 入力[データ]誤差. ~ of the first [second] kind 第一[二]種の誤り. ~ space 誤差空間. ~ term 誤差項. inherited ~ 遺伝誤差. instrumental ~ 器(械誤)差. mean absolute ~ 平均絶対誤差. mean square ~ 平均二乗[平方]誤差. probability ~ 確率誤差. relative ~ 相対誤差. round-off ~, rounding ~ 丸め誤差. sampling ~ 抽出誤差. theory of ~s 誤差論. truncate(d) [truncation] ~ 打切り誤差.
error-correction *n.* 誤りの訂正.
escape *n.* 流出; 脱出; 逃げ道. *vi.* 流出する; 脱出する. *vt.* 免れる.
escribe *vt.* 傍接させる.
~d circle 傍接円.
escription *n.* 傍接.
essential *a.* 本質的な; 真性の. *n.* 本質的要素, 要点.
~ algebra 本質環. ~ extension 本質的拡大. ~ set 本質集合. ~ singularity 真性特異点. ~ submodule 本質部分加群. ~ supremum 本質的上限.
essentially *ad.* 本質的に.
~ bounded 本質的に有界な. ~ complete

class 本質的完備類.

establish *vt.* 確立する, 確証する.

estimable *a.* 推定可能な.

estimate *n.* 推定値; 評価(式). *vt., vi.* 評価する. ⇨ estimator.

estimation *n.* 評価, 概算; 推定.
~ space 推定空間. exact ~ 精確な評価. interval ~ 区間推定. maximum likelihood ~ 最尤推定法. point ~ 点推定. precise [sharp] ~ 精確な評価. statistical ~ 統計的推定.

estimator *n.* 推定量.
consistent ~ 一致推定量. efficient ~ 有効推定量. invariant ~ 不変推定量. maximum likelihood ~ 最尤推定量. regression ~ 回帰推定量. unbiased ~ 不偏推定量.

eta *n.* イータ, エータ. ギリシア語アルファベットの7番目の文字【H, η】.

et al. (*L.*) et alibi (=and elsewhere); et alii (=and others).

etc., &c. (*L.*) et cetera (=and so on [forth]). など, その他.

et s(e)q. (*L.*) et sequence (=and the following). およびその次[行・ページなど](参照).

Euclid (330? -275? B. C.) [(365? -300? B. C.)] ユークリッド, エウクレイデス.

Euclidean, euclidean *a.* ユークリッドの; ユークリッド的な; ユークリッド幾何学の.
~ algorithm ユークリッドの互除法. ~ axiom(s) ユークリッドの公理. ~ cell complex ユークリッド[幾何学的]胞複体. ~ domain ユークリッド整域. ~ geometry ユークリッド幾何(学). ~ metric ユークリッド計量. ~ ring ユークリッド環. ~ space ユークリッド空間. locally ~ 局所ユークリッド的な. non-~ 非ユークリッド的な.

Euclidian =Euclidean.

Eudoxus of Knidos (408-355 B.C.) クニドスのエウドクソス.

Eukleides =Euclid.

Euler, Leonhard (1707-1783) オイラー.

Eulerian *a.* オイラーの.
~ angle オイラーの角. ~ function [number] オイラーの関数[数].

evaluate *vt.* 評価する; 数値を求める, 数的に表現する.

evaluation *n.* 評価; 数値を求めること, 数的表現; 計算.

even *a.* 偶数の; 平らな; 平等な. *ad.* 平等に; さえ. ⇨ odd.
~ function 偶関数. ~ if たとえ…としても. ~ number 偶数. ~ permutation 偶置換, 偶順列. ~ point 偶点.

event *n.* 事象; 結果.
complementary ~ 余事象. compound ~ 複(合)事象. dependent ~(s) 従属事象. elementary ~ 素[根元]事象. empty ~ 空事象. exclusive ~(s) 排反事象. fundamental ~ 基本[根元]事象. independent ~(s) 独立事象. limit ~ 極限事象. measurable ~ 可測事象. probability ~ 確率事象. product ~ 積事象. regular ~ 正則事象. simple ~ 単(純)事象. sum ~ 和事象. whole ~ 全事象.

eventual *a.* いつかは起こる, 時には起こりうる, 可能な.

eventuality *n.* 起こり得ることがら, 偶然性.

every *a.* すべての, いずれの…も皆; …ごとに.
(at) ~ point どの点でも. ~ three days 三日ごとに.

everywhere *ad.* どこでも, いたるところ.
almost ~ 殆んどいたるところ. ~ dense いたるところ稠密な. ~ on *E* *E* 上のどこでも.

evidence *n.* 証拠; 明白さ.

evident *a.* 明らかな, 明白な.

evidently *ad.* 明らかに.

evolute *n.* 縮閉線. *a.* 縮閉した.

evolution *n.* 発展; 開方【開平・開立など】.
~ equation 発展方程式.

evolve *vt.* 発展させる; 展開させる. *vi.* 開く; 展開する.

ex. example; exception; exercise.

ex- *pref.*「外」の意. ⇨ in-.

exact *a.* 完全な; 正確な, 精確な, 厳密な.

～ differential 完全微分. ～ differential equation 完全微分方程式. ～ differential form 完全微分形式. ～ estimation 精確な評価. ～ functor 完全関手. ～ sequence 完全(系)列. ～ test 正確検定. left ～ 左完全な. long [short] ～ sequence 長[短]完全(系)列. right ～ 右完全な.

exactitude *n.* 正確さ, 精密度, 厳密性.

exactly *ad.* 正確に; 厳密に; ちょうど.
～ divisible 割り切れる.

exactness *n.* 完全性; 精度; 厳密性.
～ axiom 完全性公理. ～ of reproduction 再生精度.

exaequali *n.* 等比の理.

examination *n.* 吟味, 考察; 試験; 検査.
oral [written] ～ 口頭[筆記]試験.

examine *vt.* 吟味する, 考察する; 試験する; 検査する.

example *n.* 例, 実例; 例題.
counter(-) ～ 反例. for ～ たとえば. give ～ 例をあげる.

exceed *vt., vi.* 越[超]える, 超過する.

excenter *n.* 傍心.

except *prep.* を除いて.

exceptional *a.* 例外の, 除外の.
～ curve 例外曲線; 除外曲線. ～group 例外群. ～ point 除外点. ～ set 除外集合. ～ value 除外値.

excess *n.* 超過, 過剰.
approximation in ～ 過剰近似. ～ of nines 九去法. ～-three code 三増しコード. spherical ～ 球面過剰.

excessive *a.* 超過的な; 極端な.

exchange *n.* 交換. *vt., vi.* 交換する.
sorting by ～(-ging) 交換(分類)法.

excircle *n.* 傍接円.

excise *vt.* 切除する; 削除する.

excision *n.* 切除; 削除.
～ axiom 切除公理. ～ isomorphism 切除同型[形].

excitation *n.* 刺激; 励起; 興奮.

excite *vt.* 刺激する, 励起する; 興奮させる.

exclude *vt.* 除外する.
law of ～d middle 排中律.

exclusion *n.* 除外; 排他.
principle of inclusion and ～ 包除原理.

exclusive *a.* 排反の; 排他的な. ⇨ inclusive.
～ events 排反事象. ～ or 排他的論理和.

exclusively *ad.* 排他的に; もっぱら.

execute *vt.* (計画などを)実行する; 遂行する.

exercise *n.* 練習, 演習; 練習問題, 応用問題.
practical ～ 応用問題.

exhaust *vt.* 空にする; 使い尽くす. *vi.* 排出する.

exhaustion *n.* 取り尽すこと, 汲み尽し.
～ method, method of ～ 悉尽法, 汲み[取り]尽し法. regular ～ 正則近似列.

exhaustive *a.* すべてを尽くす; 悉皆の.
～ search 全数探索.

exist *vi.* 存在する; 実在する.
there ～(s) …が存在する.

existence *n.* 存在.
～ proof 存在証明. ～ region 存在領域. ～ theorem 存在定理.

existent *a.* 存在している; 実在の.

existential *a.* 存在の; 存在上の.
～ proposition 特称命題. ～ quantifier 存在作用素[記号]【∃】.

exit *n.* 流出. ⇨ entrance.
～ boundary 流出境界.

exogen(e)ous *a.* 外生の. ⇨ endogenous.
～ variable 外生変数.

exotic *a.* 異種の.
～ sphere 異種[異常]球面.

exp. exponent, exponential.

expand *vt.* 展開する; 拡張する.
～ed form 展開形. ～ into series 級数に展開する.

expansion *n.* 展開; 展開式; 膨張, 拡大.
asymptotic ～ 漸近展開. binomial ～ 二項展開. ～ coefficient 展開係数. cofactor ～ 余因子展開. ～ into series 級数展開. ～ of determinant [function] 行列式[関数]の展開. Fourier ～ フーリエ展開. Laurent ～ ローラン展開. series ～ 級数展開. Taylor ～ テイラー展開.

expectation *n.* 期待; 期待値, 平均値.
conditional ~ 条件つき期待値. ~ of life 平均余命.

expected *a.* 予期された; 期待された.
~ amount of inspection 平均検査量. ~ quality level 期待品質水準. ~ value 期待値.

experiment *n.* 実験, 試験. *vi.* 実験する.
design of ~ 実験計画(法).

experimental *a.* 実験の.
~ analysis 実験(的)解析(法). ~ design 実験計画. ~ formula 実験式, 経験式. ~ probability 経験的確率.

explain *vt.* 説明する; 明白にする.

explanatory *a.* 説明的な.
~variable 説明変数.

explicit *a.* 明白な, 陽(的)な, 明示的な.
~ expression 明らさまにかかれた式. ~ formula 明示公式. ~ function 陽関数.

explicitly *ad.* 明白に; 陽[あらわ]に.

exploratory *a.* 探索的な.
~ data analysis 探索的データ解析.

exponent *n.* 指数, ベキ指数
convergence ~, ~ of convergence 収束指数. fractional ~ 分数指数[ベキ]. law(s) of ~s 指数法則. ramification ~ 分岐指数.

exponential *a.* 指数の.
double ~ formula 二重指数関数型公式. ~ curve 指数曲線. ~ distribution 指数分布. ~ equation 指数方程式. ~ family 指数型分布族. ~ function 指数関数. ~ generating function 指数的母関数. ~ inquality 指数不等式. ~ law(s) 指数法則. ~ mapping 指数写像. ~ series 指数級数. ~ smoothing 指数平滑法. ~ valuation 指数付値.

exponentiate *vt.* 指数のベキにあげる【a から e^a をつくる】.

exposition *n.* 説明, 解説.

expository *a.* 解説的な.

express *vt.* 表す, 表現[示]する.

expression *n.* 式; 表示.
algebraic ~ 代数式. alternating ~ 交代式. analytic ~ 解析(的)式. approximate ~ 近似式. congruence [congruent] ~ 合同式. divergence ~ 発散式. elementary alternating ~ 基本交代式. elementary alternating ~ 基本対称式. integral ~ 整式. linear ~ 一次式. mathematical ~ 数式. rational ~ 有理式. regular ~ 正規表現. symmetric ~ 対称式.

extend *vt.* 拡張する, 拡大する.
~ed field 拡大体. ~ed figure 拡大図.

extendable *a.* 拡大[張]可能な.

extensible *a.* 拡大[張]可能な.

extension *n.* 拡張; 拡大; 延長; 外延.
algebraic ~ 代数(的)拡大. central ~ 中心拡大. continuous ~ 連続拡大. cyclic ~ 巡回拡大. ~ degree 拡大次数. ~ field 拡大体. ~ theorem 拡張定理. finite ~ 有限(次)拡大. Galois ~ ガロア拡大. group ~ 群拡大. Hermitian ~ エルミート拡大. infinite ~ 無限(次)拡大. Kummer ~ クンマー拡大. linear ~ 線形拡大. normal ~ 正規拡大. simple ~ 単純拡大. strong ~ 強拡張. transcendental ~ 超越(的)拡大. weak ~ 弱拡張.

extensive *a.* 広い, 広範な; 外延の; 大量の, 多数の.
~ form 展開形. ~ quantity 外延量.

extent *n.* 程度; 範囲; 外延.

exterior *n.* 外部. *a.* 外部の; 外-. ⇨ outer; interior.
~ algebra 外積代数, 外積多元環. ~ angle 外角. ~ angles on the same side 同傍外角. ~ derivation 外微分. ~ derivative 外微分. ~ differential form 外微分形式. ~ differentiation 外微分. ~-interior angles 同位角. ~ point 外点. ~ power (operator) 外積ベキ(作用素). ~ problem 外部問題. ~ product 外積, 交代積.

external *a.* 外部の. ⇨ internal.
~ angle 外角. ~ common tangent 共通外接線. ~ division 外分. ~ language 外部言語. ~ product 外部積. ~ ratio 外分比. ~ term 外項. ~ variable 外生変数. ~ variance 外分散.

externally *ad.* 外で[に].
divide ~ 外分する. ~ dividing point 外分点.

extinction *n.* 消滅.
~ probability 消滅確率. ~ time 消滅時刻[時間].

extr. extremum, extremal.

extra *a.* 余分の, 特別の. *n.* 余分なもの. *ad.* 余分に.

extract *n.* 抽出(物). *vt.* 抽出する; (根を)求める; 開く.

extraction *n.* (根の)開方.
~ of cubic root 開立. ~ of square root 開平.

extraneous *a.* 無関係な; 外来の, 固有でない.
~ root 無縁根.

extraordinarily *ad.* 異常に.

extraordinariness *n.* 異常性.

extraordinary *a.* 異常な.

extrapolate *vt.* 外挿する; 補外する.

extrapolation *n.* 外挿(法), 補外(法). ⇨ interpolation.
~ formula 補外[外挿]公式

extremal *a.* 極値的な, 極値-; 端の. *n.* 極値; 極値関数[曲線].
conditional ~ 条件つき極値. constrained ~ problem 条件つき極値問題. ~ curve 極値曲線. ~ distance 極値的距離. ~ function 極値関数. ~ length 極値的長さ. ~ method 極値的方法. ~ metric 極値計量. ~ point 端点. ~ problem 極値問題. ~ surface 極値曲面. ~ value 極値.

extreme *a.* 最端の, 最後の; 極値の; 極端な. *n.* (比例・比または級数の)外項【初項または末項】.
~ and mean ratio 外中比. ~ point 極値点; 端点. ~ term (比の)外項. ~ value 極値.

extremely *ad.* 極端に.

extremum (*pl.* -ma) *n.* 極値.
constrained [conditional] ~ 条件つき極値. local ~ (局所的)極値.

F

face *n.* 面, 表面; 表; 辺; 字形[体].
bold ~ ボールド体, 太字. ~ angle 面角. ~-centred lattice 面心格子. ~ operator 面作用素.
facet *n.* (小)面.
facial *a.* 面の.
~ angle 面角.
faciend *n.* 被乗数.
facilitate *vt.* 容易にする, 楽にする.
fact *n.* 事実; 事件.
in ~ じっさい(に)
factor *n.* 因子, 因数; 要因. *vt.* 因数に分解する.
common ~ 共通因子, 公因子. composition ~ 組成因子. constant ~ 定数因子[数]. ~ analysis 因子分析(法). ~ group 因子群, 剰余群; 商群. ~ loading 因子荷重. ~ module 剰余加群, 因子加群. ~ representation 因子表現. ~ score 因子スコア. ~ set 因子団. ~ space 商空間; 剰余空間. ~ theorem 因数定理. integrating ~ 積分因子. prime ~ 素因子[数]. scale ~ 倍率, 尺度係数.
factorability *n.* 因子[因数]分解可能性.
factorial *a.* 階乗の; 要因の. *n.* 階乗.
~ design 要因配置計画[実験]. ~ experiment 要因分析(法). ~ function 階乗関数. ~ moment 階乗モーメント. ~ series 階乗級数. ~ unit 要因単位. *n* ~, ~ of *n* *n* (の)階乗.
factorization *n.* 因数[因子]分解.
~ into prime factors 素因数[因子]分解. ~ method 因数[因子]分解法; 昇降演算子法. ~ theorem (因子[因数])分解定理. unique ~ domain (UFD) 一意分解整域.
factorize *vt.* 因数(に)分解する.
faculty *n.* 学部.
~ of science 理学部.
failure *n.* 失敗; 不足.
faisceau *n.* (*F.* =sheaf.) 層.
faithful *a.* 忠実な.
~ representation 忠実な表現.
faithfully *ad.* 忠実に.
~ flat 忠実に平坦な.
fallacy *n.* 虚偽; 謬論.
false *a.* 偽の; 誤った.
~ proposition 偽の命題.
falsity *n.* 偽, 虚偽.
truth and ~ 真偽.
Faltung *n.* (*G.* =convolution.) たたみ込み; 合成積.
family *n.* 族. ⇨ class.
~ of curves, curve ~ 曲線族. ~ of functions, function ~ 関数族. ~ of sets 集合族. normal ~ 正規族. one-parameter ~ 一パラメータ族.
fascicle, -cule *n.* 分冊.
FATOU, Pierre (1878–1929) ファトゥ.
fault *n.* 欠点; 過失; 故障.
favo(u)rable *a.* 好都合な; 有利な.
~ case 好都合な場合.
F-distribution F分布.
feasible *a.* 実行しうる[可能な].
~ region 実行可能領域, 許容領域. ~ solution 実行可能解.
feedback *n.* 帰還[饋還], フィードバック. *a.* フィードバックの. *vt., vi.* フィードバックする.
FEFFERMAN, Charles Louis (1949–) フェファーマン.

Fejér, Leopold (1880–1959) ファイエ, フェエール.

Fekete, Michael (1886–1957) フェケテ.

Feller, Wilhelm (1906–1970) フェラー.

Fenchel, Werner (1905–1988) フェンヒェル.

Fermat, Pierre de (1641–1665) フェルマ.

Ferrari, Ludovico (1522–1565) フェラリ.

Ferro, Scipione dell (1465–1526) フェロ.

fetch *n.* 取出し. *vt.* 取り出す.

fiber *n.* ファイバー. ⇨ fibre.
~ bundle ファイバー. バンドル[束]. ~ space ファイバー空間.

Fibonacci (1170–1250) フィボナッチ(= Leonardo of Pisa, Leonardo Pisano, ピサのレオナルド).

fibre =fiber.
~d space ファイバー空間.

fictitious *a.* 虚構の.

fiducial *a.* フィドゥーシアル, 基準の, 起点の.
~ distribution フィドゥーシアル分布. ~ interval 信頼区間. ~ limit(s) 信頼限界. ~ point 起点, 基準点.

field *n.* 体; 場; 野, 広場.
algebraic function ~ 代数関数体. algebraic number ~ 代数的数体, 代数体. algebraically closed ~ 代数的閉体. coefficient ~ 係数体. commutative ~ 可換体. electric ~ 電界, 電場. extended ~, extension ~ 拡大体. extension of ~ 体の拡大. ~ of force 力の場. ~ of lines 線野. ~ of points 点野. ~ of quotients, quotient ~ 商の体, 商体. ~ of rational functions, rational function ~ 有理関数体. ~ supervisor 現地指導員. ~ theory 場の理論; 体論. ~ work 現地作業. finite ~ 有限体. fixed ~ 固定体, 不変体. Galois ~ ガロア体. imperfect ~ 非完全体. inertia ~ 惰性体. inertial ~ 慣性場. infinite ~ 無限体. invariant ~ 不変体. magnetic ~ 磁界, 磁場. minimal splitting ~ 最小分解体. non-commutative ~ 非可換体, 斜体. over ~ 拡大体. perfect ~ 完全体. prime ~ 素体. quaternion ~ 四元数体. ramification ~ 分岐体. rational number ~, ~ of rational numbers, ~ of rationals 有理数~. real closed ~ 実閉体. real number ~, ~ of real numbers 実数体. scalar ~ スカラー場. skew ~ 斜体, 非可換体. splitting ~ 分解体. tensor ~ テンソル場. underlying ~ 基礎体. unified ~ 統一場. valuation ~ 付値体. vector ~ ベクトル場.

Fields, John Charles (1863–1932) フィールズ.
Fields medal フィールズ牌[メダル].

FIFO フィフォ, ファイフォ.
first-in first-out 先入れ先出し.

fifth *a.* 第五(番目)の. *n.* 第五; 五分の一.
~ postulate 第五公準. ~ problem of Hilbert ヒルベルトの第五問題. two ~s 五分の二.

fig. =figure.

figure *n.* 図. 図形, 図式; 数字; (*pl.*) 計算. *vt.* 図で示す; 数字で示す; 計算する. *vi.* 計算する.
attached ~ 付図. cross ~ 断面図. double ~ 二桁(の数). geometric(al) ~ 幾何図形. plane ~ 平面図形. solid ~ 立体図形.

file *n.* ファイル. *vt.* ファイルする.
sub~ 副ファイル.

filter *n.* フィルター.
~ base フィルター基. maximal ~ 極大フィルター. ultra~ 極大フィルター.

filtering *n.* フィルタリング.
stochastic ~ 確率フィルタリング.

filtration *n.* フィルターづけ; フィルトレーション.
~ degree フィルター次数.

final *a.* 最終の; 最後の.
~ object 終対象. ~ set 終集合. ~ velocity (最)終速度.

find *vt.* 見出す; 計算して出す, 求める; 分かる, 気づく.

fine *a.* 細かい; 細-.
~ classification 細かい類別. ~ convergence 細収束. ~ topology 細位相.

finger *n.* (手)指.
~ alphabet 指文字.

finis *n.* (*L.*) 終末; 限界.

～ inferior 下限. ～ superior 上限.

finish *vt.* 終える; 完成する. *vi.* 終わる. *n.* 終局; 仕上げ.
～ed product 完成品.

finite *a.* 有限な.
closure ～ 閉包有限な. ～ covering 有限被覆. ～ decimal 有限小数. ～ difference 有限差分. ～(-)dimensional 有限次元の. ～ element method 有限要素法. ～ extension 有限(次)拡大. ～ field 有限体. ～ graph 有限グラフ. ～ group 有限群. ～ interval 有限区間. ～ mathematics 有限数学. ～ multiplier 有限修正(乗数). ～ order 有限位(数). ～ population 有限母集団. ～ series 有限級数. ～ set 有限集合. ～ sum 有限和; (差分法の)和分. ～-valued function 有限値関数. locally ～ 局所有限な. of ～ type 有限型の.

finitely *ad.* 有限(的)に.
～ additive 有限加法的な. ～ additive class [family] 有限加法族. ～ additive measure 有限加法的測度. ～ distinguishable 有限的分離可能な. ～ generated 有限生成の.

finiteness *n.* 有限性.
～ condition 有限(性)条件.

Finsler, Paul (1894-1970) フィンスラー.

first *a.* 第一の; 最初の, *ad.* 第一に; 最初に.
～ class 第一類. ～ cosine rule [formula] 第一余弦法則. ～ countability axiom 第一可算公理. ～ fundamental form 第一基本形式. ～-in first-out 先入れ先出し. ～ integral 第一積分. ～ kind 第一種. ～ quadrant 第一象限. ～ term 初項. ～ variation 第一変分.

Fischer, Charles Albert (1884-1922) フィッシャー.

Fisher, Ronald Aylmer (1890-1962) フィッシャー.

fit *vt., vi.* に適合する. 適合させる; 当てはめる. *n.* 適合.
goodness of ～ 当てはまりの良さ, 適合度.

fitting *n.* 適合させること.
curve-～ 曲線の当てはめ.

five *n.* 5. *a.* 5の.
～ centroids of triangle 三角形の五心. ～ lemma 5 補題, 5 項補題. ～ number summary 5 数要約.

fix *vt.* 固定する.

fixed *a.* 固定した, 固定の.
～ branch point 動かない分岐点. ～ effect 固定効果. ～ field 固定体, 不変体. ～ group 固定群, 不変群. ～ point 固定点, 不動点; 固定小数点. ～ point representation 固定小数点表示. ～ point theorem 不動点定理. ～ scale 固定尺. ～ vector 束縛ベクトル.

flabby *a.* ゆるんだ; 軟弱な.
～-scattered 散布的な.

flag *n.* 旗; 標識.
～ manifold 旗多様体.

flat *a.* 平らな, 平坦な. *ad.* 平坦に. *n.* 平地.
concircularly ～ 共円的に平坦な. conformally ～ 共形的に平坦な. ～ function 平坦な関数. ～ module 平坦な加群. ～ point 平坦点. ～ space 平坦な空間. ～ subspace 平坦な部分空間. locally ～ 局所平坦な. projectively ～ 射影的に平坦な.

flatness *n.* 平坦さ.
point of ～ 平坦点.

flection *n.* 屈曲.

flexion ＝flection.

flexure *n.* ひずみ; 屈曲.

flight *n.* 飛行; 逃走.
random ～ 乱歩, 酔歩.

float *vi.* 浮かぶ; 浮動する, *n.* 浮き.
～ing point 浮動小数点.

flow *n.* 流れ. *vi.* 流れる. *vt.* 流す. ⇨ stream.
～ chart, ～ sheet 流れ図, フローチャート. ～ line 流線. geodesic ～ 測地的流れ, 測地流. laminar ～ 層流. maximum ～ 最大流, 最大フロー. multicommodity ～ 多品種フロー. Ricci ～ リッチ・フロー. steady ～ 定常流

flowchart *n.* 流れ図, フローチャート, 生産工程一覧表.

fluctuation *n.* 変動; ゆらぎ, 揺動; 振幅.
～-dissipation theorem 揺動散逸定理.

fluent *n.* 変数, 変量, 流量. *a.* なめらか

な，よどみのない．

fluid *n.* 流体． *a.* 流体の．
compressible ～ 圧縮(性)流体． ～ dynamics 流体力学(＝hydrodynamics)． ～ mechanics 流体力学(＝hydromechanics)． incompressible ～ 非圧縮(性)流体． perfect ～ 完全流体． viscous ～ 粘性流体．

fluidal *a.* 流体の．

flux *n.* フラックス; 流量; 流出; 流動; 力束．

fluxio *n.* (*L.* ＝fluxion.) 流率(法)．

fluxion *n.* 流率(法)，微分商．

fluxional *a.* 微分の; 絶えず変化する．

fluxionary ＝fluxional.

fn. ＝function.

fns. ＝functions.

focal *a.* 焦点の; 焦-.
～ axis 焦軸． ～ chord 焦弦． ～ conics 焦円錐曲線 ～ distance, ～ length 焦点距離 ～ line 焦線． ～ plane 焦面． ～ point 焦点(＝focus); 渦状点． ～ radius 焦点半径． ～ surface 焦曲面．

focus (*pl.* -cases, -ci) *n.* 焦点; 焦点距離; 渦状点． *vt.* 焦点を合わす． *vi.* 焦点が合う．
～-center 渦状渦心点． ～ of ellipse 楕円の焦点． ～ of flow 流れの渦状点．

fold *n.* 襞[ひだ]; 折り目．
～ of mapping 写像の襞．

-fold *suf.*「-重，-倍」の意．
k～ symmetry *k* 重対称(性)． two～ 二倍の[に]，二重の[に]．

foliate *vt.* 葉状にする．
～d structure 葉層構造．

foliation *n.* 葉層(構造)．
analytic ～ 解析葉層． simple ～ 単純葉層．

folium (*pl.* -lia) *n.* 葉線，葉状曲線．
～ of Descartes デカルトの葉状曲線[葉形]．

follow *vt.* 従う． *vi.* 当然…となる．
as ～s 次の通り． ～ing 次の． ～ up 徹底させる，進度管理する; 再[事後]調査．

foot (*pl.* feet) *n.* 足; フィート．
～ of perpendicular 垂線の足．

footnote *n.* 脚注．

force *n.* 力． *vt.* 強制する．
external ～ 外力． couple of ～s 偶力． ～ function 力の関数． ～ polygon 力の多角形． parallelogram of ～s 力の平行四辺形．

forced *a.* 強制的な．
～ oscillation [vibration] 強制振動．

forecast *n.* 予測; 予報． *vt., vi.* 予測する．

forecasting *n.* 予測(すること); 予報．
demand ～ 需要予測．

forked *a.* 叉状の，分岐した．

form *n.* 形; 形式，同次式． *vi.* 形づくる; 形成する．
automorphic ～ 保型[形]形式． bilinear ～ 双一次形式． definite ～ 定符号[値]形式． differential ～ 微分形式． (row) echelon ～ (行)階段行列． ～ a basis 基を形成する． harmonic ～ 調和形式． Hermitian [Hessian] ～ エルミート[ヘッセ]形式． Killing ～ キリング形式． linear ～ 一次形式． polar ～ 極形式． quadratic ～ 二次形式． quadratic integral ～ 積分二次形式． reduced row echelon ～ 簡約行階段行列． semidefinite ～ 半定値形式． tensorial ～ テンソル形式．

-form *suf.*「…形」，「…状」の意．

formal *a.* 形式的な; 形式の; 正式の．
～ degree 形式的次数，見掛けの次数． ～ expansion 形式的展開． ～ Fourier series 形式的フーリエ級数． ～ logic 形式論理(学)． ～ power series 形式的ベキ級数． ～ power series ring 形式的ベキ級数環． ～ solution 形式解．

formalism *n.* 形式主義; 形式論．

formally *ad.* 形式的に．
～ self-adjoint 形式的自己随伴な．

formation *n.* 構成．
class ～ 類構造． ～ rule 構成規則．

formula (*pl.* -las, -lae) *n.* 論理式; 公式; 式; 方式． addition ～ 加法公式． asymptotic ～ 漸近公式． Cramer's ～ クラーメルの公式． double-angle ～ 二倍角の公式． empirical ～ 実験式． expansion ～ 展開(公)式． experimental ～ 実験式，経験式． explicit ～ 明示公式． factorization ～，～ of factorization 因数分解(の)公式． ～ manipulation 数式処理． half-angle ～ 半角の公式． integral ～ 積分公式． interpolation ～

補間公式. logical ～ 論理式. multiplication ～ 乗法公式. recurrence [recursion, recurring, recursive] ～ 漸化(公)式. recursive ～ 再帰式. sine ～ 正弦公式. well-formed ～ (命題論理学の)論理式. X's ～ X(氏)の公式.

formulate *vt.* 定式化する; 式で示す.

formulation *n.* 定式化; 公式化.

forward *ad.* 前方に[へ]. *a.* 前進の, 前方の. ⇨ backward.
～ difference 前進差分. ～ elimination 前進消去. ～ equation 前進方程式. ～ substitution 前進代入.

foundation *n.* 基礎; 基礎論; 基礎づけ.
～s of mathematics [geometry] 数学[幾何学]基礎論.

four *n.* 四. *a.* 四の.
～ arithmetic operations 四則算法. ～ colo(u)r problem, problem of ～ colo(u)rs 四色問題. ～(-)dimensional 四次元の. ～ rules 四則. ～score 八十. ～ vector 四元ベクトル. ～ vertex theorem, theorem of ～ vertices 四頂点定理.

Fourier, Jean Baptiste Joseph (1768-1830) フーリエ.

four-square *a.* 正方形の, 四角な. *n.* 正方形, 四角.

fourth *a.* 第四の. *n.* 第四; 四分の一.
～ harmonic poznt 第四調和点. ～ proportional 第四比例項. ～ quadrant 第四象限. three ～s 四分の三.

four-vertex *a.* 四頂点の.
～ theorem 四頂点定理 (=theorem of four vertices).

fractal *n.* フラクタル.

fraction *n.* 分数; 端数.
common [vulgar] ～ (普通の)分数. complex ～ 繁分数. continued ～ 連分数. decimal ～ 十進小数. defective ～ 不良率. improper ～ 仮分数. irreducible ～ 既約分数. partial ～ 部分分数. proper ～ 真分数. rational ～ 有理分数. recurring continued ～ 循環連分数.

fractional *a.* 分数の; 端数の.
～ equation 分数方程式. ～ exponent 分数指数[ベキ]. ～ expression [function] 分数式[関数]. ～ ideal 分数イデアル. ～ order 非整数位. ～ power 分数ベキ. ～ replication 一部実施(法).

Fraenkel, Adolf Abraham Halevi (1891-1965) フレンケル.

fragmentary *a.* 断片的な; はんぱな.
～ data 不完全データ.

frame *n.* 枠, 標構; 座標; 構造.
affine ～ アフィン枠. ～ bundle フレーム・バンドル, 枠束. ～ of space and time 時空系座標. moving ～ 動標構, 動枠. natural ～ 自然標構. orthogonal ～ 直交枠. projective ～ 射影枠.

Fréchet, Maurice (1878-1973) フレシェ.

Fredholm, Eric Ivar (1866-1927) フレドホルム.

free *a.* 自由な; …のない. ⇨ -free.
～ boundary 自由境界. ～ game 無規準ゲーム. ～ group 自由群. ～ mobility 自由運動(性). ～ module 自由加群. ～ oscillation 自由振動. ～ product 自由積. ～ resolution 自由分解. ～ variable 自由変数. ～ vector 自由ベクトル. vortex(-) ～ 渦なしの, 無渦の.

-free *suf.* 「…のない」の意.
parameter～ 母数によらない. singularity～ 特異点のない. square～ 無平方の. torsion～ ねじれのない. umbilic～ 臍点のない. vortex～ 渦なしの, 無渦の. zero～ region 零点のない領域.

freedom *n.* 自由.
degree of ～ 自由度.

freely *ad.* 自由に.
act ～ 自由に作用する.

Frege, Gottlob (1848-1925) フレーゲ.

frequency *n.* 振動数; 周波数; 度数, 頻度.
angular [circular] ～ 角振動数. cumulative ～ 累積度数. ～ curve 度数曲線, 頻度曲線. ～ distribution 度数分布, 頻度分布. ～ modulation 周波数変調. ～ of occurrence 出現頻度. ～ polygon 度数(分布)多角形. ～ ratio 頻度比. ～ table 度数分布表. relative ～ 相対度数[頻度].

frequent *a.* しばしば起こる; 頻繁な.

FRESNEL, Augustin Jean (1788-1827) フレネル.

FROBENIUS, Georg Ferdinand (1849-1917) フロベニウス.

from *prep.* …から.
bounded ~ above [below] 上[下]に有界な. ~ A to B A から B まで.

front *n.* 前面, 正面. *a.* 正面の.
~ side 表側, 表面. wave ~ 波面.

frontier *n.* 境界. ⇨ boundary.
~ point 境界点.

frustum (*pl.* -tums, -ta) *n.* 台形; 切[截]頭体.
~ of pyramid 角錐台, 切[截]頭角錐. ~ of right cone 直円錐台.

F_σ set エフ・シグマ集合.

F-test F 検定(法).

FUBINI, Ghirin Guido (1897-1943) フビニ.

FUCHS, Lazarus (1833-1902), Maximilian Ernst Richard (1873-1944) フックス.

Fuchsian *a.* フックスの.
~ form フックス形式. ~ group フックス群. of ~ type フックス型の.

fulfill *vt.* (約束, 義務などを)果たす; (希望, 要求などを)満たす.

full *a.* 満ちた; 十分な; 全部の, 完全な. *ad.* 十分に.
~ circle 周角. ~ linear group 全線形群. ~ rank 最大階数. ~ space 全空間. ~ subcategory 充満部分圏.

fully *ad.* 十分に, 完全に.
~ faithful 充満忠実な. ~ normal 全体正規な.

func. =function.

function *n.* 関数, 函数; 機能.
algebraic ~ 代数関数. algebraic ~ field 代数関数体. analytic ~ 解析関数. Baire ~ ベール関数. Bessel ~ ベッセル関数. bounded ~ 有界関数. complex ~ 複素関数. continuous ~ 連続関数. convex ~ 凸関数. decreasing ~ 減少関数. elementary ~ 初等関数. elliptic ~ 楕円関数. exponential ~ 指数関数. extremal (~) 極値関数. ~ element 関数要素. ~ of bounded variation 有界変動関数. ~ of real variable(s) 実変数関数. ~ of several variables 多変数関数. ~ space 関数空間. ~ theory 関数論. hazard ~ 危険関数, ハザード関数. hyperbolic ~ 双曲線関数. hypergeometric ~ 超幾何関数. increasing ~ 増加関数. inverse ~ 逆関数. logarithmic ~ 対数関数. membership ~ メンバーシップ関数. Möbius ~ メビウス関数. monotone ~ 単調関数. rational ~ 有理関数. real ~ 実関数. regular ~ 正則関数. risk ~ リスク[危険]関数. sn-~ sn 関数. theory of ~s 関数論. transcendental ~ 超越関数. trigonometric ~ 三角関数. univalent ~ 単葉関数. X('s) ~ X(氏の)関数.

functional *n.* 汎関数. *a.* 関数の; 機能の.
additive ~ 加法的汎関数. ~ analysis 関数解析(学). ~ dependence 関数的従属性. ~ determinant 関数行列式. ~-differential equation 汎関数(微分)方程式. ~ equation 関数方程式. ~ Φ-operation 汎関数 Φ 作用素. ~ relation 関数関係. ~ scale 関数尺, 関数目盛. ~ space 関数空間. ~ symbol 関数記号. ~ test 機能テスト. linear ~ 線形汎関数. support ~ 支持汎関数.

functionally *ad.* 関数的に.
~ dependent 関数的(に)従属な. ~ independent 関数的(に)独立な.

function-theoretic *a.* 関数論的な.
~ null set 関数論的零集合.

functor *n.* 関手.
additive ~ 加法的関手. cohomological ~ コホモロジー関手. covariant [contravariant] ~ 共変[反変]関手. derived ~ 導来関手. exact ~ 完全関手.

functorial *a.* 関手の.
~ isomorphism 関手(間)の同型[形].

functoriality *n.* 関手性.

fundamental *a.* 基本的な; 基本の.
~ construction(s) 基本作図. ~ cycle 基本輪体. ~ event 基本[根元]事象. ~ form 基本形式. ~ function 基本関数. ~ group 基本群. ~ loci 基本軌跡. ~ neighbo(u)rhood 基本近傍. ~ period 基本周期. ~ point(s) 基本点. ~ region 基本領

域. ~ sequence 基本列. ~ solution 基本解. ~ space 基本空間. ~ system 基本系. ~ theorem 基本定理. ~ triangle 基礎三角形. ~ vector 基本ベクトル.

funnel *n.* 漏斗面.

FURTWÄNGLER, Philipp (1869-1940) フルトウェングラー.

fuzzy *a.* ファジィな, あいまいな.
~ set ファジィ集合.

G

gain *n.* 利益; 利得. *vt., vi.* 利益をえる.

GALERKIN, Boris Grigor'evič (1871-1945) ガレルキン.

GALILEI, Galileo (1564-1642) ガリレオ.

GALOIS, Évaliste (1811-1832) ガロア(ガロワ).

GAM ジー・エイ・エム.
generalized additive model 一般化加法モデル.

game *n.* ゲーム.
bimatrix ~ 双行列ゲーム. closed ~ 閉鎖ゲーム. concave [convex] ~ 凹[凸]ゲーム. cooperative ~ 協力ゲーム. ~--theoretic ゲーム理論的な. ~ theory, theory of ~s ゲームの理論. ~ tree ゲームの樹形図. political ~ 政治ゲーム. zero-sum ~ 零和ゲーム.

gamma *n.* ガンマ. ギリシア語アルファベットの3番目の文字【Γ, γ】.
di~ function ディガンマ関数. ~ distribution ガンマ分布. ~ function ガンマ[Γ]関数. ~ structure ガンマ構造.

gap *n.* 間隙, 空隙.
~ series 空隙級数. ~ theorem 空隙定理. ~ value 空隙値. Hadamard('s) ~ アダマール(の)空隙.

gate *n.* ゲート【論理回路の一種】.

GÂTEAUX, René Eugène (1889-1914) ガトー.

gauge *n.* ゲージ; 規格; 計器.
~ transformation ゲージ変換.

GAUSS, Carl Friedrich (1777-1855) ガウス.

Gaussian *a.* ガウスの.
~ curvature ガウス(の)曲率. ~ curve ガウス曲線, 正規曲線(=normal curve). ~ distribution ガウス分布, 正規分布. ~ elimination ガウスの消去法. ~ integer ガウスの整数. ~ plane ガウス平面. ~ process ガウス型(確率)過程. ~ sum ガウス(の)和.

G.C.D. Greatest Common Divisor. 最大公約数.

G.C.M. Greatest Common Measure. 最大公約数

G_δ set ジー・デルタ集合.

gear *n.* 歯車.
bevel ~ 傘歯車. ~-shaped domain 歯車型領域. geometry of ~s 歯車の幾何学. hypoid ~ ハイポイド歯車. involute ~ インボリュート歯車.

Gebilde (*G.*) *n.* 形成体; 構造(物).
analytisches ~ 解析(的)形成体.

GELFAND, Izrail Moiseevich (1913-2009) ゲルファント.

GEL'FOND, Alexandr Osipovič (1906-1968) ゲルフォント.

general *a.* 一般な, 一般的な.
~ angle 一般角. ~ coordinates 一般座標. ~ derivative 一般微分係数. ~ Dirichlet series 一般ディリクレ級数. ~ epidemics 一般伝染理論. ~ form 一般形. ~ helix 一般螺線. ~ linear group 一般線形群. ~ position 一般な位置. ~ principle of relativity 一般相対性原理. ~ set theory, ~ theory of sets 一般集合論. ~ solution 一般解. ~ term 一般項, 公項. ~ theory 一般(理)論. ~ topology 一般位相数学. ~ validity 一般妥当性. in ~ 一般に. in the ~ 概して.

generality *n.* 一般性, 普遍性.

without loss of ~ 一般性を失うことなく.

generalization *n.* 一般化.

generalize *vt.* 一般化する.

generalized *a.* 一般化された.
~ additive model (GAM) 一般化加法モデル. ~ coordinates 一般座標. ~ distance 一般距離. ~ function 一般関数. ~ inverse (matrix) 一般化逆行列. ~ linear model (GLIM) 一般化線形モデル. ~ uniform convergence 広義の一様収束. ~ valuation 一般付値. ~ variance 一般分散. ~ zeta function 一般(化された)ゼータ関数.

generate *vt.* 生成する; (点・線・面が動いてそれぞれ線・面・立体を)画く.
finitely ~d 有限生成の.

generating *a.* 生成の, 母-.
~ element 生成元. ~ function 母関数, 生成関数. ~ line 導線, 母線. ~ space 母空間.

generation *n.* (図形の)生成; 世代.
~ number 世代数. ~ tree 生成(樹)木.

generator *n.* 生成元; 母線, 母面; 生成作用素; 生成素.
~ matrix 生成行列. ~ of surface of revolution 回転面の母線. ~ polynomial 生成多項式. ~ system, system of ~s 生成系; 母線系.

generatrix (*pl.* -trices) *n.* 母量; 母点, 母線, 母面.

generic *a.* 一般的な, 生成の.
~ point 生成点.

genetic *a.* 発生の; 発生的な; 遺伝学の.

genetics *n.* 遺伝学.
human ~ 人類遺伝学. population ~ 集団遺伝学.

Gentzen, Gerhard Karl Erich (1909-1945) ゲンツェン.

genuine *a.* 真の.
~ solution 真の解.

genus (*pl.* -nera, -nuses) *n.* 種数; 示性数, ジーナス; 種; 類数.
arithmetic [geometric] ~ 算術[幾何]種数. infinite ~ 無限種数. principal ~ 主種.

geodesic *a.* 測地線の, 測地(的)-. *n.* 測地線.
closed ~ (line) 閉測地線. ~ arc 測地弧. ~ circle 測地円. ~ coordinates 測地座標. ~ correspondence 測地的対応. ~ curvature 測地(的)曲率. ~ deviation 測地的変位. ~ flow 測地流. ~ line 測地線. ~ polar coordinates 測地的極座標. ~ radius 測地半径. ~ torsion 測地(的)捩率.

geodesically *ad.* 測地的に.
~ complete 測地的(に)完備な.

geodesy *n.* 測地学.

geometric *a.* 幾何学の, 幾何(学的)-, 幾何学上の; 幾何学図形の.
~ cell complex 幾何学的胞複体, ユークリッド胞複体. ~ difference equation 幾何的差分方程式. ~ locus 軌跡. ~ mean 幾何平均, 相乗平均, 等比中項, 比例中項. ~ multiplicity 幾何的重複度. ~ probability 幾何(学的)確率. ~ progression 等比[幾何]数列[G.P.]. ~ series 等比[幾何]級数.

geometrical =geometric.
~ distribution 幾何分布. ~ figure 幾何(学的)図形. ~ optics 幾何光学. ~ quantity 幾何学的量. ~ vector 幾何ベクトル.

geometrically *ad.* 幾何学で; 幾何学的に.

geometrico- 「幾何」の意の結合辞.
~ harmonic mean 幾何調和平均.

geometry *n.* 幾何学.
affine ~ アフィン幾何学. algebraic ~ 代数幾何学. analytic (al) ~ 解析幾何学. computational ~ 計算幾何学. conformal ~ 共形幾何学. coordinate ~ 座標幾何学. descriptive ~ 画法幾何学. differential ~ 微分幾何学. Euclidean ~ ユークリッド幾何学. ~ of numbers 数の幾何学. ~ of paths 道の幾何学. ~ on sphere 球面上の幾何(学). information ~ 情報幾何学. integral ~ 積分幾何学. non-Euclidean ~ 非ユークリッド幾何学. plane ~ 平面幾何学. projective ~ 射影幾何学. Riemannian ~ リーマン幾何学. solid ~ 立体幾何学. spherical ~ 球面幾何学. synthetic ~ 総合幾何学.

geophysics *n.* 地球物理学.

germ *n.* 芽.
~ of analytic function [set] 解析関数[集合]の芽. group ~ 群芽.

Germain, Sophie (1776-1831) ジェルマン.

Gevrey, Maurice (1884-1957) ジェブレー.

Gibbs, Josiah Willard (1839-1903) ギブズ.

giga- *pref.* 「10^9」の意.

gist *n.* 要旨, 要点, 骨子.

give *vt.* 与える.
This gives the desired solution. これが求める解を与える.

g. l. b. greatest lower bound. 下限.

Gleichverteilung *n.* (*G.*) 一様分布.

GLIM グリム.
generalized linear model 一般化線形モデル.

global *a.* 大域の, 大域的な. ⇨ local.
~ analysis 大域的解析学. ~ differential geometry 大域的微分幾何学. ~ dimension 大域次元. ~ optimization 大域的最適化. ~ programming 大域的計画法. ~ property 大域的性質. ~ section 大域切断. ~ theory 大域(的)理論.

globally *ad.* 大域的に.
~ defined function 大域的に定義された関数.

globe *n.* 球; 地球, 天体; 地球儀; 天体儀.

globular *a.* 球状の; 丸い.
~ projection 球面投影(法).

glossary *n.* 用語集, 語彙.

glueing *n.* 貼り合わせ.

gnomon *n.* ノーモン【平行四辺形からその一角を含む相似四辺形を取り去った残りの形】, 周髀; 指時針(日時計の).

gnomonic *a.* ノーモンの.
~ projection 心射図法.

Gödel, Kurt (1906-1978) ゲーデル.

Godement, Roger (1921-) ゴドマン.

Goldbach, Christian (1690-1764) ゴルドバッハ.

golden *a.* (黄)金の.
~ cut 黄金分割. ~ ratio 黄金比. ~ section 黄金分割.

Golusin, Gennadi Michailovič (1906-1952) ゴルジン.

-gon *suf.* 「-角形」, 「-辺形」の意.
dodeca ~ 十二角形. penta ~ 五角形. poly ~ 多角形.

goniometer *n.* 角度計, 測角器.

goodness *n.* 良さ.
~ of fit 当てはまりの良さ, 適合度.

googol *n.* 10 の 100 乗【10^{100}】.

googolplex *n.* 10 を 10 の 100 乗々した数【$10^{10^{100}}$】.

Gordan, Paul (1837-1912) ゴルダン.

Goursat, Edouard Jean Baptiste (1858-1936) グールサ.

G.P. Geometrical Progression. 等比数列.

grade *n.* 階数; 百分度【直角の 1/100】; 等級.

graded *a.* 次数つきの.
~ *A*-module 次数つき *A* 加群. ~ ideal 次数つきイデアル. ~ Lie algebra 次数つきリー環. ~ ring 次数つき環.

gradient *n.* 勾配; 傾斜, 傾き; グラディエント. *a.* 傾斜した.
conjugate ~ method 共役勾配法. ~ function 勾配関数. ~ method 勾配法, 傾斜法. ~ vector 勾配ベクトル. reduced ~ 簡約勾配, 縮約勾配.

graduate *vi.* 卒業する. *vt.* 卒業させる. *n.* 卒業生; 学士; 大学院生. *a.* 卒業生の, 学士号を受けた; 大学院の.
~ school 大学院. ~ student 大学院生.

graduation *n.* 度, 目盛; 滑らかにすること, 平滑化; 卒業.
~ thesis 卒業論文.

graduator *n.* 分度器.

Graeffe, Karl Heinrich (1799-1873) グレッフェ.

Gram, Jørgen Pedersen (1850-1916) グラム.

Gramian *a.* グラムの, *n.* グラムの行列式.

Gram-Schmidt
~ orthonomalization グラム・シュミットの正規直交化法.

graph *n.* グラフ, 図式, 図表.
area ~ 面積グラフ. band ~ 帯グラフ. bar ~ 棒グラフ. bipartite ~ 二部グラフ.

circle [circular] ~ 円グラフ. closed ~ theorem 閉グラフ定理. complement ~ 補グラフ. complete ~ 完全グラフ. connected ~ 連結グラフ. dual ~ 双対グラフ. finite linear ~ 有限線形グラフ. ~ of function 関数のグラフ. ~ theory グラフ理論. directed ~ 有向グラフ. planar ~ 平面的グラフ. plane ~ 平面グラフ. simple ~ 単純グラフ. undirected ~ 無向グラフ.

graphic *a.* グラフ(式)の, 図式の.
~ calculation 図(式)計算. ~ differentiation 図式微分. ~ integration 図式積分. ~ interpolation 図式補間. ~ solution 図式解(法).

graphical =graphic.
~ calculation 図式計算. ~ mechanics 図式力学. ~ representation グラフ[図]表現.

graphics *n.* 製図法; 図学; 図式算法.

graphology *n.* 筆跡学.

GRASSMANN, Hermann Günther (1809–1877) グラスマン.

GRAUERT, Hans (1930–2011) グラウエルト.

gravitation *n.* 引力, 重力.
universal ~ 万有引力.

gravity *n.* 重力, (地球)引力; 重量.
center of ~ 重心. ~ acceleration 重力加速度. specific ~ 比重.

great *a.* 大きな, 大きい.
~ circle 大円.

greater *a.* より大きい.
~ than or equal to 以上の. not ~ (than) 以下の.

greatest *a.* 最大な. ⇨ least.
~ common divisor 最大公約数[元, 因子][G.C.D.]. ~ common measure 最大公約数[G.C.M.]. ~ lower bound 最大下界, 下限[g.l.b.].

Greco- 「ギリシア」の意の結合辞.
~ Latin square グレコラテン方格[方陣].

Greek *a.* ギリシアの.
~ mathematics ギリシア数学.

GREEN, George (1793–1841) グリーン.

GREGORY, James (1638–1675) グレゴリ.

GRÖBNER, Wolfgang (1899–1980) グレブナー.

groin *n.* 穹稜【二つのアーチ形天井の交差線; 交差三円筒より成る面】.

gross (*pl.* -) *n.* グロス, 十二ダース; 総計.
great ~ 大グロス, 十二グロス. ~ amount 総額. ~ area 総面積. ~ mean 総平均. small ~ 小グロス, 十ダース.

GROTHENDIECK, Alexandre (1928–2014) グロタンディク.

GRÖTZSCH, Herbert (1902–1993) グレッチュ.

ground *n.* 地表; 基礎, 基本. *vt.* 基礎におく.
~ field 基礎体. ~ form 基本形式. ~ line 基線.

group *n.* 群. *vt.* 集める. *vi.* 集まる.
Abelian ~ アーベル群. additive ~ 加法群, 加群. adèle ~ アデール群. alternating ~ 交代群. Betti ~ ベッチ群. commutative ~ 可換群. continuous ~ 連続群. cyclic ~ 巡回群. discontinuous ~ 不連続群. dodecahedral ~ (正)十二面体群. factor ~ 因子群. finite ~ 有限群. fixed ~ 固定群, 不変群. function ~ 関数群. fundamental ~ 基本群. Galois ~ ガロア群. ~ algebra 群環, 群多元環. ~ character 群指標. ~ code 群符号. ~ determinant 群行列式. ~ germ 群芽. ~ manifold 群多様体. ~ object 群対象. ~ of holonomy, holonomy ~ ホロノミー群. ~ of units, unit ~ 単数群, 単元群. ~ ring 群環. hexahedral ~ (正)六面体群. homology [homotopy] ~ ホモロジー[ホモトピー]群. icosahedral ~ (正)二十面体群. inertia ~ 惰性群. infinite ~ 無限群. Lie ~ リー群. mixed ~ 混群; 混合(アーベル)群. matrix ~ 行列群. octahedral ~ (正)八面体群. order of ~ 群の位数. orthogonal ~ 直交群. permutation ~ 置換群. polyhedral ~ (正)多面体群. quaternion ~ 四元数群. quotient ~ 商群. simple ~ 単純群. spinor ~ スピノル群. sporadic simple ~ 散在型単純群. Sylow

(sub-) ~ シロウ群. symmetric ~ 対称群. tetrahedral ~ (正)四面体群. transformation ~ 変換群. unitary ~ ユニタリ群.

grouped *a.* 集められた.
~ sequential inspection 群逐次抜取り検査.

grouping *n.* 組分け, 分類; 集めること.

groupoid *n.* 亜群.

grow *vi.* 増大する.

growth *n.* 増大; 生長.
~ curve 生長曲線. ~ of function 関数の増大.

guarantee *n.* 保証. *vt.* 保証する.

GUDERMANN, Christoph (1798–1852) グーデルマン.

Gudermannian *a.* グーデルマンの. *n.* グーデルマン(の)関数【gd】.

GULDIN, Paul (1577–1643) ギュルダン.

GÜNTER, Nicolai Macsimovič (1871–1941) ギュンター.

gyroscope *n.* ジャイロスコープ, 回転儀.

gyroscopic *a.* 回転儀の.
~ term 回転儀項.

H

HAAR, Alfred (1885–1933) ハール.

HADAMARD, Jacques Salomon (1865–1963) アダマール.

HAHN, Hans (1879–1934) ハーン.

half (*pl.* -lves) *n.* 半分. *a.* 半分の.
~-angle formula 半角の公式.

half- *pref.* 「半」の意. ⇨ semi-.

half-closed *a.* 半閉の. =semi(-)closed.
~ interval 半閉区間.

half-exact *a.* 半完全な. =semi(-)exact.

half-infinite *a.* 半無限の.

half-line *n.* 半直線.

half-open *a.* 半開の. =semi(-)open.
~ interval 半開区間.

half-period *n.* 半周期.

half-periodic *a.* 半周期的な.

half-plaid *n.* 半格子縞.
~ square 半格子縞形.

half-plane *n.* 半平面.
left ~ 左半(平)面. lower ~ 下半(平)面. right ~ 右半(平)面. upper ~ 上半(平)面.

half-space *n.* 半空間. =semispace.

half-sphere *n.* 半球(面). =hemisphere.

half-spinor *n.* 半スピノル.

HALMOS, Paul Richard (1916–2006) ハルモス.

HALPHEN, Georges Henri (1844–1889) アルファン.

halt *vi.* 停止する. *n.* 停止.
~ing problem 停止問題.

halve *vt.* 二(等)分する.

HAMBURGER, Hans Ludwig (1889–1956), Meyer (1838–1903) ハンブルガー.

HAMEL, Georg (1877–1954) ハーメル.

HAMILTON, Sir William Rowan (1805–1865) ハミルトン.

Hamiltonian *a.* ハミルトンの. *n.* ハミルトニアン, ハミルトン演算子[作用素]; ハミルトン関数.
~ mechanics ハミルトン力学. ~ operator ハミルトン作用素. ~ type ハミルトン型.

HAMMING, Richard Wesley (1915–1998) ハミング.

handle *n.* 把手, ハンドル.
~-attaching 把手つけ操作. ~ body ハンドル(把手)体.

handle body *n.* ハンドル(把手)体.

HANKEL, Hermann (1839–1873) ハンケル.

hardware *n.* ハードウェア.

HARDY, Godfrey Harold (1877–1947) ハーディ.

harmonic *a.* 調和な.
~ analysis 調和解析. ~ analyzer 調和解析機. ~ boundary 調和境界. ~ conjugate 調和共役な; 共役調和関数. ~ continuation 調和接続. ~ division 調和分割. ~ form 調和形式. ~ function 調和関数. ~ integral 調和積分. ~ majorant 調和優関数. ~ mean 調和平均. ~ measure 調和測度. ~ minorant 調和劣関数. ~ pencil of lines [planes] 調和線[面]束. ~ progression 調和数列[H.P.], 調和級数. ~ range of points 調和列点. ~ series 調和級数. ~ space 調和空間. simple ~ motion 単(弦)振動[S.H.M.]. solid ~ function 球調和関数, 体球(調和)関数. spherical ~ function 球面(調和)関数.

harmonically *ad.* 調和に.

divide 〜, separate 〜 調和に分かつ.
harmonicity *n.* 調和性.
harmonics (*pl.*) *n.* 調和関数.
ellipsoidal 〜 楕円体調和関数. solid 〜 球調和関数, 体球(調和)関数. spherical 〜 球面(調和)関数. spheroidal 〜 回転楕円体調和関数. surface 〜 球面(調和)関数. tesseral 〜 縞球[方球]調和関数. zonal 〜 帯球調和関数.
harmonization *n.* 調和化.
harmonize *vt.* 調和化する,
HARNACK, Axel (1855-1888) ハルナック.
HARTOGS, Friedrich (1874-1943) ハルトクス.
HASSE, Hermut (1898-1979) ハッセ.
HAUSDORFF, Felix (1868-1942) ハウスドルフ.
haversine *n.* 半正矢【hav】.
hazard *n.* ハザード, 危険.
〜 function ハザード[危険]関数.
heat *n.* 熱.
〜 conduction 熱伝導. 〜 equation 熱方程式. 〜 source 熱源.
HEAVISIDE, Oliver (1850-1925) ヘビサイド.
hebdomadal *a.* 毎週の.
HECKE, Erich (1887-1947) ヘッケ.
hecto- *pref.*「百」の意.
〜 gram(me) 百グラム.
height *n.* 高さ, 高度, 標高.
Krull's 〜 [altitude] theorem クルルの標高定理.
HEINE, Heinrich Eduard (1821-1881) ハイネ.
helicoid *n.* 螺旋体, 螺旋面, ヘリコイド. *a.* 螺旋状の.
ordinary [right] 〜 常螺旋面.
helicoidal *a.* 螺旋の, 螺旋状の.
〜 motion 螺旋運動. 〜 surface 螺旋面.
helix *n.* つるまき線, 螺線; 螺旋.
general 〜 一般螺線. ordinary 〜 常螺線.
HELLINGER, Ernst (1883-1950) ヘリンガー.
HELMHOLTZ, Hermann Ludwig Ferdinand von (1821-1894) ヘルムホルツ.
hemi- *pref.*「半」の意. ⇨ demi-, semi-.
hemicycle *n.* 半円形.
hemisphere *n.* 半球(面).
hemispherical *a.* 半球の.
hence *ad.* ゆえに; よって; それゆえに.
HENSEL, Kurt (1861-1941) ヘンゼル.
hepta- *pref.*「七, 七倍」の意.
heptagon *n.* 七角[辺]形.
hereditary *a.* 遺伝的な.
〜 ring 遺伝環. left [right] 〜 左[右]遺伝的な.
heredity *n.* 遺伝.
HERGLOTZ, Gustav (1881-1953) ヘルグロッツ.
HERMITE, Charles (1822-1901) エルミート, エルミット.
Hermitian *a.* エルミートの, エルミート的な.
anti-〜 反エルミート的な. 〜 extension エルミート拡大. 〜 form エルミート形式. 〜 manifold エルミート多様体. 〜 matrix エルミート行列. 〜 metric エルミート計量. 〜 product エルミート積. 〜 transformation エルミート変換.
HERO(N) of Alexandria (2〜3 世紀?) アレクサンドリアのヘロン.
HESSE, Ludwig Otto (1811-1874) ヘッセ.
Hessian *n.* ヘッセ行列(式); ヘッセ形式; ヘッセ汎関数. *a.* ヘッセの. 〜 form ヘッセ形式.
hetero- *pref.*「異」の意. ⇨ homo-.
heterogeneity *n.* 異質性.
〜 test, test of 〜 異質性の検定.
heteroskedastic *a.* 異分散性の; 不均一分散な.
heuristic *a.* 発見的な.
〜 method 発見的方法. 〜 program 発見的プログラム.
hexa- *pref.*「六, 六倍」の意. ⇨ sex-.
hexadecimal *a.* 十六進(法)の.
〜 digit 十六進数字. 〜 notation 十六進法.
hexagon *n.* 六角形, 六辺形.
hexagonal *a.* 六角[辺]形の.
〜 number 六角数. 〜 prism 六角柱. 〜 system 六方晶系. 〜 web 六角形織.
hexagram *n.* 六線形, 六芒星形.
hexahedral *a.* 六面体の.

～ group (正)六面体群.

hexahedron (*pl.* -rons, -dra) *n.* 六面体. regular ～ 正六面体, 立方体.

hexaspherical *a.* 六球の.
～ coordinates 六球座標.

hierarchical *a.* 階層制の, 聖職者位階性の.
～ model 階層モデル. ～ subdivision 階段的分類.

hierarchy *n.* 階層.
analytic [arithmetic] ～ 解析[算術]的階層. ～ theorem 階層(の)定理. theory of ～chies 階層(の)理論.

high *a.* 高い.
～ precision computation 高精度計算. ～ (-)speed 高速(の).

higher *a.* より高い; 高階の, 高位の; 高等な. ⇨ lower.
equation of ～ degree 高次方程式. ～ algebra 高等代数(学). ～ mathematics 高等数学. ～ order infinitesimal, infinitesimal of ～ order 高位の無限小. ～ order infinity 高位の無限大. ～ transcendental function 高等超越関数.

highest *a.* 最高の, 最も高い.
～ weight 最高の重み. term of ～ degree 最高次の項.

Hilbert, David (1862–1943) ヒルベルト.

Hill, George William (1838–1914) ヒル.

Hille, Karl Einar (1894–1980) ヒレ.

Hinčin, Aleksandr Jakovlevič (1894–1959) ヒンチン.

histogram *n.* ヒストグラム, 柱状図(表).

hit *vt.* 的中する, 当たる. *vi.* 打つ, 的中する; 成功する. *n.* 的中, あたり.
chance ～ まぐれ当り.

hitting *a.* 到達する; 的中する.
～ measure 到達測度. ～ time 到達時刻.

Hobson, Ernest William (1856–1933) ホブソン.

Hochschild, Gerhard Paul (1915–2010) ホッホシルト.

Hodge, Sir William Vallance Douglas (1903–1975) ホッジ.

hodograph *n.* ホドグラフ.
～ method ホドグラフ法. ～ plane ホドグラフ面. ～ transformation. ホドグラフ変換.

hold *vi.* 成り立つ; 適用される; 続く. *vt.* 持つ; 開催する. *n.* 把握.

Hölder, Otto (1859–1937) ヘルダー.

hole *n.* 穴; 孔

holo- *pref.* 「全-」の意.

holomorphic *a.* 正則な.
～ function 正則関数. ～ part 正則部分. ～ sectional curvature 正則断面曲率.

holomorphical ＝holomorphic.
～ convexity 正則凸性.

holomorphically *ad.* 正則(的)に.
～ convex 正則凸な.

holomorphy *n.* 正則(性), 解析(性).
domain of ～ 正則領域. envelope of ～ 正則包.

holonomic *a.* ホロノームな.
～ system ホロノーム(な)系. ～ space ホロノーム空間.

holonomy *n.* ホロノミー.
～ group ホロノミー群. homogeneous ～ group 同次ホロノミー群. restricted ～ group 制限ホロノミー群.

homentropic *a.* 等エントロピーの.
～ flow 等エントロピー流.

homeo- *pref.* 「類似-」の意.

homeomorphic *a.* 位相同型[形]な, 同相な.
～ mapping 同相写像. ～ to *A* *A* に同相な.

homeomorphism *n.* 同相; 同相写像.
linear ～ 線形同相(写像).

homo- *pref.* 「同」の意. ⇨ hetero-.

homogeneity *n.* 同次性, 斉次性; 等質性, 均一性.
～ test 等質性の検定.

homogeneous *a.* 同次の, 斉次の; 等質な.
～ coordinates 同[斉]次座標. ～ equation 同[斉]次方程式. ～ expression 同次式. ～ form 同[斉]次形. ～ holonomy group 同次ホロノミー群. ～ ideal 同[斉]次イデアル. ～ part 同[斉]次部分. ～ polynomial 同[斉]次多項式. ～ product 同[斉]次積. ～ space 等質空間. ～ turbulence

一様乱流.

homolog *a.* (*G.*) ホモローグな.

homological *a.* ホモロジーの, ホモロジー的な.

~ algebra ホモロジー代数. ~ dimension ホモロジー次元. ~ mapplng ホモロジー写像.

homologous *a.* ホモローグな; 対応する.

~ element 対応元. ~ to 0 ゼロにホモローグな.

homologue *n.*, *a.* (*F.*) ホモローグ(な).

homology *n.* ホモロジー.

~ class ホモロジー類. ~ exact sequence ホモロジー完全系列. ~ group ホモロジー群.

homomorphic *a.* 準同型[形]な.

~ image 準同型[形]像. ~ mapping 準同型[形]写像.

homomorphism *n.* 準同型[形]; 準同型[形]写像.

algebra ~ 多元環準同型[形]. canonical ~ 標準的準同型[形](写像). connecting ~ 連結準同型[形](写像). ~ theorem 準同型[形]定理. local ~ 局所準同型[形](写像). open ~ 開準同型[形].

homoscedastic =homothedastic.

homothecy =homothety.

homothedastic *a.* 等分散[散布]の.

hornothetic *a.* 相似伸縮の.

~ center 相似中心. ~ figure(s) 相似伸縮図形. ~ ratio 相似比. ~ transformation 相似変換.

homothety *n.* 相似伸縮; 相似変換.

center of ~ 相似(の)中心. central ~ 中心相似. ratio of ~ 相似比.

homotop (*G.* =homotopic.) *a.* ホモトープな.

homotopic *a.* ホモトープな.

~ relative to *A* *A* に関してホモトープな. ~ to 0 ゼロにホモトープな.

homotopy *n.* ホモトピー.

~ class ホモトピー類. ~ equivalence ホモトピー同値. ~ equivalent ホモトピー同値な. ~ exact sequence ホモトピー完全系列. ~ group ホモトピー群. ~ invariant ホモトピー不変量, ホモトピー不変な. ~ operator ホモトピー作用素. ~ sphere ホモトピー球面. ~ type ホモトピー型. restricted ~ 制限ホモトピー.

Hopf, Eberhard (1902-1983), Heinz (1894-1971) ホップ.

horizon *n.* 水平線, 地平線.

horizontal *a.* 水平な.

~ component 水平成分. ~ line 水平線, 地平線. ~ plane 水平面, 地平面; 平画面. ~ projection 水平投象. ~ trace 水平跡.

Hörmander, Lars Valter (1931-2012) ヘルマンダー.

Horn, Jacob (1867-1946) ホルン.

Horner, William George (1768-1837) ホーナー.

horocycle *n.* (非ユークリッド幾何学の)ホロサイクル, 界線.

horosphere *n.* ホロ球面.

Hostinsky, Bohuslav (1884-1951) ホスティンスキ.

H.P. Harmonic Progression. 調和数列.

hull *n.* 莢, 閉包, 包.

closed convex ~ 閉凸包. convex ~ 凸包. injective ~ 単射[移入]包絡.

hull-kernel *n.* 莢核, 包核.

~ closure 莢[包]核閉包. ~ topology 莢[包]核位相.

human *a.* 人の, 人間の. *n.* 人, 人間.

~ engineering 人間工学. ~ genetics 人類遺伝学.

Hurewicz, Witold (1904-1957) フレビッチ.

Hurwitz, Adolf (1859-1919) フルウィッツ.

Huygens, Christiaan (1629-1695) ホイヘンス.

hybrid *a.* 混合形の, 混血の; 混性の, 混成の; 雑種の.

~ tensor 混合テンソル.

hydrodynamics *n.* 流体(動)力学.

hydromechanics *n.* 流体力学.

hydrostatic *a.* 静水の; 流体静力学の.

hydrostatics *n.* 流体静力学.

hyper- *pref.* 「上, 超越, 過度」の意. ⇨

hypo-.

hyperabelian *a.* 超アーベル的な.
～ function 超アーベル関数.

hyperbola *n.* 双曲線.
confocal ～s 共焦双曲線, conjugate ～ 共役双曲線. equilateral ～ 等辺双曲線. rectangular ～ 直角双曲線.

hyperbolic *a.* 双曲線の; 双曲型の, 双曲的な.
～ cosine 双曲(線)余弦. ～ cylinder 双曲柱. ～ domain [region] 双曲的領域, 双曲型領域. ～ function 双曲線関数. ～ geometry 双曲(的)幾何学. ～ metric 双曲(的)計量. ～ paraboloid 双曲放物面. ～ point 双曲的点. ～ quadrics 双曲型二次曲面. ～ sine 双曲(線)正弦. ～ space 双曲型空間. ～ spiral 双曲渦[螺]線, 逆匝線. ～ type 双曲型.

hyperbolical ＝hyperbolic.

hyperbolicity *n.* 双曲型性.

hyperboloid *n.* 双曲面.
canfocal ～s 共焦双曲面. conjugate ～ 共役双曲面. ～ of one sheet 一葉双曲面. ～ of revolution 回転双曲面. ～ of two sheets 二葉双曲面.

hyperboloidal *a.* 双曲面の.

hyperboloidic *a.* 双曲面的な.
～ position 双曲面的位置.

hypercohomology *n.* 超コホモロジー.

hypercornplex *a.* 多元の.
～ number 多元数. ～ system 多元環 (＝algebra).

hyperelliptic *a.* 超楕円の; 超楕円的な.
～ curve 超楕円曲線. ～ integral 超楕円積分. ～ Riemann surface 超楕円的リーマン面.

hyperfunction *n.* (佐藤)超関数.

hypergeometric *a.* 超幾何の, 超幾何-.
～ (differential) equation 超幾何(微分)方程式. ～ distribution 超幾何分布. ～ function 超幾何関数. ～ polynominal 超幾何多項式. ～ series 超幾何級数.

hypergroup *n.* 超群.

hyperplane *n.* 超平面.
～ at infinity 無限遠超平面. ～ coordinates 超平面座標. regression ～ 回帰超平面. separating ～ 分離超平面. supporting ～ 支持超平面. tangent ～ 接超平面.

hypersphere *n.* 超球.
～ geometry 超球幾何学. imaginary ～ 虚超球. point ～ 点超球. proper ～ 常超球.

hyperspherical *a.* 超球の.
～ coordinate 超球座標. ～ differential equation 超球微分方程式. ～ function 超球関数.

hypersurface *n.* 超曲面.
characteristic ～ 特性超曲面. quadratic ～ 二次超曲面. totally geodesic [umbilical] ～ 全測地[臍点]超曲面.

hypo- *pref.*「下, 以下, 少ない」の意. ⇨ hyper;- epi-.

hypocontinuous *a.* 亜連続な.

hypocycloid *n.* 内(転)サイクロイド, 内擺線.

hypocycloidal *a.* 内サイクロイドの.

hypoelliptic *a.* 準楕円的な, 準楕円型の.
～ operator 準楕円型作用素.

hypofunction *n.* 下関数.

hypograph *n.* ハイポグラフ.

hypoid *n.* ハイポイド.
～ gear ハイポイド歯車.

hypotenuse *n.* (直角三角形の)斜辺.

hypothesis (*pl.* -ses) *n.* 仮定; 仮説.
alternative ～ 対立仮説. composite ～ 複合仮説. continuum ～ 連続体仮説. linear ～ 線形仮説. null ～ 帰無仮説. simple ～ 単純仮説. statistical ～ 統計(的)仮説.

hypothetical *a.* 仮定の; 仮説の; 仮言の.
～ syllogism 仮言三段論法.

hypotrochoid *n.* 内トロコイド, 内余擺線.

hypotrochoidal *a.* 内トロコイドの.

hysteresis [hìstərí:sis] *n.* (磁気)履歴, ヒステリシス.

I

I, i （ローマ数字の）1.

i.i.d アイ・アイ・ディー.
independent and identically disitributed 独立で同一な分布に従う.

i *n.* アイ【虚数単位 $\sqrt{-1}$】.

ib., ibid. ibidem.

ibidem *ad.* (*L.* ＝in the same place.) 同じ箇所[本，掲載誌，章など]に.

-ible *suf.* 「…できる」の意: expressible, reducible. ⇨ -able.

ICM International Congress of Mathematicians. 国際数学者会議.

icosahedral *a.* 二十面体の.
～ group (正)二十面体群.

icosahedron (*pl.* -rons, -dra) *n.* 二十面体.
regular ～ 正二十面体.

id. identity; (*L.*) idem.

ideal *n.* イデアル. *a.* 理想的な，仮想上の.
admissible ～ 許容イデアル. contracted ～ 縮約イデアル. fractional ～ 分数イデアル. ～ boundary 理想境界. ～ class イデアル類. ～ group イデアル群. homogeneous ～ 同[斉]次イデアル. ～ point 理想[仮想]点. ～ quotient イデアル商. irrelevant ～ 無縁イデアル. left [right] ～ 左[右]イデアル. maximal ～ 極大イデアル. prime ～ 素イデアル. principal ～ 単項[主]イデアル. principal ～ domain (PID) 単項[主]イデアル整域. two-sided ～ 両側イデアル. unit ～ 単位イデアル.

idèle *n.* (*F.*) イデール.
～ group イデール群. principal ～ 主イデール.

idem *pron., a., ad.* (*L.* ＝same.) 同一(の); 同書(の，で).

idemfactor *n.* 単位ダイアディック.

idempotency *n.* ベキ等(性).

idempotent *a.* ベキ等の. *n.* ベキ[べき，冪]等元.
～ element ベキ等元. ～ matrix ベキ等行列. primitive ～ 原始ベキ等元.

identical *a.* 同一の; 恒等-.
～ equation 恒等式. ～ transformation 恒等変換.

identically *ad.* 恒等的に.
～ equal 恒等的に等しい.

identification *n.* 等化，同一視; 識別.
～ space 等化空間. ～ topology 等化位相.

identify *vt.* 等化する，同一視する; 識別する.
～ *A* with *B* *A* を *B* と同一視する.

identity *n.* 恒等; 恒等式; 恒等[単位]元.
～ component 単位(元)成分. ～ element 恒等元，単位元. ～ mapping 恒等写像. ～ matrix 単位行列. ～ operator 恒等作用素. ～ permutation 恒等置換. ～ transformation 恒等変換. theorem of ～ 一致(の)定理.

id est (*L.* ＝that is.) すなわち.

idle *a.* 無用の; 遊休の.

i.e. id est.

if and only if のときかつそのときに限って.

iff ＝if and only if.

if-then 含意.

ill-conditioned *a.* 不良条件の，たちの悪い.
～ case 不良条件の場合.

ill-posed *a.* 非適切な.
【well-posed の反対語.】

illustration *n.* 説明，例証; 図解.

im- *pref.* ＝in-【b，m，p の前で用いる】.

image *n.* 像.
homomorphic ～ 準同型[形]像. ～ by inversion 鏡像. ～ curve 像曲線. ～ domain 像領域. ～ measure 像測度. ～ point 像点. ～ set 像集合. inverse ～ 原像，逆像. inversion ～ 鏡像. mirror ～ 鏡像. reflected ～ 鏡像. virtual ～ 虚像.

imaginary *a.* 虚の.
～ axis 虚軸. ～ circle 虚円. ～ element 虚元素. ～ ellipse 虚楕円. ～ ellipsoid 虚楕円面. ～ geometry 虚幾何学. ～ hypersphere 虚超球. ～ number 虚数. ～ part 虚(数)部(分). ～ root 虚根. ～ unit 虚数単位. purely ～ number 純虚数. totally ～ 全虚の.

imbed ＝embed.

imbedding ＝embedding.

im Grossen *ad.* (*G.* ＝in the large; globally.) 大局的に.

im Kleinen *ad.* (*G.* ＝in the small; locally.) 局所的に.

immanency *n.* 内在性.

immanent *a.* 内在する.

immanently *ad.* 内在的に; 本質的に.

immediate *a.* 直接の; 近接の; 即時の.
～ payment 即時払い.

immediately *ad.* 直接に; すぐに.
～ adjacent すぐ隣りの. ～ after 直後. ～ before 直前.

immerse *vt.* 浸む; はめこむ; 陥らせる.
～d submanifold はめこまれた部分多様体.

immersion *n.* はめこみ，埋め込み.
differentiable ～ 微分可能なはめこみ. topological ～ 位相的はめこみ.

impedance *n.* インピーダンス.
～ matrix インピーダンス行列.

imperfect *a.* 非完全な; 不完全な.
～ field 非完全体.

impersonation *n.* なりすまし.

implication *n.* 含意，内包.

implicational *a.* 含意の.
～ proposition 含意[条件]命題.

implicit *a.* 陰の，陰的な; 間接的な.
～ definition 間接的定義. ～ function 陰関数.

imply *vt.* 含む; 意味する; 含意する，導く.
A ～lies *B A* ならば *B* である.

impose *vt.* 賦課する.
～ further condition(s) 更に条件を課する.

impossibility *n.* 不可能性; 不可能なこと.

impossible *a.* 不可能な，不能な.
of ～ construction 作図不(可)能な.

impredicative *a.* 非可述的な.
～ number 非可述的数.

impression *n.* 刷，版; 印象.

imprimitive *a.* 非原始的な.
～ group 非原始群. ～ period 複循環節.

improper *a.* 非固有な，仮性の; 特異な，変格の.
～ fraction 仮分数. ～ function 擬関数. ～ integral 広義積分，仮性積分，変格積分，非固有積分. ～ orthogonal matrix 変格直交行列. ～ rotation 非固有回転. ～ singular point [singularity] 仮性特異点.

impulse *n.* 衝撃; 力積.

impulsive *a.* 衝撃の，衝撃的な.
～ force 瞬間力. ～ function 衝撃関数.

imputation *n.* 転嫁; 配分; 補完.

in- *pref.*「内，中」の意 ⇨ ex-;「無，不」の意【il-，im-，ir- ともなる】.

inaccessible *a.* 到達不(可)能な.
～ boundary point 到達不能境界点. ～ cardinal 到達不能基数.

inaccurate *ad.* 不正確な.

inbreeding *n.* 近親交配.
～ coefficient 近交係数.

Ince, Edward Lindsay (1891–1941) インス.

incenter *n.* 内心.

incidence *n.* 結合; 発生，生起; 投射; 接続，隣接.
angle of ～ 入射角. ～ axiom 結合公理. ～ matrix 生起行列; 接続行列. ～ number 結合係数.

incircle *n.* 内接円.

inclination *n.* 傾度，傾角; 傾斜; 俯角，伏角.
angle of ～ 傾角.

incline *vt.* 傾ける. *vi.* 傾く. *n.* 傾斜面; 勾

配.
~d plane 斜面.
inclose =enclose.
inclosure =enclosure.
include *vt.* 含む; 包含する.
~d angle 夾角.
inclusion *n.* 包含.
~ mapping 包含写像. ~ relation 包含関係. principle of ~ and exclusion 包除原理. relation of ~ 包含関係.
inclusive *a.* 含めた. ⇨ exclusive.
incommensurability *n.* 通約不能性, 不通約性.
incommensurable *a.* 通約不能な, 公約数をもたない.
incompatibility *n.* 非両立性; 相容れないこと.
~ condition 非両立条件.
incompatible *a.* 両立しない, 矛盾した【with】.
~ events 排反事象.
incomplete *a.* 不完全な.
~ block 不完備計画. ~ beta [gamma] function 不完全ベータ[ガンマ]関数. ~ elliptic integral 不完全楕円積分.
incompleteness *n.* 不完全性; 不完備性.
incompressible *a.* 縮まない, 非圧縮性の.
~ fluid 非圧縮(性)流体. ~ stationary flow 非圧縮(な)定常流.
incongruence *n.* 不合同.
incongruent *a.* 不合同な.
incongruous *a.* 不適当な; とつじつまが合わない【with】; 不合同な.
inconsistency *n.* 不一致; 矛盾.
inconsistent *a.* 不能な; 不一致の, 不成立の, 矛盾する.
~ algebraic equation 不能な代数方程式. ~ statistic 一致性のない統計量.
increase *n.* 増加, 増大. *vt., vi.* 増す, 増加する. ⇨ decrease.
~ and decrease 増減. ~ by one 一だけ増す.
increasing *a.* 増加の. ⇨ decreasing.
~ function 増加関数. ~ process 増加過程. ~ sequence 増加列. ~ state 増加の状態. monotone ~ 単調増加の. slowly ~ 緩増加の. strictly ~ 狭義に増加の.
increment *n.* 増分. ⇨ decrement.
~ function 増分関数.
incremental *a.* 増加の; 増分の.
~ cost 増分費用.
indecomposable *a.* 分解不(可)能な; 直既約な.
directly ~ 直既約な. ~ continuum 分解不能な連続体. ~ group 直既約群. ~ module 直既約加群. ~ representation 直既約表現.
indeed *adv.* 実際に, 実は, 現実に.
indefinite *a.* 不定値の; 不定符号の; 不定な.
~ form 不定符号形式. ~ integral 不定積分. ~ quadratic form 不定符号二次形式.
indefinitely *ad.* 不定に.
~ divergent 不定発散の.
indent *vt.* ひっこませる.
indentation *n.* ひっこみ; 陥入; 字さげ.
indention =indentation.
independence *n.* 独立(性).
linear ~ 線形[一次]独立(性).
independency *n.* 独立性.
independent *a.* 独立な.
algebraically [analytically] ~ 代数[解析]的(に)独立な. functionally ~ 関数的(に)独立な. ~ and identically disitributed (i.i.d) 独立で同一な分布に従う. ~ event(s) 独立事象. ~ of と独立な. ~ trial(s) 独立試行. ~ variable 独立変数. linearly ~ 一次[線形]独立な.
independently *ad.* 独立に; 無関係に.
indeterminate *a.* (量が)不定な; 不確定な, *n.* 不定元.
~ coefficient 未定係数. ~ element 不定元. ~ equation 不定方程式. ~ form 不定形. method of ~ coefficients 未定係数法. polynomial in one ~ 一元[変数]多項式.
indetermination *n.* 不確定.
index (*pl.* -dexes, -dices) *n.* 指数; 添数, 添字; 指標; 標高; 索引, インデックス. *vt.* 添数をつける; 索引をつける.
author ~ 著者索引. contravariant [cova-

riant] ~ 反[共]変指標. dummy [umbral] ~ 無効添数. ~ed plan 標高平面図. ~ing set 添数集合. ~ line 指示線. ~ number of prices 物価指数. ~ of dispersion 散布指数. ~ of inertia 慣性指数. ~ set 添数[添字]集合. ~ theorem 指数定理. ramification ~ 分岐指数. subject ~ 事項索引. three ~ symbol 三添字記号.

Indian *a.* インドの.
~ mathematics インド数学.

indicator *n.* 指示するもの, 標識, 表示するもの; 定義関数.
~ function 定義関数.

indicatrix (*pl.* -trixes, -trice) *n.* 標形, 標構; 指示曲線; 基準面.
binormal ~ 従[陪]法線標形. Dupin ~ デュパン標構. ~ of tangent, tangent ~ 接線標形.

indicial *a.* 指数の.
~ equation (指数)決定方程式.

indifferent *a.* 無関心な; 重要でない, どちらでもよい【to】.

indirect *a.* 間接の.
~ continuation 間接接続. ~ measurement 間接測定. ~ proof 間接証明. ~ transcendental singularity 間接超越特異点.

indirectly *ad.* 間接に.
~ conformal 逆等角な.

individual *n.* 個体. *a.* 個別の, 個々の.
~ body 個品. ~ ergodic theorem 個別エルゴード定理. ~ risk theory 個別的危険論. ~ symbol 対象記号.

indivisibility *n.* 割り切れないこと; 非整除性, 不可分性.

indivisible *a.* 割り切れない, 整除できない, 不可分な. *n.* 不可分者.

induce *vt.* 誘導する.

induced *a.* 誘導された, 誘導の.
~ Cartan connection 誘導されたカルタン接続. ~ character 誘導指標. ~ equation 誘導方程式. ~ homomorphism 誘導準同形[型]写像. ~ metric 誘導計量. ~ relation 誘導[導来]関係. ~ representation 誘導表現. ~ topology 誘導位相.

inductance *n.* インダクタンス.

induction *n.* 帰納法, 帰納; 誘導.
complete ~ 完全帰納法. double ~ 二重帰納法. ~ equation 誘導方程式. magnetic ~ 磁気誘導. mathematical ~ 数学的帰納法. multiple ~ 多重帰納法.

inductive *a.* 帰納的な.
~ assumption 帰納法の仮定. ~ definition 帰納的定義. ~ dimension 帰納的次元. ~ hypothesis (数学的)帰納法の仮定. ~ limit 帰納的極限(=direct limit 直[順]極限). ~ method 帰納法. ~ statistics 推計学. ~ system 帰納系(=direct system 直[順]系).

inductively *ad.* 帰納的に.
~ ordered set 帰納的順序集合.

inelastic *a.* 非弾性の.
~ body 非弾性体. ~ scattering 非弾性散乱.

inequality *n.* 不等式.
absolute ~ 絶対不等式. conditional ~ 条件つき不等式. differential ~ 微分不等式. ~ condition 不等式条件. ~ sign 不等号. integral ~ 積分不等式. irrational ~ 無理不等式. logarithmic ~ 対数不等式. logarithmic Sobolev ~ 対数的ソボレフ不等式. simultaneous ~ties 連立不等式. strict ~ 狭義の不等式. triangle ~ 三角不等式. trigonometric ~ 三角不等式. unconditional ~ 絶対不等式. variational ~ 変分不等式. X's ~ X(氏)の不等式.

inequivalent *a.* 同値でない, 非同値な.

inertia *n.* 慣性, 惰性.
ellipsoid of ~ 慣性楕円体. ~ field 惰性体. ~ group 惰性群. law of ~ 慣性法則. moment of ~ 慣性能率.

inertial *a.* 慣性の, 惰性の.
~ field 慣性場. ~ sysem 慣性系.

inessential *a.* 非本質的な.

inexact *a.* 不正確な, 厳密でない.
~ Newton method 非厳密ニュートン法.

inf =infimum. 下限. ⇨ sup.

infeasible *a.* 実行可能でない.

inference *n.* 推論, 推理; 結論.
statistical ~ 統計的推論.

inferior *a.* 下級の, 劣の; 低い, 下位の

【to】. ⇨ superior.
~ angle 劣角. ~ arc 劣弧. ~ limit 下極限. ~ limit event 下極限事象. limes ~ 下極限.

infimum (*pl.* -ma) *n.* (*L.*) 下限【inf】. ⇨ supremum.

infinite *a.* 無限な. *n.* 無限.
countably [denumerably, enumerably] ~ 可算無限の. ~ continued fraction 無限連分数. ~ decimal 無限小数. ~ determinant 無限行列式. ~ dimension 無限次元. ~(-)dimensional 無限次元の. ~ field 無限体. ~ group 無限群. ~ integral 無限積分. ~ interval 無限区間. ~ matrix 無限行列. ~ order 無限位数. ~ population 無限母集団. ~ product 無限乗積. ~ sequence 無限数列. ~ series 無限級数. ~ set 無限集合. ~ sum 無限和.

infinitely *ad.* 無限に; 無数に.
~ differentiable 無限回微分可能な. ~ divisible 無限分解可能な. ~ many 無限に多くの, 無数の.

infinitesimal *n.* 無限小. *a.* 無限小の, 微小な.
~ calculus 無限小算法; 微(分)積分(法). ~ change 微小[無限小]変化. ~ deformation 無限小変形. ~ isometry 無限小等長変換. ~ motion 無限小運動. ~ of higher order 高位の無限小. ~ rotation 無限小回転. ~ transformation 無限小変換.

infinitum *n.* (*L.* =infinity.)
ad ~ =to infinity 無限に.

infinity *n.* 無限; 無限大【∞】.
axiom of ~ 無限公理. line at ~ 無限遠直線. order of ~ 無限大の位数. point at ~ 無限遠点. positive [negative] ~ 正[負]の無限大. space at ~ 無限遠空間. to ~ 無限に(まで).

inflation *n.* 膨張写像; インフレーション.

inflection *n.* 変曲.
~ point, point of ~ 変曲点.

inflexion =inflection.

inflow *n.* 流入. ⇨ outflow.

influence *n.* 影響.
domain [region] of ~ 影響領域.

influx *n.* 流入. ⇨ efflux.

informatics *n.* 情報科学.

information *n.* 情報.
amount of ~ 情報量. ~ channel 情報路. ~ geometry 情報幾何学. ~ matrix 情報行列. ~ polynomial 情報多項式. ~ processing 情報処理. ~ retrieval 情報検索[IR]. ~ science 情報科学. ~ source 情報源. ~ stability 情報安定性. ~ theory 情報理論.

informative *a.* 情報のある, 有情報の.
~ prior distribution 有情報事前分布. more ~ いっそう情報に富んだ.

ingenious *a.* 技巧的な, 巧妙な.

inherit *vt.* 遺伝的に受け継ぐ.
~ed character [quality] 遺伝形質.

inheritance *n.* 相続; 遺伝.

inhibit *vt.* 抑制する, 禁止する.

inhomogeneous *a.* 非同次の, 非斉次の.
~ coordinates 非同[斉]次座標. ~ equation 非同[斉]次方程式. ~ lattice 非同次格子.

initial *a.* 初期の, 最初の. *n.* かしら文字.
~ block 初ブロック. ~ condition 初期条件. ~ data 初期データ. ~ distribution 初期分布. ~ line 始線, 原線, 首線. ~ phase 初期位相. ~ point 始点, 起点. ~ side (角の)始辺. ~ term 初項. ~ value 初期値. ~ value problem 初期値問題. ~ velocity 初速.

injection *n.* 単射(写像); 1対1写像.
canonical ~ 標準[規準]的単射. natural ~ 自然な単射.

injective *a.* 単射的な.
~ homomorphism 単射準同型[形]. ~ mapping 単射写像. ~ module 移入[単射, 入射]加群. ~ resolution 移入分解, 単射[入射]的分解.

injectivity *n.* 単射性.

inner *a.* 内の, 内部の. ⇨ interior; outer.
~ area 内面積. ~ automorphism 内部自己同型[形](写像). ~ capacity 内容量. ~ cluster set 内集積値集合. ~ derivation 内部微分. ~ function 内関数. ~ measure 内測度. ~ point 内点. ~ product 内積. ~ transformation 内部変換. ~ volume 内

体積.

input *n.* 入力, インプット. ⇨ output.
～ area 入力域. ～ data 入力データ.

input-output *n.* 入出力[I/O].
～ analysis 産業連関分析, 投入産出分析.

inquiry *n.* 問合せ; 調査; 研究.

inrevolvable *a.* 内転(形)の.
～ oval 内転卵形.

inscribe *vt.* 内接させる.
～d angle (円)周角. ～d circle 内接円.
～d polygon 内接多角形.

inscription *n.* 内接.

inseparable *a.* 非分離的な.
～ degree 非分離次数. ～ element 非分離元. ～ extension 非分離拡大. ～ function 分離不能な関数. ～ polynomial 非分離的多項式. purely ～ 純非分離的な.

in sich dicht (*G.* =dense in itself.) 自己稠密な.

inside *n.* 内部. *a.* 内部の. *ad.* 内部に. *prep.* の内部に. ⇨ outside.
～ diameter 内径.

insignificant *a.* 些細な, 無意味な.
～ component [root] 無縁成分[根].

in solido *ad.* (*L.*) 立体[空間; 三次元]で.

insolvable *a.* 解けない; 非可解な.

inspection *n.* 検査.
acceptance [check] ～ 受入[監査]検査. control ～ 管理検査. destructive ～ 破壊検査. 100% ～ 全数検査. lot ～ 仕切検査. sampling ～ 抜取り検査. spot ～ つまみとり検査. total ～ 全数検査.

instability *n.* 不安定(性).

instable *a.* 不安定な.
～ equilibrium 不安定平衡. ～ solution 不安定解.

instal(l)ment *n.* 分割払; 賦払金.
accumulation by annual ～ 年金積立.

instance *n.* 例.
for ～ たとえば.

instantaneous *a.* 即時の; 瞬間の.
～ description 時点表示. ～ rotation 瞬間回転. ～ state 瞬間状態. ～ velocity 瞬間速度.

instead *ad.* その代りに.
～ of の代りに.

institute *n.* 研究所; 学会, 協会; 大学.

instruction *n.* 命令; 教育; 説明; 指示.

instrumental *a.* 器械の; 器具の.
～ drawing 用器画. ～ error 器(械誤)差.

insufficiency *n.* 不十分(性); 不足.

insufficient *a.* 不十分な; 不足な; 不適当な.

insurance *n.* 保険.
～ mathematics 保険数学. ～ rate 保険率. life ～ 生命保険.

integer *n.* 整数.
algebraic ～ 代数的整数. ～ programming 整数計画法. negative ～ 負の整数. positive ～ 正の整数, 自然数. rational ～ 有理整数; (ふつうの)整数.

integrability *n.* 積分可能性, 可積性.
～ condition, condition of ～ 積分可能[可積]条件.

integrable *a.* 積分可能な; 可積な.
absolutely ～ 絶対可積な. ～ function 積分可能[可積]関数. locally ～ 局所可積な. quadratically ～ 二乗可積な. square (-) ～ 平方可積な.

integral *n.* 積分. *a.* 整数の; 積分の.
abstract ～ 抽象積分. action ～ 作用積分. contour ～ 路に沿う積分. curvilinear ～ (曲)線積分. Darbourx ～ ダルブー積分. definite ～ 定積分. double ～ 二重積分. first ～ 第一積分. improper ～ 広義[変格, 仮性]積分. indefinite ～ 不定積分. ～ basis 整数基. ～ calculus 積分学. ～ closure 整閉包. ～ curvature 総曲率. ～ curve 積分[解]曲線. ～ domain 整域. ～ element 整元; 積分要素. ～ equation 積分方程式. ～ expression 整式. ～ extension 整拡大. ～ formula 積分公式. ～ function 整関数. ～ geometry 積分幾何学. ～ inequality 積分不等式. ～ invariant 積分不変式. ～ manifold 積分多様体. ～ number 整数. ～ operator 積分作用素. ～ part 整数部分. ～ quotient 整商. ～ representation 積分表示. ～ root 整(数)根. ～ sine 積分正弦. ～ surface 積分曲面. ～ theorem 積分定理. intermediate ～ 中間積分.

Lebesgue ~, (L) ~ ルベーグ積分. multiple ~ 重(複)積分. path-~ 経路積分. repeated ~ 累次積分. Radon ~ ラドン積分. Riemann ~, (R) ~ リーマン積分. simple ~ 単積分. sine ~ 正弦積分. Stieltjes ~ スティルチェス積分. stochastic ~ 確率積分. surface ~ 面積分. triple ~ 三重積分. volume ~ 体積分.

integrally *ad.* 整的に.
~ closed 整閉の. ~ dependent 整従属の.

integrand *n.* 被積分関数.

integrate *vt.* 積分する; 統合する.
~d science 総合科学.

integrating *a.* 積分の.
~ factor 積分因子.

integration *n.* 積分すること; 積分(法); 統合.
complex ~ 複素積分. contour ~ 路に沿う積分. domain of ~ 積分領域. graphical ~ 図式積分. ~ by parts 部分積分. ~ by substitution 置換積分. ~ constant, constant of ~ 積分定数. ~ operator 積分演算; 積分作用素. interval of ~ 積分区間. numerical ~ 数値積分(法). path of ~ 積分路. repeated ~ 累次積分. termwise ~ 項別積分. variable of ~ 積分変数.

integrator *n.* 求積器.
Amsler's ~ アムスラーの面積計.

integrity *n.* 整; 完全(性).
~ domain, domain of ~ 整域.

integro- 「積分」の意の結合辞.
~ differential equation 積分微分方程式.

intelligence *n.* 知能.
artificial ~ 人工知能. ~ index 知能指数. ~ test 知能検査.

intelligible *a.* 分かり易い; 理解できる, 明瞭な.

intelligibly *ad.* 分かり易く; 明瞭に.

intension *n.* 内包.

inter- *pref.* 「間, 中, 相互」の意.

interaction *n.* 交互作用, 相互作用.
simple ~ 二因子交互作用.

intercept *vt.* (面・線を二線[点]によって)切り取る. *n.* 切片.
y-~ y 切片.

interchange *vt.* 交換する; 交代する. *vi.* 交代する. *n.* 交換; 交代.

interchangeability *n.* 互換性.

interclass *n.* 級内.
~ correlation 級内相関.

interdisciplinary *a.* 学際的な.

interest *n.* 利子, 利息; 関心.
annual ~ 年利. calculation of ~ 利息算. compound [simple] ~ 複[単]利. daily ~ 日歩. principal and ~ 元利. rate of ~ 利率. yielding ~ 利回り.

interface *n.* 接合点, インタフェース.

interference *n.* 干渉, 妨害.
~ detection 妨害検出.

interior *n.* 内部. *a.* 内の, 内部の; 内-. ⇨ inner; exterior.
~ angle 内角. ~ division 内分. ~ opposite angle(s) 内対角. ~ point 内点. ~ problem 内部問題. ~ product 内部積.

intermediate *a.* 中間の; 中級の. *n.* 中間物.
~ convergent 中間近似分数. ~ field 中間体. ~ integral 中間積分. ~ value theorem 中間値の定理.

internal *a.* 内部の, 内部的な. ⇨ external.
~ angle 内角. ~ center 内心. ~ common tangent 共通内接線. ~ division 内分. ~ law of composition 内部算法. ~ ratio 内分比. ~ state 内部状態. ~ symmetry 内部対称性. ~ term 内項. ~ variance 内分散.

internally *ad.* 内で[に].
divide ~ 内分する. ~ dividing point 内分点.

international *a.* 国際的な.
~ congress [conference] 国際会議.

interpolate *vt.* 補間する.
~ting sequence 補間列.

interpolation *n.* 補間(法), 内挿(法). ⇨ extrapolation.
~ formula 補間公式. ~ polynomial 補間多項式. ~ set 補間集合. ~ space 補間空間. inverse ~ 逆補間. Lagrange's ~ ラグランジュ補間. linear ~ 線形補間. polynomial ~ 多項式補間.

interpolatory *a.* 補間的な; 補間(法)の.
interpret *vt.* 説明する; 解釈する.
interpretation *n.* 説明; 解釈.
interpreter *n.* 通訳.
interpretive *a.* 通訳の; 解釈の.
interquartile *a.* 四分位間の.
～ range 四分位範囲.
interrelation *n.* 相互関係.
interrupt *vt., n.* 割り込む; 中断する.
interruption *n.* 中断; 割りこみ.
intersect *vt., vi.* 交わる; 交差[交叉]する.
～ orthogonally 直交する. ～ transversally 横断的に交わる.
intersection *n.* 交点, 交線; 交差[交叉]; 共通部分, 交わり.
complete ～ 完全交線[交差]. finite ～ property 有限交差性. ～ angle 交角. ～ chart 共点図表. ～ multiplicity (交わりの) 重複度. ～ number 交点数, 交わりの数. ～ product 交差積. ～ theorem 交点定理. line of ～ 交線. oblique ～ 斜交.
intersectional *a.* 交わりの.
～ angle 交角.
interval *n.* 区間.
admissible ～ 許容区間. closed ～ 閉区間. confidence ～ 信頼区間. finite ～ 有限区間. infinite ～ 無限区間. ～ analysis 区間解析. ～ arithmetic 区間演算. ～ estimate 区間推定. ～ function 区間関数. ～ of integration 積分区間. method of diminishing [nested] ～s 区間縮小法. open ～ 開区間. semi-open ～ 半開区間. tolerance ～ 許容区間.
into *prep.* …の中に; 割る.
～-mapping 中への写像. six ～ eighteen is three 18÷6=3.
intraclass *n.* 級内. ⇨ interclass.
intransitive *a.* 非推移(的)な.
～ group 非推移群.
intransitivity *n.* 非推移性.
intrinsic *a.* 内在的な.
～ accuracy 内在精度. ～ equation 自然方程式. ～ geometry 内的幾何学. ～ homology 内在的ホモロジー. ～ property 内(在)的性質.
intrinsically *ad.* 内在的に; 本質的に.
introduction *n.* 序論, 序文; 入門; 紹介.
introductory *a.* 序論的な; 序文の.
intuition *n.* 直観.
intuitionism *n.* 直観主義.
intuitionistic *a.* 直観主義的な.
～ logic 直観主義論理(学).
intuitive *a.* 直観的な.
～ geometry 直観幾何(学).
invalid *a.* 不当な; 無効な.
invariable *a.* 一定の; 定数の; 不変な. *n.* 定数, 不変量.
invariance *n.* 不変性.
～ principle 不変性(の)原理.
invariant *n.* 不変式; 不変元; 不変系; 不変量. *a.* 不変な.
absolute ～ 絶対不変式. basic ～ 基本不変式. conformal ～ 等角不変量; 共形不変量. differential ～ 微分不変式. fundamental ～ 基本不変式. integral ～ 積分不変式. ～ differential form 不変微分形式. ～ distribution 不変分布. ～ estimator 不変推定量. ～ field 不変体. ～ integral 不変積分. ～ measure 不変測度. ～ statistic 不変統計量. ～ subgroup 不変部分群. ～ submanifold 不変部分多様体. ～ system 不変系. ～ test 不変検定. left [right] ～ 左[右]不変な. maximal ～ 最大不変量. relative ～ 相対不変式[量; 元]. topological ～ 位相不変量. vector ～ ベクトル不変式.
inventory *n.* 在庫品.
buffer ～ 緩衝用在庫. ～ control 在庫管理. ～ management 在庫管理. optimal ～ 最適在庫.
inverse *a.* 反対の, 逆の. *n.* 逆, 反形.
additive ～ 反数. complete ～ image 全逆像. generalized ～ (matrix) 一般化逆行列. ～ calculation 逆算. ～ cosine 逆余弦. ～ curve 反形曲線. ～ element 逆元. ～ figure 反形. ～ function 逆関数. ～ image 原像, 逆像. ～ interpolation 逆補間. ～ limit 逆極限(=projective limit 射影的極限). ～ mapping 逆写像. ～ matrix 逆行列. ～ operator 逆作用素. ～ order 逆順.

～ path 逆の道. ～ permutation 逆置換. ～ point 反点. ～ problem 逆問題. ～ proportion 反比例, 逆比例. ～ ratio 反比. ～ relation 逆関係. ～ rotation 逆回転. ～ sine 逆正弦. ～ system 逆系(＝projective system 射影(的)系). ～ transformation 逆変換. ～ trigonometric function 逆三角関数. ～ vector 逆ベクトル. left [right] ～ element 左[右]逆元.

inversely *ad.* 逆に.
～ proportional 反比例の. ～ well-ordered set 逆整列集合.

inverser ＝inversor.

inversible (*F.*) ＝invertible.

inversion *n.* 反転, 鏡像; 転倒; 転位.
～ formula 反転公式. ～ geometry 反転幾何学. ～ image 鏡像. ～ principle, principle of ～ 鏡像[反転]の原理.

invert *vt.* 逆転させる; 反転させる.

invertible *a.* 可逆な.
～ element 可逆元. ～ matrix 可逆行列. ～ sheaf 可逆層. ～ transformation 可逆変換.

investigate *vt.* 吟味する.

investigation *n.* 吟味.

involute *n.* 伸開線, インボリュート.
～ gear インボリュート歯車.

involution *n.* 対合, 対合変換.
elliptic [hyperbolic] ～ 楕円[双曲]的対合. ～ system, system of ～ 包合系, 縮閉系. ～ transformation 対合変換.

involutive *a.* 対合的な.
～ automorphism 対合的自己同型[形]. ～ correlation, ～ correspondence 対合的相反変換.

involve *vt.* 含む; 包含する.

inward *a.* 内を向いた; 内への. *ad.* 内側へ. *n.* 内部, 内面. ⇨ outward.
～ normal 内向き(の)法線.

I/O Input/Output. 入出力.

iota *n.* イオタ. ギリシア語アルファベットの9番目の文字【*I*, *ι*】.

IR Information Retrieval. 情報検索.

ir- *pref.* ＝in-【r の前で用いる】.

irrational *a.* 無理の. *n.* 無理数.
～ cut 無理切断. ～ equation 無理方程式. ～ expression 無理式. ～ function 無理関数. ～ inequality 無理不等式. ～ number 無理数; 不尽根数. ～ quantity 無理量.

irrationality *n.* 無理性; 不合理性.

irreducibility *n.* 既約性.

irreducible *a.* 既約な.
absolutely ～ 絶対既約な. directly ～ 直既約な. ～ character 既約指標. ～ component 既約成分. ～ decomposition 既約分解. ～ element 既約元. ～ fraction 既約分数. ～ module 既約加群. ～ polynomial 既約多項式. ～ representation 既約表現. ～ residue class 既約剰余類. ～ system of residues 既約剰余系.

irredundant *a.* 余分のない, むだがない.
～ representation 余分のない表現.

irreg. irregular, irregularly.

irregular *a.* 非正則な, 特異な.
～ point 非正則点, 特異点. ～ singularity, ～ singular point 不確定特異点.

irregularity *n.* 非[不]正則性.

irrespective *a.* どちらでもよい【of】.

irreversibility *n.* 不可逆性.

irreversible *a.* 不可逆な, 非可逆な.
～ process 非可逆過程.

irrotational *a.* 渦なしの, 無渦の, 非回転的な.
～ motion 無渦運動. ～ vector field 非回転ベクトル場.

island *n.* 島.

iso- *pref.* 「等しい, 同じ」の意.

isochrone *n.* 等時(性)曲線.

isochronous *a.* 等時(性)の.

isoclinal 等傾の; 定角の.
～ line 等傾(曲)線. ～ spiral 等傾渦[螺]線. ～ trajectory 定角軌道.

isocline *n.* 等傾線.

isogenous *a.* 同種な.

isogeny *n.* 同種写像.

isogon *n.* 等角多角形.

isogonal *a.* 等角の.
～ conjugate point 等角共役点. ～ transformation 等角変換. ～ trapezoid 等角台形.

isolate *vt.* 孤立させる; 分離させる.
isolated *a.* 孤立した, 孤立の.
~ ordinal number 孤立順序数. ~ point 孤立点. ~ prime divisor 孤立素因子. ~ set 孤立集合. ~ singularity 孤立特異点.
isometric *a.* 等距離の, 等長の; 距離同型[形]な.
~ mapping 等長写像. ~ operator 等長作用素. ~ projection 等測投[射]影(法). ~ transformation 等長変換.
isometry *n.* 等長写像[変換].
infinitesimal ~ 無限小等長変換.
isomorphic *a.* 同型[形]の.
~ image 同型[形]像. ~ mapping 同型[形]写像. spatially ~ 空間的同型[形]な. spectrally ~ スペクトル同型[形]な.
isomorphism *n.* 同型[形]; 同型[形]写像.
admissible ~ 許容同型[形]. affine [algebra] ~ アフィン[多元環]同型[形]. canonical ~ 標準的同型[形]. group ~, ~ of group(s) 群(の)同型[形]. ~ theorem 同型[形]定理. operator ~ 作用同型[形].
isoparametric *a.* 等径の.
~ function 等径関数. ~ hyperplane 等径超曲面.
isoperimetric *a.* 等周の.
~ inequality 等周不等式. ~ problem 等周問題.
isopleth *n.* 等値(曲)線.
isoplethic *a.* 等値の.
~ curve 等値(曲)線.
isosceles [aisάsəlì:z/-sɔ́s-] *a.* 二等辺の; 等脚の.
~ trapezoid 等脚台形. ~ triangle 二等辺三角形.
isospin *n.* アイソスピン.
isothermal *a.* 等温の.
~ coordinates 等温座標. ~ parameter 等温媒介変数.
isotomic *a.* 等分(割)の.
~ conjugate points 等線分共役点.
isotopic *n.* イソトープな, 同位な.
isotopy *n.* イソトピー, 同位.
~ type イソトピー型.
isotropic *a.* 等方的な, 等方性の.
~ turbulence 等方乱流. totally ~ 全等方的.
isotropy *n.* 等方性.
~ group 等方群.
italic *n., a.* イタリック字体[斜体](の).
item *n.* 項目, 個条; 単位体.
~-by-~ sequential inspection 各個[個別]逐次検査. ~ response theory 項目反応理論.
iterate *vt.* 反復する; くりかえす. *n.* 反復(でえられたもの).
iterated *a.* 累次の, 反復した.
~ integral 累次積分. ~ kernel 反復核. ~ limit 累次極限. ~ logarithm 累次[反復; 重復]対数. ~ series 累次級数. ~ sine 反復正弦.
iteration *n.* 反復(法); 逐次代入.
~ method 反復法.
iterative *a.* 反復の.
~ly reweighted least saquares 反復重み付き最小二乗法. ~ method 反復法. ~ solution 反復解(法).
-ity *suf.* =-ty: generality.
-ize *suf.*「…にする, …化する」の意の動詞をつくる: factorize, generalize, harmonize.

J

JACOBI, Cari Gustav Jacob (1804–1851) ヤコビ.

Jacobian *a.* ヤコビの. *n.* ヤコビ行列式, 関数行列式.
~ criterion (of regularity) (正則性の)ヤコビ判定法. ~ matrix ヤコビ行列. ~ variety ヤコビ多様体.

JACOBSON, Nathan (1910–1999) ジャコブソン.

JENSEN, Johann Ludwig Wilhelm Waldemar (1859–1925) イェンゼン.

jet *n.* ジェット.
~ space ジェット空間.

job *n.* 作業, ジョブ.
~-shop scheduling 職種機種別作業計画.

JOHN, Fritz (1910–1994) ジョン.

join *vt., vi.* 結ぶ. *n.* 接合点[線・面]; 結び, 合併(集合). ⇨ cup; meet.
reduced ~ 約結.

joint *n.* 結合; 継ぎ目. *a.* 同時の; 共同の.
~ author 共著者. ~ director circle 連合準円. ~ directrix 連合準線. ~ distribution 結合分布, 同時分布. ~ random variable 結合確率変数. ~ work [paper] 共著.
rule ~ 折り尺.

JORDAN, Marie Ennemond Camille (1838–1922) ジョルダン.

JOUKOWSKI, Nikolai Jegorovič (1847–1921) ジュコフスキ.

journal *n.* (定期刊行の)雑誌.

J-shape *n.* J 型.
~(d) distribution J 型分布.

judgment,【英】judgement *n.* 判定; 判断.
~ sample 有意標本.

JULIA, Gaston (1893–1978) ジュリア.

jump *n.* 跳躍; 飛越. *vi., vt.* 飛越する.
~ point 跳躍点.

junction *n.* 接合(点).

JUNG, Heinrich Wilhelm Ewald (1876–1953) ユンク.

juxtaposition *n.* 並置, 並列.

K

KÄHLER, Erich (1906-2000) ケーラー.

Kählerian *a.* ケーラーの.
~ manifold ケーラー多様体. ~ metric ケーラー計量.

Kalkül *n.* (*G.* =calculus.)

KAMKE, Erich (1890-1961) カムケ.

KAPLANSKY, Irving (1917-2006) カプランスキ.

kappa *n.* カッパ. ギリシア語アルファベットの10番目の文字【K, κ】.

KELLOGG, Oliver Dimon (1878-1932) ケロッグ.

KEPLER, Johannes (1571-1630) ケプラー, ケプレル.

KERÉKJÁRTÓ, Béla von (1898-1946) ケレキャルト.

kernel *n.* 核.
adjoint [conjugate] ~ 随伴[共役]核. convolution ~ 合成核. degenerate ~ 退化核. difference ~ 差核. Hermitian ~ エルミート核. iterated ~ 反復核. ~ differential 核微分式. ~ function 核関数. ~ of homomorphism 準同型[形]の核. ~ of polynomial type 多項式型の核. measurable ~ 等測核. open ~ 開核. reproducing ~ 再生核. resolvent ~ 解核. singular ~ 特異核. symmetric ~ 対称核. symmetrizable ~ 対称化可能な核.

key *n.* 見出し; キー; 鍵.
~ word キーワード, キー[手掛かりとなる]語.

keyword *n.* キーワード. =key word.

KHINTCHINE =HINČIN.

killing *n.* 消滅.
~ measure 消滅測度. ~ method 消滅法. ~ operator 消滅作用素

KILLING, Wilhelm Karl Joseph (1847-1923) キリング.
Killing form キリング形式.

KILLING, Walter (1847-1923) キリング.

kilo- *pref.* 「千」の意.
~gram キログラム. ~meter キロメートル.

kind *n.* 種類.
of the first [second] ~ 第一[二]種の.

kinematic *a.* 運動学の.

kinematics *n.* 運動学.

kinetic *a.* 運動の.
~ energy 運動エネルギー. ~ potential 運動ポテンシャル.

kinetics *n.* 動力学.

Klassenkalkül *n.* (*G.* =class calculus). (論理)集合算.

KLEIN, Felix (1849-1925) クライン.
~ four group クラインの四元群.

Kleinian *a.* クラインの.
~ group クライン群.

KNESER, Adolf (1862-1930) クネザー.

KNOPP, Konrad (1882-1957) クノップ.

knot *n.* 結び目; 結び糸; 節. *vt.* 結ぶ.
alternating ~ 交代結び目. clover-leaf ~ 三葉結び目. ~ group 結び目群. ~ projection 結び目射影. ~ theory 結び目理論. ~ type 結び目型. tame ~ 順結び.

knotted *a.* 結ばれている.

known *a.* 既知の.
~ quantity 既知数[量].

KOEBE, Paul (1882-1945) ケーベ.

KOLMOGOROV, Andrei Nikolaevič (1903-1987) コルモゴロフ.

Koszul, Jean-Louis (1921-) コスツル, コジュール, コシュール, コスール.

Kovalevskaja, Sof'ja [Sophia] Vasilévna (1850-1891) コワレフスカヤ, コワレフスキ.

Kowalewski =Kovalevskaja.

Kreisverwandtschaft *n.* (*G.* =circle-to--circle correspondence.) 円々対応.

Kronecker, Leopold (1823-1891) クロネッカー.

Krull, Wolfgang (1899-1971) クルル.

***K*-sample** *n.* 多標本.

***K*-theory** *n.* *K*理論.

Kummer, Ernst Eduard (1810-1893) クンマー.

Kuratowski, Casimir (1896-1980) クラトフスキ.

kurtosis *n.* 尖度. とがり.

L

L, l. (ローマ数字の) 50.

label *n.* ラベル. *vt.* ラベルをつける, 分類する.

lack *n.* 欠如, 不足. *vi., vt.* 欠く.
~ of control 管理欠如. ~ of memory 無記憶性.

lacunary *a.* 空白[隙]の. ⇨ gap.
~ series 空隙級数.

ladder *n.* はしご; 昇降演算子.
down-~ 下降演算子. ~ method 昇降演算子法. ~ operation 昇降演算. up-~ 上昇演算子.

lag *n.* おくれ; 遅延.
~ correlation おくれ相関. time ~ 時刻のおくれ.

LAGRANGE, Joseph Louis (1736-1813) ラグランジュ.

Lagrangian *a.* ラグランジュの.
~ [Lagrange's] bracket ラグランジュの括弧式. ~ function ラグランジュ(の)関数.

LAGUERRE, Edmond Nicolas (1834-1886) ラゲール.

lambda *n.* ラムダ. ギリシア語アルファベットの 11 番目の文字【Λ, λ】.
~ function ラムダ[λ]関数. ~ matrix ラムダ行列.

LAMBERT, Johann Heinrich (1728-1777) ランベルト.

Lambertian *a.* ランベルトの.

LAMÉ, Gabriel (1795-1870) ラメ.

lamella (*pl.* -lae, -las) *n.* 薄い層.

lamellar *a.* 薄い層の; 層状の, 渦なしの, 無渦の.
~ stream [flow] 層流.

lamina (*pl.* -nae, -nas) *n.* 薄板; 薄層.

laminar *a.* 薄層の.
~ flow 層流.

LANDAU, Edmund Georg (1877-1938) ランダウ.

LANDEN, John (1719-1790) ランデン.

language *n.* 言語, 語.
computer ~ 計算機言語. context-free ~ 文脈自由言語. context-sensitive ~ 文脈依存言語. definite ~ 限定言語. external ~ 外部言語. machine ~ 機械語. programming ~ プログラム言語. regular ~ 正規言語.

LAPLACE, Pierre Simon (1749-1827) ラプラス.

Laplacian *n.* ラプラシアン, ラプラス演算子[作用素]【Δ】. *a.* ラプラスの.
~ operator ラプラス演算子[作用素].

large *a.* 大きな; 多数の; 広い. *n.* 大きなこと. ⇨ small.
in the ~ 大域的に. ~ number 大数. ~ sample 大標本. ~ sample theory 大標本論. law of ~ numbers 大数の法則.

largest *a.* 最大の, 最大な.

last *a.* 終りの, 最終の【late の最上級】.
~-in first-out 後入れ先出し. ~ multiplier 最終乗式. ~-off sampling 途中抜取り. ~ place 末位. ~ term 末項. ~ theorem 最終定理; 最後の定理.

latency *n.* 潜在.

latent *a.* 潜在的な.
~ root [value] 固有値. ~ variable 潜在変数.

lateral *a.* 横の; 側面の; 傍系の.
~ area 側面積. ~ edge 側稜. ~ face 側面.

latest *a., ad.* 最後の[に]; 最新の[に].

Latin *a.* ラテン(人)の. *n.* ラテン語; ラテン民族.

Greco-~ square グレコラテン方格[方陣]. ~ square ラテン方格[方陣].

latitude *n.* 緯度, 緯線.

~ circle 緯(度)線, 緯円. ~ distance 緯距. parallel of ~ 緯線.

latitudinal *a.* 緯度の.

lattice *n.* 格子; 束.

Banach ~ バナッハ束. complemented ~ 可補束, 相補束. complete ~ 完備束. critical ~ 臨界格子. cubic ~ 立方格子. distributive ~ 分配束. double ~ 複格子. dual ~ 双対格子. ~ design 格子計画. ~ distribution 格子分布. ~ group 格子群. ~ homomorphism 束(順序)準同型[形]. ~ isomorphism 束(順序)同型[形]. ~-ordered 束順序の. ~ point 格子点. ~-point formula 格子点公式. ~ space 束空間. ~ square (method) 格子方格(法). ~ theory 束論. linear ~ 線形束. quadruple ~ 四重格子. quotient ~ 商束. simple ~ 単格子. space ~ 空間格子; 空間束. vector ~ ベクトル束.

lattice-ordered *a.* 束順序の.

~ group 束順序群, 束群. ~ linear space 束順序線形空間. ~ set 束順序集合.

latus (*pl.* latera) *n.* (*L.*) 側辺.

~ rectum (*pl.* latera recta) (円錐曲線の) 通径.

LAURENT, P. A. (1813-1854) ローラン.

law *n.* 法則; 律, 規則.

absorptive [absorption] ~ 吸収法則. associative ~ 結合法則[律]. cancellation ~ 簡約法則. commutative ~ 交換法則[律]. cosine ~ 余弦法則. differential ~ 微分法則. distributive ~ 分配法則[律]. ~ of composition 算法. ~ of excluded middle 排中律. ~ of exponents 指数法則. ~ of inertia 慣性法則. ~ of large numbers 大数の法則. ~ of naught or one 0-1 法則, 零または一の法則, 悉無律. ~ of non(-)contradiction 無矛盾律. ~ of parallelogram 平行四辺形の法則. ~ of reaction 反作用の法則. ~ of symmetry 対称律. reflexive ~ 反射法則[律]. sine ~ 正弦法則. transitive ~ 推移法則[律]. zero-one ~ 0-1 法則, 零または一の法則, 悉無律.

LAX, Peter David (1926-　) ラックス.

layer *n.* 層.

boundary ~ 境界層. double [simple, single] ~ 二重[単一]層. thin ~ 薄層.

layout *n.* 割りつけ, レイアウト; 配列(表); 設計.

orthogonal ~ 直交配列表.

l. c. =loco citato.

L.C.M. Least Common Multiple. 最小公倍数[元].

lead *vt.* 導く. *n.* 先導.

~ time 調達期間.

leading *a.* 先導する; 主要な.

~ coefficient 主係数, 最高次の係数. ~ term 主要項.

leaf (*pl.* leaves) *n.* 葉; 葉体.

leap *n.* 跳躍.

~ function 跳躍関数.

learn *vt., vi.* 学習する.

learner *n.* 学習者; 初学者.

learning *n.* 学習; 学習すること.

~ model 学習模型.

least *a.* 最小[少]な. *ad.* 最少に. *n.* 最小のもの. ⇨ most; greatest.

at ~ 少なくとも. ~ action 最小作用. ~ common denominator 最小公分母. ~ common multiple 最小公倍数[元] [L. C. M.]. ~-square estimator 最小二[自]乗推定量. ~ upper bound 最小上界, 上限 [l. u. b.]. method of ~ squares 最小二[自]乗法.

LEBESGUE, Henri Léon (1875-1941) ルベーグ.

lecture *n.* 講義, 講演. *vt., vi.* 講義[講演]する.

~ note 講義ノート. ~ theater 階段教室.

LEFSCHETZ, Solomon (1884-1972) レフシェッツ.

left *a.* 左の. *ad.* 左に. *n.* 左. ⇨ right.

~ *A*-module 左 *A* 加群. ~ annihilator 左零化元[イデアル]. ~ continuous 左連続

な. ~ coset 左剰余類. ~ endpoint 左端(点). ~ exact 左完全な. ~ hereditary 左遺伝的な. ~ ideal 左イデアル. ~ identity [unit] element 左単位元. ~ invariant 左不変な. ~ inverse element 左逆元. ~ order 左整環. ~ resolution 左分解. ~ regular 左正則な. ~ shunt 左通過点. ~ translation 左移動.

left-hand *a.* 左の; 左側の.
~ed system 左手系. ~ lower [upper] derivative 左下[上]導関数[微分係数]. ~ side [member] 左辺.

left-handed *a.* 左回りの.
~ system 左手系.

leg *n.* 脚; 辺.
~s of right triangle 直角三角形の直角を夾む二辺.

Legendre, Adrien Marie (1752-1833) ルジャンドル.

Leibniz, Gottfried Wilhelm (1646-1716) ライプニッツ.

lemma (*pl.* -mas, -mata) *n.* 補助定理, 補題.
five ~ 5 補題, 5 項補題. fundamental ~ 基本補助定理. Schwarz(') ~ シュワルツの補助定理.

lemniscate *n.* レムニスケート, 連珠形.

length *n.* 長さ.
arc ~ 弧長. code ~ 符号長. extremal ~ 極値的長さ. focal ~ 焦点距離. minimal ~ 最短[極小]距離. unit ~ 単位の長さ. wave ~ 波長. word ~ 語長.

lens (*pl.* lenses) *n.* レンズ.
~ space レンズ空間.

Leonard of Pisa =Fibonacci.

leptokurtic *a.* 狭幅の.
~ distribution 狭幅[幅ほそ; 急尖]分布.

Leray, Jean (1906-1998) ルレー.

Lerch, Matyáš (1860-1922) レルヒ.

less *a.* より少ない. *ad.* より少なく.【little の比較級】. ⇨ more.
~ than or equal to 以下の. not ~ (than) 以上の.

-less *suf.* 名詞につけて「…のない; 無限の, 無数の」の意の形容詞をつくる: memoryless, endless. 動詞につけて「…できない」の意の形容詞をつくる: countless.

let *vt.*「仮定」…せよ【命令形】.

letter *n.* 文字; 手紙.
capital ~ 大[頭]文字.

level *n.* 高度, 水準; 段階. ⇨ niveau.
confidence ~ 信頼水準. ~ curve [surface] 等位線[面], 水準線[面], 等高線[面]. ~ of control 管理水準. significant ~, ~ of significance 危険率, 有意水準.

level(l)ing *n.* 水準測量.

leverage *n.* てこ作用; てこ比.

Levi, Beppo (1875-1961) レビ.

Levi-Civita, Tullio (1873-1941) レビ・チビタ.

Lévy, Paul Pierre (1886-1971) レビ.

lexical *a.* 辞書の.
~ order 辞書式順序.

lexicographic(al) *a.* 辞書式の.
~ order 辞書式順序.

lexicography *n.* 辞書編集(法).

lexicon *n.* 辞書.

L'Hôpital, Guillaume François Antoine de (1661-1704) ロピタル.

Lichnérowicz, André Léon Jean Maurice (1915-1998) リヒ[シュ]ネロビッチ.

lie *vi.* …にある; 位置する.

Lie, Marius Sophus (1842-1899) リー.

life *n.* 生命; 寿命; 生涯.
~ annuity 終身年金. ~ curve 寿命曲線. ~ insurance 生命保険. ~ interest 終身年金. ~ science 生命科学. ~ table 死亡生残表, 生命表. ~ time 生存時間.

LIFO ライフォ.
last-in first-out 後入れ先出し.

lift *n.* (曲線・ベクトル場の)持上げ, リフト; 膨張写像.

light *n.* 光.
~ source 光源. ~-year 光年.

like *a.* 同様な, 類似な, 等しい.
~ figures 相似形. ~ instance 類似な例. ~ quantities [signs] 等量[号]. ~ surds 同類根数, ~ terms 同類項.

-like *suf.* -型の.
star~ domain 星型領域.

likelihood *n.* 尤度.
～ equation 尤度方程式. ～ function 尤度関数, 公算関数. ～ ratio 尤度比, 公算比, 確度. maximum ～ 最大尤度; 最尤の. maximum ～ estimation 最尤推定法. maximum ～ estimator 最尤推定量.

likely *a.* ありそうな, 確からしい.
equally ～ 同様[同程度]に確からしい.

limaçon *n.* (*F.*) 蝸牛線, リマソン.

limes (*pl.* limites) *n.* (*L.* =limit.) 極限.
～ inferior 下極限. ～ superior 上極限. limites principales 主極限.

limit *n.* 極限; 限界. *vt.* 限る.
check ～ 検査限界. confidence ～ 信頼限界. control ～ 管理限界. direct ～ 直[順]極限, 帰納的極限(=inductive limit). inductive ～ 帰納的極限. inverse ～ 逆極限(= projective ～ 射影的極限). ～ cycle 極限周期軌道. ～ distribution 極限分布. ～ event 極限事象. ～ function 極限関数. ～ing circle 極限円. ～ing point (極)限点. ～ normal 限界法線. ～ number 極限数. ～ of class 級端. ～ of error 誤差(の)限界. ～ point 極限点; 集積点. ～ relation 極限関係. ～ set 極限集合. ～ theorem 極限定理. ～ value 極限値. lower [upper] ～ 下[上]限; 下[上]端. order ～ 順序極限. projective ～ 射影的極限. unilateral ～ 片側極限.

LINDELÖF, Ernst Leonard (1870–1946) リンデレフ.

LINDEMANN, Karl L. Ferdinand (1852–1939) リンデマン.

line *n.* 線; 直線.
broken ～ 破線. dotted ～ 点線. horizontal ～ 水平線. initial ～ 始線, 原線. ～ at infinity 無限遠直線. ～ bundle 直線束. ～ complex 直線棒. ～ congruence (直)線叢. ～ element 線素. ～ graph 折れ線グラフ. ～ of curvature 曲率線. ～ of force 力線. ～ of regression 反帰線. ～ of torsion 捩率線, ねじれの線. ～ relaxation 線状緩和. ～ segment 線分. normal ～ 法線. real ～ 数直線. solid ～ 実線. straight ～ 直線. tangent ～ 接線. vertical ～ 垂直[鉛直]線. wavy ～ 波線. whole ～ 全直線.

linear *a.* 一次の, 線形[型]の.
complete ～ system 完備線形系. general ～ group 一般線形群. ～ algebra 線形代数(学). ～ approximation 線形近似. ～ code 線形符号. ～ combination 一次[線形]結合. ～ connection 線形接続. ～ correlation 線形相関. ～ dependence 一次[線形]従属性. ～ differential equation 線形微分方程式. ～ equation 一次[線形]方程式. ～ equivalence 線形同値. ～ expression 一次式. ～ extension 線形延長. ～ form 線形[一次]形式. ～ fractional function 一次分数関数. ～ function 一次関数. ～ homogeneous equation 一次同次方程式. ～ hypothesis 線形仮説. ～ independence 一次[線形]独立性. ～ inequality 一次不等式. ～ interpolation 線形[一次]補間. ～ mapping 線形写像. ～ network 線形回路. ～ operator 線形作用素. ～ order 線形順序. ～ programming 線形計画(法)[LP]. ～ regression 線形回帰. ～ relation 線形関係. ～ space 線形空間. ～ subspace 線形部分空間. ～ system 線形系. ～ transformation 一次[線形]変換. piecewise ～ 区分的線形な.

linearity *n.* 線形性.

linearization *n.* 線形化.

linearly *ad.* 一次的に, 線形的に.
～ compact 線形コンパクトな. ～ dependent 一次[線形]従属な. ～ disjoint 線形離別[無関連]の. ～ equivalent 線形同値な. ～ independent 一次[線形]独立な. ～ ordered set 線形順序集合.

linguistic *a.* 言語の.
～ algebra 言語代数.

linguistics *n.* 言語(学); 語学.
computational ～ 計算言語学. mathematical ～ 数理言語学.

link *n.* まつわり; リンク; (連鎖の)輪; 連結; つながり. *vi.* まつわる; 連接する; つながる. *vt.* つなぐ.
～ complex まつわり複体. ～ function 連結関数. ～ing coefficient 纏絡係数, まつわり数. ～ing number まつわり数, 絡数.

～ relatives 連環比率，連環指数.

linkage *n.* 連結; 連係; リンク; 連関.
～ coefficient 絡数. ～ value 連関率.

LINNIK, Jurii Vladimirovič（1915–1972）リンニク.

LIONS, Pierre-Louis（1956– ）リオンス，ピエール＝ルイ.

LIONS, Jacques-Louis（1928–2001）リオンス，ジャック＝ルイ.

LIOUVILLE, Joseph（1809–1882）リウヴィル.

LIPSCHITZ, Rudolf Otto Sigismund（1832–1903）リプシッツ.

liquid *n.* 液体，流体. *a.* 流体の. ⇨ fluid.

list *n.* リスト; 一覧表. *vt.* 列挙する.

literal *a.* 文字の; 文字で表した.
～ coefficient 文字係数. ～ equation 文字方程式. ～ expression 文字式.

literally *ad.* 文字通りに; 逐語的に.

literature *n.* 文献.
～ search 文献探索.

LITTLEWOOD, John Edensor（1885–1977）リトルウッド.

lituus *n.* リチュウス.

LJAPUNOV, Aleksandr Mihairovič（1857–1918）リャプノフ.

LOBAČEVSKIĬ, Nicolai Ivanovič（1793–1856）ロバチェフスキ.

LOBACHEVSKI ＝LOBAČEVSKIĬ.

local *a.* 局所の，局所的な; 軌跡の. ⇨ global.
～ basis 局所基. ～ chart 局所座標. ～ connectivity 局所連結度[性]. ～ coordinates 局所座標. ～ field 局所体. ～ homomorphism 局所準同型[形]（写像）. ～ maximum 極大. ～ minimum 極小. ～ parameter 局所(媒介)変数. ～ property 局所的性質. ～ ring 局所環. ～-ringed space 局所環つき空間. ～ search 局所探索. ～ theory 局所(的)理論. ～ transformation 局所変換. ～ uniformization 局所一意化. ～ value 位価. regular ～ ring 正則局所環.

localization *n.* 局所化.
～ theorem 局所化定理.

locally *ad.* 局所的に.
～ absolute *p*-valent 局所絶対 p 葉の. ～ bounded 局所有界な. ～ closed set 局所閉集合. ～ compact 局所コンパクトな. ～ connected 局所連結の. ～ convex 局所凸な. ～ Euclidean 局所ユークリッド的な. ～ finite 局所有限な. ～ flat 局所(的に)平坦な. ～ integrable 局所可積な. ～ isomorphic 局所同型[形]な. ～ solvable 局所(的に)可解な. ～ symmetric space 局所対称空間. ～ trivial 局所自明な.

local-ringed *a.* 局所環つきの.
～ space 局所環つき空間.

location *n.* 場所; 所在; 位.
～ parameter 位置母数. ～ test ずれの検定.

loc. cit. loco citato.

loco citato（*L.*）*ad.* 上記引用文中[個所]に.

locus（*pl.* loci [lóusai]）*n.* 軌跡; 軌道; 場所，位置.
fundamental ～ [loci] 基本軌跡. geometric ～ 軌跡. optimal ～ 最適軌道.

LOEWY, Alfred（1873–1935）レービ.

log. logarithm, logarithmic.

logarithm *n.* 対数.
anti ～ 真数. common ～ 常用対数. discrete ～ 離散対数. iterated ～ 累次[反復]対数. natural ～ 自然対数. table of ～s 対数表.

logarithmic *a.* 対数の.
～ branch point 対数分岐点. ～ capacity 対数容量. ～ convexity 対数的凸性. ～ curve 対数曲線. ～ decrement 対数減衰率. ～ derivative 対数(的)導関数. ～ differentiation 対数微分. ～ equation 対数方程式. ～ function 対数関数. ～ inequality 対数不等式. ～ modulus 対数的モジュラス. ～ paper 対数方眼紙. ～ potential 対数ポテンシャル. ～ residue 対数的留数. ～ scale 対数目盛，対数尺. ～ scheme ログ概型. ～ series 対数級数. ～ singularity 対数(的)特異点. ～ Sobolev inequality 対数的ソボレフ不等式. ～ spiral 対数螺線. ～ structure ログ構造.

logarithmically *ad.* 対数的に.

differentiate ~ 対数(的に)微分する. ~ convex 対数的(に)凸な.

logic *n.* 論理学; 論理; 論法.
first order predicate ~ 一階述語論理. ~ (al) element 論理素子. many-valued [multiple-valued] ~ 多値論理. predicate ~ 述語論理. propositional ~ 命題論理. symbolic ~ 記号論理(学).

logical *a.* 論理的な; 論理の.
~ axiom 論理的な公理. ~ design 論理設計. ~ element 論理素子. ~ expression 論理式. ~ formula 論理式. ~ operator 論理演算子. ~ product 論理積. ~ proposition 論理命題. ~ sum 論理和. ~ symbol 論理記号.

logicism *n.* 論理主義.

logistic *a.* ロジスティックな.
~ curve ロジスティック曲線, 生長曲線, 算定曲線. ~ equation ロジスティック方程式.

long *a.* 長い.
~ exact sequence 長完全(系)列.

longitude *n.* 経度, 経線.
~ circle 経(度)線. ~ distance 経距.

longitudinal *a.* 縦の; 経度の; 経時的な.
~ data 経時的データ. ~ vibration 縦振動. ~ wave 縦波.

loop *n.* 自閉線, ループ, 閉道; ループ【群の一つの一般化】.
~ed curve 自閉線. ~ space ループ空間, 閉道空間. ~ theorem ループ定理.

Lorentz, Hendrik Antoon (1853-1928) ローレンツ.

loss *n.* 損失; 喪失.
~ function 損失関数. ~ of information 情報の損失. ~ probability 呼損率. without ~ of generality 一般性を失うことなく.

lot *n.* 仕切, ロット; 一山.
~ inspection 仕切検査. ~ number 仕切番号. ~ quality protection 仕切品質保護. ~ tolerance percent defective 仕切許容不良率[LTPD].

lottery *n.* 福引, 富籤[とみくじ].
compound ~ 複合くじ.

lower *a.* 下の, 下方の; より低い; 下層の. ⇨ upper; higher. *vt., vi.* 下げる, 下がる; 減じる, 減る.
~ approximate value 過小近似値. ~ base 下底. ~ bound 下界. ~ central series 降中心列. ~ derivative 下微分係数. ~ envelope 下包. ~ half-plane 下半(平)面. ~ integral 下積分. ~ limit 下限; 下端. ~ order 下位数. ~ quartile 下側四分位数. ~ semi-continuous 下(に)半連続な. ~ sum 不足和. ~ triangular matrix 下三角行列. of ~ order 低位の.

Löwner, Karl [Loewner, Charles] (1893-1968) レブナー.

loxodrome *n.* 航海線, ロクソドローム.
~ curve 航海線.

loxodromic *a.* 斜航的な; 等方位の.
~ curve 航海線. ~ transformation 斜航(的)変換.

lozenge *n.* ひし形, 菱形.

LP Linear Programming. 線形計画(法).

L.S.D. Least Significant Digit. 最下位の(有効)数字.

LTPD Lot Tolerance Percent Defective. ロット[仕切]許容不良率.

l. u. b. least upper bound. 上限.

lucid *a.* 分かり易い, 明解な.

lucidly *ad.* 分かり易く, 明解に.

Ludolf, van Ceulen (1540-1610) ルドルフ.

lunar *a.* 月(形)の.

lune *n.* 弓形, 弓月形; 円弧二角形.
spherical ~ 球面月形.

Lüroth, Jacob (1844-1910) リューロート.

Luzin, Nikolai Nikolaevič (1883-1950) ルージン.

-ly *suf.* 形容詞に添えて副詞をつくる: analytically, continuously. 名詞に添えて形容詞をつくる: timely.

M

M, m (ローマ数字の) 1000.

MACAULAY, Francis Sowerby (1862–1937) マコーレー.

MACHIN, John (1685–1751) マチン.

machine *n.* 機械.
account ~ 会計機. ~ configuration (機械の)内部状態. ~ language 機械語. ~ learning 機械学習. Turing ~ チューリング機械.

MACLANE, Saunders (1909–2005) マクレーン.

MACLAURIN, Colin (1698–1746) マクローリン.

macro- *pref.*「長い, 大きい, 大規模な」の意.

magic *n.* 魔法. *a.* 魔法の.
~ square 魔方陣.

magnetic *a.* 磁気の.
~ field 磁界, 磁場. ~ induction 磁気誘導.

magnetism *n.* 磁気; 磁気学.

magnification *n.* 倍率; 拡大.
~ ratio 倍率.

magnify *vt.* 拡大する.

magnitude *n.* 大きさ, 量.
geometric(al) ~ 幾何学的量【長さ, 面積, 体積, 角など】.

mail *n.* 郵便物; 郵便.
~ survey 郵送調査.

main *a.* おもな, 主要な; 第一の.
~ effect 主効果. ~ theorem 主(要)定理.

major *a.* (二つの中で)大きい方の; 主要な; 優角の. *n.* 大前提. ⇨ minor.
~ angle 優角. ~ arc 優弧. ~ axis (of ellipse) (楕円の)長軸[径]. ~ defect 重欠点. ~ function 優関数. ~ premise 大前提.

majorant *n.* 優級数; 優関数.
harmonic ~ 調和優関数. ~ series 優級数. ~ test 優級[関]数判定法.

MALMQUIST, Johannes (1822–1952) マルムキスト.

management *n.* 経営; 管理.
inventory ~ 在庫管理. ~ science 経営科学.

MANDELBROJT, Szolem (1899–1983) マンデルブロイト.

MANDELBROT, Benoît B. (1924–2010) マンデルブロート.

manifold *n.* 多様体. ⇨ variety.
algebraic ~ 代数多様体. center ~ 中心多様体. complex ~ 複素多様体. covering ~ 被覆多様体. differentiable ~ 可微分多様体. Hermitian ~ エルミート多様体. integral ~ 積分多様体. Kählerian ~ ケーラー多様体. ~ with boundary 境界のある多様体. open ~ 開多様体. Riemannian ~ リーマン多様体. Stein ~ シュタイン多様体. topological ~ 位相多様体.

manipulation *n.* 操作.
formula ~ 数式処理.

manner *n.* 方法, やり方.

MANOVA Multivariate ANalysis Of VAriance. 多変量分散分析.

mantissa *n.* (対数の)仮数.

manual *n.* マニュアル, 手引; 便覧.

manuscript *n.* 原稿.

many *a.* 多数の. *n.* 多数.
infinitely ~ 無限に多くの. ~-body problem, problem of ~(-)bodies 多体問題. ~

points 多くの点.

many-sided *a.* 多辺の.

many-valued *a.* 多価の, 多値の.
~ function 多価関数. ~ logic 多値論理.

map *n.* 写像; 地図. *vt.* 写す; 移す.
canonical ~ 標準(的)写像. conformal ~ 等角写像. ~ ping function 写像関数. ~ projection 地図投影[象](法). restriction ~ 制限写像.

mapping *n.* 写像.
barycentric ~ 重心写像. closed ~ 閉写像. composed [composite] ~ 合成写像. conformal ~ 等角写像. constant ~ 定値写像. continuous ~ 連続写像. contraction ~ 縮小写像. convex ~ 凸写像. degree of ~ 写像度. identity [identical] ~ 恒等写像. injective ~ 単射写像. inverse ~ 逆写像. linear ~ 線形写像. ~ cone [cylinder] 写像錐[柱]. ~ degree 写像度. ~ function 写像関数. ~ radius 写像半径. ~ space 写像空間. ~ theorem 写像定理. one-to-one ~ 一対一写像. open ~ 開写像. perspective ~ 配景写像. reduced ~ cone 約写像錐, 被約写像錐. schlicht ~ 単葉写像. star(-)like ~ 星型写像. surjective ~ 全射写像. topological ~ 位相写像. transposed ~ 転置写像. univalent ~ 単葉写像.

MARCINKIEWICZ, Józef (1910–1940) マルチンキェヴィッチ.

marginal *a.* 縁の; 周辺の; 限界の.
~ checking 限界検査. ~ distribution 周縁[周辺]分布. ~ frequency 周縁[周辺]度数. ~ utility 限界効用.

mark *n.* 標識, 印, マーク; 得点. *vt., vi.* 印をつける.
average ~ 平均点. class ~ 階級値. ~ reader 得点読取機.

market *n.* 市場.
~ research 市場分析, マーケット・リサーチ. ~ survey 市場調査.

MARKOV, Andrej Andrevitch (1856–1922) マルコフ.

martingale *n.* マルチンゲール.

MASCHERONI, Lorenzo (1750–1800) マスケロニ.

mass *n.* 質量; 多数, 多量.
center of ~ 質量中心, 重心. ~ distribution 質量分布. ~ point 質点. ~ production 大量生産, 量産. point ~ 点質量; 質点. unit ~ 単位質量.

master *n.* 基本; 主.
~ equation マスター方程式. ~ sample 親標本, マスター・サンプル.

matching *a.* 釣合う. *n.* 符合; マッチング.
~ number 符合数. ~ problem 符合(の)問題. maximal ~ 極大マッチング. maximum ~ 最大マッチング. perfect ~ 完全マッチング.

math. mathematics; mathematician; mathematical.

mathematical *a.* 数学の, 数理的な, 数学上の.
~ analysis 数学解析. ~ axiom 数学的な公理. ~ biology 数理生物学. ~ economics 数理経済学. ~ expression 数式. ~ finance 数理ファイナンス. ~ induction 数学的帰納法. ~ linguistics 数理言語学. ~ logic 数理論理学. ~ physics 数理物理学, 物理数学. ~ probability 数学的確率. ~ programming 数理計画(法). ~ science 数理(的)科学. ~ society 数学会. ~ statistics 数理統計学. ~ structure 数学的構造. ~ system 数学的体系.

mathematically *ad.* 数学上, 数学的に.

mathematician *n.* 数学者.

mathematics *n.* 数学.
advanced ~ 高等数学. ancient ~ 古代数学. applied ~ 応用数学. classical ~ 古典数学. discrete ~ 離散数学. elementary ~ 初等数学. finite ~ 有限数学. higher ~ 高等数学. Japanese ~ 和算. ~ for programming 計画数学. modern ~ 現[近]代数学. pure ~ 純粋数学.

MATHIEU, Emile Leonard (1835–1890) マシュー.

matrix (*pl.* -trices, -trixes) *n.* マトリックス, 行列.
adjacency ~ 隣接行列. adjoint ~ 随伴行列, 共役転置行列. alternating ~ 交代行列. bordered ~ 縁どった行列. coefficient

〜 係数行列. cofactor 〜 余因子行列. dense 〜 密行列. diagonal 〜 対角行列. (row) echelon 〜 (行)階段行列. full 〜 ring 全行列環. generalized inverse 〜 一般化逆行列. generator 〜 生成行列. Hermitian 〜 エルミート行列. improper orthogonal 〜 変格直交行列. incidence 〜 生起行列; 接続行列. infinite 〜 無限行列. inverse 〜 逆行列. invertible 〜 可逆行列. lambda 〜 ラムダ[λ]行列. 〜 element 行列要素. 〜 equation 行列方程式. 〜 group 行列群. 〜 representation 行列表現. 〜 unit 行列単位. negative-definite 〜 負定値行列. normal 〜 正規行列. orthogonal 〜 直交行列. period 〜 周期行列. permutation 〜 置換行列. positive-definite 〜 正定値行列. reduced row echelon 〜 簡約行階段行列. regular 〜 正則行列. representation 〜 表現行列. sparse 〜 疎行列, スパース行列. square(d) 〜 正方行列. symmetric 〜 対称行列. total 〜 ring 全行列環. transposed 〜 転置行列. triangular 〜 三角行列. unit 〜 単位行列. unitary 〜 ユニタリ行列.

matroid *n.* マトロイド.
〜 lattice マトロイド束.

matroidal *a.* マトロイドの.

max. maximum, maximal.

maxim *n.* 公理.

maximal *a.* 極大の; 最大の. ⇨ minimal.
〜 condition 極大条件. 〜 deficiency 最大不足数. 〜 ideal 極大イデアル. 〜 operator 最大作用素. 〜 order 極大整環; 最大位数. 〜 solution 最大解. 〜 term 最大項. 〜 value 極[最]大値.

maximality *n.* 極大性; 最大性.

maximally *ad.* 極大(的)に.
〜 almost periodic group 極大概周期群.

maximin *n.* 最大最小, マクシミン.
〜 principle [value] マクシミン原理[値].

maximize *vt.* 最大にする.

maximizer *n.* 最大解.
global 〜 大域的最大解. local 〜 局所的最大解.

maximizing *a.* 最大にする.
〜 sequence 最大化列.

maximum (*pl.* -ma, -mums) *n.* 最大; 極大. ⇨ minimum.
absolute 〜 最大. 〜 coefficient 最大係数. 〜 condition 極大条件. 〜 flow 最大流, 最大フロー. 〜 likelihood 最大尤度; 最尤の. 〜 likelihood decoding 最尤復号法. 〜 likelihood estimation 最尤推定法. 〜 modulus principle 最大(絶対)値の原理. 〜 principle 最大値(の)原理; 極大の原理. relative 〜 極大.

max-min *n.* 最大最小, マクシミン.
〜 principle [value] マクシミン原理[値].

MAXWELL, James Clerk (1831-1879) マクスウェル.

MAZURKIEWICZ, Stefan (1888-1945) マズルキーウィッチ.

mean *a.* 平均の. *n.* 平均, 平均値; 中項.
arithmetic 〜 算術[相加]平均. convergence in the 〜 平均収束. geometric 〜 幾何[相乗]平均. harmonic 〜 調和平均. 〜 absolute error 平均絶対誤差. 〜 convergence 平均収束. 〜 curvature 平均曲率. 〜 deviation 平均偏差. 〜 error 平均誤差. 〜 motion 平均運動. 〜 oval 平均卵形(線). 〜 proportional 比例中項; (数表の)比例部分. 〜 p-valent 平均 p 葉の. 〜 sojourn 平均訪問回数. 〜 square 平均平方, 不偏分散. 〜 square error 平均二乗[平方]誤差. 〜-unbiased 平均不偏な. 〜 value 平均値. 〜 value theorem 平均値の定理; 中間値の定理. 〜 velocity 平均速度. population 〜 母(集団)平均. progressive 〜 累加平均. running 〜 移動平均. sample 〜 標本平均. trimmed 〜 刈り込み平均. weighted 〜 荷重平均.

means *n.* (*pl.*) 手段; 方法.
by 〜 of によって.

measurability *n.* 可測性.

measurable *a.* 可測な; 面積確定の.
Borel(-) 〜 ボレル可測な. Lebesgue(-) 〜 ルベーグ可測な. 〜 cover 等測包. 〜 event 可測事象. 〜 function 可測関数. 〜 kernel 等測核. 〜 set 可測集合. 〜 space 可測空間. 〜 variable 計量. μ-〜 ミュー可

測な.

measure *n.* 測度; 約数. *vt., vi.* 測定する; はかる.
Borel ~ ボレル測度. bounded ~ 有界測度. circular ~ 弧度(法). common ~ 公約数. convergence in ~ 測度収束. cubic ~ 体積, 容量. greatest common ~ 最大公約数[G.C.M.]. Haar ~ ハール測度. harmonic ~ 調和測度. image ~ 像測度. inner ~ 内測度. invariant ~ 不変測度. killing ~ 消滅測度. Lebesgue ~ ルベーグ測度. ~d value 測定値. ~ preserving transformation 保測変換. ~ space 測度空間. ~ theory 測度論. ~ vector 尺度ベクトル. of ~ zero 測度ゼロの. outer ~ 外測度. probability ~ 確率測度. product ~ 直積測度. quotient ~ 商測度. σ-additive [-finite] ~ シグマ加法的[有限]測度. solid ~ 体積.

measurement *n.* 測定; 測量.
actual ~ 実測. conditional ~ 条件つき測定. direct [indirect] ~ 直[間]接測定. photo(-) ~ 写真測量. trigonometrical ~ 三角測量.

measure-theoretic *a.* 測度論的な.
~ method 測度論的方法.

measuring *a.* 測定の.
~ line 側線. ~ point 測点.

mechanical *a.* 器械的な.
~ construction 器械的作図. ~ integration 器械的積分. ~ quadrature 器械的求積.

mechanics *n.* 力学.
analytical ~ 解析力学. celestial ~ 天体力学. graphical ~ 図式力学. Hamiltonian ~ ハミルトン力学. Newtonian ~ ニュートン力学. non(-)linear ~ 非線形力学. quantum ~ 量子力学. statistical ~ 統計力学. wave ~ 波動力学.

mechanism *n.* 機械; 機構.

medial *a.* 中点の.
~ triangle 中点三角形.

median *a.* 中央の, 中位の; 中位数の. *n.* 中線; 中点; 中央値, 中位数, メジアン.
~ line 中線. ~-unbiased estimator 中央値不偏推定量. population ~ 母(集団)中央値. sample ~ 標本中央値.

mediant *n.* 中間数.

meet *n.* 交わり. *vt., vi.* 出会う, 交わる. ⇨ cap; join. *vt.* を満たす.

mega- *pref.* 「大; 百万」の意.
~ cycle メガ[百万]サイクル. ~ watt メガ[百万]ワット.

Mehler, Ferdinand Gustav (1835-1895) メーラー.

Mellin, Robert Hjalmar (1854-1933) メリン.

member *n.* 項; 辺; 元, 構成要素.
a ~ of equation 方程式の一項. left [right]-hand ~ 左[右]辺.

membership *n.* 会員, メンバーシップ.
~ function メンバーシップ関数.

membrane *n.* (薄)膜.
vibrating ~ 振動膜.

memoir *n.* 研究報告; 論文.

memorize *vt., vi.* 記憶する; 暗記する.

memoryless *a.* 無記憶の.

Menelaus [**Menelaos**] of Alexandria (100頃) アレクサンドリアのメネラウス.

Menge *n.* (*G.* =set.) 集合.

meniscus (*pl.* -cuses, -ci) *n.* 新月図形.

mensuration *n.* 測定(法), 測量(法), 求積(法).
~ by parts 区分求積(法).

mental *a.* 暗算の; 知的な.
~ arithmetic [calculation] 暗算.

Méray, Hugues Charles Robert (1836-1911) メレー.

Mercator, Gerhardus (1512-1594) メルカトル.

merging *n.* 併合; 混合.
sorting by ~ 併合整列法.

meridian *n.* 子午線, 経線. *a.* 子午線の.
~ circle 子午線.

meromorphic *a.* 有理形の, 有理型の.
~ curve 有理型[形]曲線. ~ function 有理型[形]関数. ~ mapping 有理型[形]写像.

meromorphy *n.* 有理型[形](性).
radiusof ~, ~ radius 有理型[形]半径.

MERSENNE, abbé Marin (1588-1648) メルセンヌ.

mesh *n.* 網目.
～ of covering 被覆の目(の大きさ).

message *n.* メッセージ; 伝言; 通信(文).

meta- *pref.* 「メタ-, 超-」の意.
～abelian [metabelian] group メタアーベル群. ～theory メタ理論.

metamathematics *n.* 超数学.

metaphysics *n.* 形而上学.

method *n.* 方法, 手段.
damped Newton ～ 減速ニュートン法. diagonal ～ 対角線論法. direct [indirect] ～ 直[間]接法. divide and conquer ～ 分割統治法. gradient ～ 勾配法, 傾斜法. iterative ～ 反復法. ～ of averaging 平均法. ～ of diminishing [nested] intervals 区間縮小法. ～ of elimination 消去法. ～ of inversion 反転法. ～ of least squares 最小二[自]乗法. ～ of maximum likelihood 最尤法. ～ of randomized block 乱塊法. ～ of steepest descent 最急降下法. ～ of successive approximation [substitution] 逐次近似[代入]法. multiplier ～ 乗数法. nearest neigbo(u)r ～ 最近傍法. Newton's ～ ニュートン法. numerical ～ 数値解法. relaxation ～ 緩和法. simplex ～ 単体法. steepest descent ～ 最急降下法. summation [summability] ～ 総和法. trust region ～ 信頼領域法.

metric *a.* 計量の; 距離の. *n.* 計量, 距離; 距離関数.
almost ～ structure 概計量構造. Euclidean [Hermitian] ～ ユークリッド[エルミート]計量. hyperbolic ～ 双曲(的)計量. ～ connection 計量接続. ～ property 計量的性質. ～ space 距離空間. ～ tensor 計量テンソル. ～ vector space 計量ベクトル空間. non-Euclidean ～ 非ユークリッド計量. Riemannian ～ リーマン計量. spherical ～ 球面計量. supporting ～ 支持計量.

metrical *a.* 計量の, 計量的な.
～ transitivity 測度推移[可遷]性.

metrically *ad.* 計量的に.
～ transitive 測度推移[可遷]的な.

metrizable *a.* 距離づけ可能な, 計量化可能な.

metrization *n.* 計量化, 距離づけ.

MEUSNIER, Jean Baptiste Marie Charles (1754-1793) ムーニエ.

micro- *pref.* 「小, 微小」の意; 「百万分の一」の意.

microbundle *n.* マイクロ束.

micro-local *a.* 超局所的な.

micron (*pl.* -crons, -cra) *n.* ミクロン[百万分の一メートル].

middle *a.* 中央の. *n.* 中央.
～ point 中点. ～-square method 二乗[平方]採中法.

midperpendicular *n.* 垂直二等分線.

mid(-)point *n.* 中点; 中点値; 範囲の中央.
～ rule 中点法則. perpendicular at ～ 垂直二等分線.

mid(-)range *n.* 中点値; 範囲の中央.

milestone *n.* 主要管理点; 里程標.

millesimal *n., a.* 千分の一(の).

milli- *pref.* 「千分の一」の意.
～ meter ミリ[千分の一]メートル.

millile *n.* 千分位数.

million *n., a.* 百万(の).
hundred ～s 億.

MILNOR, John Willard (1931-) ミルナー.

min. minimum, minimal.

minimal *a.* 最小の, 極小の; 最短の. ⇨ maximal.
～ condition 極小条件. ～ curve 極小曲線. ～ ideal 極小イデアル. ～ polynomial 最小[低]多項式. ～ splitting field 最小分解体. ～ surface 極小曲面. ～ value 局[最]小値.

minimality *n.* 極小性; 最小性.

minimally *ad.* 極小(的)に.
～ almost periodic group. 極小概周期群. ～ imbedded submanifold 極小的に埋めこまれた部分多様体.

minimax *n.* 最小最大, ミニマックス.
～ principle ミニマックス原理, 鞍(部)点原理. ～ solution ミニマックス解. ～ test ミニマックス検定. ～ value ミニマックス値. point of ～ 鞍(部)点.

minimization *n.* 最小にすること，最小化.
minimize *vt.* 最小にする.
minimizer *n.* 最小解.
global ~ 大域的最小解. local ~ 局所的最小解.
minimizing *a.* 最小にする.
~ sequence 最小化列.
minimum (*pl.* -ma, -mums) *n.* 極小; 最小. ⇨ maximum.
~ chi(-)square estimator 最小カイ二乗推定量. ~ condition 極小条件. ~ point 最小点. ~ solution 最小解. ~ value 最小値. relative ~ 極小.
MINKOWSKI, Hermann (1864-1909) ミンコフスキ.
minor *a.* 小さい方の，より少ない. *n.* 小行列式; 小前提. ⇨ major.
~ angle 劣角. ~ arc 劣弧. ~ axis 短軸[径]. ~ defect 軽欠点. ~ determinant 小行列式. ~ function 劣関数. ~ premise 小前提. principal ~ 主(座)[首座]小行列式.
minorant *n.* 劣関数; 劣級数.
harmonic ~ 調和劣関数.
minuend *n.* 被減数.
minus *prep.* マイナス，…を減じて. *a.* マイナスの，負の. *n.* 負号，マイナス記号; 負数. ⇨ plus.
~ sign 負号，減号. plus and ~ 正負.
minute *n.* 分【時間・角度の単位】. *a.* 微小な，極微な.
mirror *n.* 鏡，ミラー.
~ image 鏡像.
miscalculate *vt., vi.* 計算違いをする，誤算する.
miscalculation *n.* 計算違い，誤算.
miscellaneous *a.* 種々な，雑多な.
~ exercises 雑(問)題.
MISES, Richard Edler von (1883-1953) ミーゼス.
missing *a.* 見落しの; 欠けている.
~ data 欠測値. ~ plot, ~ value 欠測値.
MITTAG-LEFFLER, Gösta Magnus (1846-1927) ミッタク-レフラー.
mix *vt., vi.* 混ぜる，混ざる. *n.* 混合.
~ing property 混合性.
mixed *a.* 混合した.
~ boundary value problem 混合境界値問題. ~ decimal 帯小数. ~ difference 混合差分. ~ fraction 帯分数. ~ group 混群; 混合(アーベル)群. ~ partial derivative 混合偏導関数. ~ recurring decimal 混循環小数. ~ strategy 混合戦略. ~ tensor 混合テンソル. ~ type 混合型.
mixture *n.* 混合.
~ distribution 混合分布.
M.K.S. Meter-Kilogram-Second.
mobility *n.* 可動性.
free ~ 自由運動[可動性].
MÖBIUS, Augustus Ferdinand (1790-1868) メビウス.
mod, mod. module, modulo, modulus.
~ *n* *n*を法として.
modular *a.* モジュラー; 母数の.
~ character モジュラー指標. ~ curve モジュラー曲線. ~ form モジュラー形式. ~ function 母数関数，モジュラー関数. ~ group 母数群，モジュラー群. ~ lattice モジュラー束. ~ representation モジュラー表現.
modal *a.* 様相の; モードの.
~ class 最頻級. ~ logic 様相論理. ~ proposition 様相命題.
mode *n.* モード，並数，最頻値; 方式.
sample ~ 標本モード.
model *n.* 模型; ひな型，モデル.
allocation ~ 配分模型. deterministic ~ 決定論的モデル[模型]. enumerable ~ 可算モデル. hidden Markov Model 隠れマルコフモデル. ~ theory モデル理論. statistical linear ~ 統計的線形モデル. stochastic ~ 確率(論)的モデル[模型]. topological ~ 位相(的)モデル.
model(l)ing *n.* 模型作製，モデル化.
modern *a.* 現代の; 近世の; モダンな.
~ mathematics 現[近]代数学.
modification *n.* 変更; 修正.
analytic ~ 解析的改変. proper ~ 固有改変. spherical ~ 球面改変.
modified *a.* 変形された，変形の; 変更した，修正した．
～ Bessel function 変形ベッセル関数．

modify *vt.* 変形する，修正する．

modulation *n.* 変調．
frequency ～ 周波数変調．～ rate 変調速度．～ space モデュレーション空間．

module *n.* 加群; モジュール．
dual ～ 双対加群．factor ～ 剰余加群，因子加群．free ～ 自由加群．indecomposable ～ 直既約加群．injective ～ 移入[単射，入射]加群．irreducible ～ 既約加群．left [right] *A*-～ 左[右] *A* 加群．～ of boundaries [cycles] 境界[輪体]加群．Noetherian ～ ネーター加群．projective ～ 射影加群．simple ～ 単純加群．topological ～ 位相加群．

modulo *prep.* (*L.*) を法として【modulus の従格】．
～ *n* *n* を法として(＝to modulus *n*)．～-*n* residue モジュロ *n* 剰余．

modulus (*pl.* -li) *n.* 係数; (合同式の)法【mod】; (実数，複素数の)絶対値; (楕円積分の)母数; モジュラス．
complementary ～ 補母数．logarithmic ～ 対数的モジュラス．～ of continuity 連続度[率]．periodicity ～ 周期母数．

modus *n.* (*L.*) 規則; 様式; (推論様)式，(推論形)式．
～ ponendo tollens 肯定否定式．～ ponens 肯定式．～ tollendo ponens 否定肯定式．～ tollens 否定式．

mollifier *n.* 軟化子．

mollify *vt.* 軟化させる．

moment *n.* 能率; モーメント，積率; 力率．
absolute ～ 絶対能率[積率]．～ generating function 積率母関数．～ matrix 能率行列．～ method 積率法．～ of inertia 慣性能率．～ problem モーメント問題．second ～ 二次能率．

momentum (*pl.* -ta, -tums) *n.* 運動量．
angular ～ 角運動量，運動量の能率．

MONGE, Gaspard (1746–1818) モンジュ．

monic *a.* モニックな．
～ polynomial モニック多項式．

mono- *pref.*「単一，一つの」の意．⇨ poly-.

monoclinic *a.* 単斜晶の．
～ system 単斜晶系．

monodromy *n.* モノドロミー; 一価性．
～ group モノドロミー群．～ matrix モノドロミー行列．～ theorem 一価性の定理．

monogeneity *n.* 一元性．

monogenic *a.* モノジーンな; 単生[成]的な; 一遺伝子性の; 一元の．
～ function (複素)解析関数; 単生的関数．

monoid *n.* 単位(的)半群．

monoidal *a.* モノイダルな，単項的な．
～ transformation モノイダル[単項的]変換．

monolithic *a.* モノリシックな; 一方的な; 単一基板上の．

monomial *a.* 単項の．*n.* 単項式．
～ factor 単項因子．～ representation 単項表現．

monomorphism *n.* 単射，単射準同型[形]．

monotone *n.* 単調．*a.* 単調な．*ad.* 単調に．
～ convergence 単調収束．～ decreasing [increasing] 単調減少[増加]の．～ function 単調関数．～ likelihood ratio 単調尤度比．～ sequence 単調(数)列．strictly ～ 狭義(に)単調な．

monotoneity *n.* 単調性．

monotonic *a.* ＝monotone.

monotonicity ＝monotoneity.

monotonous *a.* ＝monotone.

Monte Carlo モンテカルロ．
～ method モンテカルロ法．

MONTEL, Paul Antoine (1876–1975) モンテル．

MOORE, Eliakim Hastings (1862–1932) ムーア．

more *a.* より多い．*ad.* もっと多く[大きく]【many, much の比較級】．⇨ less. *n.* それ以上のもの．
～ than or equal to 以上の．much ～, still ～ なおさら．not ～ (than) 以下の．

MORERA, Giacinto (1856–1909) モレラ．

MORITA, Kiichi (1915–1995) 森田紀一．

morphism *n.* 射，準同型[形]．
diagonal ～ 対角射．inverse ～ 逆射．

proper ~ 固有射. structure ~ 構造射.

MORSE, Harold Kelvin Marston (1892-1977) モース.

mortality *n.* 廃却率; 死亡率.
~ rate 死亡率. ~ table 死亡率統計表.

MOSER, Jürgen (1928-1999) モーザー.

most *a.* 最も多くの; たいがいの. *ad.* 最も. *n.* 最大(数, 量); の大部分【of】. ⇨ least.
at ~ 多くとも, 高々. ~ powerful test 最有力検定. ~ probable value 最確値.

motion *n.* 運動; 合同変換.
Brownian ~ ブラウン運動. circular [elliptic] ~ 円[楕円]運動. equation of ~ 運動方程式. infinitesimal ~ 無限小運動. ~ group 運動群. relative ~ 相対運動. simple harmonic ~ 単(弦)振動 [S. H. M.]. uniform ~ 等速運動. vortex ~ 渦動.

movable *a.* 可動の.
~ branch point 可動(の)[動く]分岐点. ~ singularity 可動(の)[動く]特異点.

move *vt., vi.* 動かす. *n.* 手番.

movement *n.* 運動.

moving *a.* 動く; 移動する.
~ average 移動平均. ~ coordinate system 動座標系. ~ frame 動枠, 動標構. ~ trihedral 動三面角.

M.S.D. Most Significant Digit. 最上位の数字.

M-test =majorant test.

mu *n.* ミュー. ギリシア語アルファベットの12番目の文字【M, μ】.

MÜLLER, Johann (es) =REGIOMONTANUS. ミュラー.

multangular *a.* 多角の.

multi- *pref.* 「多い」の意.

multi-collinearity *n.* 多重共線性.

multidimensional *a.* 多[高]次元の.

multi-index *n.* 多重指数.

multilinear *a.* 多重線形の.
~ form 多重線形形式. ~ mapping 多重線形写像.

multinomial *a.* 多項の. *n.* 多項式.
⇨ polynomial.
~ coefficient 多項係数. ~ distribution 多項分布. ~ expansion [theorem] 多項展開[定理].

multi-phase *n.* 多相.
~ sampling 多相抽出(法).

multiplanar *a.* 多面の.
~ coordinates 多面座標.

multiple *a.* 倍数の; 重複した, 多重の. *n.* 倍数, 倍元.
common ~ 公倍数[元]. least [lowest] common ~ 最小公倍数[L.C.M.]. ~ comparison 多重比較. ~ correlation 重相関. ~ covariant 多重共変式. ~ integral 重(複)積分, 多重積分. ~ mathematical induction 多重数学的帰納法. ~ point 重複点. ~ ratio 相乗比. ~ regression 重回帰. ~ root 重(複)根. ~ sampling 多(数)回抽出. ~ sequence 多重数列. ~ series 多重級数. ~ stratification 多重層化. ~ suffix 多重添数. ~-valued 多値の. scalar ~ スカラー倍.

multiplicable *a.* 相乗可能な.

multiplicand *n.* 被乗数.

multiplication *n.* 乗法, 掛け算.
complex ~ 虚数乗法. continued ~ 連乗(積). ~ table 掛け算九々表, 乗法表. ~ theorem 乗法定理. scalar ~ スカラー乗法.

multiplicative *a.* 掛け算の, 乗法の; 乗法的な.
~ congruence 乗法合同. ~ factor 乗法[数](的)因子. ~ function 乗法的関数. ~ functional 乗法的汎関数. ~ group 乗法群. ~ process 分枝過程. ~ set (=~ly closed set) 乗法的閉集合, 積閉集合. ~ valuation 乗法付値.

multiplicatively *ad.* 乗法的に.
~ closed set 乗法的閉集合, 積閉集合.

multiplicator *n.* 乗法子; 乗数, 乗法因子.

multiplicity *n.* 重複; 重複度.
algebraic ~ 代数的重複度. geometric ~ 幾何的重複度.

multiplier *n.* 乗数, 乗式, 乗法因子.
characteristic ~ 特性乗数. Fourier ~ フーリエ乗算作用素. Lagrange ~ ラグランジュ乗数. last ~ 最終乗式. ~ algebra 乗法子環. ~ effect 乗数効果. ~ method,

method of ～(s) 乗数法.

multiply *vt., vi.* 乗じる，掛ける．*ad.* 多重に，重複して; より多く; 複数にある.
～(-)connected (重)複[多重]連結の．～ 5 by [into] 3 5 に 3 を掛ける．～ transitive group 多重推移群.

multipolar *a.* 多極の.
～ coordinates 多極座標.

multi(-)stage *a.* 多段(式)の; 多段階の.
～ allocation 多段配分．～ choice process 多段選択過程．～ sampling 多段抽出.

multi(-)valued *a.* 多価の，多値の.
～ function 多価関数.

multivalence, multivalency *n.* 多葉(性).

multivalent *a.* 多葉な; 多価の.
～ function 多葉関数.

multivariate *a.* 多変量の.
～ analysis 多変量解析．～ analysis of variance 多変量分散分析[MANOVA]．～ data 多変量データ．～ linear model 多変量線形モデル．～ normal distribution 多変量正規分布.

multivector *n.* 多重ベクトル.

Mumford, David Bryant (1937-) マンフォード.

mutatis mutandis *ad.* (*L.* ＝necessary changes being made.)「必要な変更を加えれば」,「必要に応じて変更を加えて」

mutual *a.* 相互の.
～ division 互除(法)．～ energy 相互エネルギー.

mutually *ad.* 互いに，相互に.
～ disjoint (集合が)互いに素な．～ exclusive 互いに排反な．～ orthogonal 互いに直交の．～ prime (整数，イデアルが)互いに素な．～ symmetric 相互対称な.

N

n 《記号》不定整数[量].

nabla *n.* ナブラ【∇】. ⇨ atled.

nadir *n.* 天底.

Nakayama, Tadashi (1912–1964) 中山正.

namely *ad.* すなわち.

nano- *pref.* ナノ【10^{-9} の意】.

Napier, John (1550–1617) ネイピア.

Napierian *a.* ネイピアの.

~ logarithm ネイピア(の)対数, 自然対数.

nappe *n.* 面葉, 面包.

narrow *a.* 狭い, 細い.

~ convergence 狭収束.

n.a.s.c. =necessary and sufficient condition. 必要十分条件.

Nash Jr., John Forbes (1928–2015) ナッシュ.

nat *n.* ナット【自然対数による情報量やエントロピーの単位】.

natural *a.* 自然の, 自然な.

~ boundary 自然境界. ~ equation 自然方程式. ~ frame 自然標構. ~ geometry 自然幾何学. ~ homomorphism 自然(な)準同型[形]. ~ injection 自然な単射. ~ logarithm 自然対数. ~ model 自然なモデル. ~ number 自然数. ~ projection 自然射影. ~ scale 自然尺度. ~ transformation 自然変換.

naturality *n.* 自然性.

naught *n.* 零; 無. =nought.

law of ~ or one 0–1 法則, 零または一の法則, 悉無律.

Navier, Claude-Louis (1785–1836) ナヴィエ.

N.B. Nota Bene. 注意(せよ).

near *ad.* 近く. *prep.* の近くに. *a.* 近い. *vt.*, *vi.* (に)近づく.

~est neighbor method 最近傍法.

nearly *ad.* 殆んど, ほぼ.

~ equal ほぼ等しい.

necessarily *ad.* 必ず, 必然的に.

necessary *a.* 必要な.

~ and sufficient condition 必要十分条件.

~ condition 必要条件.

necessity *n.* 必要性; 必要.

necklace *n.* 首飾り, ネックレス.

~ permutation 数珠順列. ~ problem 首飾りの問題.

negation *n.* 否定.

double ~ 二重否定.

negative *a.* 負の, マイナスの; 否定の, 否定的な. *n.* 負数, 反数; 否定. ⇨ positive.

double ~ 二重否定. ~ angle 負(の)角. ~ binomial distribution 負の二項分布. ~ correlation 負(の)相関. ~-definite, ~ definite 負(定)値の, 負の定符号の. ~ direction 負の向き. ~ domain [region] 負領域. ~ infinity 負の無限大. ~ number 負(の)数. ~ orientation 負の向き(づけ). ~ proposition 否定命題. ~ root 負(の)根. ~ semi-definite 半負値の, 負の半定符号の. ~ sense 負の向き. ~ sign 負号. ~ term 負項. ~ value 負(の)値. ~-valued 負値の. ~ variation 負変動.

negatively *ad.* 負に, 負の向きに; 否定的に.

~ asymptotically stable 負の向きに漸近安定な. ~ infinite マイナス無限大の.

negativity *n.* 負値性.

negligible *a.* 無視できる, 無視してもよい.

~ quantity [term] 無視できる量[項].

negligibly *ad.* 無視してもよく.
~ small 無視できるほど小さい.

neighbo(u)rhood *n.* 近傍.
analytic ~ 解析的近傍. convex ~ 凸近傍. coordinate ~ 座標近傍. ε-~ ε 近傍. fundamental ~ (system) 基本近傍(系). ~ system 近傍系. open ~ 開近傍. spherical ~ 球近傍. sufficiently small ~ 十分小さい近傍.

neighbo(u)ring *a.* 隣接している.

nephroid *n.* 腎曲線.

nerve *n.* 脈体; 神経.

-ness *suf.* 形容詞などに添えて「性質，状態」を表す名詞をつくる: boundedness, completeness.

nest *n.* 巣; 入れ子構造. *vi., vt.* 巣ごもる[らせる]. *vt.* 入れ子にする.
~ed design 入れ子型計画. sequence of ~ed intervals 縮小区間列【次第に前者に含まれる区間の列】.

net *a.* 正味の. *n.* 有向点族; 正価; 網.
~ present worth 正味現価. ~ price 正価.

network *n.* ネットワーク; 回路; 網目; 網系.
bilateral ~ 相反回路; 双方向性ネットワーク. contact ~ 接点回路. electric ~ 電気回路. ~ structure 網目構造. neural ~ ニューラルネットワーク. reciprocal ~ 相反回路.

NEUMANN, Carl Gottfried (1832-1925), Franz Ernst (1798-1895) ノイマン.

neutral *a.* 中立の; 中性の.
~ element 単位元，中立元. ~ type 中立型.

NEVANLINNA, Rolf Herman (1895-1980) ネヴァンリンナ.

NEWTON, Sir Isaac (1642-1727) ニュートン.

Newtonian *a.* ニュートンの.
~ mechanics ニュートン力学. ~ potential ニュートン・ポテンシャル.

next *a.* 次の. *ad.* 次に.

nice *a.* 適当な.
~ function 好適な関数.

NIKODYM, Otto Martin (1887-1974) ニコディム.

nil *n.* 零. *a.* 零の.
~ ideal ベキ零元イデアル. ~ radical ベキ零元根基.

nilalgebra *n.* ベキ零元代数[多元環].

nil(-)factor *n.* 零因子.

nilpotency *n.* ベキ零(性).

nilpotent *a.* ベキ零の.
~ component ベキ零成分. ~ element ベキ零元. ~ group ベキ零群. ~ ideal ベキ零イデアル. ~ matrix ベキ零行列. ~ radical ベキ零根基. ~ ring ベキ零環. ~ transformation ベキ零変換.

nilradical *n.* ベキ零元根基.

nine-point *a.* 九点の.
~ circle 九点円.

NIRENBERG, Louis (1925-) ニーレンバーグ.

niveau *n.* (*F.*) 水準. ⇨ level.
~line [curve] 等位線. ~ surface 等位面.

No., no. (*pl.* Nos., nos.) (*L.*) numero.

nodal *a.* 結節状の，交軌点の.
~ line [curve] 節線. ~ point 節点. ~ region 結節状領域.

node *n.* 結節点，節点，ノード【曲線の交点】; 節(ふし).
~-locus 結節点の軌跡. tac~ 二重尖点.

NOETHER, Amalie Emmy (1882-1935), Max (1844-1921) ネーター.

Noetherian *a.* ネーターの，ネーター的な.
~ module ネーター加群. ~ ring ネーター環. ~ space ネーター空間.

noise *n.* 雑音.
white ~ 白色雑音.

noiseless *a.* 雑音のない.
~ channel 雑音のない情報路.

noisy *a.* 雑音のある.
~ channel 雑音のある情報路.

nominal *a.* 名目上の.
~ scale 名義尺度.

nomogram *n.* 計算図表，ノモグラム.

nomograph *n.* 計算図表，ノモグラフ.

nomographic *a.* 計算図表(学)の.
~ chart 共線図表.

nomography *n.* 計算図表学.
non- *pref.*「非-, 無-, 不-」の意.
nona *pref.* ノナ【九の意】.
non-abelian *a.* 非アーベル的な, 可換でない, 非可換な.
~ group 非可換群.
nonagon *n.* 九辺形, 九角形.
non-Archimedean *a.* 非アルキメデス的な.
~ geometry 非アルキメデス幾何学. ~ valuation 非アルキメデス(的)付値.
non-arithmetic *a.* 非算術的な.
~ use 非算術的応用[利用].
nonary *a.* 九進法の. *n.* 九個の組.
~ scale 九進法.
non-associative *a.* 非結合的な.
~ algebra 非結合的代数.
nonbasic *a.* 非基底の.
~ variable 非基底変数.
non-central *a.* 無心の.
~ conics 無心二次[円錐]曲線. ~ quadrics 無心二次曲面.
non-centrality *n.* 非[無]心(性).
~ parameter 非[無]心母数.
non-commutative *a.* 非可換な, 可換でない.
~ field 非可換体, 斜体. ~ ring 非可換環.
non-compact *a.* 非コンパクトな.
~ set 非コンパクト(な)集合. of ~ type 非コンパクト型の.
non(-)contradiction *n.* 無矛盾(性).
law of ~ 無矛盾律.
non(-)contradictory *a.* 無矛盾な.
non-cooperative *a.* 非協力の.
~ game 非協力ゲーム.
non-correlated *a.* 無相関の.
non-countable *a.* 非可算の.
~ set 非可算集合.
non-degenerate *a.* 非退化の, 退化しない.
~ critical point 非退化臨界点. ~ kernel 非退化核. ~ quadratic form 非退化二次形式.
non-dense *a.* 疎な.
~ set 疎集合.
non-Desarguesian *a.* 非デザルグ的な.
~ geometry 非デザルグ幾何学.
non-destructive *a.* 非破壊的な.
~ inspection 非破壊検査.
non-deterministic *a.* 非決定論的な.
~ computation 非決定性計算. ~ polynomial time (NP) 非決定性多項式時間.
non-empty *a.* 空でない.
~ set 空でない集合.
non-Euclidean, non-euclidean *a.* 非ユークリッド的な.
~ angle 非ユークリッド角. ~ distance 非ユークリッド距離. ~ geometry 非ユークリッド幾何学. ~ metric 非ユークリッド計量. ~ space 非ユークリッド空間.
non-expansive *a.* 非拡大の.
~ mapping 非拡大写像.
non(-)finiteness *n.* 非有限性.
non-holonomic *a.* 非ホロノームな.
~ relation [system] 非ホロノーム(な)関係(式)[系].
non-homogeneous *a.* 非同次の, 非斉次の. ⇨ inhomogeneous.
~ chain 非同次鎖. ~ coordinates 非斉[同]次座標.
noninformative *a.* 情報のない, 無情報の.
~ prior distribution 無情報事前分布.
nonion *n.* 九元数.
~ form 九元数の形.
non(-)linear *a.* 非線形[型]の.
~ equation 非線形方程式. ~ functional 非線形汎関数. ~ mechanics 非線形力学. ~ oscillation 非線形振動. ~ phenomenon 非線形現象. ~ prediction 非線形予報. ~ problem 非線形問題. ~ programming 非線形計画.
non(-)singular *a.* 非特異な; 退化しない.
~ matrix 非特異行列, 正則行列. ~ point 非特異点; 単純点; 正則点.
non-negative *a.* 非負の.
~ integer 非負の整数, 自然数(と零). ~ number 負でない数.
non-orientable *a.* 向きづけ不(可)能な, 不可符号の.

〜 surface 不可符号曲面.

non-parametric *a.* ノンパラメトリック, 母数によらない.
〜 estimation ノンパラメトリック推定. 〜 method ノンパラメトリック法, 非母数法. 〜 test ノンパラメトリック検定.

non-Pascalian *a.* 非パスカル的な.
〜 geometry 非パスカル幾何学.

non-periodic *a.* 非循環の; 非周期的な.
〜 part 非循環節.

non-positive *a.* 非正の.
〜 integer 非正の整数, 負の整数と零.

non-prime *a.* 非素の.
〜 number 非素数.

non-primitive *a.* 非原始の.
〜 character 非原始指標.

non(-)redundant *a.* 余分のない, むだがない.

non-residue *n.* 非剰余.
quadratic 〜 平方非剰余.

non-sampling *a.* 非標本の.
〜 error 非標本誤差.

non-signed *a.* 無符号の.
〜 number 無符号数.

non-stable *a.* 非安定な.

non-tangential *a.* 接しない.
〜 limit 接しない路に沿う極限(値).

non-trivial *a.* 自明でない.
〜 solution 自明でない解, 非自明解.

non-void *a.* 空でない.
〜 set 空でない集合.

non(-)zero *a.* 零でない.
〜 element 零でない元. 〜(-)sum game 非零和ゲーム. 〜 zero 原点と一致しない零点.

NÖRLUND, Niels Erick (1885-1981) ネールンド.

norm *n.* ノルム. *vt.* ノルムづける.
absolute 〜 絶対ノルム. l^p 〜 L^p ノルム. 〜-residue ノルム剰余. 〜 [〜ed] space ノルム空間. operator 〜 作用素ノルム. reduced 〜 被約ノルム. relative 〜 相対ノルム. sup 〜 上限ノルム. trace 〜 トレース・ノルム.

normable *a.* ノルムづけ可能な.

normal *a.* 標準の, 正規な; 法線の, 垂直な. *n.* 法線, 垂直線[面]; 平均量.
affine 〜 アフィン法線. bisecting 〜 垂直二等分線. inward 〜 内(向き)法線. 〜 basis 正規基. 〜 bundle 法束. 〜 chain 正規鎖. 〜 component 法線成分. 〜 continued fraction 正規連分数. 〜 coordinates 標準座標. 〜 curvature 法曲率. 〜 curve 正規曲線. 〜 derivative 法線導関数. 〜 distribution 正規分布. 〜 equation 標準[正規]方程式. 〜 extension 正規拡大. 〜 family 正規族. 〜 form 標準形. 〜 function 正規関数. 〜 line 法線. 〜 matrix 正規行列. 〜 operator 正規作用素. 〜 plane 法平面. 〜 population 正規母集団. 〜 random numbers 正規乱数. 〜 regression 正規回帰. 〜 ring 正規環. 〜 sampling 規準抜取(り). 〜 scheme 正規スキーム. 〜 section 直截口. 〜 space 正規空間. 〜 subgroup 正規部分群. 〜 time 標準時間. 〜 valuation 正規付値. 〜 vector 法(線)ベクトル. 〜 vibration 規準振動. outward 〜 外(向き)法線. principal 〜 主法線. unit 〜 単位法線.

normality *n.* 正規性.
〜 test 正規性検定.

normalization *n.* 正規化.
Noether's 〜 theorem ネーターの正規化定理. 〜 condition, condition of 〜正規化条件. ortho〜 正規直交化.

normalize *vt.* 正規化する.

normalized *a.* 正規化された, 規格化された, 標準化された.
〜 contrast 標準化対比. 〜 form 正規形. 〜 function 正規化された関数. 〜 system 正規系. 〜 valuation 正規化された付値. 〜 vector 正規化されたベクトル.

normalizer *n.* 正規化群.

normally *ad.* 法線的に; 正規に.
〜 algebraic 正規代数的な. 〜 flat 法線的に平坦な; 正規に平坦な.

normed *a.* ノルムを備えた.
〜 linear space ノルム線形空間. 〜 ring ノルム環.

normic *a.* ノルムの.

～ form ノルム形式.

north *n.* 北. *a.* 北の.

～(ern) hemisphere 北半球. ～ pole 北極.

Northcott, Douglas Geoffrey (1916–2005) ノースコット.

northern *a.* 北の; 北向きの.

not *ad.* …でない.

～ greater [less] than 大きく[小さく]ない, 以下[以上].

nota bene (*L.* =mark well.) 注意; 注意せよ.

notation *n.* 記号; 記号法, 表示法, 記法; 記数法.

abridged ～ 略記法[記号]. binary ～ 二進法. decimal ～ 十進法. factorial ～ 階乗記号. positional ～ 位取り記数法. scale of ～ 記数法(の基). system of ～ 記数法; 記号法.

nothing *n.* 無; 零.

notice *vt.* 注意する.

notion *n.* 観念; 概念.

nought *n.* 零; 無. =naught.

nowhere *ad.* どこにも…ない.

～ dense (いたるところ)疎の, 全疎の.

NP non-deterministic polynomial time 非決定性多項式時間(計算可能な問題のクラス).

～ complete NP 完全. ～ hard NP 困難.

nu *n.* ニュー. ギリシア語アルファベットの 13 番目の文字【N, ν】.

nuclear *a.* 核の.

～ operator 核(型)作用素.

null [nəl] *a.* 零の, 空の. *n.* 零.

～ boundary 零境界. ～ complex 零複体. ～ distribution 帰無分布. ～ event 空事象. ～ function 零関数. ～ hypothesis 帰無仮説. ～ recurrent 零再帰の. ～ sequence 零列. ～ set 空集合. ～ space 零空間. ～-system 零系.

null-homotopic *a.* 零にホモトープな.

nullity *n.* 無; 無であること, ナリティ; 退化次数.

number *n.* 数; 数字; 個数; 号.

absolute [abstract] ～ 無名数. acceptance ～ 合格判定個数. algebraic ～ 代数的数. algebraic ～ field 代数的数体; 代数体. Betti ～ ベッチ数. cardinal ～ 基数; 濃度. Cayley ～ ケーリー数. chromatic ～ 染色数, 彩色数. clique ～ クリーク数. complex ～ 複素数. composite ～ 合成数. compound ～ 複名数, 諸等数. condition ～ 条件数. cubic ～ 立方数. deficient ～ 不足数, 輸数. imaginary ～ 虚数. integral ～ 整数. irrational ～ 無理数; 不尽根数. Lebesgue ～ ルベーグ数. Ludolf ～ ルドルフ数, 円周率. negative ～ 負(の)数. ～ field 数体. ～ line 数直線. ～ of cases 場合の数. ～ π [pi] 数「パイ」, 円周率. ～ theory (整)数論. ～ vector 数ベクトル. opposite [opposed] ～ 反数. ordinal ～ 序数. positive ～ 正(の)数. prime ～ 素数. rational ～ 有理数. real ～ 実数. round(-ed) ～ 丸められた数. square ～ 平方数. theory of ～s 整数論, 数論. transcendental ～ 超越数. transfinite ～ 超限数. wave ～ 波数. whole ～ 整数.

number-theoretic *a.* (整)数論的な.

～ function (整)数論的関数.

numeral *a.* 数の. *n.* 数字; 数詞.

Arabic ～ アラビア数字. Roman ～ ローマ数字.

numerary *a.* 数の.

numeration *n.* 計数; 計算(法); 記数法.

～ system 記数法. ～ table 数表. Roman ～ ローマ記数法; 命数法.

numerator *n.* (分数の)分子. ⇨ denominator.

partial ～ 部分分子.

numeric *n.* 数; 分数.

a. =numerical.

numerical *a.* 数の; 数値の; 数を表す, 数字の.

～ analysis 数値解析. ～ calculation, ～ calculus, ～ computation 数値計算. ～ coefficient 数値[字]係数. ～ control 数値制御. ～ differentiation 数値微分. ～ equation 数値[字]方程式. ～ integration 数値積分. ～ method 数値解法. ～ range 数域. ～ solution 数値解(法). ～ value 数値; 絶対値. ～ vector 数ベクトル.

numerically *ad.* 数値的に.
 ～ equivalent 数値的に同値な.
numero *n.*（*L.* ＝number.）号，番.
numerous *a.* 多数の.
nutation *n.* 章動.

O

object *n.* 対象; 目標物; 目的.
control ~ 制御対象. final ~ 終対象. group ~ 群対象. initial [cofinal] ~ 始対象. ~ domain 対象領域. ~ variable 対象変数. quotient ~ 商対象.

objective *a.* 目的の; 客観的な. *n.* 目的; 目的物.
~ expression 目的式. ~ function 目的[目標]関数.

objectivity *n.* 客観性.

oblate *a.* 扁平な, 扁球の.
~ spheroid 扁平回転楕円体[面]; 偏球.

oblique *a.* 斜めの, 傾いた, 斜角の; 斜面の.
~ angle 斜角【鋭角又は鈍角】. ~ axis 斜交軸. ~ circular cone 斜円錐. ~ (Cartesian) coordinates 斜交座標. ~ intersection 斜交. ~ line 斜線. ~ perspective 有角透視図. ~ prism 斜角柱. ~ projection 斜投影[象]. ~ symmetry 斜対称(変換). ~ triangle 非直角三角形.

obliqueness *n.* 傾角; 傾斜.

oblong *n., a.* 長方形(の), 矩形(の); 長扁の; 長円[楕円]形(の).
~ spheroid 長偏球.

observation *n.* 観察; 観測.
direct [indirect] ~ 直接[間接]観測. ~ equation [vector] 観測方程式[ベクトル].

observe *vt.* 観察する; 観測する; 注意する.
~d value 観測値.

obsolescence *n.* 廃退, 退化; 萎縮.
~ loss 陳腐化損失.

obstruction *n.* 障害, 妨害.
~ cocycle 障害(的)双対輪体. ~ to extension 拡張への障害. primary [secondary] ~ 第一[二]障害. theory of ~s 障害理論.

obtain *vt.* 得る; 達成する. *vi.* 広まっている; 成り立っている.

obtuse *a.* 鈍角の.
~ angle 鈍角. ~ triangle 鈍角三角形.

obverse *n.* (コインなどの)表; 換質命題.

obvert *vt.* (命題を)換質する.

obvious *a.* 明らかな, 明白な.
~ reason 明白な理由.

obviously *ad.* 明らかに.

occur *vi.* 起こる, 生ずる, 発生する, 現れる.

OC curve Operating Characteristic curve. OC 曲線, 検査特性曲線.

oct-, octa-, octo- *pref.*「八」の意.

octagon *n.* 八角[辺]形.

octagonal *a.* 八角[辺]形の.

octahedral *a.* 八面体の; 八面の.
~ group (正)八面体群.

octahedron (*pl.* -roes, -dra) *n.* 八面体.

octal *a.* 八進法(の).
~ notation 八進法.

octant *n.* 八分円.

octave *n.* オクターブ.

octenary *a.* 八進法の.

octillion *n.*【米・仏】千の九乗;【英・独】千の十六乗.

octonary *a.* 八進法の. *n.* 八進数.

octonion *n.* 八元数.

octuple *a.* 八倍の; 八重の. *n.* 八倍. *vt.* 八倍にする.

odd *a.* 奇数の, 奇の; 奇妙な. ⇨ even.
~ function 奇関数. ~ number 奇数. ~ permutation 奇置換, 奇順列. ~ point 奇点. ~ triangle 奇三角形.

odds *n.* オッズ.
〜 ratio オッズ比.

offprint *n.* (雑誌・論集などに掲載した論文の)抜き刷り, 別刷.

ogive *n.* 累積分布[累積度数]曲線図.

-oid *suf.* 「様な(もの), 状の(もの)」の意: cycloid, Fuchsoid.

omega *n.* オメガ. ギリシア語アルファベット最後[24番目]の文字【Ω, ω】.

omicron *n.* オミクロン. ギリシア語アルファベットの15番目の文字【O, o】.

omission *n.* 切捨て; 省略.
〜 of fractions (端数の)切捨て.

omit *vt.* 省く; 抜かす; 切捨てる.

one *a.* 一つの; 一方の, *n.* 一; 一つ.
〜 third 三分の一.

one-parameter *n.*, *a.* 一パラメータ(の).
〜 family 一パラメータ族. 〜 group 一パラメータ群.

one-point *n.*, *a.* 一点(の).
〜 compactification 一点コンパクト化.

one-sided *a.* 単側の; 片側の.
〜 Chi-square test 片側カイ二乗検定. 〜 ideal 片側イデアル. 〜 limit 片側極限. 〜 stable process 片側安定過程. 〜 surface 単側曲面. 〜 test 片側検定.

one-to-one *a.* 1対1 [一対一]の.
〜 correspondence 1対1 [一対一]対応. 〜 mapping 1対1 [一対一]写像.

one-valued *a.* 一価の.
〜 function 一価関数.

one-way *n.*, *a.* 単向(の).

onto *prep.* …の上に.
〜 (-)mapping 上への写像, 全射.

open *a.* 開いた, 開いている; 未決定の. ⇨ closed.
〜 arc 開弧. 〜 ball 開球. 〜 base 開基. 〜 circle, 〜 circular disc [disk] 開円板. 〜 cover(ing) 開被覆. 〜 game 開放ゲーム. 〜 interval 開区間. 〜 kernel 開核. 〜 manifold 開多様体. 〜 mapping 開写像. 〜 neighbo(u)rhood 開近傍. 〜 problem, 〜 question 未解決の問題. 〜 Riemann surface 開リーマン面. 〜 set 開集合. 〜 sphere 開球. 〜 surface 開曲面; 開いた面. semi-〜 半開の.

operand *n.* 被演算数; オペランド.

operate *vi.* 作用する. *vt.* 作用させる.

operating *a.* 操業の, 稼動の, オペレーティングの.
〜 characteristic 動作[検査]特性. 〜 characteristic curve OC曲線. 〜 cost 操業費用. 〜 time 稼動時間.

operation *n.* 運算, 演算, 算法; 操作, 動作; 作用; オペレーション, 命令.
algebraic [analytic] 〜 代数[解析]的演算. binary 〜 二項演算. elementary 〜 基本演算. four 〜s 四則(演算), 加減乗除. ladder 〜 昇降演算. 〜 on set 集合への作用. 〜s research オペレーションズ・リサーチ【OR】.

operational *a.* 演算の; 作用の.
〜 calculus 演算子法. 〜 research オペレーショナル・リサーチ【OR】.

operator *n.* 作用素, 演算子; 操作員, オペレータ.
adjoint 〜 随伴作用素, 共役作用素. boundary 〜 境界作用素. coboundary 〜 双対境界作用素. conjugate 〜 共役作用素. degeneracy 〜 退化作用素. differential 〜 微分作用素[演算子]. dispersive 〜 散逸作用素. integral 〜 積分作用素. inverse 〜 逆作用素. Laplace 〜 ラプラス作用素[演算子]. 〜 algebra 作用素代数. 〜 domain 作用域. 〜 isomorphism 作用同形[型]. 〜 norm 作用素ノルム. projection 〜 射影作用素. ring of 〜s; 〜 algebra 作用素環. shape 〜 型(かた)作用素, シェイプ作用素, 形(かたち)作用素. step-down [-up] 〜 下降[上昇]演算子. translation 〜 移動演算子.

opportunity *n.* 機会.
〜 cost 機会費用.

oppose *vt.* 対立させる.
〜d number 反数.

opposite *a.* 向かい合った; 向い側の; 逆の; 正反対の.
〜 dihedral angle 対稜二面角. 〜 number 反数. 〜 orientation 逆の向き. 〜 point 対点. 〜 side 反対側, 向い側; 対辺. 〜

sign 反対[逆]符号. vertically ~ angle(s) 対頂角.

optical *a.* 光学の.
~ axis 光軸. ~ distance 光学的距離.

optics *n.* 光学.
geometric(al) ~ 幾何光学. wave ~ 波動光学.

optimal *a.* 最善の, 最適の.
~ allocation 最適割当[配分]. ~ control 最適制御. ~ inventory 最適在庫. ~ linear predictor 最良線形予報値. ~ locus 最適軌道. ~ policy 最適政策. ~ solution 最適解. ~ stopping problem 最適停止問題. ~ stopping-time problem 最適停止時刻問題. ~ strategy 最適戦略.

optimality *n.* 最適性.
principle of ~ 最適性の原理.

optimization *n.* 最適化.
combinatorial ~ 組合せ最適化. constrained ~ 条件つき最適化, 制約付き最適化. discrete ~ 離散最適化. global ~ 大域的最適化. ~ theory 最適化理論.

optimum (*pl.* -ma, -mums) *n.* 最適.
~ allocation 最適割当[配分]. ~ estimate 最適推定量. ~ programming 最適計画(法). ~ solution 最適解.

optional *a.* 任意の; 随意の.
~ sampling 任意抽出.

OR Operations Research, Operational Research.

or *conj.* または; すなわち. *n.* 論理和【∨, +】.
and/~ 非排他的論理和. exclusive ~ 排他的論理和.

orbit *n.* 軌道.
~ closure 軌道閉包. ~ determination 軌道決定. ~ space 軌道空間. periodic ~ 周期軌道.

orbital *a.* 軌道の.
~ curve 軌道曲線. ~ plane 軌道面. ~ stability 軌道安定性.

order *n.* 順序; 順位; 位数; 階; 次数; 整環; 度; 命令; 発注. *vt.* 順序づける; 命令する.
alphabetical ~ アルファベット順. axiom(s) of ~ 順序公理. depression of ~ 階数低下. derivative of the second ~ (第)二階[次]導関数. infinitesimal of higher [lower] ~ 高[低]位の無限小. left ~ 左整環. lexicographical ~ 辞書式順序. lower ~ 低位数. maximal ~ 極大整環; 最大位数. of higher [lower] ~ 高[低]位の. ~ condition 位数条件. ~ function 位数関数. ~ ideal 位数イデアル. ~ isomorphism 順序同形[型]. ~ limit 順序極限. ~ of contact 接触位数. ~ of element 元の位数. ~ of group 群の位数. ~ relation 順序関係. ~ statistic 順序統計量. ~ topology 順序位相. ~ type 順序型. partial ~ 半順序. reverse ~ 逆の順序. right ~ 右整環. same ~ 同順; 同位(数). semi-~ 半[準]順序. total ~ 全順序.

ordered *a.* 順序づけられた.
linearly ~ set 線形順序集合, 全順序集合. ~ basis 順序基底. ~ complex 順序複体. ~ field 順序体. ~ group 順序群. ~ module 順序加群. ~ pair 順序対. ~ product 順序積. ~ set 順序集合. ~ sum 順序和. partially [partly] ~ 半順序の. totally ~ additive group 全順序加群. well-~ set 整列集合.

ordering *n.* 順序づけること.
~ cycle [interval] 発注間隔.

ordinal *n.* 序数, 順序数. *a.* 序数の; 順序の.
constructive ~ 構成的順序数. ~ number (順)序数. ~ scale 順序尺度.

ordinary *a.* 通常の; 正常の.
~ curve 通常曲線. ~ differential equation 常微分方程式. ~ double point 結節点. ~ helicoid 常螺旋面. ~ helix 常螺線. ~ point 通常点. ~ scale 普通尺. ~ singularity 通常[通性]特異点. ~ solution 正常解.

ordinate *n.* 縦座標; 縦軸. ⇨ abscissa.
~ set 縦線集合.

or else *conj.* さもなければ.

Oresme, Nicole (1323?–1382) オレーム.

orient *vt.* 方向を定める.

orientable *a.* 可符号の, 向きが付けられる.

~ manifold 可符号(の)[向きづけ可能な]多様体. ~ surface 可符号曲面.

orientation *n.* 方位; 方向づけ, 向き.
negative [positive] ~ 負[正]の向き. ~ preserving [reversing] mapping 向きを保つ[変える]写像.

oriented *a.* 有向の, 向きづけられた. ⇨ directed.
coherently ~ 同調に向きづけられた. ~ arc 有向弧. ~ circle 有向円. ~ distance 有向距離. ~ manifold 向きづけられた多様体. ~ plane 有向平面. ~ segment 有向線分. ~ simplex 有向単体. ~ surface 有向曲面. ~ system of coordinate neighborhoods 向きづけられた座標近傍系. ~ volume 有向体積.

origin *n.* 原点; 起点, 始点; 起源.

original *a.* もとの; 原物の; 独創的な. *n.* 原物; 原文.
~ paper 原論文, 原著. ~ permutation 基準順列. ~ point 始点, 起点. ~ table 原表.

orthant *n.* 象限.
positive ~ 正の象限.

orthic *a.* 垂線の.
~ triangle 垂足三角形.

orthocenter *n.* 垂心.

orthocentric *a.* 垂心の.
~ tetrahedron 垂心四面体.

orthocomplement *n.* 直交補空間.

orthogonal *a.* 直角の; 直交の.
~ array 直交配列. ~ axes 直交軸. ~ axonometric projection 正軸測投影. ~ basis 直交基. ~ complement 直交補空間. ~ component 直交成分. ~ coordinates 直交座標. ~ cross section 直切面. ~ decomposition 直交分解. ~ frame 直交枠. ~ group 直交群. ~ layout 直交配列表. ~ matrix 直交行列. ~ polynomials 直交多項式. ~ projection 正射影, 直交射影, 正射影作用素; 正投影[象]. ~ series 直交級数. ~ system 直交系. ~ tetrahedron 直稜四面体. ~ trajectory 直交切[截]線. ~ transformation 直交変換. triply ~ 三重直交の.

orthogonality *n.* 直交性.
~ relation 直交関係.

orthogonalization *n.* 直交化.
~ theorem 直交化定理.

orthogonally *ad.* 直角に; 直交して.
cut [intersect] ~ 直交する.

orthograph *n.* 正射影.

orthographic *a.* (3 次元空間における平面への)正射影の.
~ projection 正射影.

orthographical =orthographic.

orthography *n.* (3 次元空間における平面への)正射影法.

orthonormal *a.* 正規直交の. [orthogonal and normal.]
~ basis 正規直交基. ~ system 正規直交系.

orthonormalization *n.* 正規直交化.
Gram-Schmidt ~ グラム・シュミットの正規直交化法.

orthonormalize *vt.* 正規直交化する.

orthopolar *a.* 直極の.
~ circle 直極円.

orthopole *n.* 直極点.

oscillate *vi.* 振動する.
~ting sequence 振動数列.

oscillation *n.* 振動; 振幅. ⇨ vibration.
autonomous ~ 自励振動. damped ~ 減衰振動. eigen-~ 固有振動. forced ~ 強制振動. free ~ 自由振動. harmonic ~ 調和振動. limit ~ 極限振幅. ~ theorem 振動定理. proper ~ 固有振動. relaxation ~ 緩和振動.

oscillatory *a.* 振動的な.
~ sequence [series] 振動数列[級数].

osculate *vt., vi.* 接触する, 共通点をもつ.

osculating *a.* 接触している.
~ circle 接触円. ~ conics 接触円錐曲線. ~ plane 接触平面. ~ sphere 接触球.

osculation *n.* 接触.
best ~ 最大接触. ~ of *n*th order *n* 位の接触. point of ~ 接点.

Osgood, William Fogg (1864–1943) オスグッド.

Ostrogradski, Michail Wassiljevič (1801–

1862) オストログラドスキ.

OSTROWSKI, Alexandre Markovič (1893-1986) オストロフスキ.

other *a.* 他の; (2つのうち)もう一つの. *pron.* 他のもの, (2つのうちで)もう一方のもの.
each ~ 互いに, 相互に.

otherwise *ad.* さもなければ; 他の点では.

OUGHTRED, William (1574-1660) オートレッド.

-ous *suf.* 「…性の, …様の」の意の形容詞をつくる: autonomous, continuous.

outer *a.* 外の; 外部の. ⇨ exterior; inner.
~ area 外面積. ~ automorphism 外部自己同型[形]. ~ capacity 外容量. ~ control limit 外側管理限界. ~ function 外関数. ~ measure 外測度. ~ point 外点. ~product 外積. ~ volume 外体積.

outflow *n.* 流出. ⇨ inflow.

outlier *n.* 外れ値.

outline *n.* 要旨; 概要; 梗概; アウトライン.

out of *ad.* の外で; はずれて.
~ control 管理はずれの.

output *n.* 出力, アウトプット. ⇨ input.
~ area 出力域. ~ data 出力データ. ~ rate 消費速度.

outside *n.* 外側, 外部. *a.* 外側の, 外部の. *ad.* 外に. *prep.* の外に. ⇨ inside.
~ diameter 外径.

outward *a.* 外を向いた; 外への. *ad.* 外へ. *n.* 外部, 外界. ⇨ inward.
~normal 外向き(の)法線.

oval *a.* 卵形線の; 長円形の. *n.* 卵形線; 長円体.
inrevolvable ~ 内転卵形. mean ~ 平均卵形(線).

ovaloid *n.* 卵形面.

over *prep.* の上に; をこえて; に関して. *ad.* 上に; 越えて; 向う側へ.
algebra ~ K K上の多元環. a little ~ 10 percent 一割強. a ~ b b分の$a\left[\frac{a}{b}\right]$. ~(-)relaxation 加速緩和(法).

overconverge *vi.* 超収束する.

overconvergence *n.* 超収束. ⇨ ultraconvergence.

overconvergent *a.* 超収束の.
~ power series 超収束ベキ級数.

overcross *vi.* 上交差[叉]する, 上で交わる.
~ing point 上交差[叉]点.

overdetermined *a.* 優決定の, 過剰決定の.
~ system 優決定系, 過剰決定系.

overdispersion *n.* 超過変動; 過分散.

overfield *n.* 拡大体.

overflow *n.* 桁あふ[溢]れ, オーバーフロー. *vi.* 桁あふれする; あふれる.
line ~ 行あふれ.

overlap *n.* 重複.

over(-)relaxation *n.* 加速緩和(法).
successive ~ 逐次緩加速法[SOR].

overring *a.* 拡大環.

ovoid *n.* 卵形体. *a.* 卵形体の.

P

p. (*pl.* pp.) page; part.
pack *n.* パック. *vt.* パックする, 詰め込む, 充填する.
packing *n.* パッキング, 詰め込み, 充填.
closest ~ 最密充填. ~ density 充填密度.
***p*-adic** *a.* p進の.
~ fraction p進分数. ~ integer p進整数. ~ (number) field, ~ system p進体. ~ valuation p進付値.
Painlevé, Paul (1863–1933) パンルベ.
pair *n.* 対; 組; 一対. *vt.* 対にする.
ordered ~ 順序対. ~ed comparison 一対比較(法). ~ test 二点比較法. ~wise disjoint 対ごとに素な. two ~s 二対, 二組.
pairing *n.* 対になる[する]こと.
Palais, Richard S. (1931–) パレ.
Paley, Raymond Edward Alan Christopher (1907–1933) ペーリー.
panharmonic *a.* 汎調和の.
~ series 汎調和級数.
panmixia *n.* 汎混合.
pantograph *n.* パント[タ]グラフ; 写図[拡図, 縮図]器.
paper *n.* 紙; 論文.
joint ~ 共著(の)論文. logarithmic ~ 対数方眼紙. originat ~ 原論文, 原著. probability ~ 確率(方眼)紙. ruled ~, section ~ 方眼紙. semi-logarithmic ~ 半対数方眼紙. stochastic ~ 推計紙.
Pappus [**Pappos**] of Alexandria (320 頃) アレクサンドリアのパップス.
papyrus (*pl.* -ri) *n.* パピルス【写本】.
parabola *n.* 放物線, 抛物線.
asymptotic ~ 漸近放物線. confocal ~s 共焦放物線.
parabolic *a.* 放物線の; 放物型の, 放物的な.
~ coordinates 放物線座標. ~ cusp 放物的尖点. ~ cylinder 放物柱[筒]. ~ domain [region] 放物的領域. ~ equation 放物型方程式. ~ geometry 放物的幾何学. ~ point 放物的点. ~ quadrics 放物型二次曲面. ~ spiral 放物螺[渦]線. ~ transformation 放物的変換. ~ type 放物型.
parabolical ＝parabolic.
parabolicity *n.* 放物型性.
paraboloid *n.* 放物面[体].
elliptic [hyperbolic] ~ 楕円[双曲]放物面.
~ of revolution 回転放物面.
paraboloidal *a.* 放物面の.
~ coordinates 放物面座標.
paracompact *a.* パラコンパクトな.
~ space パラコンパクト空間.
paradox *n.* 逆説, パラドックス; 逆理, 背理; 二律背反.
~ of Russell, Russell's ~ ラッセルの逆理.
paradoxical *a.* 逆説的な, 矛盾した.
parallactic *n.* 変位の; 視差の.
parallax *n.* 視差.
parallel *a.* 平行な; 並列の. *ad.* (と)平行して. *n.* 平行線; 緯(度)線; 類似; 並列.
axiom of ~s [~ lines] 平行線公理. ~ computation 並列計算. ~ coordinates 平行座標. ~ curves 平行曲線. ~ displacement 平行移動. ~ lines 平行線. ~ of latitude 緯線. ~ perspective figure 平行透視図. ~ projection 平行射[投]影. ~ ring 平行環. ~ surface 平行曲面. ~ translation 平行移動.
parallelepiped [pæ̀rəlèləpáipid, -ped]

n. 平行六面体.
rectangular ～ 直六面体.

parallelism *n.* 平行; 平行性.
absolute ～, tele ～ 遠隔平行性. angle of ～ 平行角.

parallelizable *a.* 平行化可能な，平行性をもつ.

parallelogram *n.* 平行四辺形.
～ identity 中線[平行四辺形]定理. ～ of forces 力の平行四辺形. period ～ 周期平行四辺形.

parallelopiped ＝parallelepiped.

parallelotope *n.* 平行体.

parameter *n.* 媒介変数，パラメータ，助変数，径数; 母数.
canonical ～ 標[規]準助変数. interval of ～ パラメータ区間. one [two]-～ family 一[二]パラメータ族. ～ disc [disk] パラメータ円板. ～(-) free 母数によらない. ～(s) of connection 接続の径数. ～ space 母数空間. population ～ 母数. system of ～s パラメータ系. transformation ～ 変換母数. uniformizing ～ 一意化(媒介)変数.

parametric *a.* 媒介変数の，助変数の，径数の; 母数の.
～ equation 媒介変数[パラメータ]方程式. ～ function 母数関数. ～ representation 媒介変数[パラメータ]表示.

parametrix *n.* パラメトリックス.

parametrization *n.* パラメータ(表示)を定めること.

parametrized *a.* パラメータ[径数]のつけられた.
～ curve [surface] パラメータのつけられた曲線[曲面].

parenthesis (*pl.* -ses) *n.* 括弧【(　)】.

parity *n.* パリティ; 奇偶性，偶奇性.
～ check 奇偶検査，パリティ・チェック. ～ transformation パリティ変換.

part *n.* 部分，一部;【無冠詞で】整除数; (*pl.*) 部分分数.
adjacent ～ 隣接部分. common ～ 共通部分. decimal ～ 小数部(分). imaginary ～ 虚(数)部(分). integral ～ 整数部分. integration by ～s 部分積分. principal ～ 主要部. real ～ 実(数)部(分). singular ～ 特異部分.

partial *a.* 部分の，部分的な; 偏-.
～ confounding 部分交絡(法). ～ correlation 偏相関. ～ denominator 部分分母. ～ derivative 偏導関数. ～ differential coefficient 偏微分係数. ～ differential equation 偏微分方程式. ～ differentiation 偏微分. ～ fraction 部分分数. ～ function 部分関数. ～ numerator 部分分子. ～ order 半順序. ～ product 部分積. ～ quotient 部分商. ～ regression coefficient 偏回帰係数. ～ sum 部分和.

partially *ad.* 部分的に.
～ balanced design 一部釣合い型計画法. ～ differentiable 偏微分可能な. ～ differentiate 偏微分する. ～ ordered set 半順序集合.

particle *n.* 粒子; 極(小)量; 質点.
elementary ～ 素粒子. material ～ 質点. system of ～s 質点系.

particular *a.* 特殊な，特殊の，特別な; 個別の; 特定の. *n.* 事項.
in ～ 特に. ～ solution 特殊解.

partition *n.* 分割; 区分け.
number of ～s 分割数. ～ function 分配関数. ～ of number 数の分割. ～ of unity 一[単位]の分割，一[単位]の分解. ～ problem 分割問題.

partout *ad.* (*F.* ＝everywhere.) いたるところ.
presque ～ 殆んどいたるところ[p.p.].

Pascal, Blaise (1623-1662) パスカル.

pass *vi., vt.* 通る; 横切る. *n.* 通過; 合格; 峠.
mountain ～ theorem 峠の定理. ～ through 通る，通過する.

passim *ad.* (*L.*) いたるところに; 諸所に.

passing *n.* 合格，通過，経過; 消滅，終わり.

passive *a.* 受動的な; 消極的な.
～ boundary point 受動境界点.

patch *n.* 細片. *vt.* つぎあわせる.
surface ～ 曲面片.

path *n.* 経路，路，道; 軌道.
asymptotic ～ 漸近路. closed ～ 閉路[道].

geometry of ~s 道の幾何学. ~ algebra 道多元環. ~ of contact 接触点の軌跡. ~ of integration 積分路. ~ space 道の空間. Stolz ~ シュトルツの路.

path-dependent *a.* 試行列に従属な.

path-independent *a.* 試行列に独立な.

path-integral *n.* 経路積分.

pathological *a.* 病(理学)的な; 野生的な.
~ example 病的な例.

path-polygon *n.* 軌道多角形.

pattern *n.* 模型; パターン.
~ analysis パターン解析. ~ recognition パターン認識.

pay-back *n.* 返済.
~ period 回収[返済]期間.

pay-off *n.* ペイオフ; 利得; 支払い.
~ function 利得関数. ~ matrix 支払行列, 利得行列.

pd pushdown.

pe ペー【℘】.
~ function ペー関数.

peak *n.* 峰, 峯.
~ point 峰点. ~ set 峰集合.

Peano, Giuseppe (1858-1932) ペアノ.

pearl *n.* 真珠.
~ curve 真珠曲線.

Pearson, Karl (1857-1936) ピアソン.

pedal *n.* 垂足線; 垂足面. *a.* 垂足線[面]の.
~ circle 垂足円. ~ curve 垂足曲線. ~ transformation 垂足変換. ~ triangle 垂足三角形.

pedestal *n.* 足.

Pellian *a.* ペルの.
~ equation ペル方程式.

pencil *n.* 束; 束線, 線束.
axial ~ 平面束. harmonic ~ 調和線束. linear ~ 線形束. ~ of conics 円錐[二次]曲線束. ~ of lines (直)線束, 束線. ~ of planes (平)面束, 束平面. ~ of quadrics 二次曲面束.

pendulum *n.* 振子.
circular ~ 円振子. cycloid ~ サイクロイド振子. reversible ~ 可逆振子. simple ~ 単振子. spherical ~ 球面振子.

penetration *n.* 貫通.

penta- *pref.* 「五」の意.

pentad *n.* 五.

pentadecagon *n.* 十五角[辺]形.

pentagamma *n.* ペンタガンマ.
~ function ペンタガンマ関数.

pentagon *n.* 五角形; 五辺形.
regular ~ 正五角[辺]形.

pentagonal *a.* 五角(形)の; 五辺形の.
~ number 五角数.

pentagram *n.* ペンタグラム, 星形.

pentahedral *a.* 五面の; 五面体の.

pentahedron (*pl.* -rons, -dra) *n.* 五面体.

penta(-)spherical *a.* 五球の.
~ coordinates 五球座標.

per *prep.* (*L.*) につき, 当り.
~ annum 一年につき[当り]. ~ capita 頭割りで. ~ cent 百分率. ~ centum 百につき[当り].

percent *n.* 百分率; パーセント【%】. *a.* 百分の. *ad.* 百につき.
100 ~ inspection 全数検査. ~ defective 不良(百分)率.

percentage *n.* 百分率; 割合.
~ error 百分率誤差.

percentile *n.* 百分位数, パーセント点.

perfect *a.* 完全な.
~ correlation 完全相関. ~ cube 完全立方. ~ field 完全体. ~ fluid 完全流体. ~ matching 完全マッチング. ~ number 完全数. ~ set 完全集合. ~ square 完全平方.

perfectly *ad.* 完全に.
~ normal space 完全正規空間. ~ separable 完全可分な.

perform *vt.* (約束, 義務などを)果たす.

perigon *n.* 周角【360°】.

perimeter *n.* 周囲; 周辺; 周辺の長さ.

period *n.* 周期; 循環節.
busy ~ 稼動期間. fundamental ~ 基本周期. imprimitive ~ 複循環節. ~ inequality 周期不等式. ~ matrix 周期行列. ~ parallelogram 周期平行四辺形. ~ relation 周期関係式. ~ strip 周期帯. primitive ~ (s) 基本周期.

periodic *a.* 周期的な; 周期の.

almost ~ function [system] 概周期関数[系]. doubly ~ 二重周期の. ~ function 周期関数. ~ group 周期群. ~ orbit 周期軌道. ~ solution 周期解. ~ system 周期系. simply ~ 単周期の.

periodical *a.* 定期刊行の; ＝periodic. *n.* 定期刊行物.

periodicity *n.* 周期性; 周期数.
~ modulus 周期母数.

periodogram *n.* ペリオドグラム.

peripheral *a.* 周辺の.
~ singularity (収束円)周上の特異点.

periphery *a.* 周, 周囲; 円周.

permanence *n.* 不変(性).
~ of functional relation(s) 関数関係の不変(性). ~ of sign 符号の継続(性).

permanent *n.* 不変な.

permil *n.* 千分率【‰】.

permillage *n.* 千分率.

permissible *a.* 許容できる; さしつかえない程度の.

permit *vt., vi.* 許す.

permutable *a.* 並べかえできる; 交換できる.
~ functions 可換関数.

permutation *n.* 順列; 置換.
circular ~ 円順列. cyclic ~ 巡回置換, 巡換. even ~ 偶順列[置換]. identity ~ 恒等置換. inverse ~ 逆置換. necklace ~ 数珠順列. odd ~ 奇順列[置換]. original ~ 基準順列. ~ group 置換群. ~ matrix 置換行列. ~ representation 置換表現. repeated ~ 重複順列.

permute *vt.* 並べかえる; 交換する.

perpendicular *a.* 直角をなす【to】; 垂直の, 直立した. *n.* 垂線; 垂直面; 垂直.
be ~ to 直交する. common ~ 共通垂線. ~ at midpoint 垂直二等分線. ~ bisector 垂直二等分線[面]. ~ displacement 垂直変位.

perpendicularity *n.* 垂直(性).

perpendicularly *ad.* 垂直に.

Perron, Oscar (1880–1975) ペロン.

perspective *n.* 透視画[図]; 投象; 遠近(画)法. *a.* 配景的な; 透視の.
angular ~ 有角透視図. central ~ drawing 中心投象画法. oblique ~ 有角透視図. ~ correspondence 配景対応. ~ mapping 配景写像. ~ position 配景の位置. ~ projection 透視図(法). ~ relation 配景的関係. ~ surface 透視面. ~ transformation 配景変換.

perspectivity *n.* 配景.
center of ~ 配景の中心.

PERT *n.* Program Evaluation and Review Technique. パート.
~/COST パート・コスト.

perturbation *n.* 摂動.
~ method, method of ~ 摂動法. regular ~ 正則摂動. secular ~ 永年摂動. singular ~ 特異摂動.

Petrovskiĭ, Ivan Georgievič (1901–1973) ペトロフスキ.

Pfaff, Johann Friedrich (1765–1825) パッフ.

Pfaffian *a.* パッフの.
~ (form) パッフ形式.

phase *n.* 相, 位相.
initial ~ 初期位相. ~ angle 位相角. ~ average 相平均. ~ group 相群. ~ shift 位相のずれ. ~ space 相空間. ~ velocity 位相速度.

phasic *a.* 相の; 位相の.

phenomenon (*pl.* -mena) *n.* 現象
Gibbs ~ ギブズ現象. natural [social] ~ 自然[社会]現象.

phi *n.* ファイ, フィー. ギリシア語アルファベットの 21 番目の文字【Φ, φ, ϕ】.
~ function ファイ関数.

Phragmén, Lars Edvard (1863–1937) フラグメン.

physical *a.* 物質の; 自然な; 物理的な; 物理の.
~ distribution management 物的流通管理.
~ mathematics 物理数学.

physico- 「物理」の意の結合辞.
~ mathematical society 数学物理学会.

physics *n.* 物理学.
mathematical ~ 数理物理学. theoretical ~ 理論物理学.

PI Programmed Instruction. プログラム学習.

pi *n.* パイ. ギリシア語アルファベットの16番目の文字【$\Pi, \pi; \varpi$】; 円周率.

PI-algebra *n.* PI [Polynomial Identity] 多元環【多項式恒等関係をもつ多元環】.

PICARD, Charles Emile (1856-1941) ピカール.

pico- *pref.* ピコ【10^{-12}】.

pictogram *n.* 絵画図表, 絵グラフ.

pictograph ＝pictogram.

picture *n.* 画像; 絵; 図.
attached ~ 付図. ~ plane 画面.

PID, P.I.D. Principal Ideal Domain. 単項[主]イデアル整域.

piece *n.* 片, 一片.
~ of curve [surface] 曲線[曲面]片.

piecewise *ad.* 区分的に.
~ continuous function 区分的(に)連続(な)関数. ~ linear 区分的線形の. ~ smooth 区分的に滑らかな.

pierce *vt., vi.* 貫通する.
~cing point 貫通点.

pilot *n.* 案内人. *a.* 試験的な.
~ survey 試験[予備]調査. ~ test 試験調査.

pinch *vt.* 一点に縮める; 締めつける.
~ed space 一点に縮められた空間. ~ing problem 締めつけの問題.

PINCHERLE, Salvatore (1853-1936) ピンケルレ.

PIR, P.I.R. Principal Ideal Ring. 単項イデアル環.

pitch *n.* 傾斜度, 勾配; 刻み, ピッチ.
~ surface ピッチ面. reduced ~ 準ピッチ.

pivot *n.* 枢軸, ピボット.
~ing, ~ operation ピボット[枢軸]演算.

pivotal *a.* 中枢の; 枢軸の.
~ element 中心要素. ~ line 要[かなめ]線.

pivoting *n.* ピボット[枢軸]演算.

place *n.* 場所; 位, 桁; 順序; 座.
decimal ~ 小数位. in ~ of の代りに. three ~ number; number of three ~s 三桁の数.

placement *n.* 位置.
~ problem 位置の問題.

plaid *n.* 格子縞.

plain *a.* はっきりした, 明白な; 平易な; 率直な; 平らな, 平坦な. *ad.* はっきりと; 平易に.

plainly *ad.* 分かり易く; 平易に; 明白に; 簡単に.

plan *n.* 平面図; 水平図; 計画.
indexed ~ 標高平面図.

planar *a.* 平面の; 平面的な; 単葉型の.
~ character 単葉型. ~ curve 平面曲線. ~ domain [region] 平面領域. ~ point 平面的(な)点. ~ polygon 平面多角形. ~ quadrilateral 平面四辺形.

PLANCHEREL, Michael (1885-1967) プランシュレル.

plane *n.* 平面; 面. *a.* 平面の.
complex ~ 複素平面. coordinate ~ 座標平面. cutting ~ 切除平面. diametral ~ 径面. Euclidean ~ ユークリッド平面. Gaussian ~ ガウス平面. normal ~ 法平面. osculating ~ 接触平面. ~ area 平面積. ~ coordinates 平面座標. ~ curve 平面曲線. ~ domain [region] 面分; 平面領域. ~ figure 平面図形. ~ geometry 平面幾何学. ~ graph 平面グラフ. ~ of symmetry 対称面. ~ trigonometry 平面三角法. ~ wave 平面波. principal ~ 主平面. projective ~ 射影平面. tangent ~ 接平面.

planet *n.* 惑星, 遊星.

planetary *a.* 惑星の, 遊星の.
~ motion 惑星運動.

planimeter *n.* プラニメーター, 面積計.

planning *n.* (手順)計画. *a.* 計画する.
~ mathematics 計画数学. ~ of experiment 実験の計[企]画.

plano-concave *a.* 平凹の【一面平らで一面凹の】.

plano-convex *a.* 平凸の【一面平らで一面凸の】.

plate *n.* (薄)板.
vibrating ~ 振動板.

platykurtic [plati'kərtik] *a.* 広幅の.

～ distribution 広幅[幅ひろ; 緩尖]分布.

plausible *a.* ありそうな，もっともらしい.

play *vi., vt.* 遊ぶ; 行う; 働く. *vt.* (役割を)果たす. *n.* 遊戯; 競技; 勝負事.

P.L.K. method Poincaré-Lighthill-Kuo method. P.L.K. 法.

plot *n.* プロット; 計画; 区画地; 図面. *vt.* 計画する; 図面をかく; 地図[グラフ]上にマークする.

box-whisker ～ 箱ひげ図. curve (-) ～ting 曲線をえがくこと.

PLÜCKER, Julius (1801-1868) プリュッカー.

plural *a.* 複数の; 二つ以上の. *n.* 複数，複数形. ⇨ singular.

pluri- *pref.* 「複数，複」の意.

pluriharmonic *a.* 多重調和な.

～ function 多重調和関数.

plurisubharmonic *a.* 多重劣調和な.

～ function 多重劣調和関数.

plus *prep.* プラス. *a.* プラスの，加号の; 正の. *n.* 正号，プラス記号; 正数. ⇨ minus.

～ and minus 正負. ～ sign 正の記号，プラス記号; 加号【+】.

-ply *suf.* 「-重」の意.

k～ transitive k 重推移的な. n～ connected n 重連結の. p～ symmetric p 重対称な.

POCHHAMMER, Leo (1841-1920) ポッホハンマー.

POINCARÉ, Jules Henri (1854-1912) ポアンカレ.

point *n.* 点; 小数点; 要旨.

accumulation ～ 集積点. boundary ～ 境界点. Brocard ～ ブロカール点. cluster ～ 集積点. concircular ～s 共円点. corner ～ 角点. corresponding ～s 対応点. critical ～ 臨界点，危点. double ～ 二重点. end [extreme; extremal] ～ 端点. exterior ～ 外点. fixed ～ 固定点，不動点; 固定小数点. harmonic range of ～s 調和列点. interior ～ 内点. isolated [isolate] ～ 孤立点. Lemoine ～ ルモアーヌ点. multiple ～ 重複点. non(-)singular ～ 非特異点; 単純点; 正則点. ～ at infinity 無限遠点. ～ derivation 点微分. ～ equation 点方程式. ～ estimation 点推定. ～ function 点関数. ～ graph 点グラフ. ～ hypersphere 点超球. ～ mass 点質量. ～ of condensation 凝集点. ～ of contact 接点. ～ of density 密集点. ～ of discontinuity 不連続点. ～ of inflection; inflection ～ 変曲点. ～ of intersection 交点. ～ of sight 視点. ～ of symmetry 対称点. ～ of tangency 接点. ～ range 点域. ～ sampling 点抽出(法). ～ sequence, sequence of ～s 点列. ～ set 点集合. ～ symmetry 点対称. ～ transformation 点変換. range of ～s 列点. regular [singular] ～ 正則[特異]点. saddle ～ 鞍点. set of ～s 点集合. stationary ～ 定常[停留]点. triple ～ 三重点. unit ～ 単位点. zero ～ 零点.

pointed *a.* 尖った，鋭い.

pointwise *a., ad.* 点別の[に]，各点ごとの[に].

～ convergent 各点収束の.

POISSON, Siméon Denis (1781-1844) ポアソン.

polar *a.* 極の; 極線の. *n.* 極線.

～ angle (極座標の)角座標. ～ axis 原線. ～ circle 極円. ～ cone 極錐. ～ conics 極円錐曲線. ～ coordinates 極座標. ～ curve 極曲線. ～ distance 極距離. ～ element 極要素. ～ equation 極方程式. ～ form 極形式. ～ line 極線. ～ plane 極平面. ～ projection 極射[投]影. ～ set 極集合. ～ singularity 極. ～ surface 極曲面. ～ system 極系. ～ tangential coordinates 接線極座標. ～ tetrahedron 極四面体. ～ triangle 極三角形. ～ vector 極性ベクトル.

polarity *n.* 極関係，極反(性); 極性.

polarization *n.* 偏り; 偏光; 偏極，分極，極化.

～ identity 極化恒等式.

polarize *vt., vi.* 偏光させる; 極性を与える.

～d half-plane [variety] 偏極半平面[多様体].

pole *n.* 極; 極点.

logarithmic ～ 対数的極. north ～ 北極. ～ of order n n 位の極. simple ～ 一位の極. south ～ 南極.

policy *n.* 政策，ポリシー.

optimal ~ 最適政策.

poly- *pref.*「多, 複」の意. ⇨ mono-.

Pólya, György (1887-1985) ポーヤ.

polyadic *a., n.* ポリアディック.

polyconic *a.* 多円錐の.
~ projection 多円錐投影[射], 多円錐図法.

polydisc, polydisk *n.* 多重円板.

polygamma *n.* ポリガンマ.
~ function ポリガンマ関数.

polygon *n.* 多角形, 多辺形.
circumscribed ~ 外接多角形. concave [convex] ~ 凹[凸]多角形. force ~ 力の多角形. frequency ~ 度数(分布)多角形. inscribed ~ 内接多角形. path-~ 軌道多角形. planar ~ 平面多角形. regular ~ 正多角[辺]形. summability ~ 総和多角形. twisted ~ 捻れ多角形.

polygonal *a.* 多角形の, 多辺形の; 多角の.
~ domain [region] 多角形領域. ~ line 折れ線, 屈折線. ~ number 多角数.

polyharmonic *a.* 多(重)調和な.
~ function 多(重)調和関数.

polyhedral *a.* 多面体の; 多面の.
~ angle 多面角. ~ cone 多面錐. ~ group 正多面体群. ~ projection 多面体図法.

polyhedron (*pl.* -rons, -dra) *n.* 多面体.
concave [convex] ~ 凹[凸]多面体. regular ~ 正多面体.

polymorphic *a.* 多形[型]な.

polynomial *a.* 多項式の. *n.* 多項式, 整式.
alternating ~ 交代(多項)式. approximate ~ 近似多項式. characteristic ~ 特性多項式. constant ~ 定数多項式. elementary alternating ~ 基本交代(多項)式. elementary symmetric ~ 基本対称多項式. generator ~ 生成多項式. homogeneous ~ 斉[同]次多項式. information ~ 情報多項式. irreducible ~ 既約多項式. Legendre('s) ~ ルジャンドル(の)多項式. minimal ~ 最小多項式. of ~ type 多項式型の. ~ approximation 多項式近似. ~ function 多項式関数. ~ of degree n n 次の多項式. ~ predicate 多項式述語. ~ representation 多項式表現. ~ ring 多項式環. ~ series 整[多項]式級数. ~ solution 多項式解. ~ theorem 多項定理. ~ time algorithm 多項式時間アルゴリズム. pseudo-~ 擬多項式. quadratic ~ 二次(の多項)式. reducible ~ 可約多項式. separable ~ 分離多項式. skew ~ 歪(わい)多項式. strongly ~ 強多項式の. symmetric ~ 対称多項式.

polynomially *ad.* 多項式的に.
~ convex 多項式的に凸な.

polytropic *a.* ポリトロープな.
~ curve ポリトロープ曲線. ~ equation ポリトロープ方程式.

Pompeiu, Dimitrie (1873-1954) ポンペイウ.

Poncelet, Jean Victor (1788-1867) ポンスレ.

Pontryagin, Lev Semenovic (1908-1988) ポントリャーギン.

pool *vt.* 共同計算にする.
~ed estimate こみにした推定.

population *n.* 母集団; 人口.
normal ~ 正規母集団. ~ characteristic 母集団特性値. ~ correlation 母(集団)相関. ~ distribution 母(集団)分布. ~ genetics 集団遺伝学. ~ mean [median] 母(集団)平均[中央値]. ~ parameter 母数. ~ problem 人口問題. ~ statistics 人口統計(学). ~ variance 母分散.

porism *n.* 不定命題, (ギリシア幾何学の)系, 系論.

porismatic *a.* 不定命題の.

porous *a.* 多孔質の.
~ media 多孔性媒質.

portion *n.* 一部, 部分.

position *n.* 位置.
general ~ 一般な位置. hyperbolic ~ 双曲的位置. ~ of similarity 相似の位置. ~ of symmetry 対称の位置. ~ vector 位置ベクトル. skew ~ 捻れの位置. standard ~ 標準の位置. symmetric ~ 対称な位置.

positional *a.* 位置の.
~ notation 位取り記数法.

positive *a.* 正の, プラスの; 肯定的な. *n.*

実在; 正量. ⇨ negative.
of ~ type 正型の. ~ correlation 正(の)相関. ~-definite, ~ definite 正(定)値の, 正の定符号の. ~ direction 正の向き. ~ divisor 正因子. ~ domain [region] 正領域. ~ infinity 正の無限大. ~ integer 正の整数, 自然数. ~ number 正(の)数. ~ orientation 正の向き. ~ quadrant 正の象限. ~ root 正(の)根. ~ semi-definite 半正(定)値の, 正の半定符号の. ~ sense 正の向き. ~ sign 正号. ~ term series 正項級数. ~ value 正(の)値. ~-valued 正値の. ~ variation 正変動.

positively *ad.* 正に, 正の向きに; 肯定的に.
~ related 正に結ばれている.

positivity *n.* 正値性.

possess *vt.* 所有する.

possibility *n.* 可能性.

possible *a.* 可能な.
best ~ 最良の.

post- *pref.* 時間・時代. 時期的に「後-」の意. ⇨ pre-.

posterior *a.* 後の; 事後の. ⇨ anterior.
~ distribution 事後分布. ~ probability 事後確率. ~ risk 事後危険.

postgraduate *a.* 大学院の; 大学卒業後の. *n.* 大学院生.
~ course 大学院課程. ~ school 大学院.

postoptimization *n.* 再最適化.

postulate *vt.* 要求する; (自明として)仮定する. *n.* 公準; 要請.
~ for construction 作図の公法, 作図公準.

postulation *n.* 仮定.

potency *n.* 濃度; 基数.
~ of set 集合の濃度.

potential *n.* ポテンシャル; 可能性. *a.* 潜在的な; 位置の; ポテンシャルの.
bipolar ~ 双極ポテンシャル. complex ~ 複素ポテンシャル. dipole ~ 双極ポテンシャル. kinetic ~ 運動ポテンシャル. logarithmic ~ 対数ポテンシャル. Newtonian ~ ニュートン・ポテンシャル. ~ energy ポテンシャル[位置]エネルギー. ~ function ポテンシャル関数. ~ scattering ポテンシャル散乱. ~ theory ポテンシャル論. retarded ~ 遅延ポテンシャル. scalar ~ スカラー・ポテンシャル. vector ~ ベクトル・ポテンシャル. velocity ~ 速度ポテンシャル.

power *n.* 累乗, ベキ[冪]; 方ベキ; 濃度; 力.
~ function ベキ関数; 検出力関数. ~ method ベキ乗法. ~ of set 集合のベキ. ~ of test 検出力. ~-residue ベキ剰余. ~ root ベキ根. ~ series ベキ級数. ~ set ベキ集合. ~-sum ベキ和. ~ theorem 方ベキ定理. raise ~ to the second 二乗する. symbolic ~ 記号ベキ.

powerful *a.* 有力な, 強力な.
most ~ 最強力な.

p. p. (*F.* =a.e.) prersque partout. 殆んどいたるところ.

p. p. proportional part. 比例部分.

practical *a.* 実際(的)な; 実用の; 実質的な.
~ exercise 応用問題. ~ mathematics 実用数学.

practically *ad.* 実際的に; 実用的に; 実質的に.

practice *n.* 実算; 練習.

Prandtl, Ludwig (1875-1953) プラントル.

pre- *pref.* 時間・時代・時期的に「前-」の意. ⇨ post-.
~Hilbert space 前ヒルベルト空間.

prealgebraic *a.* 前代数的な.
~ variety 前代数多様体.

precession *n.* 歳差; 歳差運動.

precise *a.* 精確な, 厳密な.
~ estimation 精確な評価.

precision *n.* 精度; 正しさ.
~ attained, ~ reached 達成精度. ~ prescibcd 要求精度.

precompact *a.* 前コンパクトな, 全有界な.

precondition *n.* 前処理.

predator *n.* 捕食動物.
~-prey equation 捕食者-被食者方程式.

predecessor *n.* 直前(のもの); 先行(のもの). ⇨ successor.

predicate *n.* 述語.

polynomial ～ 多項式述語. ～ calculus 述語計算. ～ logic 述語論理. ～ symbol 述語記号. ～ variable 述語変数.

predicative *a.* 可[叙]述的な. *n.* 叙述語.
～ logic 述語論理.

prediction *n.* 予測.
linear [nonlinear] ～ 線形[非線形]予測.

predictor *n.* 予測値, 予測子.
best ～ 最良予測値. optimal [optimum] linear ～ 最良線形予測値. ～-corrector method 予測子修正子法.

preimage *n.* 原像, 逆像.

preliminary *a.* 予備の, 準備の.

premise *n.* 前提.
major ～ 大前提. minor ～ 小前提.

prenex *n.* 冠頭.
～ normal form 冠頭標準形.

preorder *n.* 前順序.

preparation *n.* 準備.
～ theorem 予備定理.

prerequisite *n., a.* 予備知識(の).

prescheme *n.* 前概型.

present *a.* 現在の; 居合わせる, 出席している. *vt.* 示す; 発表する; 与える.
～ value 現価.

preservation *n.* 保存.
measure-～ 保測. theorem of domain ～ 領域保存の定理.

preserve *vt.* 保存する; 維持する.
measure-～ ving transformation 保測変換. ～ ving function 保存関数.

presheaf *n.* 準層; 前層.

presque *ad.* (*F.* ＝almost.) 殆んど.
～ partout 殆んどいたるところ[p.p.](＝almost everywhere; a.e.).

pressure *n.* 圧力.
～ equation 圧力方程式.

presumption *n.* 推定; 仮定.

pretest *n.* 予備調査[試験].

price *n.* 価格. *vt.* 値をつける; 評価する.
buying ～ 買価. net ～ 正価. ～cing vector 評価ベクトル. selling ～ 売価. unit ～ 単価.

primal *a.* 最初の; 主要な.
～ problem 主問題. ～ variable 主変数.

primary *a.* 主要な; 第一の, 最初の; 準素の.
～ group 準素群. ～ ideal 準素イデアル. ～ (ideal) decomposition 準素(イデアル)分解. ～ obstruction 第一障害[碍]類. ～ sampling unit 一次抽出単位. ～ solution 主要解.

prime *a.* 第一の; 素(数)の. *n.* 素数, 素元; プライム記号【′】.
mutually ～ 互いに素な. ～ divisor 素因子. ～ elemcnt 素元. ～ factor 素因子, 素因数. ～ field 素体. ～ formula 素論理式. ～ ideal 素イデアル. ～ number 素数. ～ spectrum プライムスペクトル. ～ spot 素点. relatively ～ 互いに素な.

primitive *a.* 原始の; 素朴な. *n.* 原線; 原始関数; 基
～ element 原始元. ～ equation 原始方程式. ～ function 原始関数. ～ idempotent 原始ベキ等元. ～ notion 無定義概念. ～ n-th root of unity 1 の原始 n 乗根. ～ period 基本周期. ～ period parallelogram [strip] 基本周期平行四辺形[帯]. ～ polynomial 原始多項式. ～ ring 原始環. ～ root 原始根. ～ solution 原始解, 固有解.

principal *a.* 主な; 主要な; 主-, 主座の. *n.* 元金.
～ adèle 主アデール. ～ and interest 元利. ～ axis 主軸. ～ branch 主枝. ～ bundle 主束. ～ component 主成分. ～ convergent 主近似分数. ～ curvature 主曲率. ～ diagonal 主対角線. ～ direction 主方向. ～ genus 主種. ～ ideal 単項[主]イデアル. ～ ideal domain (PID) 単項[主]イデアル整域. ～ matrix 主行列. ～ minor 主(座)[首座]小行列式. ～ normal 主法線. ～ part 主(要)部. ～ plane 主平面. ～ point 主点. ～ root 主要根. ～ solution 主要解. ～ theorem 主要定理. ～ value 主値.

principle *n.* 原理; 法則; 律.
argument ～ 偏角の原理. continuity ～ 連続性原理. maximum (modulus) ～ 最大(絶対)値の原理. ～ of conservation of form [calculation] 形式[算法]不易の原理. ～ of duality, duality ～ 双対原理. ～ of

inclusion and exclusion 包除原理. ~ of optimality 最適性の原理. ~ of reflection, reflection ~ 鏡像の原理. substitution ~ 代入の原理. ~ of superposition 重ね合わせの原理, 重畳原理. symmetry ~ 対称[鏡像]の原理.

PRINGSHEIM, Alfred (1850-1941) プリンクスハイム.

prior *a.* 前の, 事前の. *ad.* 先立って.
~ distribution 事前分布. ~ probability 事前確率.

priority *n.* 優先(権).
~ method 優先法.

prism *n.* 角柱; プリズム.
quadrangular ~ 四角柱. rectangular ~ 直方体. regular ~ 正角柱. right ~ 直角柱. triangular ~ 三角柱. truncate(d) ~ 截[切]頭角柱.

prismatic *a.* 角柱の; プリズムの.
~ surface 角柱面.

prismatoid *n.* 擬角柱.

prismoid *n.* 角錐台, 角台.

prob. probable, probability, probabilistic.

probabilistic *a.* 確率(論)的な.

probability *n.* 確率; 公算; 蓋然性.
conditional ~ 条件つき確率. convergence in ~ 確率収束. mathematical ~ 数学的確率. ~ a posteriori, aposteriori ~ 事後確率. ~ a priori, apriori ~ 事前確率. ~ density 確率密度. ~ distribution 確率分布. ~ element 確率エレメント, 確率素分. ~ error 確率誤差. ~ event 確率事象. ~ limit 確率限界. ~ measure 確率測度. ~ of causes 原因の確率, 事後確率. ~ paper 確率紙. ~ sample 確率標本. ~ space 確率空間. ~-theoretic 確率論的な. subjective ~ 主観確率. theory of ~ 確率論. upper tail ~ 上側確率.

probable *a.* 確からしい; 蓋然的な.
equally ~ 同様に[同程度に]確からしい. most ~ value 最確値. ~ error 確率誤差, 蓋然誤差.

problem *n.* 問題; 課題; 作図題.
boundary value ~ 境界値問題. closure ~ 閉形問題. coefficient ~ 係数問題. complementarity ~ 相補性問題. Dirichlet ~ ディリクレ[ディリシュレ]問題. dual ~ 双対問題. Fermat's ~ フェルマーの問題. four colo(u)r ~ 四色問題. geometric ~ 幾何学的問題. open ~ 未解決の問題. primal ~ 主問題. ~ of [for] construction 作図問題. ~ of many bodies 多体問題. singular ~ 特異問題. traveling salesman ~ 巡回セールスマン問題. unsolvable ~ 解けない問題. unsolved ~ 未解決の問題.

procedure *n.* 方法; 手順.

proceedings (*pl.*) *n.* 会報; 議事録.

process *n.* 過程; 工程; プロセス; 方法, 手順.
additivc ~ 加法過程. birth [death] ~ 出生[死亡]過程. counting ~ 計数過程. decision ~ 決定過程. deterministic ~ 確定的過程. differential ~ 加法過程. manufacturing ~ 生産工程. Markov ~ マルコフ過程. reversed ~ 逆進過程. stationary ~ 定常過程. stochastic ~ 確率過程. transition ~ 推移過程.

processing *n.* 処理(すること).
batch ~ 一括処理. information ~ 情報処理. ~ error 処理過程(の)誤差.

produce *vt.* 生成する; 延長する.

producer *n.* 生産者. ⇨ consumer.

producible *a.* 生成できる; 延長できる.

product *n.* 積; 生成; 産物, 製品.
bracket ~ 括弧積. canonical ~ 規準乗積. Cartesian ~ カルテシアン積, デカルト積. composition ~ 合成積. cross ~ 交差[叉]積. crossed ~ 接合積. difference ~ 差積. direct ~ 直積. exterior ~ 外積. finite ~ 有限(乗)積. infinite ~ 無限(乗)積. inner ~ 内積. interior ~ 内(部)積. outer ~ 外積. ~ event 積事象. ~ formula 積公式. ~ manifold 積多様体. ~ measure 直積測度. ~ metric 積計量. ~ notation 乗積記号【Π】. ~ of propositions 論理積【∧】. ~ series (乗)積級数. ~ set 積集合. ~ space (直)積空間. ~ sum 積和. ~ topology (直)積位相. scalar ~ スカラー積. slant ~ スラント積, 斜積. tensor ~ テンソル積. vector ~ ベクトル積. wedge ~ 楔積.

production *n.* 延長, 延長線; 生産, 生成.
～ control 工程[生産]管理. ～ planning 生産計画(法).

profile *n.* 縦断面図; 側面(図).

pro-finite *a.* 副有限な, 射(影)有限な.
～ group 副有限群, 射有限群.

profit *n.* 利益, 利潤.
～ function 利潤関数.

program(me) *n.* プログラム. 計画表. *vt.* プログラムをつくる.
Erlangen ～ エルランゲン(の)目録. ～-med instruction プログラム学習.

programmer *n.* プログラマ.

programming *n.* プログラミング, 計画法.
convex ～ 凸計画(法). dynamic ～ 動的計画法. global ～ 大域的計画法. linear ～ 線形計画(法)[LP]. mathematical ～ 数理計画(法). semi-infinite ～ 半無限計画(法).

progression *n.* 数列.
arithmetic ～ 等差[算術]数列. geometric ～ 等比[幾何]数列. harmonic ～ 調和数列. ～ of differences 階差数列.

progressional *a.* 数列の.

progressive *a.* 前進する; 進行性の.
～ mean 累加平均.

project *vt.* 投影する, 射影する. *n.* 計画.

projectile *n.* 投[放]射物, 発射体. *a.* 投放[射]する; 推進する.
height [range] of ～ 投[放]射高度[距離]. ～ force 推進力.

projection *n.* 射影, 投影, 投象; 射影作用素; 平面図法; 投影法, 投象法; 突起.
cabinet ～ キャビネット投象法. central ～ 中心射影. conformal conic ～ 正角円錐図法. natural ～ 自然射影. oblique ～ 斜投影[象]. orthogonal ～ 正射影, 直交射影, 正射影作用素; 正投影[象]. perspective ～ 透視図法. ～ chart 投影図. ～ map 射影. ～ matrix 射影行列. ～ operator 射影作用素. stereographic ～ 立体射影.

projective *a.* 射影の, 射影的な.
～ closure 射影閉包. ～ collineation 射影的共線変換. ～ connection 射影接続. ～ coordinates 射影座標. ～ correspondence 射影的対応. ～ cover 射影被覆. ～ curvature tensor 射影曲率テンソル. ～ geometry 射影幾何学. ～ limit 射影的極限(＝inverse limit 逆極限). ～ line 射影直線. ～ module 射影加群. ～ plane 射影平面. ～ property 射影的性質. ～ relation 射影的関係. ～ resolution 射影(的)分解. ～ space 射影空間. ～ symplectic group 射影シンプレクティック群. ～ system 射影(的)系(＝inverse system 逆系). ～ transformation 射影(的)変換.

projectively *ad.* 射影的に.
～ flat 射影的に平坦な.

projectivity *n.* 射影関係.

projectometry *n.* 投射測定.

projector *n.* 投影線; 射影作用素.

prolate *a.* 扁長の.
～ spheroid 扁長回転楕円体[面].

prolong *vt.* 延長する; 引き延ばす; 接続する.

prolongable *a.* 接続可能な.

prolongation *n.* 接続; 延長. ⇨ continuation.
analytic ～ 解析接続. harmonic ～ 調和接続.
analytically ～ 解析接続可能な.

proof *n.* 証明; 検算; 論証; 校正刷.
alternative ～ 別証(明). direct ～ 直接証明. galley ～ ゲラ刷. indirect ～ 間接証明. ～ by contradiction 背理法. ～ by induction 帰納法による証明. ～ reading 校正(すること). reductio ad absurdum ～ 背理法による証明.

prop. property, proposition.

propagation *n.* 伝播.
～ of errors, error ～ 誤差(の)伝播. wave ～ 波(動)の伝播.

propagative *a.* 伝播性の.

propagatory *a.* 伝播性の.

proper *a.* 真の; 固有の; 本来の; 通常の; 適正な.
～ class 本来の類. ～ conics 固有な円錐曲線. ～ convex function 真凸関数, 適正(な)凸関数. ～ convex set 真凸集合. ～

equation 固有方程式. ~ fraction 真分数. ~ hypersphere 常超球. ~ integral 通常の積分. ~ mapping 固有写像. ~ oscillation 固有振動. ~ polynomial 固有多項式. ~ product 固有積. ~ quadrics 固有な二次曲面. ~ rotation 固有回転. ~ solution 固有解. ~ subgroup 真部分群. ~ subset 真部分集合. ~ subspace 真部分空間. ~ value 固有値. ~ vector 固有ベクトル.

properly *ad.* 全く; 真に.
~ discontinuous 真性不連続な. ~ *n*-dimensional 固有 *n* 次元の. ~ posed 適切な.

property *n.* 性質; 属性.
affine ~ アフィン的性質. alternating ~ 交代性. Baire ~ ベールの性質. geometric(al) ~ 幾何学的性質. global [local] ~ 大域[局所]的性質. qualitative [quantitative] ~ 定性[量]的性質. universal ~ 普遍的性質.

proportion *n.* 割合, 比率; 比例; 比; 比例式, 比例計算.
constant of ~ 比例定数. continued ~ 連比例. direct ~ 正比例. in ~ to [as] に比例して. inverse ~ 反[逆]比例. ~ defective 不良率. reciprocal ~ 逆比例.

proportional *a.* 比例の; 比例した【to】. *n.* 比例項.
fourth ~ 第四比例項. mean ~ 比例中項. ~ compasses 比例コンパス. ~ constant 比例定数. ~ distribution 比例配分, 按分比例. ~ expression 比例式. ~ part 比例部分[p.p.]. ~ sampling 比例抽出(法). ~ scale 比例尺. third ~ 第三比例項.

proportionality *n.* 比例関係.
factor of ~ 比例因子.

proportionate *a.* 比例している. *vt.* 比例させる.
~ sampling 比例抽出(法).

proposition *n.* 命題; 定理; 提案.
affirmative ~ 肯定命題. categorical ~ 範疇命題. composite ~ 複合命題. conditional ~ 条件命題. conjunction ~ 連合[合接]命題. converse ~ 逆[転換]命題. existential ~ 存在命題, 特称命題. logical ~ 論理命題. negative ~ 否定命題. ~ variable 命題変数[変項]. universal ~ 全称命題.

propositional *a.* 命題の; 定理の.
~ calculus 命題計算. ~ connective 命題結合記号. ~ constant 命題定項. ~ function 命題関数. ~ logic 命題論理(学). ~ value 命題値. ~ variable 命題変項.

prorate *vt., vi.* 比例配分する, 割り当てる.

prospective *a.* 将来の; 見込まれる. *n.* 展望, 見込み.
~ study 前向き研究.

protection *n.* 保護.
average [lot] quality ~ 平均[仕切]品質保護.

prototype *n.* 原型.

protraction *n.* 製図; 延長.

protractor *n.* 分度器.

protuberance *n.* 突起, こぶ, 結節.

provable *a.* 証明可能な.

prove *vt.* 証明する; 検査する.

provided *conj.* もし…ならば, …という条件で, …である限り.

proximate *a.* 近似の; 非常に近い.

proximity *n.* 接近, 近いこと.
~ function 近接関数.

Prym, Friedrich (1841–1915) プリム.

pseudo- [sú:d(oʊ), s(j)ú:dəʊ] *ref.*「仮の, 擬似の」の意. ⇨ quasi-.

pseudo-analytic *a.* 準解析的な.
~ function 準解析関数.

pseudo-code *n.* 擬似コード.

pseudo-compact *a.* 擬コンパクトな.

pseudo-convex *a.* 擬凸な.
~ domain 擬凸領域. strongly ~ 強擬凸な.

pseudo-differential *a.* 擬微分-.
~ operator 擬微分作用素.

pseudo-directed *a.* 擬有向な.
~ family of points 擬有向点族. ~ set 擬有向集合.

pseudo-distance *n.* 擬距離.

pseudo-function *n.* 擬関数.

pseudo-geometric *a.* 擬幾何的な.
~ ring 擬幾何的環.

pseudo-gradient *n.* 擬勾配.
～ vector field 擬勾配ベクトル場.
pseudo-graph *n.* 擬グラフ.
pseudo-group *n.* 擬群.
～ structure 擬群構造.
pseudo-inverse *n.* 擬(似)逆行列.
pseudo-manifold *n.* 擬多様体, 準多様体.
pseudo-metric *n., a.* 擬[準]距離(の).
～ space 擬[準]距離空間.
pseudo-metrizable *a.* 擬距離づけ可能な.
pseudo-module *n.* 擬加群.
pseudo-norm *n.* 半ノルム.
pseudo-normal *n.* 擬法線. *a.* 擬法-.
～ vectors 擬法ベクトル.
pseudo-order *n.* 擬順序.
pseudo-polynomial *n.* 擬多項式.
distinguishcd ～ 特殊擬多項式.
pseudo-random *a.* 擬似乱数の.
～ numbers 擬似乱数.
pseudo-scalar *n.* 擬スカラー.
pseudo-sphere *n.* 擬球(面).
pseudo-spherical *a.* 擬球の.
～ surface 擬球(曲)面.
pseudo-valuation *n.* 擬付値.
pseudo-vector *n.* 擬ベクトル.
psi *n.* プサイ, プシー. ギリシア語アルファベットの 23 番目の文字【Ψ, ϕ】.
～ function プサイ関数.
psychometrics *n.* 計量心理学.
pt. point.
Ptolemaeus [**Ptolemaios**] of Alexandria, Claudiu[o]s (85?-165?) アレクサンドリアのトレミー[プトレマイオス].
Ptolemy ＝Ptolemaeus.
pts. points.
public *a.* 公衆の.
～ key cryptosystem 公開鍵暗号系. ～ opinion poll 世論調査. ～ opinion survey 世論調査.
Puiseux, Victor Alexandre (1820-1883) ピュイズー.
pull *vt.* 引く, 引っ張る.
～ back 引きもどす.
pullback *n.* 引き戻し, プルバック.
pulsate *vi.* 脈動する; 振動する.
pulsation *n.* 脈動; 振動.
pulse *n.* パルス.
punctual *a.* 点の.
punctured *a.* 穴[孔]をあけた.
～ disc [plane] (一)点を除いた円板[平面].
pure *a.* 純粋な.
～ decimal 純小数. ～ geometry 純粋幾何学. ～ mathematics 純粋数学. ～ recurring decimal 純循環小数. ～ strategy 純粋戦略.
purely *ad.* 純(粋)に.
～ discontinuous 純不連続な. ～ imaginary number 純虚数. ～ inseparable 純非分離的な. ～ transcendental extension 純超越拡大.
purpose *n.* 目的.
purposive *a.* 目的のある; 故意の; 目的論的な.
～ sample 有意標本. ～ selection 有意選出.
pursuit *n.* 追跡.
curve of ～ 追跡曲線(＝tractrix).
pushdown *n.* プッシュダウン; 後入れ先出し.
～ automaton プッシュダウン・オートマトン.
pushout *n.* 押し出し, プッシュアウト.
puzzle *n.* パズル; 判じ物, 謎.
mathematical ～ 数学パズル.
p. v. principal value. 主値.
p-value *n.* p 値.
pyramid *n.* 角錐, ピラミッド.
frustum of ～ 角錐台. regular ～ 正角錐. spherical ～ 球面角錐.
pyramidal *a.* 角錐の; 錐体の.
～ surface 角錐面.
Pythagoras of Samos (572?-492? B.C.) サモスのピタゴラス.
Pythagorean *a.* ピタゴラスの.
～ number ピタゴラス数. ～ table ピタゴラスの表, 掛算九々の表. ～ theorem [proposition] ピタゴラスの定理, 三平方の定理.

Q

Q.E.D., q.e.d. (*L.*) quod erat demonstrandum (=which was to be proved).

Q.E.F., q.e.f. (*L.*) quod erat faciendum (=which was to be done).

Q.E.I., q.e.i. (*L.*) quod erat inveniendum (=which was to be found).

qua *ad.* (*L.*) として.

quadrable *a.* 等積の正方形で表し得る; 可積な; 二[自]乗し得る.

quadrangle *n.* 四角形.
complete ～ 完全四角形. twisted ～ ねじれ四角形.

quadrangular *a.* 四角形の.
～ number 四角数. ～ prism 四角柱. ～ set of six points 六点図形, 四角柱六点.

quadrant *n.* 四分円; 象限.
first ～ 第一象限. positive ～ 正の象限.

quadrantal *a.* 四分円の; 象限の.
～ spherical triangle 象限弧球面三角形.

quadrate *a.* 正方形の. *n.* 正方形.

quadratic *a.* 二次の, 平方の. *n.* 二次方程式. (*pl.*)【単数扱い】二次方程式論.
～ cone 二次錐面. ～ curve 二次曲線. ～ cylinder 二次柱面. ～ differential 二次微分. ～ equation 二次方程式. ～ field 二次体. ～ form 二次形式. ～ function 二次関数. ～ hypersurface 二次超曲面. ～ loss function 二次損失関数. ～ non-residue 平方非剰余. ～ programming 二次計画法. ～ residue 平方剰余. ～ surface 二次曲面.

quadratics *n.* 二次方程式論.

quadrature *n.* 求積(法).
mechanical ～ 器械的求積. numerical ～ 数値求積(法). ～ of circle 円積.

quadri-, quadru- *pref.* 「四」の意.

quadric *a.* 二次の. *n.* 二次曲面 (*pl.*); 二次関数.
central ～s 有心二次曲面. confocal ～s 共焦二次曲面. elliptic [hyperbolic, parabolic] ～s 楕円[双曲, 放物]的二次曲面. non-central ～s 無心二次曲面. ～ cone [cylinder] 二次錐[柱](面). ～ hypersurface 二次超曲面.

quadrilateral *a.* 四辺形の; 四辺の. *n.* 四辺形.
complete ～ 完全四辺形. curvilinear ～ 曲線四辺形. harmonic ～ 調和四辺形. ～ circumscribed [inscribed] to circle, circumscribed [inscribed] ～ 円に外[内]接する四辺形. twisted ～ ねじれ四辺形.

quadrillion *n.*【英・独】百万の四乗;【米・仏】千の五乗, 千兆.

quadrinomial *a.* 四項の. *n.* 四項式.

quadrivium *n.* 四科, 四自由学芸【中世大学の算術・幾何・天文・音楽】.

quadruple *n.* 四倍数; 四つ組. *a.* 四倍の, 四重の.
～ lattice 四重格子.

quadruplicate *a.* 四倍の, 四重の. *vt.* 四倍にする.

quadruply *ad.* 四重に, 四倍に.

qualification *n.* 資格; 条件.
constraint ～ 制約想定.

qualitative *a.* 定性的な; 質的な.
～ property 定性的性質. ～ variable 質的変数.

qualitatively *ad.* 定性的に.

quality *n.* 性質; 特性; 品質.
average ～ 平均品質. ～ characteristic 品質特性. ～ control 品質管理. ～ of design

設計の品質. ~ protection 品質保護. ~ specification 品質規格. ~ standard 品質標準.

quantic *n.* 有理斉関数, 同次多項式.

quantical *a.* 有理斉関数の, 同次多項式の.

quantification *n.* 数量化; 量化.
~ method 数量化法.

quantifier *n.* 量化記号, 量化子; 限定記号, 限定作用素; 限量記号, 限量子.
existential ~ 存在記号【∃】. universal ~ 全称記号【∀】.

quantify *vt.* 定量化する, 数量化する; 量化する.

quantile *n.* 分位数; 変位数.
~ of order *p* *p* 分位数.

quantitative *a.* 定量的な; 量的な.
~ property 定量的性質. ~ variable 量的変数.

quantitatively *ad.* 定量的に.

quantity *n.* 量; 数量; 多量, 多数.
geometric ~ 幾何学的量. known ~ 既知量. scalar ~ スカラー量. transformation ~ 変換子. unknown ~ 未知量. vector ~ ベクトル量.

quantization *n.* 量子化.

quantize *vt.* 量子化する.
~d wave function 量子化(された)波動関数.

quantum (*pl.* -ta) *n.* 量; 量子.
~ information theory 量子情報理論. ~ mechanics 量子力学.

quart. quarter, quarterly.

quarter *n.* 四分の一, 四半分; 十五分; 四半期.

quarterly *a., ad.* 毎季の[に]; *n.* 季刊誌.
~ issued [published] 季刊の.

quartic *a.* 四次の. *n.* 四次(代数)曲線; 四次関数.
~ equation 四次方程式. ~ function 四次関数.

quartile *a.* 四分位の. *n.* 四分位数.
lower ~ 下側四分位数. upper ~ 上側四分位数. ~ deviation 四分位偏差.

quasi- [kwéizai, kwéisai, kwá:si, -zi] *pref.* 「擬, 準」の意. ⇨ pseudo-.
~*p*-valent 準 *p* 葉な.

quasi-affine *a.* 準アフィンな.
~ algebraic variety 準アフィン代数的多様体.

quasi-analytic *a.* 擬[準]解析的な.
~ function 擬[準]解析関数.

quasi-bounded *a.* 準有界な.
~ function 準有界関数.

quasi-coherent *a.* 準連接な.

quasi-compact *a.* 準コンパクトな.
~ space 準コンパクト空間.

quasi-complemented *a.* 準可補的な.

quasi-complete *a.* 準完備な.
~ space 準完備空間.

quasi-completeness *n.* 準完備性.

quasiconcave *a.* 準凹の.
~ function 準凹関数.

quasi-conformal *a.* 擬等角な.
~ mapping 擬等角写像.

quasi-continuity *n.* 準連続性.

quasi-continuous *a.* 準連続な.
~ function 準連続関数.

quasiconvex *a.* 準凸の.
~ function 準凸関数.

quasi-dual *n.* 準双対. *a.* 準双対な.

quasi-equivalent *a.* 準同値な.
~ representation 準同値(な)表現.

quasi-factorial *a.* 準要因の.

quasi-Frobenius *a.* 準フロベニウス的な.
~ algebra 準フロベニウス多元環.

quasi-group *n.* 準群, 擬群.

quasi-independent *a.* 準独立な.
~ of path 試行列と準独立な.

quasi-invariant *a.* 準不変な.
~ measure 準不変測度.

quasi-inverse *n.* 準逆元, 擬逆元. *a.* 準逆の, 擬逆の.
~ element 準[擬]逆元.

quasi-invertible *a.* 準可逆な, 擬可逆な.
~ element 準[擬]可逆元.

quasi-Latin *a.* 準ラテンの.
~ square 準ラテン方格.

quasi-linear *a.* 準線形な, 擬線形な.
~ equation 準[擬]線形方程式.

quasi-local *a.* 準局所的な.

quasi-norm *n.* 準ノルム.

quasi-normal *a.* 準正規な.
~ family 準正規族.

quasi-normed *a.* 準ノルムをもつ.
~ linear space 準ノルム線形空間.

quasi-order *n.* 擬順序.

quasi(-)projective *a.* 準射影的な.
~ algebraic variety 準射影代数的多様体.

quasi(-)regular *a.* 擬正則な, 準正則な.
~ element 擬正則元. ~ ideal 擬正則イデアル. ~ sequence 準正則列.

quasi-semi(-)local *a.* 準半局所的な.

quasi-spectrum (*pl.* -tra, -trums) *n.* 擬スペクトル.

quasi-stable *a.* 準安定な.
~ distribution 準安定分布.

quasi-tonnelé *a.* (*F.*) 準樽型の.
espace ~, ~ space 準樽型空間.

quasi-uniform *a.* 擬一様な.
~ convergence 擬一様収束.

quaternary *a.* 四元の; 四元数の.
~ quantic 四元同次多項式.

quaternion *n.* 四元数; 四元数算法.
conjugate ~ 共役四元数. ~ algebra 四元数環. ~ field 四元数体. ~ group 四元数群. ~ hyperbolic space 四元数双曲型空間. ~ vector bundle 四元数ベクトル束. skew field of ~s 四元数斜体.

quaternionic *a.* 四元数の.
~ structure 四元数構造.

question *n.* 質問, 問い; 疑問; 問題. *vt., vi.* 質問する.
~ and answer 問題[質問]と答.

Quetelet, Lambert Adolphe Jacques (1796–1874) ケトレ.

queue [kjúː] *n.* 待ち行列.
bulk ~ 集団待ち行列. ~ with many servers 複数窓口の待ち行列. tandem ~ 連続待ち行列.

queuing *n.* 待ち時間; 待合せ.
~ theory 待合せ[待ち行列の]理論, Qの理論.

quiescence *n.* 静止.

quiescent *a.* 静止した, 静止の.

quinary *a.* 五の. *n.* 五進法.

quincunx *n.* 五点形.

quindecagon *n.* 十五角[辺]形.

quinquangular *a.* 五角形の.

quinque-, quinqu- *pref.* 「五」の意.

quintic *a.* 五次の. *n.* 五次(代数)曲線; 五次関数.
~ curve 五次(代数)曲線. ~ equation 五次方程式.

quintillion *n.* 【米・仏】百万の三乗; 【英】百万の五乗.

quintuple *a.* 五倍の. *n.* 五倍.

quiver *n.* 箙(えびら), クイバー; 有向グラフ.

quod *pron.* (*L.* =which) それ.
~ erat demonstrandum=q. e. d., Q. E. D. ~ erat faciendum=q.e.f., Q.E.F. ~ erat inveniendum=q.e.i., Q.E.I.

quota *n.* 分けまえ; 割当.
~ method 割当法.

quotation *n.* 引用.
~ mark 引用符.

quotient *n.* 商.
difference ~ 差商. differential ~ 微分商. ideal ~ イデアル商. integral ~ 整商. prime ~ 素商. ~ algebra 商多元環. ~category 商圏. ~ field, field of ~s 商体. ~ group 商群. ~ lattice 商束. ~ law 商法則. ~ linear space 商線形空間. ~ measure 商測度. ~ module 商加群. ~ object 商対象. ~ ring, ring of ~s 商環. ~ set 商集合. ~ space 商空間. ~ system 商系. ~ topology 商位相. Rayleigh('s) ~ レイリー商. total ring of ~s, ring of total ~s, total ~ ring 全商環.

R

rad. =radian.

RADEMACHER, Hans Adolph (1892–1969) ラーデマッヘル.

radial *a.* 半径の; 放射状の.
～ limit 半径に沿う極限(値). ～ ray 放射線. ～ slit 放射截線.

radially *ad.* 放射状に.

radian *n.* ラジアン, 弧度.
～ measure 弧度(法).

radiation *n.* 放射; 輻射.

radical *a.* 根の. *n.* 根基; ベキ根, 根; 根号.
Jacobson ～ ジャコブソン根基. nilpotent ～ ベキ零根基. ～ axis 根軸. ～ center 根心. ～ equation 無理方程式. ～ number 根数. ～ plane 根(平)面. ～ root 累乗根. ～ sign 根号【√】. ～ test ベキ根判定法.

radicand *n.* 被開平[ベキ]数.

radius (*pl.* -dii, -diuses) *n.* 半径.
convergence ～, ～ of convergence 収束半径. convexity ～, ～ of convexity 凸型[凸性]半径. mapping ～ 写像半径. ～ of curvature 曲率半径. ～ of inertia 慣性半径. ～ of inversion 反転半径. ～ of meromorphy 有理型[形]半径. ～ of torsion 捩率半径. ～ vector (*pl.* -dii -res, -s) 動径, 動径ベクトル. univalency ～ 単葉半径.

radix (*pl.* -xes, -dices) *n.* (数・対数の)基; (統計の)位取り; 基数.
～ sorting 基数(分類)法.

RADÓ, Tibor (1895–1965) ラド.

RADON, Johann Karl August (1887–1956) ラドン.

raise *vt.* 累乗する; 大きくする.
～ *a* to *n*th power *a* を *n* 乗する.

RAMANUJAN, Srinivasa (1899–1943) ラマヌジャン.

ramification *n.* 分岐, 分枝.
～ exponent 分岐指数. ～ field 分岐体. ～ group 分岐群. ～ index 分岐指数. ～ number 分岐数. ～ point 分岐点. ～ ring 分岐環. tame ～ 順分岐. wild ～ 野性[生]分岐.

ramified *a.* 分岐した.
～ covering space 分岐被覆空間. ～ element 分岐要素. totally ～ 完全分岐.

ramify *vt., vi.* 分岐する.

random *n.* 無作為; でたらめ. *a.* 無作為な; でたらめの; 確率的な, ランダムな.
at ～ 無作為に; でたらめに. continuous [discrete] ～ variable 連続[離散](型)確率変数. ～ arrangement 確率的[無作為]配置. ～ current 彷徨カレント. ～ die 乱数さいころ. ～ distribution 確率超過程. ～ effect 変量効果. ～ error 確率的誤差. ～ flight 乱歩, 酔歩. ～ measure 彷徨[確率]測度. ～ number(s) 乱数. ～ operation 確率(化)操作. ～ process 確率過程. ～ sample 無作為標本, 任意標本, 確率標本. ～ sampling 任意抜取り, 無作為抽出, 確率抽出. ～ start でたらめ(の)振出し. ～ tensor field 彷徨テンソル場. ～ variable 確率変数. ～ walk 酔歩, 乱歩.

randomization *n.* 無作為化, 確率化, ランダム化.
～ operation 確率化操作.

randomized *a.* 無作為化された, 確率化された.
～ block method, method of ～ blocks 乱塊法. ～ test 確率化された検定.

range *n.* 範囲, 値域, 限界, レンジ; 射程.

vt. 並べる. *vi.* 並ぶ; 広がる.
effective ~ 有効射程. harmonic ~ of points 調和列点. numerical ~ 数域. ~ of function 関数の値域. ~ of points (射影幾何の)列点. ~ of values 値域. sample ~ 標本範囲. stable ~ 安定域.

rank *n.* 階; 位; 階数, ランク; 順位.
maximal ~ 最大階数. ~ correlation coefficient 順位相関係数. ~ of matrix 行列の階数. ~ order 順位.

ranking *n.* 順位づけ.
~ method 順位法.

rapid *a.* 速い; 急速な, 急激な. ⇨ slow.
~ decrease 急減少.

rapidity *n.* 速さ, 速度; 急速.

rapidly *ad.* 急速に.
~ decreasing distribution 急減少超関数. ~ decreasing function 急減少関数.

rate *n.* 割合, 率, 歩合.
per capita ~ 均等割. ~ of change 変化率. ~ of interest 利率. survival ~ 生存率. transmission ~ 伝送速度.

ratio (*pl.* -s) *n.* 比, 割合.
anharmonic ~ 非調和比. common ~ 公比. continued ~ 連比. correlation ~ 相関比. cross ~ 交差[叉]比, 非調和比. deformation ~ 変形率. dilatation [dilation] ~ 拡大率; 膨張率. direct ~ 正比. double ~ 複比. external [internal] ~ 外[内]分比. inverse [reciprocal] ~ 反比, 逆比. odds ~ オッズ比. ~ criterion 比判定法. ~ estimate 比推定量. ~ of the circumference of a circle to its diameter 円周率. ~ scale 比尺度. ~ test 比判定法. similitude ~ 相似比. simple ~ 単比.

rational *a.* 有理の; 有理数の, 有理的な; 合理的な.
~ curve 有理曲線. ~ differential equation 有理的微分方程式. ~ element 有理要素. ~ expression 有理式. ~ fraction 有理分数. ~ function 有理関数. ~ function field 有理関数体. ~ integer 有理整数. ~ integral function 有理整関数. ~ number 有理数. ~ number field; field of ~ numbers; field of ~s 有理数体. ~ operation 有理演算. ~ point 有理点. ~ rank 有理階数. ~ root 有理根. ~ surface 有理曲面.

rationalization *n.* 有理化.
~ of denominator 分母の有理化. ~ of integrand 被積分関数の有理化.

rationalize *vt.* 有理化する.

rationally *ad.* 有理的に.
~ equivalent 有理同値な.

ray *n.* 半径; (放)射線; 半直線.
radial ~ 放射線.

Rayleigh, Lord [Strutt, John William] (1842–1919) レイリー.

re- *pref.* 「再」の意.

reaction *n.* 反作用.
law of ~ 反作用の法則.

read *vi., vt.* 読む.
~ out 読み出す.

real *a.* 実数の, 実の. *n.* 実数.
~ analysis 実解析. ~(-)analytic 実解析的な. ~ axis 実軸. ~ closed field 実閉体. ~ field 実体. ~ form 実形. ~ function 実関数. ~ line 数直線. ~ linear space 実線形空間. ~ number 実数. ~ number field, field of ~ numbers 実数体. ~ part 実(数)部(分). ~ projective space 実射影空間. ~ quadratic field 実二次体. ~ root 実根. ~ space 実空間. ~ structure 実構造. ~-valued 実数値の. ~ variable 実変数. ~ vector space 実ベクトル空間.

reality *n.* 実質, 実在; 真実性; 事実.

realizability *n.* 実現可能性.

realizable *a.* 実現しうる, 実現可能な.

realization *n.* 実現.

realize *vt.* 実現させる; (はっきりと)理解する.

real-time *a.* 実時間の, リアルタイムの.

real-valued *a.* 実数値の.
~ function 実数値関数.

rearrange *vt.* 再配列する; 配列し直す.

rearrangement *n.* 再配列, 配列がえ.

reason *n.* 理由, 根拠; 前提; 小前提. *vi., vt.* 推論する, 論証する.

reasoning *n.* 推論, 推理; 論法. *a.* 推論の.
demonstrative ~ 論証的推理. plausible ~

もっともらしい論法.

receiver *n.* 受取人, 受報者, 受信者. ⇨ destinator.

receptor *n.* 感覚器官.

recessive *a.* 劣の; 劣性の. ⇨ dominant.
~ game 劣ゲーム.

reciprocal *a.* 相反の, 逆の; 相互の. *n.* 逆数, 反数.
~ difference 逆差分. ~ equation 相反[逆数]方程式. ~ figure 相反図形. ~ formula 反転公式. ~ linear representation 反線形表現. ~ network 相反回路. ~ number 逆数; 反数. ~ proportion 反比例. ~ ratio 反比, 逆比. ~ scale 逆数目盛. ~ spiral 逆匝線. ~ system 相反系.

reciprocally *ad.* 相反的に.

reciprocity *n.* 相反(性); 相互性, 相互作用.
law of ~, ~ law 相反法則, 相互法則. ~ relation 相反関係.

recognition *n.* 認識.
character ~ 文字認識. pattern ~ パターン認識.

record *n.* 記録, レコード. *vt.* 記録する.
block ~ ブロック・レコード. ~ing point 発注点.

rect- *pref.* 「直, 正」の意.

rectangle *n.* 長方形, 矩形.

rectangular *a.* 長方形の, 矩形の; 直角の.
~ [Cartesian] coordinates 直角[交]座標. ~ distribution 長方形分布. ~ equilateral triangle 直角二等辺三角形. ~ graph 棒グラフ. ~ hyperbola 直角双曲線. ~ hyperbolic coordinates 直角双曲線座標. ~ matrix 長方[矩]形行列. ~ parallelepiped 直六面体, 直方体. ~ prism 直方柱. ~ spherical triangle 直角球面三角形. ~ triangle 直角三角形.

rectifiable *a.* 長さの有限な, 長さのある, 求長可能な.
~ curve 長さの有限な曲線.

rectification *n.* 求長(法).

rectify *vt.* (曲線の)長さを求める, 求長する; 訂正する.
~ing plane 展直平面. ~ing surface 展直曲面.

rectilineal =rectilinear.

rectilinear *a.* 直線の.
~ complex ユークリッド複体. ~ coordinate system ユークリッド[直線]座標系.

recur *vi.* 循環する; 戻る.

recurrence *n.* 循環; 再帰; 反復; 回帰.
~ formula 漸化(公)式. ~ relation 漸化関係. ~ theorem 再帰定理.

recurrent *a.* 循環する; 再帰的な.
~ chain 再帰連鎖. ~ event 再帰事象. ~ geodesic 再帰測地線. ~ sequence 回帰数列.

recurring *a.* 循環する; 再帰的な. ⇨ circulating.
~ continued fraction 循環連分数. ~ decimal 循環小数. ~ formula 漸化(公)式. ~ sequence 循環数列.

recursion *n.* 再帰; 帰納法, 再帰法.
~ formula 漸化[再帰](公)式. ~ theorem 再帰定理.

recursive *a.* 再帰的な; 帰納的な.
~ definition 再帰[帰納]的定義. ~ formula 漸化(公)式; 再帰式. ~ function 再帰[帰納]的関数. ~ set 再帰[帰納]的集合.

recursively *ad.* 再帰的に; 帰納的に.
~ enumerable 再帰[帰納]的に可算な.

reduce *vt.* 減らす; 帰着させる; 換算する; 変形する; 通分[通約]する; 還元する. *vi.* 減る; 帰着する.

reduced *a.* 被約な, 既約の; 縮小[退]した.
~ algebra 被約代数. ~ cone 約錐, 被約錐. ~ drawing 縮図. ~ gradient 縮約勾配. ~ mapping cone 約写像錐, 被約写像錐. ~ norm 被約ノルム. ~ pitch 準ピッチ. ~ quadratic form 被約二次形式. ~ representation 被約表現. ~ residue class 既約剰余類. ~ ring 被約な環. ~ sampling inspection 緩和抜取り検査. ~ scale 縮尺. ~ suspension 約懸垂. ~ trace 被約跡[トレース].

reducible *a.* 換算[通約]できる, 可約な; 還元できる.
completely ~ 完全可約な. ~ element 可約元. ~ fraction 可約分数. ~ polynomial

可約多項式.

reductio *n.* (*L.* =reduction.)
~ ad absurdum (=reduction to absurdity) 背理法, 帰謬法.

reduction *n.* 通約, 約分; 換算; 還元; 減少; 縮図; 縮分.
data ~ データ整理. ~ formula 還元[換算]公式; 漸化式. ~ of data データ整理. ~ of order 階数低下. ~ to absurdity 背理法, 帰謬法. ~ to common denominator 通分.

reductive *a.* 簡約可能な, 通約できる.
~ absurdity 背理法, 帰謬法. ~ action 簡約可能(な)作用.

redundance (=redundancy) *n.* 余分; 冗長; 重複.

redundancy *n.* 余分; 冗長; 冗長度; 重複.
~ check 冗長(度)検査. ~ code 冗長符号.

redundant *a.* 余分な; 冗長な; 重複した.
~ code 冗長符号. ~ number 過剰数.

reentering *a.* 再び入る.
~ polygon 凹多角形.

reentrant *n.* 凹角. *a.* 凹角の.

refer *vt.* 参照させる. *vi.* 参照する.

referee *n.* 査読者, レフェリー. *vt.*, *vi.* 査読する.

reference *n.* 参照, 参考; 参考文献.
cross ~ 相互参照. ~ mark 参照符, 引用記号. ~ measure 標準測度.

refine *vt.*, *vi.* 純化する, 洗練する; 細分する.

refinement *n.* 細分.
~ theorem 細分定理. star ~ 星型細分.

reflect *vt.* 反射する.

reflected *a.* 反射した.
~ binary code 交番二進コード. ~ image 鏡像, 反射像. ~ ray 反射光線.

reflecting *a.* 反射する.
~ barrier 反射壁.

reflection *n.* 反射; 反転, 鏡映; 裏返し, 折返し.
~ principle 反射原理, 反転原理. ~ surface, surface of ~ 鏡映面.

reflective *a.* 反射的な.
~ barrier 反射壁.

reflex *a.* 反射の; 優角の. *n.* 反射.
~ angle 優角.

reflexion =reflection.

reflexive *a.* 反射的な; 再帰の.
~ law 反射法則[律].

refract *vt.* 屈折させる.
~ed ray 屈折光線.

refraction *n.* 屈折; 大気差.

refractive *a.* 屈折の.
~ index 屈折率.

reg. region; regular(ly).

Regiomontanus (1436-1476) レギオモンタヌス. =Johannes Müller.

region *n.* 範囲; 領域; 面分. ⇨ domain.
acceptance ~ 採択域. circular ~ 円領域. closed ~ 閉領域. confidence ~ 信頼域. critical ~ 棄却域. feasible ~ 実行可能領域, 許容領域. fundamental ~ 基本領域. negative ~ 負領域. open ~ 開面分[領域]. positive ~ 正領域. ~ of concern [influence] 影響領域. trust ~ method 信頼領域法.

regional *a.* 地域的な; 領域の.

regionally *ad.* 地域的に.

regression *n.* 復帰; 回帰; 折返し.
edge of ~ 反帰曲線. multiple ~ 重回帰. ~ analysis 回帰解析. ~ coefficient 回帰係数. ~ curve 回帰曲線. ~ equation 回帰方程式. ~ estimate 回帰推定値. ~ function 回帰関数. ~ hyperplane 回帰超平面. ~ line 回帰線.

regula falsi (*L.*) 誤差調整(法)【零点の挟撃】.

regular *a.* 正則な, 正規の; 等辺等角の; 正.
left [right] ~ 左[右]正則な. ~ arc 正則弧. ~ boundary 正則境界. ~ curve 正則曲線. ~ element 正則元. ~ embedding 正則な埋め込み. ~ event 正則事象. ~ exhaustion 正則近似列. ~ expression 正規表現. ~ extension 正則拡大. ~ function 正則関数. ~ local ring 正則局所環. ~ mapping 正則写像. ~ matrix 正則行列. ~ point 正則点. ~ polygon 正多角形. ~ polyhedral angle 正多面角. ~ polyhedron 正多面体. ~ position 正則な位置. ~

prism 正角柱. ~ pyramid 正角錐. ~ representation 正則表現. ~ sequence 正則列. ~ singular point, ~ singularity 確定特異点. ~ space 正則空間. ~ surface 正則曲面. ~ system of parameters 正則パラメータ系. ~ triangle 正三角形.

regularity *n.* 正則性; 正規性.
point of ~ 正則点. ~ condition 正則性(の)条件.

regularization *n.* 正則[規]化.

regularize *vt.* 正則[規]化する.

regularly *ad.* 正則的に, 正規的に; 規則正しく.
~ hyperbolic 正規双曲型の.

regulator *n.* 単数基準.

reject *vt.* 棄却する; 却下する. *n.* 不合格品.

rejection *n.* 棄却; 却下; 不合格にすること.
~ limit 棄却限界. ~ number 棄却(個)数, 不合格判定個数. ~ region [range] 棄却域. ~ test 棄却検定.

relate *vi.* 関係する. *vt.* 関係させる.

related *a.* 関係のある; 同族の.
positively ~ 正に結ばれている. ~ equation 標準[関連]方程式. ~ topic 関連する話題.

relation *n.* 関係; 関連.
binary ~ 二項関係. equivalence ~ 同値関係. functional ~ 関数関係. in [with] ~ to に関(連)して. limit ~ 極限関係. linear ~ 線形(的)関係. order ~ 順序関係. perspective [projective] ~ 配景[射影]的関係. ~ algebra 関係環. symmetric ~ 対称な関係. symmetry ~ 対称関係.

relational *a.* 関係の(ある).

relationship *n.* 関係; 関連.
~ algebra 関係環.

relative *a.* 相対的な.
~ boundary 相対境界. ~ consistency 相対無矛盾性. ~ degree 相対次数. ~ error 相対誤差. ~ frequency 相対度数[頻度]. ~ homological algebra 相対ホモロジー代数(学). ~ invariant 相対不変式. ~ maximum [minimum] 極大[小]. ~ norm 相対ノルム. ~ probability 条件つき確率. ~ risk 相対危険度. ~ tensor 相対テンソル. ~ topology 相対位相. ~ vector 相対ベクトル. ~ velocity 相対速度.

relatively *ad.* 相対的に.
~ compact 相対コンパクトな. ~ minimal curve 相対最短線. ~ prime 互いに素な.

relativistic *a.* 相対論的な.

relativity *n.* 相対性(理論).
theory of ~, ~ theory 相対性理論.

relativization *n.* 相対化.

relativize *vt.* 相対化する.
~d theorem 相対化された定理.

relaxation *n.* 緩和.
line ~ 線状緩和. over-~ 加速緩和(法). ~ method 緩和法. ~ oscillation 緩和振動. ~ table 緩和表.

relay *n.* 交替. *vt., vi.* 中継する.

relevant *a.* 関連した.

reliability *n.* 信頼性.

Rellich, Franz (1906–1955) レリッヒ.

remainder *n.* 剰余, 余り; 残り.
chinese ~ theorem 中国式剰余定理. cubic ~ 開立剰余. quadratic ~ 平方剰余. ~ term 剰余項. ~ theorem 剰余定理. square ~ 開平剰余.

remark *n.* 注意; 注. *vt., vi.* 注意する.

removable *a.* 除去可能な, 除ける.
~ singularity 除去可能(な)特異点.

removal *n.* 除去.

remove *vt.* 除去する.

rencontre *n.* (*F.* =rencounter.)

rencounter *n.* 遭遇; めぐり合い.
probability of ~ めぐり合いの確率.

renewal *n.* 再生, 更新.
~ process 再生過程.

repeat *vt., vi.* くり返す; 反復する. *n.* 反復.

repeated *a.* くり返された; 累次の.
~ combination 重複組合せ. ~ differentiation 累次微分. ~ integral 累次積分. ~ limit 累次極限. ~ permutation 重複順列. ~ series 累次級数.

repeater *n.* くり返すもの; 循環小数.

repeating *a.* くり返す.
~ decimal 循環小数.

repère *n.* (*F.* =frame.)
～ mobile 動標構，動枠.
repetition *n.* 反復; 重複.
replace *vt.* 置き換える，置換する.
replacement *n.* 置換.
replica *n.* 複製; レプリカ.
report *vt., vi.* 報告する; 記録する．*n.* 報告; 記録; (*pl.*) 議事録.
represent *vt.* 表現[表示]する; 示す; のべる.
representability *n.* 表現可能性.
representable *a.* 表現可能な.
～ event 表現可能(な)事象
representation *n.* 表現，表示; 写像.
complex ～ 複素表現．conformal ～ 等角写像．integral ～ 積分表示; 整数表現．irreducible ～ 既約表現．matrix ～ 行列表現．modular ～ モジュラー表現．ordinary ～ 通常表現．parametric ～ 媒介変数表示．Poisson ～ ポアソン表示．quotient ～ 商表現．regular ～ 正則表現．～ by components 成分表示．～ matrix 表現行列．～ module 表現加群．～ problem 表現問題．～ space 表現空間．～ theory 表現論．spherical ～ 球面表示．vector ～ ベクトル表示[表現].
representative *a.* 代表的な; 表現する．*n.* 代表(値); 代表元.
～ element 代表元．～ function 表現関数．～ ring 表現環．system of ～s 代表系.
representing *a.* 表現する.
～ function 表現関数．～ measure 表現測度.
reproduce *vt.* 再生する．*vi.* 繁殖する.
reproducibility *n.* 再生性.
reproducing *a.* 再生する; 再生の.
～ kernel 再生核．～ property 再生性.
reproductive *a.* 再生的な.
～ property 再生性.
reproductivity *n.* 再生性.
repulsion *n.* 斥力.
repulsive *a.* 斥ける，反発する.
～ force 斥力.
requirement *n.* 要求; 要請; 要件.
consistency ～ 無矛盾の要請．～ space 条件空間．～ vector 条件ベクトル.
research *n.* 研究; 調査．*vt., vi.* 研究する; 調査する.
operations ～, operational ～ オペレーションズ[ナル]・リサーチ，OR; 運営研究．～ fellow 研究員.
reset *vt.* ゼロにする; 元にもどす; 破算する; 置き直す; 払う．*n.* 置き直すこと.
residual *a.* 余りの，剰余の; 残留の; 等終の．*n.* 剰余; 誤差; 残差.
～ error 残留誤差．～ quantity 残量．～ spectrum 剰余スペクトル．～ subset 等終部分集合．～ term 剰余項．～ variance 残差分散.
residue *n.* 剰余; 留数.
logarithmic ～ 対数的留数．quadratic ～ 平方剰余．～ calculus 留数解析，留数計算．～ character 剰余指数．～ class 乗余類．～ class group [ring] 剰余(類)群[環]．～ system 剰余系．～ theorem 留数(の)定理.
residuum (*pl.* -dua) *n.* 余り，剰余; 残留誤差; 留数.
resolution *n.* 分解; 解消.
free ～ 自由分解．injective ～ 移入分解，単射[入射]的分解．projective ～ 射影(的)分解．～ of identity 単位の分解．～ of singularities 特異点の解消．spectral ～ スペクトル分解．standard ～ 標準分解.
resolutive *a.* 可解な.
～ compactification 可解コンパクト化.
resolve *vt., vi.* 分解する.
～ into partial fractions 部分分数に分解する.
resolvent *n.* 逆核，解核，レゾルベント; 分解方程式.
cubic ～ 三次分解(方程)式．～ equation 分解方程式，レゾルベント方程式．～ kernel 解核．～ operator レゾルベント作用素．～ set レゾルベント集合.
resonance *n.* 共鳴，共振.
～ energy 共鳴エネルギー．～ scattering 共鳴散乱．～ theorem 共鳴定理.
respect *n.* 関心; 関係，関連.
in ～ of に関して，について．with ～ to に

関して.

respective *a.* それぞれの.

respectively *ad.* それぞれ(に).

respond *vi.* 応答する【to】.

response *n.* 応答.
~ time 応答時間.

rest *n.* 静止; 残余, 残高.
~ point 静止点, 休止点.

restitution *n.* 復元.

restitutive *a.* 復元の.
~ force 復元力.

restraint *n.* 拘束.

restrict *vt.* 制限する, 限定する.

restricted *a.* 制限された, 限定された.
in the ~ sense 狭義に. ~ direct product 制限直積. ~ holonomy group 制限ホロノミー群. ~ mapping 制限された写像. ~ root 制限根. ~ type 限定型.

restriction *n.* 制限, 限定; 制約, 拘束; 縮小.
~ map 制限写像. ~ of mapping 写像の制限.

result *n.* 結果; 答. *vi.* 帰着する【in】; 結果として生じる.
as a ~ それゆえ. ~ table 結果表.

resultant *n.* 終結式; 合力. *a.* 結果の; 合成的な.
~ force 合力.

résumé *n.* (*F.*) 摘要, 要約; 要旨.

retard *vt.* 遅らせる. *n.* 遅れ.
~ed potential 遅延ポテンシャル. ~ed type 遅れ型.

retardation *n.* 遅れ.
difference-differential equation of ~ type 遅れ型差分微分方程式.

retract *n.* レトラクト; 牽縮.
absolute neighbo(u)rhood ~ 絶対近傍レトラクト. deformation ~ 変位レトラクト.

retraction *n.* 引き込み, 左逆写像.

retrieval *n.* 探索, 検索.
information ~ 情報検索[IR].

retrograde *a.* 逆進の. *vi.* 逆行する; 退化する.

retrospective *a.* 回顧的な; 後ろ向きの.
~ study 後ろ向き研究.

return *n.* 復帰; 返答; 往復. *vi.* 帰る. *vt.* 返す.

Reuleaux, Franz (1829–1905) ルーロー.

revenue *n.* 収益.

reverse *n.* 逆, 裏. *vt.* 逆にする; 反転する. *a.* 逆の.
~d process 逆進過程. ~(d) series 反転級数. ~ inequality 逆向きの不等式. ~ order 逆の順序. ~ side 裏側.

reversibility *n.* 可逆性.

reversible *a.* 可逆な.
~ pendulum 可逆振子. ~ process 可逆過程.

reversion *n.* 反転.
~ of series 級数の反転.

review *vt., vi.* 評論する; 査読する. *n.* 評論, レビュー; 査読.

reviewer *n.* 評論者; 査読者.

revise *vt.* 修正する; 校訂する.
~d edition 改訂版.

revolution *n.* 回転; 周期.
axis of ~ 回転軸. cone of ~ 回転錐(面). cylinder of ~ 回転柱(面). ellipsoid of ~ 回転楕円面[体]. paraboloid of ~ 回転放物面[体]. solid of ~ 回転体. surface of ~ 回転面.

revolve *vi.* 回転する; 公転する. *vt.* 回転させる.

rheonomic *a.* レオノームな.
~ system レオノーム系.

rheonomous =rheonomic.

rho *n.* ロー. ギリシア語アルファベットの17番目の文字【P, ρ】.

rhomb =rhombus.

rhombic *a.* 菱形の; 斜方晶系の.
~ system 斜方晶系.

rhombohedral *a.* 菱面体晶系の.
~ system 菱面体晶系.

rhombohedron (*pl.* -rons, -dra) *n.* 菱面体, 斜方六面体.

rhomboid *a.* 偏菱形の; 長斜方形の. *n.* 偏菱形; 長斜方形.

rhomboidal *a.* =rhomboid.

rhombus (*pl.* -buses, -bi) *n.* 菱形, 斜方形.

rhumb *n.* 羅針方位; 航程線.
 ~ line 航程線; 航海線.
RIBAUCOUR, Albert (1845-1893) リボークール.
RICCATI, Jacopo Francesco (1676-1754) リッカチ.
RICCI-CURBASTRO, Gregorio (1853-1925) リッチ.
 ~ curvature リッチ曲率. ~ flow リッチ・フロー. ~ tensor リッチ・テンソル.
RIEMANN, Georg Friedrich Bernhard (1826-1866) リーマン.
Riemannian *a.* リーマンの.
 ~ connection リーマン接続. ~ curvature リーマン曲率. ~ geometry リーマン幾何学. ~ manifold リーマン多様体. ~ metric リーマン計量. ~ product リーマン積. ~ space リーマン空間. ~ symmetric space 対称リーマン空間.
RIESZ, Frédéric (1880-1956), Marcel (1886-1969) リース.
right *a.* 直角の; 右の, 右側の; 正しい. *n.* 右, 右側. ⇨ left.
 ~ angle 直角. ~ circular cone [cylinder] 直円錐[柱]. ~ conoid 正コノイド. ~ continuous 右連続な. ~ coset 右剰余類. ~ differentiable 右微分可能な. ~ exact 右完全な. ~ helicoid 常螺旋面. ~ ideal 右イデアル. ~ invariant 右不変な. ~ inverse element 右逆元. ~ order 右整環. ~ prism 直角柱. ~ regular 右正則な. ~ translation 右移動. ~ triangle 直角三角形. ~ unit element 右単位元.
right-angled *a.* 直角の.
 ~ triangle 直角三角形.
right-hand *a.* 右の, 右側の.
 ~ member, ~ side 右辺.
right-handed *a.* 右回りの.
 ~ system 右手系.
rigid *a.* 固定している; 剛直な; 硬い.
 ~ algebra 剛代数. ~ body 剛体. ~ game 規準ゲーム. ~ motion 剛体運動.
rigidity *n.* 剛性.
rigidness =rigidity.
ring *n.* 環; 環面. *vt.* 環にする.
 anchor ~ 輪環面, トーラス. catenary ~ カテナリー環. circular ~ 円環. commutative ~ 可換環. differential ~ 微分環. division ~ 斜体. formal power series ~ 形式的ベキ級数環. group ~ 群環. local ~ 局所環. power series ~ ベキ級数環. principal ideal ~ 主[単項]イデアル環. quotient ~, ~ of quotients 商環. reduced ~ 被約な環. regular local ~ 正則局所環. ~ domain 環状領域. ~ homomorphism 環準同形[型]. ~ isomorphism 環同形[型]. ~ of operators 作用素環. ~ of sets 集合環. total ~ of quotients, ~ of total quotients, total quotient ~ 全商環. ~ of valuation vectors 付値ベクトル環. semilocal ~ 半局所環. simple ~ 単純環. spectrum of a (commutative) ~ (可換)環のスペクトル. spherical ~ 球環. unit of a ~ 環の単元.
ringed *a.* 環つきの; 環にされた.
 ~ space 環つき空間.
risk *n.* 危険, リスク, 危機.
 ~ function リスク[危険]関数.
RITZ, Walther (1878-1909) リッツ.
ROBIN, Gustave (1855-1897) ロバン.
ROBINSON, Abraham (1918-1974) ロビンソン.
robustness *n.* 頑健性.
ROCHE, Edouard-Albert (1820-1883) ロシュ.
rod *n.* 棒.
 Napier ~ ネイピアの計算棒.
RODRIGUES, O. (1794-1851) ロドリーグ.
roll *vi.* 転る. *vt.* 転す. *n.* 回転.
ROLLE, Michel (1652-1719) ロール.
Roman, roman *a.* ローマの; ローマ字体[立体]の. *n.* ローマ字体[立体].
 ~ character ローマ字. ~ numeral ローマ数字. ~ numeration ローマ記数法.
root *n.* 根; ベキ根, 累乗根; ルート.
 characteristic ~ 特性根. common ~ 共通根. conjugate ~ 共役根. cube ~, cubic ~ 立方根. double ~ 二重根. equal ~ 等根, 重根. imaginary ~ 虚根. multiple ~ 重(複)根. negative ~ 負(の)根. positive ~

正(の)根. primitive n-th ~ of unity 1 の原始 n 乗根. radical ~ 累乗根, ベキ根. rational ~ 有理根. real ~ 実根. ~ of unity 1 のベキ根[累乗根]. ~ sign 根号【√】. ~ system 根系, ルート系. ~ test ベキ根判定法. simple ~ 単根. square ~ 平方根. triple ~ 三重根.

rolling *a.* 転る. *n.* 回転; 横ゆれ.
~ curve 転(跡)曲線, 輪転曲線.

rot *n.* 回転(演算)子. ⇨ curl.

rotate *vi.* 回転する; 交替する. *vt.* 回転させる.
~ting sample (逐次)交代標本. ~ting wing 回転翼.

rotation *n.* 回転. ⇨ rot.
axis of ~ 回転軸. improper ~ 非固有回転. proper ~ 固有回転. ~ group 回転群. ~ index, ~(-) number 回転(指)数. ~ theorem 回転定理.

rotational *a.* 回転の; 渦のある.

rotationless *a.* 渦なしの.

Rouché, Eugène (1832–1910) ルーシェ.

rough *a.* 粗い; 概略の.
~ calculation 目の子算. ~ classification 粗い類別. ~ estimate 粗い評価. ~ sketch 見取図.

roulette *n.* 転跡線, 輪転曲線; ルーレット.

round *a.* 丸い, 丸めた. *n.* 丸いもの; 円, 環, 球. *vt.* 丸める; 回る; 一周する.
~ angle 周角. ~(ed) number 丸められた数, はんぱのない数; およその数; 概数. ~-off [-up] error 丸め誤差; 切捨て[切上げ]誤差.

round-down *vt.* 切捨てて丸める.

rounding *n.* 丸めること.
~ error 丸め誤差. ~ of number 数の丸め.

round-off *vt.* 切捨てて丸める.
~ error 丸め誤差.

round-up *vt.* 切上げて丸める.

routine *n.* ルーチン; おきまりの手順.

row *n.* 行; 並び.
elementary ~ operation 行基本変形. ~ echelon matrix [form] 行階段行列. ~-finite matrix 行有限行列. ~ rank 行階数. ~ space 行空間. ~ vector 行[横]ベクトル.

rudimentary [rù:dəméntəri] *a.* 基本の; 初歩の.
~ operation(s) 基本的演算.

Ruffini, Paolo (1765–1822) ルフィニ.

ruin *n.* 破産. *vt.* 破産させる.
~ probability 破産確率.

rule *n.* ものさし; 定規; 規則, 法則; 解式. *vt.* 定規で線を引く.
chain ~ 連鎖律. cosine ~ 余弦法則. four ~s 四則. ~ joint 折り尺. ~ of inference 推論. ~ of supposition 仮定法. sine ~ 正弦法則. slide ~ 計算尺. three-eights ~ 八分の三公式. trapezoid(al) ~ 台形公式. T-~ ティ定規.

ruled *a.* 線織の, 直線から成る.
~ paper 方眼紙. ~ surface 線織(曲)面.

ruler *n.* 定規, 定木.

run *vi.* 走る; 動く, わたる. *vt.* 運転する.
~ning mean 移動平均.

Runge, Carl David Tolmé (1856–1927) ルンゲ.

running *a.* 動く. *n.* 運転.
~ mean 移動平均. ~ suffix 動く添字.

Russell, Bertrand Earl (1872–1970) ラッセル.

S

's 名詞の所有格をつくる【省略されることもある】. Legendre's function, Möbius's (むしろ Möbius') function; 例: x's は x_1, x_2, … などの総称.

SACCHERI, Girolamo (1667–1733) サッケリ.

saddle *n.* 鞍.
~ point 鞍点, とうげ[峠]点. ~ region 鞍状領域.

safety *n.* 安全性.
~ stock 安全在庫.

sagitta [sədʒítə] *n.* 矢【円弧とその弦との中点間の線分または距離】.

sake *n.* 理由, 目的.
for the ~ of simplicity [brevity] 簡単のために.

SAKS, Stanisław (1897–1942) サクス.

SALEM, Raphael (1898–1963) サレム.

salience *n.* 突起.

salient *a.* 凸角の; 突起した. *n.* 凸角.
~ point 凸角点.

saltus *n.* 振幅; 跳躍.
~ function 跳躍関数. ~ of function 関数の振幅. ~ point 跳躍点.

same *a.* 同じ, 同一の; 変わらない.
infinitesimal of the ~ order 同位の無限小. ~ sign 同符号.

sample *n.* 見本; 標本, 試料, 資料. サンプル.
additional ~ 追加(の)資料. balanced ~ 釣合いのとれた標本. judgement ~ 有意標本. large ~ 大標本. master ~ 親標本, マスター・サンプル. random ~ 無作為標本. ~ characteristic 標本特性値. ~ correlation coefficient 標本相関係数. ~ design 標本設計. ~ distribution 標本分布. ~ function 見本関数. ~ mean 標本平均. ~ median 標本中央値. ~ point 見本点. ~ process 見本過程. ~ range 標本範囲. ~ space 標本空間. ~ survey 標本調査. ~ value 標本値. ~ variance 標本分散. small ~ 小標本. stratified ~ 層化標本.

sampling *n.* 標本抽出; 抽出, 抜取り, 試験採集, サンプリング.
areal ~ 地域抽出. cluster ~ 集落抜取り. last-off ~ 途中抜取り. ~ distribution 標本分布. ~ error 抽出誤差. ~ fraction 抽出比. ~ inspection 抜取り検査, 標本調査. ~ interval 抽出間隔. ~ method 標本調査法. ~ procedure 抽出方式. ~ ratio 抽出比. ~ theory 標本論. ~ unit 抽出単位. ~ with probability proportional size 確率比例抽出法. ~ with replacement 復元抽出. ~ without replacement 非復元抽出. single ~ 一回抽出.

satellite *n.* 衛星.
~ (functor) 衛星関手.

satisfaction *n.* みたすこと.

satisfiability *n.* 充足可能性.

satisfiable *a.* 満足させうる.

satisfy *vt.* みたす; 満足させる.
~ condition [equation] 条件[方程式]をみたす.

saturated *a.* 染み込んだ; 飽和した.
~ model 飽和モデル.

saturation *n.* 飽和.
~ of approximation 近似の飽和.

say *vt., vi.* いう, のべる; (たとえば)…とせよ.

scalar *n.* スカラー, 数量. *a.* スカラーの.
ring of ~s 係数環. ~ change 係数変更.

～ curvature スカラー曲率. ～ extension 係数拡大. ～ field スカラー場. ～ matrix スカラー行列. ～ multiple スカラー倍. ～ multiplication スカラー乗法. ～ potential スカラー・ポテンシャル. ～ product スカラー積. ～ quantity スカラー量. ～ restriction 係数制限. ～ triple product スカラー三重積.

scale *n.* 比率; 記(数)法, -進法; 目盛, 比例尺, 縮尺; 比例.
binary ～ 二進法. B-～ B尺. decimal [denary] ～ 十進法. functional ～ 関数目盛. logarithmic [ordinary] ～ 対数[普通]目盛; 対数[普通]尺. ratio ～ 比尺度. reduced ～ 縮尺. ～ factor 倍率, 尺度係数. ～ off 目外れ. ～ of notation 記数法(の基). ～ parameter 尺度母数. sliding ～ 滑尺. uniform ～ 一様[普通]目盛.

scalene *a.* (三角形が)不等辺の; (円錐体で)軸が斜めの. *n.* 不等辺三角形.
～ triangle 不等辺三角形.

scaling *n.* 尺度構成; スケーリング.

scanned *a.* 注目されている; 精査された.
～ symbol 注目されている記号.

scatter *n.* 散布; 散在; 散乱. *vt.* 散乱させる. *vi.* 四散する.
～ diagram 散布図, 点図表.

scattered *a.* 散在する; 局所有限の; 散乱された.
～ family 局所有限族. ～ set 分散集合. ～ sheaf 散布的層.

scattergram *n.* 散布図.

scattering *n.* 散乱; 散布.
degree of ～ 散布度. elastic [inelastic] ～ 弾性[非弾性]散乱. potential ～ ポテンシャル散乱. resonance ～ 共鳴散乱. ～ length 散乱の長さ. ～ operator 散乱作用素.

scedastic *a.* 散布の.
～ curve 散布線.

SCHAUDER, Pawel Juliusz (1899-1943) シャウダー.

schedule *n.* 予定表; 計画. *vt.* 予定表に入れる.

scheduling *n.* 日程計画.
complete [start] ～ function 完了[開始]計画関数.

schema (*pl.* -mata) *n.* 図解, 図式; (独)シェーマ.

scheme *n.* 概型, スキーム; 計画; 図表; 概要; 模型.
affine ～ アフィンスキーム. algebraic ～ 代数的概型. logarithmic ～ ログ概型. normal ～ 正規スキーム.

SCHLÄFLI, Ludwig (1814-1895) シュレフリ.

schlicht *a.* (*G.*) 単葉な. ⇨ univalent.
～ domain 単葉領域. ～ function 単葉関数. ～ mapping 単葉写像.

schlichtartig *a.* (*G.*) 単葉型の.

SCHLÖMILCH, Oskar (1823-1901) シュレーミルヒ.

SCHMIDT, Erhard (1876-1959) シュミット.

Schmiegungsfunktion *n.* (*G.* =proximity function.) 近接関数.

SCHOENFLIES, Arthur (1853-1928) シェーンフリース.

SCHOTTKY, Friedrich (1851-1935) ショットキ.

SCHREIER, Otto (1901-1929) シュライアー.

school *n.* 学校; 学派.
abacus ～ 算盤派. algoristic ～ 筆算派.

SCHRÖDER, Ernst (1841-1902) シュレーダー.

SCHRÖDINGER, Erwin Rudolf Josef Alexander (1887-1961) シュレーディンガー.

SCHUR, Isaai (1875-1941) シューア.

SCHWARTZ, Laurent (1912-2002) シュワルツ, シュヴァルツ.

SCHWARZ, Karl Hermann Amandus (1843-1921) シュワルツ, シュヴァルツ.

Schwarzian *a.* シュワルツの.
～ derivative シュワルツの導関数.

sci. science; scientific.

science *n.* 科学; 理学.
computer ～ コンピュータ科学. faculty of ～ 理学部. information ～ 情報科学. integrated ～ 総合科学. mathematical ～ 数理科学. natural ～ 自然科学. social ～ 社会科学.

scleronomic *a.* スクレロノームな.

～ system スクレロノーム系.

scleronomous ＝scleronomic.

score *n.* 得点; 二十.
eight ～ 百六十. four ～ 八十. ～ test スコア検定. T(-)～ ティ-スコア, 偏差値.

screen *vt.* 選別する. *n.* 仕切, 幕.
～ing inspection 選別検査.

screw *n.* 螺旋体; ねじ.
～ motion 螺旋運動.

scrutinization *n.* 吟味.

scrutinize,【英】scrutinise *vt.* 吟味する.

SD Standard Deviation. 標準偏差.

search *n.* 探索. *vt.* 探す.
binary [dichotomizing] ～ 二分探索. breadth first ～ 幅優先探索. depth first ～ 深さ優先探索. exhaustive ～ 全数探索. literature ～ 文献探索. local ～ 局所探索. ～ direction 探索方向.

searcher *n.* 探索者.

seasonality *n.* 季節変動.

secant *a.* 割る; 交差する, 分つ. *n.* 割線; 正割, セカント.

second *a.* 第二の, 二番目の. *ad.* 第二に.
～ boundary value problem 第二境界値問題. ～ category set 第二類集合. ～ cosine formula 第二余弦公式. ～ countability axiom 第二可算公理. ～ fundamental form 第二基本形式. ～ isomorphism theorem 第二同型[形]定理. ～ kind 第二種. ～ mean value theorem 第二平均値(の)定理. ～ order differential equation 二階微分方程式. ～ quadrant 第二象限. ～ variation 第二変分. set of ～ category 第二類集合.

second *n.* 秒【時刻・時間・角度の単位】.

secondary *a.* 第二位[次]の; 次の.
～ composition 二次合成. ～ obstruction 第二障害.

section *n.* 区分; 断面; 切断; 節; 切り口; 右逆写像.
conic ～ 円錐曲線. cross ～ (横)断面. global ～ 大域切断. normal ～ 直截口. orthogonal ～ 直断面. plane ～ 平断面. ～ paper 方眼紙.

sectional *a.* 断面の; 区分的な.
～ curvature 断面曲率.

sectionally *ad.* 区分的に. ⇨ piecewise.
～ smooth 区分的に滑らかな.

sector *n.* 扇形; 関数尺.
～ graph 扇形グラフ.

sectoral *a.* 扇形の.

sectorial *a.* 扇形の.
～ domain [region] 扇形領域.

secular *a.* 永年の, 長年の.
～ equation 永年方程式, 固有(値)方程式.
～ perturbation 永年摂動.

security *n.* 安全; 保証.
～ level 保証水準.

sedenion *n.* 十六元数.

see *vt.* 見る. *vi.* 見える; わかる.

segment *n.* 線分; (円の)弓形; 切片.
directed (line) ～ 有向線分. line ～ 線分. oriented ～ 有向線分. spherical ～ 球台. *y*-～ *y* 切片.

segmental *a.* 線分の; 弓形の.

segmentary ＝segmental.

Sᴇɢʀᴇ, Corrado (1863-1924) セグレ.

select *a.* 選んだ. *vt.* 選ぶ.
～ [～ed] works 選集.

selection *n.* 選出, 選択; 選定.
～ theorem 選択[出]定理. ～ parameter 選定母数. ～ statistic 選定統計量.

self *n.* 自己, 自身. *a.* 同じ, 単一の.

self- *pref.*「自己, 自動」の意.

self-adjoint *a.* 自己共役な, 自己随伴な.
～ differential equation 自己随伴微分方程式. ～ information 自己情報量. ～ operator 自己共役[自己随伴]作用素.

self-conjugate *a.* 自(己)共役な.
～ point 自(己)共役点.

self-consistent *a.* 自己矛盾のない; 首尾一貫した.

self-contained *a.* 自己充足的な.

self-dense *a.* 自己稠密な.

self-dual *a.* 自己双対な.

self-embedding *n.* 自己埋め込み, 自己埋蔵.

self-excited, self-exciting *a.* 自励の.
～ system 自励系. ～ vibration 自励振動.

self-injective *a.* 自己移入的な, 自己入射

的な.

self-intersection *n.* 自己交叉.
～ number 自己交点数.

self-linking *a.* 自己まつわり[纏絡]の.
～ coefficient 自己まつわりの[纏絡]数.

self-mapping *n.* 自己写像.

self-polar *a.* 自極の.
～ tetrahedron 自極四面体. ～ triangle 自極三角形.

self-reciprocal *a.* 自己相反な.
～ function 自己相反関数.

self-symmetric *a.* 自己対称な.
～ figure 自己対称図形.

self-weighting *a.* 自動加重の.
～ sample 自動加重標本.

semantics *n.* 意味論.

semi- *pref.* 「半」の意. ⇨ demi-, hemi-.

semicircle *n.* 半円(形).
upper [lower] ～ 上[下]半円.

semicircular *a.* 半円(形)の.

semi(-)closed *a.* 半閉の.
～ interval 半閉区間.

semi(-)continuity *n.* 半連続性.

semi(-)continuous *a.* 半連続な.
lower [upper] ～ function 下[上](に)半連続(な)関数.

semi-convergence *n.* 半収束.

semi-convergent *a.* 半収束の.
～ series 半収束級数.

semi-definite *a.* 半定符号の, 半定値な.
negative ～ 半負値な, 負の半定符号の.
positive ～ 半正値な, 正の半定符号の.

semidirect *a.* 半直の.
～ product 半直積.

semidisc, semidisk *n.* 半円板.

semi(-)exact *a.* 半完全な.
～ differential 半完全微分.

semifinite *a.* 半有限な.

semigroup *n.* 半群, 準群.
～ algebra 半群環. unitary [unital] ～ 単位的半群.

semi-infinite *a.* 半無限な.
～ interval 半無限区間. ～ programming 半無限計画(法). ～ strip 半無限帯(状面分).

semi-intuitionism *n.* 半直観主義.

semi-invariant *a.* 半不変な. *n.* 半不変係数, 半不変式; 半不変元.

semilinear *a.* 半線形[型]の.
～ mapping 半線形写像.

semilocal *a.* 半局所的な.
～ ring 半局所環.

semi-logarithmic *a.* 半対数(的)の.
～ paper 半対数方眼紙.

semi-martingale *n.* 半マルチンゲール.

seminar *n.* セミナー, ゼミナール; 研究室.

seminorm *n.* 半ノルム.

semi(-)open *a.* 半開の.
～ interval 半開区間.

semiorder *n.* 半順序, 準順序.

semiordered *a.* 半順序の.
～ set 半順序集合.

semiperimeter *n.* 半周長.

semi-polar *a.* 半極-.
～ coordinates 半極座標.

semiprimary *a.* 半準素の.
～ ring 半準素環.

semiprime *a.* 半素の.
～ ideal 半素イデアル. ～ ring 半素環.

semiprimitive *a.* 半原始の.
～ ring 半原始環.

semireductive *a.* 半簡約可能な.

semireflexive *a.* 半反射的な.

semiregular *a.* 半正則な.
～ point 半正則点. ～ transformation 半正則変換.

semiscalar *n., a.* 半スカラー(の).
～ product 半スカラー積.

semisimple *a.* 半単純な.
～ algebraic group 半単純代数群. ～ component 半単純成分. ～ module 半単純加群. ～ ring 半単純環.

semisimplicial *a.* 半単体の.
～ complex 半単体複体.

semi(-)space *n.* 半空間.

semi(-)sphere *n.* 半球(面). ＝hemisphere.

semistable *a.* 半安定な.
～ distribution 半安定分布.

sender *n.* 送信者.

sense *n.* 意味; 認識; 向き，方向.
in the restricted [wider] ~ 狭[広]義に. negative [positive] ~ 負[正]の向き.

sense-preserving *a.* 向きを保存する.
~ mapping 向きを保存する写像.

sense-reversing *a.* 向きを逆にする.
~ mapping 向きを逆にする写像.

sensitivity *n.* 感度.
~ analysis 感度解析.

sensory *a.* 知覚の.
~ test 官能検査.

sentinel *n.* 標識，しるし; 見張(人).

separability *n.* 分離性; 可分性.

separable *a.* 分離可能な，分離的; 可分な.
perfectly ~ 完全可分な. ~ algebra 分離的多元環. ~ closure 分離閉包. ~ degree 分離次数. ~ element 分離元. ~ extension 分離拡大. ~ metric space 可分距離空間. ~ polynomial 分離多項式. ~ transcendence basis 分離超越基底.

separably *ad.* 分離的に.
~ generated 分離生成の.

separate *vt.* 分離する. *a.* 分離した.
~ harmonically 調和に分かつ. ~ print 別[抜]刷. ~ting hyperplane 分離超平面.

separated *a.* 分離された.
~ topological group 分離位相群.

separately *ad.* 個々に，各個に.
~ continuous 各個連続な.

separation *n.* 分離.
degree of ~ 分離度. ~ axiom, axiom of ~ 分離公理. ~ of roots 根の分離. ~ of variables 変数分離. ~ theorem 分離定理.

separative *a.* 分離性の; 分離的な.

septangle *n.* 七角形.

septangular *a.* 七角(形)の.

septillion *n.* 【米・仏】千の八乗【1 に 0 を二十四個つけた数】; 【英・独】百万の七乗.

septimal *a.* 七の.

septuple *n.*, *a.* 七倍(の). *vt.* 七倍する.

seq. sequentia, sequentes (*L.* ＝the following).
p. 23 et ~ 23 ページ以後[下].

sequence *n.* 列，系列; 数列.
admissible ~ 許容列. Cauchy ~ コーシー列，基本列. convergent ~ 収束列. convex ~ 凸数列. decreasing [increasing] ~ 減少[増加] (数)列. diagonal ~ 対角列. double ~ 二重(数)列. exact ~ 完全(系)列. fundamental ~ 基本列. infinite ~ 無限(数)列. normal ~ 正規列. minimizing [maximizing] ~ 最小化[最大化]列. monotone ~ 単調(数)列. regular ~ 正則列. ~ of diminishing intervals 縮小区間列. ~ of events 事象列. ~ of functions 関数列. ~ of nested intervals 縮小区間列. ~ of numbers 数列. ~ of points, point ~ 点列. ~ of positive type 正型数列. ~ of sets 集合列. ~ of trials, trial ~ 試行(の)列.

sequential *a.* 続く，逐次の; (系)列の; 点列的な.
~ analysis 逐次分析. ~ game 逐次ゲーム. ~ limit 列極限. ~ sampling 逐次抜取り[抽出(法)]. ~ test 逐次検定.

sequentially *ad.* 逐次に; (点)列的に.
~ compact 点列コンパクトな. ~ complete 列的完備な. ~ definable 逐次定義可能な.

serial *a.* 直列の; 逐次の.
~ correlation 系列相関. ~ operation 直列操作. ~ processing 逐次処理.

series (*pl.* -) *n.* 級数; 列; 系列.
alternating ~ 交代級数. arithmetic ~ 算術[等差]級数. asymptotic ~ 漸近級数. central ~ 中心列. complementary ~ 補系列. composition ~ 組成列. convergent ~ 収束級数. divergent ~ 発散級数. double ~ 二重級数. finite ~ 有限級数. Fourier ~ フーリエ級数. gap ~ 空隙級数. geometric ~ 幾何[等比]級数. harmonic ~ 調和級数. infinite ~ 無限級数. major ~, majorant ~ 優級数. multiple ~ 多重級数. normal ~ 正規列. panharmonic ~ 汎調和級数. polynomial ~ 多項式級数. power ~ ベキ[整]級数. power ~ ring ベキ級数環. reversion of ~ 級数の反転. semiconvergent ~ 半収束級数. ~ development [expansion] 級数展開. ~ solution, solution by series 級数解. ~ with [of] positive terms 正項級数. Taylor ~ テイラー級数.

time ～ 時系列. trigonometric ～ 三角級数.

serpentine *a.* 曲がりくねった, エス[S]字形の. *n.* エス[S]字形.
～ curve エス[S]字(状)曲線.

Serre, Jean-Pierre (1926-) セール.

Serret, Joseph Alfred (1819-1885) セレ.

service *n.* 奉仕; 事務; サービス. *vt.* 奉仕作業をする.
bulk ～ 集団サービス. ～ life 耐用期間. ～ time 応対時間.

servo *n.* サーボ.
～-mechanism サーボ機構. ～ theory サーボ理論.

sesqui- *pref.* 「倍半[1½]の意」.
～ centennial 百五十年の.

sesquilinear *a.* 半双線形な.
～ form 半双線形形式.

set *n.* 集合; 組. *vt.* 置く; 配置する. *vi.* 沈む.
annihilator ～ 零化集合. Baire ～ ベール集合. border ～ 縁集合. Borel ～ ボレル集合. Cantor ～ カントル集合. closed ～ 閉集合. cluster ～ 集積値集合. compact ～ コンパクト集合. complementary ～ 補[余]集合. convex ～ 凸集合. countable [denumerable, enumerable] ～ 可算集合, 可付番集合. derived ～ 導集合. directed ～ 有向集合. empty ～ 空集合. finite ～ 有限集合. function ～, ～ of functions 関数(の)集合. fuzzy ～ ファジィ集合. infinite ～ 無限集合. Lebesgue measurable ～ ルベーグ可測集合. nondense ～ 疎集合. open ～ 開集合. ordered ～ 順序集合. point ～, ～ of points 点集合. proper convex ～ 真凸集合. ring of ～s 集合環. ～ function 集合関数. ～ operation 集合(演)算. ～ theory, theory of ～s 集合論. singular ～ 特異集合. singularity ～ 特異点集合. ternary ～ 三進集合. universal ～ 普遍集合. void ～ 空集合. whole ～ 全体集合.

set-theoretic(al) *a.* 集合論的な.
～ topology 集合論的位相幾何学.

set-valued *a.* 集合値の.
～ function 集合値関数.

several *a.* 数個の. *pron.* いくつか, 数個.
function of ～ variables 多変数(の)関数. ～ complex variables 多複素変数.

Severi, Francesco (1879-1961) セベリ.

sew *vt.* 縫う.
～ing theorem 縫合定理.

sex- *pref.* 「六」の意. ⇨ hexa-.

sexadecimal ⇨ hexadecimal.

sexagesimal *a.* 六十の, 六十進法の.
～ measure 六十分法. ～ scale 六十進法.

sexangle *n.* 六角形.

sexangular *a.* 六角形の.

sexi- *pref.* 「六」の意.

sexillion ＝sextillion.

sextant *n.* 円の六分の一, 六分円.
～ surveying 六分儀測量.

sextic *a.* 六次の.
～ curve 六次曲線.

sextillion *n.* 【米・仏】千の七乗; 【英】百万の六乗.

sextuple *a.* 六倍の. *n.* 六倍.

shade *n.* 陰.
～ and shadow 陰影.

shadow *n.* 影.
～ price 影の価格, 潜在価格.

Shannon, Claude Elwood (1916-2001) シャノン.

shape *n.* 形, 外形; 型; シェイプ. *vt.* 形づくる.
barrel ～ 樽形. J-～ ジェー字形. L-～ エル字形. L-～d エル字形の. ring-～d 輪状の. ～ operator 型(かた)作用素, シェイプ作用素, 形(かたち)作用素. star-～ 星形.

sharp *a.* 鋭い; 急な; 鋭敏な.
～ estimation 精確な評価.

sheaf (*pl.* sheaves) *n.* 層; 束.
ample ～ 豊富層. analytic ～ 解析的層. canonical ～ 標準層. coherent ～ 連接層. conormal ～ 余法線層. constant ～ 定数層. invertible ～ 可逆層. scattered ～ 散布的層. ～ of germs of continuous functions 連続写像[関数]の芽の層. ～ of planes 平面束. ～ space 層空間. trivial ～ 自明な層.

sheafification *n.* 層化.

shear *vt., vi.* 剪断する; ずれを起す. *n.* 剪断変形.
~ing stress ずれ応力.

sheet *n.* 葉; 枚; シート.
balance ~ 貸借対照表. hyperboloid of one ~ 一葉[単葉]双曲面. hyperboloid of two ~s 二葉[複葉]双曲面. number of ~s 葉数.

shell *n.* 殻.
spherical ~ 球殻.

shift *n.* (桁)移動. *vt.* 変える, 移す; 置き替える. *vi.* 移動する.
phase ~ 位相のずれ. ~ transformation ずらしの変換. unilateral ~ 片側シフト.

S.H.M. Simple Harmonic Motion. 単(弦)振動.

shock *n.* 衝撃.
~ wave 衝撃波.

short *a.* 短い; 簡潔な.
~ division 短除法. ~ exact sequence 短完全系列. ~ proof 短い[簡単な]証明.

shortest *a.* 最短の.
~ breadth 短幅. ~ distance 最短距離. ~ path 最短路. ~ representation 最短表現.

show *vt.* 示す, 立証する; 明らかにする.

shrink *vi.* 縮む, 収縮する.
~ing sequence 収縮列.

shunt *n.* 通過点; 分路.
left [right] ~ 左[右]通過点.

side *n.* 辺; 面; 側. *a.* 副の.
adjacent ~ 隣辺. angle-~ table 角辺表. corresponding ~s 対応辺. opposite ~ 対辺; 反対側. ~-angle table 辺角表. ~ condition 付帯条件. ~ elevation 側面図. ~ face 側面. ~ plane 側画面. ~ trace 側跡.

SIEGEL, Carl Ludwig (1896–1981) ジーゲル.

SIERPIŃSKI, Wacław (1882–1969) シェルピンスキ.

sieve *n.* 篩. *vt.* ふるう.
~ method 篩法. ~ of Eratosthenes, Eratosthenes' ~ エラトステネスの篩.

sift *vt.* 吟味する.

sight *n.* 視覚; 視界; 視距離.
point of ~ 視点.

sigma *n.* シグマ. ギリシア語アルファベットの 18 番目の文字【$\Sigma, \sigma, \varsigma$】.
co(-) ~ functions コシグマ関数. ~ compact シグマ・コンパクトな. ~ function シグマ関数. three ~ method 三シグマ法.

σ-additivity *n.* シグマ加法性, 可算加法性.

σ-finite *a.* シグマ有限な.
~ measure シグマ有限測度.

sigmoid *a.* エス字形の.
~ curve エス字形[状]曲線.

sign *n.* 符号; 記号; 信号.
change ~ 符号を変える. double ~ 複号. equality [inequality] ~ 等[不等]号. law of ~s 符号の法[規]則. minus [negative] ~ 負(の)符号. plus [positive] ~ 正(の)符号. ~-test 符号検査.

signal *n.* 信号; 記号.

signature *n.* 記号; 符号, 符号(定)数; 署名.
digital ~ デジタル署名. ~ of quadratic form 二次形式の符号数.

signed *a.* 符号のついた.
~ measure 符号のついた測度. ~ number 符号のついた数. ~ rank 符号付き順位.

significance *n.* 有意性; 重要性.
~ test 有意性(の)検定. statistical ~ 統計的有意性.

significant *a.* 有意の, 有意な; 有効な.
~ condition 有意状態. ~ digit, ~ figure 有効数字.

sign-test *n.* 符号検定.

signum *n.* (*L.* =sign.) 符号【sgn】.
~ function 符号関数. ~ of x x の符号【$\mathrm{sgn}\,x$, $x/|x|$】.

similar *a.* 相似な; 同様な, 同類の【to】.
~ correspondence 相似的対応. ~ figure(s) 相似図形. ~ matrix 相似な行列. ~ representation 相似[同形]な表現. ~ term(s) 同類項. ~ test 相似検定. ~ transformation 相似変換. ~ translation 相似移動.

similarity *n.* 相似; 相似変換.
~ transformation 相似変換.

similarly *ad.* 同様に.

similitude *n.* 相似; 相似変換.
center of ~ 相似の中心. external [internal] center of ~ 相似の外[内]心. ~ ratio

相似比.

simple *a.* 単純な; 簡単な; 単一の; 分かり易い.

~ algebra 単純多元環. ~ arc 単純弧. ~ character 単純指標. ~ closed curve 単純[単一]閉曲線. ~ component 単純成分. ~ convergence 単純収束. ~ correlation 単相関. ~ curve 単純曲線. ~ distribution 単一(層)分布. ~ event 単(純)事象. ~ extension 単純拡大. ~ function 単関数; 単葉関数. ~ graph 単純グラフ. ~ group 単純群. ~ harmonic motion 単(弦)振動[S.H.M.]. ~ hypothesis 単純仮説. ~ integral 単(一)積分. ~ interaction 二因子交互作用. ~ interest 単利. ~ lattice 単格子. ~ layer 単一層. ~ module 単純加群. ~ pole 一位の極. ~ polygon 単純多角形. ~ proof 簡単な証明. ~ ratio 単比. ~ ring 単純環. ~ root 単根. ~ series 単一級数. ~ statistical hypothesis 単純統計仮説. ~ zero 一位の零点.

simplest *a.* 最簡の.

~ alternating polynomial [expression] 最簡交代式; 差積. ~ orthogonal polynomial 最簡直交多項式.

simplex (*pl.* -lexes, -lices) *n.* 単体, シンプレックス.

boundary ~ 辺単体. continuous ~ 曲単体. *n*-~ *n* 次元単体. oriented ~ 有向単体. ~ criterion 単体規準. ~ method 単体法, シンプレックス法. ~ tableau 単体表, シンプレックス表.

simplicial *a.* 単体の.

~ approximation 単体近似. ~ complex 単体(的)複体. ~ decomposition 単体分割. ~ map 単体写像.

simplification *n.* 簡約(化); 単純化.

simplify *vt.* 簡約する; 簡単にする; 単純化する.

simply *ad.* 単純に; 単一に.

~ (-) connected 単(一)連結の. ~ (-) periodic function 単(一)周期関数. ~ transitive 単純推移的な.

SIMPSON, Thomas (1710-1761) シンプソン.

simulate *vt.* シミュレートする; 模擬[擬態]をする.

simulation *n.* シミュレーション, 模擬実験.

analog(ue) [system] ~ アナログ[システム]・シミュレーション. numerical ~ 数値シミュレーション.

simulator *n.* シミュレータ.

simultaneous *a.* 同時の, 同時に起る; 連立の.

~ distribution 同時分布. ~ equations 連立方程式. ~ inequalities 連立不等式.

since *ad.* それ以来. *pref.* …以来. *conj.* …のゆえに, …であるから.

sine *n.* 正弦, サイン【sin】.

coversed ~ 余矢. hyperbolic ~ 双曲正弦. integral ~ 積分正弦. inverse ~ 逆正弦. ~ curve 正弦曲線. ~ formula 正弦公式. ~ integral 正弦積分. ~ law, ~ rule 正弦法則. ~ theorem 正弦定理. ~ transform 正弦変換. ~ wave 正弦波. versed ~ 正矢.

SINGER, Isadore Manuel (1924-) シンガー.

single *a.* ただ一個の, 単一の.

~ sampling inspection 一回抜取検査. ~-valued function 一価関数. ~ word 単語.

singleton *n.* 一個; 一枚札; 単集合.

singular *a.* 特異な, 異常な. ⇨ regular. *n.* 単数, 単数形. ⇨ plural.

left [right] ~ 左[右]特異な. non-~ matrix 正則行列. ~ integral 特異積分. ~ integral equation 特異積分方程式. ~ kernel 特異核. ~ part 特異部分. ~ point 特異点. ~ problem 特異問題. ~ series 特異級数. ~ set 特異集合. ~ simplex 特異単体. ~ solution 特異解. ~ subspace 特異部分空間. ~ value 特異値.

singularity *n.* 特異性; 特異点.

algebraic ~ 代数(的)特異点. essential ~ 真性特異点. improper ~ 仮性特異点. isolated ~ 孤立特異点. logarithmic ~ 対数特異点. ordinary ~ 通常[通性]特異点. polar ~ 極. regular ~ 確定特異点. removable ~ 除去可能(な)特異点. resolution of ~ies 特異点の解消. ~ set 特異点集合. transcendental ~ 超越特異点.

sinistrorse *a.* 左巻きの. ⇨ dextrorse.
sink *n.* 吸込み. *vi., vt.* 沈む[める].
sinusoid *n.* シヌソイド. *a.* 正弦状の.
sinusoidal *a.* シヌソイドの, 正弦(状)の.
situation *n.* 状況.
sixtactic *a.* 六点接触の.
size *n.* 大きさ, 寸法.
memory ～ 記憶容量. ～ of population 母集団の大きさ. ～ of sample 標本の大きさ.
skeleton *n.* 骨格.
～ of simplex 単体の骨格.
skew *a.* 斜めの; 歪[非]対称の. *vt., vi.* ゆがめる, ゆがむ.
～ cissoid 歪疾走線. ～ coordinates 斜交座標. ～ distribution 非対称分布. ～ field, ～-field 非可換体, 斜体. ～ polynomial 歪(わい)多項式. ～ position 捩れの位置. ～ quadrilateral 捩れ四辺形. ～ surface 斜[歪]曲面.
skewfield *n.* 非可換体, 斜体.
skew-Hermitian *a.* 歪エルミートの.
～ form 歪エルミート形式. ～ matrix 歪エルミート行列.
skewness *n.* 歪度, 非対称度; ゆがみ, ひずみ.
skew-symmetric *a.* 歪[逆]対称の; 交代の.
～ matrix 歪[反; 逆]対称行列, 交代行列.
～ part 交代部分. ～ tensor 交代テンソル.
skew-symmetry *n.* 逆[歪]対称.
Skolem, Thoralf (1887–1963) スコレム.
slack *a.* 余裕のある; ゆるい.
～ variable 緩和変数, 余裕変数. ～ vector 緩和ベクトル.
slant *a.* 斜めの. *n.* 傾き; 斜面[線].
～ height 斜高. ～ product スラント積, 斜積.
slide *vi.* すべる. *vt.* すべらせる. *n.* すべること; 滑走.
～ rule, ～ding rule 計算尺. ～ding scale 滑尺.
slit *n.* 截線; 切れ目. *vt.* 切れ目を入れる.
circular ～ 円弧截線. parallel ～ 平行截線. radial ～ 放射截線. ～ disc [disk] 截線(のはいった)円板. ～ mapping 截線写像. ～ plane [domain] 截線(のはいった)平面[領域].
slope *n.* 傾き; 勾配, 傾斜.
cost ～ 費用増加率. ～ function 勾配関数.
slow *a.* 遅い, 緩慢な. ⇨ rapid.
～ increase 緩増加. ～ wave 遅い波.
slowly *ad.* 緩慢に.
～ increasing function 緩増加関数.
Smale, Stephen (1930–) スメール.
small *a.* 小さい. *n.* 小さな部分. ⇨ large.
in the ～ 局所的に, 小域的に. law of ～ numbers 小数の法則. ～ circle 小円. ～ inductive dimension 小さな帰納的次元. ～ sample 小標本. ～ submodule 小部分加群.
smooth *a.* 滑らかな. *vt.* 滑らかにする.
piecewise ～ 区分的に滑らかな. ～ curve [surface] 滑らかな曲線[曲面]. ～ function 滑らかな関数. ～ manifold 滑らかな多様体. ～ structure 平滑構造.
smoothing *n.* 平滑化. *a.* 平滑化する.
exponential ～ 指数平滑法. ～ problem 平滑化(の)問題. ～ effect 平滑化効果.
smoothness *n.* 滑らかさ.
～ condition 滑らかさの条件. ～ test 平滑検定.
SMSG School Mathematics Study Group.
Sobolev, Sergei L'vovič (1908–1989) ソボレフ.
socle *n.* 底(てい), ソクル, ソークル.
software *n.* ソフトウェア.
sojourn *n.* 滞在.
mean ～ 平均訪問回数.
solenoid *n.* ソレノイド, 円筒.
solenoidal *a.* 管状の, 層状の, ソレノイダル.
～ stream [flow] 管状流, 層流.
solid *a.* 立体の. *n.* 立体.
～ angle 立体角. ～ content 容積, 体積. ～ figure 立体図形. ～ geometry 立体幾何学, 空間幾何学. ～ harmonic function, ～ harmonics 体球(調和)関数, 球調和関数. ～ line 実線. ～ measure 体積. ～ number 立体数. ～ of revolution 回転体. ～ sphere 球体. ～ torus 円環体.
solidity *n.* 立体性; 体積, 容積.

solidus *n.* 斜線【/】.

soluble ＝solvable.

solution *n.* 解; 解答; 解法.
algebraic ～ 代数的解(法). approximate ～ 近似解. elementary ～ 素解; 基本解. formal ～ 形式解. fundamental ～ 基本解. general ～ 一般解. numerical ～ 数値解(法). optimal ～ 最適解. particular ～ 特殊解. periodic ～ 周期解. primitive ～ 原始解. proper ～ 固有解. series ～ 級数解. singular ～ 特異解. ～ curve [surface] 解曲線[面]. ～ of triangle 三角形の解法. ～ space 解空間. strong ～ 強解. trivial ～ 自明な解. unique ～ 唯一の解. weak ～ 弱解.

solvability *n.* 解決可能性; 可解性.

solvable *a.* 可解な; 解ける.
algebraically ～ 代数的に解ける. ～ algebra 可解代数. ～ by radicals ベキ根で解ける. ～ group 可解群.

solve *vt.* 解く; 解答する.

some *a.* いくらかの, ある【限定】.

Sommerfeld, Arnold Johann Wilhelm (1868–1951) ゾンマーフェルト.

sonic *a.* 音(波)の; 音速の.

Sonin(e), Nikolai Jakovlevič (1849–1915) ソニン.

SOR Successive Over-Relaxation. 逐次加速緩和法.

sorites *n.* 連鎖式(三段論法), 連鎖論法.

sort *n.* 種類. *vt.* 分類する.
a ～ of ある種の. ～ file 分類用ファイル.

sorting *n.* 分類(すること), 順序づけ.
radix ～ 基数法. ～ by exchanging 交換法. ～ by insertion 挿入整列法. ～ by merging 併合整列法.

source *n.* 源泉, 源, 原始; 湧出し; 始点.
heat [light] ～ 熱[光]源. ～-free 湧出しのない. ～ language 原始言語, 原語. ～ material 原始資料.

south *n.* 南. *a.* 南の.
～(ern) hemisphere 南半球. ～ pole 南極.

southern *a.* 南の; 南向きの.

space *n.* 空間; 間隔. *vt., vi.* 間隔をおく.
abstract ～ 抽象空間. affine ～ アフィン空間. Baire ～ ベール空間. Banach ～ バナッハ空間. base ～ 底空間, 基礎空間. Besov ～ ベゾフ空間. covering ～ 被覆空間. dual ～ 双対空間. equally ～d 等間隔の. euclidean [Euclidean] ～ ユークリッド空間. function ～ 関数空間. Hermitian ～ エルミート空間. Hilbert ～ ヒルベルト空間. L^p ～ L^p 空間. lens ～ レンズ空間. linear ～ 線形空間. measurable ～ 可測空間. measure ～ 測度空間. metric ～ 距離[計量]空間. modulation ～ モデュレーション空間. *n*-dimensional ～, *n*-～ *n* 次元空間. Noetherian ～ ネーター空間. normed ～ ノルム空間. null ～ 零空間. projective ～ 射影空間. quotient ～ 商空間. Riemannian ～ リーマン空間. ～ at infinity 無限遠空間. ～ curve 空間曲線. ～ form 空間形. ～ geometry 空間幾何学. ～ lattice 空間格子. ～ of constant curvature 定曲率空間. ～ of elementary events 基礎空間. ～ reflection 空間反転. Stein ～ シュタイン空間. symmetric ～ 対称空間. topological ～ 位相空間. underlying ～ 基底空間. unitary ～ ユニタリ空間.

space-like *a.* 空間的な.

space-time *n.* 時空.
～ Brownian motion 時空ブラウン運動.

spacial ＝spatial.

span *n.* スパン. *vi., vt.* 張る, 渡す.
make ～ 作業期間幅. space ～ned by で張られた空間.

sparse *a.* 疎な, まばらな.
～ matrix 疎行列.

spatial *a.* 空間の; 空間的な.
～ distribution 空間(的な)分布. ～ lattice 空間格子.

spatially *ad.* 空間的に.
～ homogeneous 空間的に一様な. ～ isomorphic 空間的同型[形]な.

special *a.* 特殊な, 特殊の.
～ curve(s) 特殊曲線. ～ function(s) 特殊関数. ～ linear group 特殊線形群. ～ orthogonal group 特殊直交群. ～ surface(s) 特殊曲面. ～ theory of relativity 特殊相対(性理)論. ～ unitary group 特殊ユニタリ

群.

speciality *n.* 特殊性; 特別品.
~ index, index of ~ 特殊指数.

specialization *n.* 特殊化.

specific *a.* 明確な; 特定の; 特有な; 特殊な; 特異な.

specification *n.* 仕様.

specificity *n.* 特異度.

specified *a.* 指定された.

spectral *a.* スペクトルの.
~ analysis スペクトル解析. ~ decomposition スペクトル分解. ~ integral スペクトル積分. ~ measure スペクトル測度. ~ multiplicity スペクトルの重複度. ~ radius スペクトル半径. ~ resolution スペクトル分解. ~ sequence スペクトル系列. ~ synthesis スペクトル総合.

spectrally *ad.* スペクトル的に.
~ isomorphic スペクトル同型[形]な.

spectrum (*pl.* -tra, -trums) *n.* スペクトル.
continuous ~ 連続スペクトル. discrete ~ 離散スペクトル. point ~ 点スペクトル. prime ~ プライムスペクトル. ~ decomposition [resolution] スペクトル分解. ~ of a (commutative) ring (可換)環のスペクトル.

speed *n.* 速さ.
average ~ 平均の速さ. high (-) ~ 高速. instantaneous ~ 瞬間の速さ. ~ per second 秒速.

sphere *n.* 球, 球面.
complex ~ 複素球面. concentric ~s 同心球. homological ~ ホモロジー球面. open ~ 開球. Riemann ~ リーマン球(面), 数球面. ~ bundle 球面束. ~ congruence 球叢. ~ theorem 球面定理.

spherical *a.* 球の, 球面の.
~ angle 球面角. ~ cone 球錐. ~ coordinates 球座標. ~ curve 球面曲線. ~ derivative 球面導関数. ~ distance 球面距離. ~ excess 球面過剰. ~ function 球関数. ~ geometry 球面幾何学. ~ lune 球面月形. ~ representation 球面表示. ~ ring 球環. ~ segment 球欠, 球台. ~ shell 球殻. ~ space 球面空間. ~ surface 球面. ~ harmonic function, ~ harmonics 球面(調和)関数. ~ triangle 球面三角形. ~ trigonometry 球面三角法. ~ wave 球面波.

spherics *n.* 球面幾何学; 球面三角法.

spheroid *n.* 回転楕円体[面].
oblate ~ 扁平回転楕円体[面]. prolate ~ 扁長回転楕円体[面].

spheroidal *a.* 回転楕円体[面]の.
~ function 回転楕円体関数. ~ wave function 回転楕円体波動関数.

spin *n.* スピン.
~ bundle スピン束. ~ mapping スピン写像. ~ representation スピン表現.

spinode *n.* 尖点.

spinor *n.* スピノル.
contravariant [covariant] ~ 反変[共変]スピノル. dotted ~ 付点スピノル. mixed ~ 混合スピノル. ~ group スピノル群. undotted ~ 無点スピノル.

spinorial *a.* スピンの.
~ norm スピン・ノルム.

spiral *a.* 螺旋形の; 渦巻線の. *n.* 螺線; 渦巻線, 螺旋, 匝線.
algebraic ~ 代数螺線. Archimedean ~ アルキメデス螺線. Cornu's ~ コルニュの螺線[⇨ clothoid]. equiangular ~ 等角螺線. hyperbolic ~ 双曲渦[螺]線. isoclinal ~ 等傾螺線. logarithmic ~ 対数螺線. reciprocal ~ 逆匝線. ~ motion 螺旋運動. ~ point 渦心. ~ surface 螺旋面.

spiro- *pref.*「螺線形, 螺旋状; 渦巻」の意.

spline *n.* スプライン.
~ function スプライン関数.

split *vi.* 分裂[分解]する. *vt.* 割る, 裂く; (から)裂き取る【off, from】. *a.* 分裂[分解]した, 分けた. *n.* 按分線.
~ plot experiment 分割法.

splitting *a.* 分裂[分解]している.
minimal ~ field 最小分解体. ~ field 分解体.

sporadic *a.* 散発的な; 孤立した.
~simple group 散在型単純群.

spot *n.* 点; すこし, 一口.
~ inspection つまみ取り検査.

spray *n.* スプレイ.

Spur *n.* (*G.* =trace.) 跡，シュプール，トレース.

sq. sequentia (*L.* =the following.) ⇨ seq.; square.

square *n.* 正方形，四角; 二乗，平方; (直角)定規[木]. *a.* 正方形の; 平方の，二乗の. *ad.* 直角に，四角に. *vt.* 四角にする; 二乗する. *vi.* 直角をなす.
chi-~ distribution カイ二[自]乗分布. magic ~ 魔方陣. perfect ~ 完全平方. set ~ 三角定規. ~ diagram 正方形グラフ. ~(d) matrix 正方行列. ~-free 無平方の. ~ integrable 平方可積な. ~ measure [number] 平方積[数]. ~ root 二乗根，平方根. ~ scale 平方尺，平方目盛. ~-sum 平方和，二乗和. ~ table 正方分割表. T-~ T定規. unit ~ 単位正方形.

squarely *ad.* 方形に; 直角に.

s.t. =such that.

stability *n.* 安定，安定性.
asymptotic ~ 漸近安定性. orbital ~ 軌道安定性. ~ analysis 安定性解析.

stabilizer *n.* 安定化部分群; 固定部分群; 安定化子.

stable *a.* 安定な，安定した.
absolutely ~ 絶対安定な. ~ distribution 安定分布. ~ equilibrium 安定平衡. ~ equivalence 安定同値. ~ in both directions 両側安定な. ~ population 安定人口. ~ process 安定過程. ~ range 安定域. ~ solution 安定解. ~ state 安定状態.

stably *ad.* 安定して.
~ equivalent 安定同値な. ~ parallelizable 安定平行性をもつ.

stalk *n.* 茎.
~ of presheaf 前層の茎.

standard *a.* 標準の，標準的な. *n.* 標準，基準; 規格.
~ basis 標準基. ~ complex 標準複体. ~ coordinate system 標準座標系. ~ deviation 標準偏差. ~ error 標準誤差. ~ form 標準形. ~ normal distribution 標準正規分布. ~ position 標準の位置. ~ resolution 標準分解. ~ scale 標準目盛. ~ vector space 基準ベクトル空間.

standardization *n.* 標準化，規格化; 統一.

standardize *vt.* 標準[基準，規格]に合わせる，標準[基準，規格]化する; 統一する.

star *a.* 星形[型]の，星状の. *n.* 星; 星形[型]集合，星状体; 星印【*】. *vt.* 星状にする.
principal ~ 主星領域. ~ body 星形体. ~ convergence 星形収束. ~red polygon 星状多角[辺]形. ~ region [domain] 星状領域. ~-shaped 星形の.

star-finite *a.* 星形[型]有限な.
~ property 星形有限性.

star(-)like *a.* 星形[型，状]の.
~ domain 星形[型，状]領域. ~ function [mapping] 星形[型]関数[写像].

starlikeness *n.* 星形[型]性.
radius of ~ 星形[型]半径.

star-refinement *n.* 星形[型，状]細分.

star-shaped *a.* =starlike.

start *vi.* 出発する. *vt.* 出発させる. *n.* 出発; 開始.

starting *a.* 出発する. *n.* 出発[開始]すること.
~ point 始点，振出し; 出発点. ~ value 出発値.

stat. static; stationary.

state *n.* 状態. *vt.* 陳述する，のべる.
decreasing [increasing] ~ 減少[増加]の状態. stable ~ 安定状態. ~ of control 管理状態. ~ space 状態空間. stationary ~ 定常状態.

statement *n.* 陳述，命題; 計算書.
compound ~ 合成命題.

static *a.* 静的な; 静止の.

statics *n.* 静力学.
electro ~ 静電気学. hydro ~ 流体静力学.

stationary *a.* 定常な，停留の; 動かない，定着した.
~ curve 定常[停留]曲線. ~ function 定常[停留]関数. ~ information source 定常情報源. ~ point 定常[停留]点，定留点. ~ process 定常過程. ~ state 定常状態. ~ value 定常[停留]値. ~ wave 定常波.

statist. statistics; statistical.

statistic *a.* =statistical. *n.* 統計量.

ancillary ~ 補助統計量. consistent ~ 一致統計量. sufficient ~ 十分統計量. unbiased ~ 不偏統計量.

statistical *a.* 統計(学上)の, 統計的な.
~ decision function 統計的決定関数. ~ estimation 統計的推定. ~ hypothesis 統計的仮設. ~ inference 統計的推論. ~ linear model 統計的線形モデル. ~ mechanics 統計力学. ~ probability 統計的確率. ~ quality control 統計的品質管理. ~ significance 統計的有意性.

statistics *n.* 統計学; 統計.
descriptive ~ 記述統計学. inductive ~ 推測統計学. mathematical ~ 数理統計学. population ~ 人口統計(学).

stay *vi.* とどまる; 滞在する. *n.* 滞在.
~ing time 滞在時[期]間.

steady *a.* 定常の.
~ distribution 定常分布. ~ flow 定常流. ~ state 定常状態. ~ wave 定常波.

Steenrod, Norman Earl (1910-1971) スティーンロッド.

steep *a.* 険しい; 急勾配の.
~est descent 最急降下.

Stein, Elias Menachem (1931-) スタイン.

Stein, Karl (1913-2000) シュタイン.

Steiner, Jakob (1796-1863) シュタイナー.

Steinhaus, Hugo (1887-1972) シュタインハウス.

Steinitz, Ernst (1871-1928) シュタイニッツ.

stem *n.* 幹.
~ and leaf plot 幹葉[みきは]図.

step *n.* 歩み; 階段.
backward ~ 後退段階. ~-down operator 下降演算子. ~ function 階段関数. ~ping stone method 飛石法. ~-up operator 上昇演算子.

step-formed *a.* 階段状の.
~ graph 階段状グラフ.

step-size *n.* 刻み幅; 歩み幅.

steradian *n.* ステラジアン【立体角の単位】.

stereographic *a.* 立体画法の.
~ projection 立体射影.

stereometry *n.* 求積法; 体積測定(法).

Stevin, Simon (1548-1620) ステヴィン.

Stieltjes, Thomas Jean (1856-1894) スティルチェス.

Stifel, Michael (1487-1567) シュティフェル.

stiff *a.* 硬い.
~ system 硬い系.

still *a.* 静止した, 流れのない.
~ water 静水. ~ more なおさら.

stimulus (*pl.* -li) 刺激.

Stirling, James (1692-1770) スターリング.

stochastic *a.* 確率の, 確率的な; 推計の.
~ convergence 確率収束. ~ differential equation 確率微分方程式. ~ filtering 確率フィルタリング. ~ function 確率関数. ~ integral 確率積分. ~ matrix 確率行列. ~ model 確率(論)的模型. ~ paper 推計紙. ~ process 確率過程. ~ programming 確率的計画法. ~ variable 確率変数.

stochastically *ad.* 確率的に, 推計的に.
~ large [small] 確率的に大き[小さ]い.

stochastics *n.* 推計学, 推測統計学.

stock *n.* 貯蔵, 在庫品.
safety ~ 安全在庫. ~ management 在庫管理.

Stokes, Sir George Gabriel (1819-1903) ストークス.

Stone, Marshall Harvey (1903-1989) ストーン.

stop *vt.* 止める. *n.* 止まる, 止まること; 停止.
optimal ~ping problem 最適停止問題. optimal ~ping-time problem 最適停止時刻問題. ~ping problem 停止問題. ~ping time 停止時刻.

storage *n.* 貯蔵; 蓄積.

store *n.* 貯蔵. *vt.* 貯蔵する; 蓄積する.

straight *a.* まっすぐな; 一直線の.
~ angle 平角【180°】. ~en 直化[まっすぐに]する. ~ line 直線.

strain *n.* 変形; 歪み. *vi., vt.* 引っ張る; 変形させる.
~ tensor 変形テンソル.

strategic *a.* 方略的な，戦略の．
～ variable 戦略変数．

strategy *n.* 方略，戦略，策略，術策．
mixed ～ 混合戦略．optimal ～ 最適戦略．pure ～ 純粋戦略．～ space 戦略集合[空間]．

stratification *n.* 層別，層化，層別化．

stratified *a.* 層化された．
～ sample 層化標本．～ sampling 層化抽出(法)．

stratify *vt.* 層にする，重ねる．

stratum (*pl.* -ta) *n.* 層．

stream *n.* 流れ．⇨ flow.
～ function 流れの関数．～(-)line 流線．～-line body 流線形．

strength *n.* 強さ，強度．

stress *n.* 圧力; 応力; 歪み．
normal ～ 法線応力．shearing ～ ずれ応力．～ tensor 応力テンソル．tangential ～ 接線応力．

stretch *vi.* 伸びる．
～ing and shrinking 伸縮．

strict *a.* 厳密な; 精密な; 狭義の．
～ implication 厳密な含意．～ inductive limit 狭義(の)帰納的極限．～ inequality 狭義の[等号を含まない]不等式．in the ～ sense 狭義に．

strictly *ad.* 厳密に; 精密に; 狭義に．
～ convex [concave] function 狭義(の)凸[凹]関数．～ decreasing 狭義に減少の．～ determined 確定的な．～ increasing 狭義に増加の．～ monotone 狭義(に)単調な．

string *n.* 弦, 絃; 紐．
vibrating ～ 振動絃．super ～ theory 超弦理論．

strip *n.* 帯; 帯状面分．
bicharacteristic ～ 陪特性帯．boundary ～ 境界帯．characteristic ～ 特性帯．parallel ～ 平行帯(状面分)．period ～ 周期帯．～ domain [region] 帯状領域．

strong *a.* 強い．⇨ weak.
～ convergence 強収束．～er form いっそう強い形．～ extension 強拡張．～ law of large numbers 大数の強法則．～ mixing property 強混合性．～ solution 強解．～ topology 強位相．～ variation 強変分．

strongly *ad.* 強く．⇨ weakly.
～ continuous representation 強連続表現．～ elliptic 強楕円的な[型の]．～ measurable 強可測な．～ monotone 強い意味で単調な．～ plurisubharmonic 強多重劣調和な．～ polynomial 強多項式の．～ pseudo(-)convex 強擬凸な．～ stationary process 強定常過程．

strophoid *n.* ストロフォイド．

structural *a.* 構造(上)の．
～ equation 構造方程式．～ stability 構造安定性．

structure *n.* 構造．
algebraic ～ 代数的構造．analytic ～ 解析構造．complex ～ 複素構造．conformal [differentiable] ～ 等角[可微分]構造．contact [metric] ～ 接触[計量]構造．logarithmic ～ ログ構造．quaternionic ～ 四元数構造．～ analysis 構造分析．～ constant(s) 構造定数．～ equation 構造方程式．～ formula 構造公式．～ group 構造群．～ morphism 構造射．～ space 構造空間．～ theorem 構造定理．

STRUVE, Michael (1955-　) ストルーベ．

STRUVE, Ludwig (1858-1920), Wilhelm (1793-1864) ストルーベ．

STUDENT ＝William Sealy Gosset (1876-1937) スチューデント．

study *n.* 学習; 研究．*vt., vi.* 学習する; 研究する．

STUDY, Eduard (1862-1930) ストゥディ．

STURM, Jacques Charles François (1803-1855) シュテュルム．

sub- *pref.* 「部分; 下位，劣」の意．⇨ super-.

subadditive *a.* 劣加法的な．
～ functional 劣加法(的)汎関数．

subadditivity *n.* 劣加法性．

subalgebra *n.* 部分多元環．
*-～ ＊部分環．

subbase, subbasis *n.* 部分基，準基底．

subbundle *n.* 部分バンドル．

subcategory *n.* 部分圏．
full ～ 充満部分圏．

subclass *n.* 部分族.
subcomplex *n.* 部分複体.
subcover(ing) *n.* 部分被覆.
open ～ 部分開被覆.
subdifferential *n.* 劣微分.
subdivision *n.* 細分, 分割.
barycentric ～ 重心細分.
subdomain *n.* 部分領域.
closed ～ 部分閉領域, 閉部分領域.
subduplicate *a.* 平方根の.
～ ratio 平方根比.
subfamily *n.* 部分族.
subfield *n.* 部分体.
subfile *n.* 副ファイル.
subgradient *n.* 劣勾配.
subgraph *n.* 部分グラフ.
subgroup *n.* 部分群.
cyclic ～ 巡回部分群. invariant ～ 不変部分群. normal ～ 正規部分群. proper ～ 真部分群. trivial ～ 自明な部分群.
subharmonic *a.* 劣調和な.
pluri ～ 多重劣調和な. ～ extension 劣調和拡大. ～ function 劣調和関数.
subharmonicity *n.* 劣調和性.
subindex (*pl.* -xes, -dices) *n.* 副指数, 副添字; 下付文字.
subinvariant *a.* 劣不変な.
～ measure 劣不変測度.
subject *n.* 主題; 事項; 題目, 論題. *a.* 従属する【to】.
～ index 事項索引.
subjective *a.* 主観的な.
～ probability 主観確率.
sublattice *n.* 部分束; 部分格子.
submanifold *n.* 部分多様体.
invariant ～ 不変部分多様体.
submartingale *n.* 劣マルチンゲール.
submatrix *n.* 小[部分]行列.
complementary ～ 余[補]小行列.
submersion *n.* 沈め込み.
submodular *n.* 劣モジュラー.
submodule *n.* 部分加群.
A-～ (＝sub-*A*-module) *A* 部分加群. essential ～ 本質部分加群. small ～ 小部分加群. superfluous ～ 余剰部分加群.
submultiple *n.* 約数.
subnormal *n.* 法線影. *a.* 劣正規な; 正常以下の.
～ operator 劣正規作用素. ～ subgroup 正規鎖部分群.
subobject *n.* 部分対象.
subordinate *a.* 従属の. *vt.* 従属させる.
～ function 従属関数.
subordination *n.* サブオーディネーション; 従属(性); 従属操作.
～ principle 従属原理.
subordinator *n.* 従属(化)過程.
subprogram *n.* 副プログラム.
subregion *n.* 部分領域[面分].
subrepresentation *n.* 部分表現.
subring *n.* 部分環.
sub-sampling *n.* 二段抽出.
subscript *n.* 下付きの添字.
subsequence *n.* 部分(系)列.
convergent ～ 収束部分列.
subset *n.* 部分集合.
closed ～ 閉部分集合. cofinal ～ 共終部分集合. proper ～ 真部分集合.
subsidiary *a.* 補助の.
～ condition 付帯条件. ～ equation 補助方程式.
subsonic *n.* 亜音速の.
～ wave 亜音速波.
subspace *n.* 部分空間.
complementary ～ 補部分空間. flat ～ 平坦な部分空間. invariant ～ 不変部分空間. metric ～ 部分距離空間. proper ～ 真部分空間. trivial ～ トリビアル[自明]な部分空間. vector ～ 部分ベクトル空間.
substance *n.* 実質, 実体; 要旨, 大意.
substantial *a.* 実質的な.
substantially *ad.* 実質的に.
substitute *vt., vi.* 置換する; 代入する. *n.* 代用物.
substitution *n.* 代入, 置換.
backward ～ 後退代入. forward ～ 前進代入. integration by ～ 置換積分(法). ～ principle 代入の原理. successive ～ 逐次代入.
sub-stratification *n.* 二段層化.

subsystem *n.* 部分系.
subtangent *n.* 接線影.
subtend *vt.* (対辺が弧角に)対する.
subtense *n.* 弦; 対辺.
subtheory *n.* 部分理論.
subtract *vt.* 引く, 減じる.
subtracter *n.* 減数.
subtraction *n.* 引き算, 減法【−】.
subtractive *a.* 減じる, 減ずべき, 負号【−】で表された, 負の.
subtrahend *n.* 減数.
subtriplicate *a.* 立方根の, 立方根で表される.
subvariety *n.* 部分多様体.
success *n.* 成功; 当り.
successive *a.* 逐次の.
~ approximation 逐次近似. ~ derivatives 逐次導関数. ~ differentiation 逐次微分. ~ iteration 逐次代入[反復]. ~ minima 逐次最小. ~ minimum points 逐次最小点. ~ over-relaxation 逐次加速緩和法. ~ substitution 逐次代入.
successor *n.* 後者. ⇨ predecessor.
sufficient *a.* 十分な.
necessary and ~ 必要十分な. ~ condition 十分条件. ~ statistic 十分[充足]統計量.
sufficiently *ad.* 十分に.
~ large [small] 十分大き[小さ]い. ~ many 十分多い.
suffix (*pl.* -ffices, -xes) *n.* 添数, 添字, サフィックス.
sum *n.* 和; 和分; 計. *vt.* 合計する; 和分する. *vi.* 合計して(~に)なる【into】; 要約する.
algebraic ~ 代数和. constant ~ 定和. direct ~ 直和. double ~ 二重和. finite [infinite] ~ 有限[無限]和. logical ~ 論理和. partial ~ 部分和. power-~ ベキ和. product ~, ~ of products 積和. ~ event 和事象. ~ of squares, square-~ 二乗和, 平方和. ~ up 合計する. total ~, ~ total 総和; 合計. trigonometric ~ 三角和.
summability *n.* 総和可能性.
absolute ~ 絶対総和可能性. field of ~ 総和域. logarithmic ~ 対数総和法. ~ method 総和法. ~ polygon 総和多角形. ~ region 総和領域.
summable *a.* 総和可能な; 可積な.
Cesàro-~ チェザロ総和可能な. locally ~ 局所可積な. strongly ~ 強総和可能な. ~ function (ルベーグ)可積関数.
summand *n.* 被加数; 加数, 和の一項.
direct ~ 直和因子.
summary *n., a.* 摘要(の), 要約(の); 要旨(の).
five number ~ 5 数要約.
summation *n.* 求和, 加法; 総和(法); 和分.
~ convention 総和規約. ~ formula 総和公式. ~ notation, ~ sign, ~ symbol 総和記号【Σ】; 和分記号【$\mathfrak{S}$】.
summit *n.* 頂点; 稜.
sunk *a.* 沈んだ. ⇨ sink.
~ cost 埋没費用.
sup =supremum. 上限. ⇨ inf.
super- *pref.*「上; 超」の意. ⇨ sub-.
superabundance *n.* 過多; 余数.
superficial *a.* 表面の; 面積の, 平方の.
~ content (表)面積.
superficies (*pl.* -) *n.* 表面; 面積.
superfluous [su:pə́r:fluəs] *a.* 余剰(な).
~ submodule 余剰部分加群.
superharmonic *a.* 優調和な.
~ function 優調和関数. ~ transformation 優調和変換.
superharmonicity *n.* 優調和性.
superior *a.* 上級の; 高い, 高位の【to】. ⇨ inferior.
limes ~ 上極限. ~ angle 優角. ~ arc 優弧. ~ function 優関数. ~ limit 上極限.
supermartingale *n.* 優マルチンゲール.
superosculation *n.* 超過接触.
superpose *vt.* 重ねる; 重ね合わせる.
superposition *n.* 重ね合わせ, 重畳.
principle of ~ 重ね合わせの原理, 重畳原理.
superscript *n.* 上つきの添字.
superset *n.* 超集合.
supersolvable *a.* 超可解な.
supersonic *a.* 超音波の; 超音速の.

～ wave 超音(速)波.

super(-)string *n.* 超弦.
～ theory 超弦理論.

supplement *n.* 補角; 補遺, 付録. *vt.* 補充する, 補足する.
～ed algebra 係数添加環.

supplemental ＝supplementary.
～ chord 補弦.

supplementary *a.* 補角の; 補弧の; 補足の.
～ angle 補角. ～ chord 補弦. ～ interval 劣弧.

supplementation *n.* 補足, 追加, 補完, 補充; 補遺, 増補; 付録, 増刊.

support *n.* 台; 支持(するもの). *vt.* 支持する; 支える.
be ～ed on *E E* 上に台をもつ. ～ functional 支持汎関数. ～ of function 関数の台. ～ point 支持点. ～ variety 台多様体.

supporting *a.* 支持する.
～ hyperplane 支持超平面. ～ line 支持線, 触線. ～ metric 支持計量. ～ plane 支持平面. ～ point 支持点; 示点.

suppose *vt.* 仮定する.

supposition *n.* 仮定.

supremum (*pl.* -ma) *n.* (*L.*) 上限【sup】. ⇨ infimum.
essential ～ 本質的上限.

surd *a.* 無理数の; 不尽根の. *n.* 無理数; 不尽根数.
like ～s 同類根数.

sure *a.* 確かである.

surface *n.* 面, 曲面; 表面.
algebraic ～ 代数曲面. bilateral ～ 両側曲面. boundary ～ 境界面. closed ～ 閉曲面. conical ～ 錐面. covering ～ 被覆面. curved ～ 曲面. cylindrical ～ 柱[筒, 壔]面. developable ～ 可展面, 展開可能曲面. minimal ～ 極小曲面. one-sided ～ 単側曲面. open ～ 開曲面; 開いた面. pitch ～ ピッチ面. plane [planar] ～ 平面. Riemann ～ リーマン面. ruled ～ 線織(曲)面. ～ area 曲面積, 表面積. ～ element 面素, 面積要素. ～ harmonics 球面(調和)関数. ～ integral 面積分. ～ of revolution 回転(曲)面. ～ patch 曲面片. ～ wave 表面波. transcendental ～ 超越曲面. tubular ～ 管面. two-sided ～ 両側曲面. umbilic(al) ～ 臍点曲面. unilateral ～ 単側曲面.

surgery *n.* 手術.

surjection *n.* 全射.
canonical ～ 標準的全射.

surjective *a.* 全射の.
～ homomorphism 全射準同型[形]. ～ mapping 全射写像.

surround *vt.* 囲む; 包囲する.

survey *n.* 調査; 概説; 測量.
mail ～ 郵送調査. pilot ～ 試験[予備]調査. trigonometrical ～ing 三角測量.

survival *n.* 生存; 残存; 残存者[物].
～ curve 生存曲線. ～ function 生存関数. ～ rate 生存率, 残存率.

SUSLIN, Andrei (1950-　) ススリン.

SUSLIN, Mikhail Yakovlevič (1894-1919) ススリン.

suspension *n.* 懸垂, 浮遊.
reduced ～ 約懸垂. ～ isomorphism 懸垂同型[形](写像). ～ theorem 懸垂定理.

sweep *n.* 掃過. *vt., vi.* 掃く.
～ out 掃き出す.

sweeping *n.* 掃過; 一掃. ⇨ balayage.
～-out 掃散. ～-out method 掃き出し法. ～-out principle 掃散原理.

switch *n.* 開閉器, スイッチ; 切替. *vt.* 消す, 切る【off】; つける, 通じる【on】. *vi.* 転換する; 変わる.
～ing locus [curve] 切換[替]線.

syllogism *n.* 三段論法; 演繹(法).
categorical ～ 定言三段論法. conditional ～ 仮言三段論法. disjunctive ～ 選言三段論法. hypothetical ～ 仮言三段論法.

syllogistic *a.* 三段論法(的)な; 演繹の.

SYLOW, Peter Ludwig (1832-1918) シロー.

SYLVESTER, James Joseph (1814-1897) シルベスター.

symbol *n.* 記号; 符号; 表象.
algebraic ～ 代数記号. functional ～ 関数記号. Kronecker('s) ～ クロネッカーの記号. Landau's ～ ランダウの記号. predicate ～ 述語記号. summation ～ 総和記号.

three index ～ 三添字記号.

symbolic *a.* 記号の; 符号の.
～ language 記号言語. ～ logic 記号論理(学). ～ power 記号ベキ. ～ solution 記号解法.

symbolized *a.* 符号化された.
～ pattern 符号化されたパターン.

symmedian *n.* 類似中線.
～ point 類似重心. ～ prism 類似角柱.

symmetric *a.* 対称(的)な.
axially ～, axi～ 軸対称な. binary ～ channel 二進対称通信路. elementary ～ expression 基本対称式. elementary ～ polynomial 基本対称多項式. Hermitian ～ space 対称エルミート空間. locally ～ 局所対称な. rotationally ～ 回転対称な. spherically ～ 球対称な. ～ algebra 対称代数, 対称多元環. ～ configuration 対称配置. ～ difference 対称差. ～ distribution 対称分布. ～ domain 対称領域. ～ expression 対称式. ～ figure 対称図形. ～ function 対称関数. ～ group 対称群. ～ kernel 対称核. ～ law 対称法則[律]. ～ matrix 対称行列. ～ partition 対称区分け. ～ polynomial 対称多項式. ～ position 対称な位置. ～ relation 対称な関係. ～ space 対称空間. ～ tensor 対称テンソル. ～ transformation 対称変換.

symmetrical ＝symmetric.

symmetrizable *a.* 対称化可能な.
～ kernel 対称化可能な核.

symmetrization *n.* 対称化; 対称にすること.
circular ～ 円[中心]対称化. Steiner('s) ～ シュタイナー(の)対称化.

symmetrizer *n.* 対称化作用素; 対称子.

symmetry *n.* 対称; 対称変換; 鏡像; 鏡映; 折返し.
axial ～ 軸対称(変換). axis of ～ 対称軸. central ～ 中心対称(変換). oblique ～ 斜対称(変換). plane of ～ 対称面. point of ～ 対称点. point ～ 点対称. position of ～ 対称の位置. principle of ～, ～ principle 鏡像の原理. ～ relation 対称関係.

symplectic *a.* シンプレクティックな.
～ group シンプレクティック群, 斜交群. ～ transformation シンプレクティック変換.

symposium (*pl.* -siums, -sia) *n.* シンポジウム.

synchronization *n.* 同期, 同期化.

synchronous *a.* 同時の, 同期(式)の.

synthesis (*pl.* -ses) *n.* 総合; 合成, 構成.

synthetic *a.* 総合(的)の.
～ division 組立除法. ～ geometry 総合幾何学.

system *n.* 系; 体系; 方式.
algebraic ～ 代数系. binary (number) ～ 二進法. complete ～ 完全系. complex ～ 複雑系. coordinate ～ 座標系. decimal ～ 十進法. direct ～ 直[順]系, 帰納系(＝inductive ～). generator ～ 生成系. inductive ～ 帰納系. inverse ～ 逆系(＝projective ～ 射影(的)系). involution ～ 包合系. left-handed ～ 左手系. linear ～ 線形系. orthogonal ～ 直交系. orthonormal(ized) ～ 正規直交系. projective ～ 射影(的)系. right-handed ～ 右手系. root ～ 根系, ルート系. stiff ～ 硬い系. ～ engineering システム工学. ～ of absolute units 絶対単位系. ～ of axioms 公理系. ～ of equations 方程式系, 連立方程式. ～ of generators 生成系. ～ of inequalities 連立不等式. ～ of parameters パラメータ系. ～ of rational [real; complex] numbers 有理[実; 複素]数系. ～ of representatives 代表系. ～ of transitivity 推移類. ～ simulation システム・シミュレーション. tetragonal ～ 正方晶系. zero ～ 零系.

systematic *a.* 系統的な, 体系的な, 組織的な.
～ error 系統(的)誤差. ～ sampling 系統(的)抽出法. ～ theory 体系的理論.

systematization *n.* 体系化, 組織化.

syzygy *n.* シズィジ, 合[ごう], 衝[しょう].
～ theory シズィジ理論.

Szász, Otto (1884-1952) サース.

Szegö, Gabor (1895-1985) セゲ.

T

table *n.* 表, テーブル.
addition ~ 加法九々表. analysis of variance ~ 分散分析表. contingency ~ 分割表, 偶然表. correlation ~ 相関表. difference ~ 差分表. function(al) ~ 関数表. logarithmic ~ 対数表. multiplication ~ 掛算九々表, 乗法表. numerical ~ 数表. ~ of common logarithms 常用対数表. ~ of prime numbers 素数表. ~ of random numbers 乱数表.

tableau (*pl.* -leaux) *n.* (*F.*) タブロー; 画表.
simplex ~ 単体表, シンプレックス・タブロー.

tabular *a.* 表の.
~ difference 表差.

tabulation *n.* 表作成.

tabulate *vt., vi.* 表にする; 要約する.

tac-locus *n.* タック・ローカス【曲線族の接点の軌跡】.

tacnode *n.* 二重尖点.

tag *n.* 付箋; 標識, しるし.

Takagi, teiji (1875-1960) 高木貞治.

tame *a.* 順の; 平凡な.
~ knot 順結び. ~ ramification 順分岐.

tandem *a.* 一列の; 縦に並んだ.
~ queue 連続待ち行列.

tangency *n.* 接触.
point of ~ 接点.

tangent *a.* 接線の, 接する; 正接の. *n.* 接線; 正接; タンジェント.
common ~ 共通接線. double ~ 二重接線. ~ bundle 接束; 接バンドル. ~ coefficient 正接係数. ~ cone 接錐. ~ curve 正接曲線. ~ hyperplane 接超平面. ~ line 接線. ~ plane 接平面. ~ rule 正接法則. ~ space 接空間. ~ surface 接線曲面. ~ vector 接(線)ベクトル.

tangential *a.* 接線の, 接する; 正接の; 接線方向の.
polar ~ coordinates 接線極座標. ~ component 接線成分. ~ direction 接線方向. ~ equation 接線方程式. ~ plane 接平面. ~ stress 接線応力.

tapper *n.* 盗聴者.

target *n.* 標的; 目標物, 目的物; 終点; 視準標.
~ language 目的言語. ~ variable 目標変数.

Tartaglia, Niccolò (1500?-1557) タルタリア.

Tate, John Torrence (1925-) テイト.

tau *n.* タウ. ギリシア語アルファベットの19番目の文字【T, τ】.

Tauber, Alfred (1866-1942) タウバー.

Tauberian *a.* タウバーの.
~ theorem(s) タウバー(型)の定理.

tautochrone *n.* 等時曲線.

tautological *a.* 同語反復的な.

tautology *n.* トートロジー; 同語反復.

Taylor, Brook (1685-1731) テイラー.

t-distribution t 分布.

technical *a.* 技術の; 専門の, 学術の.
~ term(s) 術語, 専門(用)語.

technic *n.* technique または technics と同意.

technics *n.* (単数扱いもある)(学)術語; 専門事項; 工学, 技術.

technique *n.* 手法, 技法, 技巧, 方法.

technology *n.* 工学; 科学技術.

Teichmüller, P. J. Oswald (1913-1943) タ

イヒミュラー.

Teilerproblem *n.* (*G.*) 約数問題.

telegraph *n.* 電信.
~ equation 電信方程式.

teleparallelism *n.* 遠隔平行性.

temporal *a.* 時(間)の; 時間的な; 一時の.

temporally *ad.* 時間的に.
~ homogeneous 時間的に一様な.

temporarily *ad.* 一時(的に).

temporary *a.* 仮の; 当座の, 一時の.
~ average 仮(の)平均. ~ quotient 仮商. ~ standard 暫定規格.

tend *vi.* 近づく, 収束する; 向かう; (する)傾向がある【to】.
~ to limit 極限に近づく[収束する].

tendency *n.* 傾向, 趨勢.
central ~ 中心傾向.

tensor *n.* テンソル, テンサー.
alternating ~ 交代テンソル. base ~ 基テンソル. connecting ~ 連絡テンソル. contracted ~ 縮約テンソル. contravariant ~ 反変テンソル. correlation ~ 相関テンソル. covariant ~ 共変テンソル. curvature ~ 曲率テンソル. deflection ~ 偏差テンソル. hybrid ~ 混合テンソル. metric ~ 計量テンソル. mixed ~ 混合テンソル. Ricci ~ リッチ・テンソル. strain ~ 変形テンソル. stress ~ 応力テンソル. symmetric ~ 対称テンソル. ~ algebra テンソル代数. ~ analysis テンソル解析. ~ bundle テンソル束. ~ calculus テンソル計算(法), テンソル解析. ~ equation テンソル方程式. ~ field テンソル場. ~ form テンソル形式. ~ product テンソル積. ~ representation テンソル表現. ~ space テンソル空間. torsion ~ 捩率テンソル.

tensorial *a.* テンソルの.
~ form テンソル形式.

tentative *a.* 試験的な; 仮の.
~ standard 仮規格.

tera- *pref.* テラ【10^{12}】.

term *n.* 項; (術)語; 論理名辞[概念].
constant ~ 定数項. in ~s of によって, を用いて. like ~s 同類項. major ~ 大名辞[概念]. middle ~ 中名辞[概念], 媒名辞[概念]. minor ~ 小名辞[概念]. technical ~(s) 術語, 専門(用)語. ~ of higher order 高位の項. undefined ~ 無定義用語.

terminal *a.* はしの, 末端の, 終りの; 期末の. *n.* はし, 終点; 端末, 端子.
~ decision 最終決定. ~ point 終点. ~ side (角の)終辺. ~ time 生存時間.

terminate *vt.* 終らせる. *vi.* 終る.
~ting decimal 有限小数.

termination *n.* 終止; 終末.

terminology *n.* 用語; 術語(学).

terminus *n.* 終点.

termwise *a.* 項別の. *ad.* 項別に.
~ differentiation 項別微分. ~ integration 項別積分.

ternary *a.* 三つの; 三個の変数をもつ, 三元の; 三進の. *n.* 三つ組.
~ logic 三値論理. ~ set 三進集合. ~ system 三進法.

terrestrial *a.* 地球の.
~ globe 地球儀. ~ gravitation 地球の引力.

tertiary *a.* 第三の.

tessera (*pl.* -rae) *n.* モザイク片.

tesseral *a.* 等軸(晶系)の; 縞球[方球]の; モザイクの.
~ harmonics 縞球[方球](調和)関数.

test *n.* 試験; 検査; 検定; 判定(法). *vt.* 試験する; 検査する.
chi-square ~ カイ二乗検定. condensation ~ 凝集判定法. consistent ~ 一致検定. convergence ~ 収束判定(法). F-~ エフ検定. majorant ~ 優級数(による)判定(法). minimax ~ ミニマックス検定. one-sided ~ 片側検定. pair ~ 二点比較法. radical ~ ベキ根判定法. randomized ~ 確率化された検定. ratio ~ 比判定法. sequential ~ 逐次検定. significance ~ 有意性(の)検定. Student ~ スチューデント検定. ~ function 検定関数. ~ing 検定. ~ of goodness of fit 適合度検定. ~ routine 検査ルーチン. ~ statistic 検定統計量. triangle ~ 三点比較法. two-sided ~ 両側検定.

tetra- *perf.* 「四」の意.
tetracyclic *a.* 四円の.
～ coordinates 四円座標.
tetradic *n., a.* テトラディック.
tetragamma *n.* テトラガンマ.
～ function テトラガンマ関数.
tetragon *n.* 四角[辺]形.
tetragonal *a.* 四角[辺]形の; 正方晶系の.
～ system 正方晶系.
tetrahedral *a.* 四面体の, 四面のある.
～ angle 四面角. ～ group (正)四面体群.
tetrahedron (*pl.* -rons, -dra) *n.* 四面体.
orthocentric ～ 垂心四面体. orthogonal ～ 直稜四面体. polar ～ 極四面体. regular ～ 正四面体. self-polar ～ 自極四面体.
TeX *n.* テフ, テック.
LaTeX ラテフ, ラテック, レイテック.
text *n.* テキスト; 本文.
～ book 教科書.
TFAE *n.* The following are equivalent. 「以下は同値」の意.
-th *suf.* 第…番目の; …次の, …倍の.
n～ order *n* 階. *n*～ power *n* 乗.
th. theorem.
Thales of Miletos (624?–548? B.C.) ミレトスのターレス.
thence *ad.* それゆえ; よって; それ以来; そこから.
theor. theorem; theoretical.
theorem *n.* 定理.
～ of X, X('s) ～ X (氏)の定理.
theorematic *a.* 定理の.
theoretic *a.* 理論(上)の, 理論的な.
function-～ 関数論的な. number-～ (整)数論的な. set-～ 集合論的な.
theoretical ＝theoretic.
～ distribution 理論的分布. ～ formula 理論式. ～ physics 理論物理学.
theory *n.* 理論.
analytic ～ of numbers 解析(的整)数論. coding ～ 符号理論. function ～, ～ of functions 関数論. Galois ～ ガロア理論. knot ～ 結び目理論. measure ～, ～ of measures 測度論. optimization ～ 最適化理論. relativity ～ 相対(性理)論. set ～, ～ of sets 集合論. ～ of complex multiplication 虚数乗法論. ～ of differential equations 微分方程式論. ～ of errors 誤差論. ～ of games ゲーム(の)理論. ～ of invariants 不変式論. ～ of large [small] samples 大[小]標本論. ～ of surfaces 曲面論.
therefore *ad.* それゆえに.
thermodynamics *n.* 熱力学.
thermometer *n.* 温度計.
Celsius [centigrade] ～ 摂氏温度計. Fahrenheit [Réaumur] ～ 華[列]氏温度計.
thesis (*pl.* -ses) *n.* 論文, 学位論文; 論題, 命題.
doctoral ～ 学位[博士]論文.
theta *n.* シータ, テータ. ギリシア語アルファベットの 8 番目の文字【$\Theta, \theta, \vartheta$】.
～ function シータ関数, テータ関数. ～ series シータ級数, テータ級数.
thickness *n.* 厚さ.
thin *a.* 細い, 尖細な; 薄い.
～ set 尖細集合. ～ wing theory 薄翼理論.
third *a.* 第三の. *n.* 第三; 三分の一.
～ fundamental form 第三基本形式. ～ proportional 第三比例項. ～ quadrant 第三象限.
Thom, René (1923–2002) トム.
Thomae, Johannes (1840–1921) トメ.
thousand *n.* 千.
ten ～ 万.
three *a.* 三の. *n.* 三.
～(-)body problem 三体問題. ～ circle [line] theorem 三円[線]定理. ～-cusped hypocycloid 三尖点内擺線. ～-dimensional 三次元(的)の. ～-eights rule 八分の三公式. ～ fourths; ～-quarters 四分の三. ～ sigma method 三シグマ法. ～-valued function 三価関数. ～-valued logic 三値論理.
three-factor *a.* 三因子の.
～ interaction 三因子交互作用.
three-quarter *a.* 四分の三の.
three-quarters *n.* 四分の三.
three-valued *a.* 三価の; 三値の.
～ function 三価関数. ～ logic 三値論理.
Threlfall, William (1888–1949) トレルフ

アル.

threshold *n.* 閾; しきい(値).
anomalous ~ 異常閾. ~ Jacobi method 閾値ヤコビ法.

tie *vt.* 縛る, 結ぶ; 同点になる.
~d rank 同順位.

TIETZE, Heinrich Franz Friedrich (1880–1964) ティーツェ.

tight *a., ad.* 緊密な[に].

tightened *a.* 緊密にされた.
~ sampling inspection 厳格抜取り検査.

TIKHONOV, Andrei Nikolaevič (1906–1993) チホノフ.

tilde *n.* (*Sp.*) チルダ, 上つき波印【スペイン語の文字 n の上につける発音符】【~】.

tilting *a.* 傾斜的な. *n.* 傾斜.
~ theory 傾(斜)理論. ~ complex 傾斜複体.

time *n.* 時, 時間, 時刻; 倍; 回.
cycle ~ サイクル・タイム. each ~ 各回. every ~ 毎回. space-~ 時空. ~ average 時間平均. ~ lag 遅れの時間. ~ parameter 時刻助変数. ~ reversal 時間[刻]反転. ~ series 時系列. two ~s 二回; 二倍. two ~s three is six 2×3=6.

time-homogeneous *a.* 斉時的(せいじてき)な.
~ Markov chain 斉時マルコフ連鎖.

time-like *a.* 時間的な.

time-sharing *n.* 時分割, タイム・シェアリング.

-tion *suf.* 行為, 状態, 結果などを表す名詞語尾 (=ion): addition, condition, reflection.

TITCHMARSH, Edward Charles (1899–1963) ティッチマーシュ.

TOEPLITZ, Otto (1881–1940) テプリッツ.

tolerance *n.* 許容; 許容差; 公差.
~ interval 許容区間. ~ limit(s) 許容限界. ~ region 許容域.

TONELLI, Leonida (1885–1946) トネリ.

tonic *a.* 音調の. *n.* 基音.

tooth (*pl.* teeth) *n.* 歯.
~ curve 歯形曲線. ~ surface 歯(形曲)面.

topological *a.* 位相的な.
~ field 位相体. ~ group 位相群. ~ invariant 位相不変量. ~ manifold 位相多様体. ~ mapping 位相写像. ~ model 位相(的)モデル. ~ module 位相加群. ~ product 位相積. ~ property 位相的性質. ~ ring 位相環. ~ space 位相空間. ~ structure 位相(的)構造. ~ subspace 部分位相空間. ~ transformation group 位相変換群. ~ union 和(位相)空間.

topologically *ad.* 位相的に.
~ complete space 位相完備空間. ~ equivalent 位相(的に)同値な. ~ invariant 位相的に不変な.

Topologie *n.* (*G.* =topology.)

topologie *n.* (*F.* =topology.)

topologize *vt.* 位相づける.

topology *n.* 位相数学; 位相幾何学; 位相; トポロジー.
algebraic ~ 代数的位相幾何学. box ~ 箱位相. combinatorial ~ 組合せ論的トポロジー. differential ~ 微分位相幾何学. direct sum ~ 直和位相. discrete ~ 離散位相. Krull ~ クルル位相. natural ~ 自然位相. product ~ (直)積位相. quotient ~ 商位相. relative ~ 相対位相. set-theoretic ~ 集合論的トポロジー. strong ~ 強位相. uniform ~ 一様位相. weak ~ 弱位相.

toric *a.* 円環体の, トーラスの; トーリック.
~ variety トーリック多様体.

toroid *n.* トロイド, 円錐曲線回転面[体].

torse *n.* 捩面.

torsion *n.* 捩率, ねじれ率; ねじれ, トーション.
radius of ~ 捩率半径. ~ coefficient ねじれ係数. ~ element ねじれ元. ~ form 捩率形式. ~ group ねじれ群. ~ product ねじれ積. ~ tensor 捩率[ねじれ率]テンソル.

torsion-free *a.* ねじれのない.
~ group [module] ねじれのない群[加群].

tortuous *a.* くねった, 彎曲した.

torus (*pl.* -ri) *n.* 円環面[体]; 輪環面[体];

トーラス.
complex ~ 複素トーラス. ~ function 円環(面)関数. ~ group 輪環群, トーラス群.

total *a.* 総計の, 全体の, 全-. *n.* 合計.
grand ~ 総計. ~ boundary operator 全境界作用素. ~ correlation 単相関, 全相関. ~ curvature 全曲率. ~ degree 総次数, 全次数. ~ differential 全微分. ~ differential equation [form] 全微分方程式[形式]. ~ inspection 全数検査. ~ order 全順序. ~ perspective 全透視図. ~ probability law 全確率の公式. ~ ring of quotients, ring of ~ quotients, ~ quotient ring 全商環. ~ space 全空間. ~ sum 総和; 合計. ~ variation 全変動.

totality *n.* 全体, 総計.

totally *ad.* 全く; 全体として; 全-; 完全な.
~ additive 完全加法的な. ~ bounded 全有界な. ~ differentiable 全微分可能な. ~ disconnected 全不連結な, 完全非連結な. ~geodesic 全測地的な. ~ imaginary 全虚な. ~ ordered set 全順序集合. ~ positive 総[全]正の. ~ ramified 完全分岐. ~ real 全実の. ~ singular 全特異な. ~ umbilical surface 全臍(曲)面.

totient *n.* トーシェント.
Euler's ~ function オイラーのトーシェント関数.

touch *vt., vi.* 接する, 接触する. *n.* 接触.

toward *prep.* の方へ, に向けて.
concave [convex] ~ pole 極に(向けて)凹[凸]な.

towards =toward.

trace *n.* 跡, トレース; 固有和, 対角和. *vt.* 追跡する.
horizontal ~ 水平跡. reduced ~ 被約跡. ~ class トレース族. ~ formula 跡公式. ~ norm トレース・ノルム. vertical ~ 直立跡.

tracing *n.* 追跡. *a.* 追跡する.
curve ~ 曲線(の)追跡.

track *n.* トラック; 航跡; 進路.

tractrix *n.* 牽引曲線, 犬曲線, 弧曲線, トラクトリックス.

traffic *n.* 交通; 交通量; 通話量.
~ intensity 呼量, トラフィック密度.

trajectory *n.* 軌道, 軌線; 弾道; 定角軌道.
orthogonal ~ 直交截線. phase ~ 相軌道.

trans- *pref.* 「変化; 移転」の意.

transactions *pl. n.* 報告書[誌].

transcendence *n.* 超越性.
~ basis 超越基底. ~ degree, degree of ~ 超越次数.

transcendency *n.* 超越(性).

transcendental *a.* 超越の, 超越的な.
elementary ~ function 初等超越関数. ~ curve 超越曲線. ~ element 超越(的)元. ~ extension 超越拡大. ~ function 超越関数. ~ number 超越数. ~ singularity 超越特異点.

transcendentality *n.* 超越性.

transfer *n.* 転送; 移送; 伝達. *vt.* 移す. *vi.* 乗り換える.
~ function 伝達関数.

transference *n.* 移すこと, 移行; 移動; 転送.

transfinite *a.* 超限の, 超越[限]的な.
~ diameter 超越直径. ~ induction 超限帰納法. ~ ordinal number 超限順序数.

transform *vt.* 変換する; 変形する. *n.* 変換(によってえられたもの).
convolution ~ 合成変換. cosine ~ 余弦変換. Fourier ~ フーリエ変換. integral ~ 積分変換. Laplace ~ ラプラス変換. sine ~ 正弦変換.

transformable *a.* 変形させることのできる, 変形可能な; 可変の.

transformation *n.* 変換; 変形.
affine ~ アフィン変換. angular ~ 角変換. birational ~ 双有理変換. conformal ~ 共形変換. congruent ~ 合同変換. coordinate ~ 座標変換. elementary ~ 初等変換; 基本変形. Fourier [Laplace] ~ フーリエ[ラプラス]変換. identity ~ 恒等変換. infinitesimal ~ 無限小変換. integral ~ 積分変換. inverse ~ 逆変換. linear ~ 一次変換. Möbius ~ メービウス変換. orthogonal ~ 直交変換. principal axis ~ 主軸変換. projective ~ 射影変換.

similar ～ 相似変換. symmetric ～ 対称変換. ～ formula 変換公式. ～ group 変換群. ～ matrix 変換行列. ～ parameter 変換母数. unitary ～ ユニタリ変換.

transgression *n.* 転入, 転入写像.

transgressive *a.* 転入的な.

transient *a.* 非再帰の; 一時の, 過渡的な.
～ component 過渡的成分. ～ problem 過渡現象の問題. ～ solution 一時解.

transition *n.* 遷移, 推移.
～ diagram 遷移図, ～ function 変換[推移]関数. ～ matrix 推移行列. ～ probability 推移確率. ～ process 推移過程.

transitive *a.* 推移的な, 可移的な, 可遷的な.
metrically ～ 測度可遷的な. *n*-ply [*n*-fold] ～ *n* 重推移的な. ～ group 推移[可移, 可遷]群. ～ law 推移法則[律]. ～ permutation group 可移[推移的, 可遷]置換群. ～ relation 推移的関係. ～ trajectory 推移軌道.

transitively *ad.* 推移的に.

transitivity *n.* 推移性.
metrical ～ 測度推移[可遷]性.

translatability *n.* (平行)移動可能性.

translate *vt., vi.* 翻訳する; 移す; (平行)移動させる.

translater, translator *n.* 翻訳者.

translation *n.* 移行; 平行移動, 移動, 並進; 翻訳.
left [right] ～ 左[右]移動. similar ～ 相似移動. surface of ～ 移動曲面. ～ number 変位数; 概周期. ～ of coordinate axes [system] 座標軸[系]の平行移動. ～ operator 移動演算子.

transmission *n.* 伝達, 伝送; 伝播.
heat ～ 熱の伝播. ～ capacity 伝送容量. ～ rate 伝送速度.

transplant *vt.* 移植する.

transport *n.* 転位; 輸送; 移送. *vt.* 輸送する.
～ equation 輸送方程式. velocity of ～ 転位速度.

transportation *n.* 輸送; 移送.
～ problem 輸送問題.

transpose *vi., vt.* 移項する, 転置する.

transposed *a.* 転置された.
～ equation 転置方程式. ～ kernel 転置核. ～ mapping 転置写像. ～ matrix 転置行列, 随伴行列. ～ representation 転置表現.

transposition *n.* 移項; 互換; 転置.

transshipment *n.* 移積(輸送), 積み換え.
～ problem 輸送問題.

transvection *n.* 移換.

transversal *n.* 横断線. *a.* 横(断線)の; 横断的な.
～ vibration 横振動. ～ wave 横波.

transversality *n.* 横断性.
～ condition 横断(性)条件.

transverse *n.* 横軸. *a.* 横断的な; 横軸の.
～ axis 横軸. ～ field 横断面の場. ～ line 截線. ～ regularity 横断正則性. ～ section 横断面.

trap *n.* 割込み; トラップ; わな.

trapezial *a.*【米】不等辺四角形の;【英】台形の, 梯形の.

trapezium (*pl.* -zia, -ums) *n.*【米】不等辺四辺形;【英】台形, 梯形.

trapezoid *n., a.*【米】台形(の), 梯形(の);【英】不等辺四辺形(の).
isogonal ～ 等角台形. isosceles ～ 等脚台形. ～ rule 台形公式.

trapezoidal *a.*【米】台形の, 梯形の;【英】不等辺四角形の.
～ rule 台形公式.

travel *vt.,vi.* 旅行する; 巡回販売する.
～ing salesman problem 巡回セールスマン問題.

traverse *vt., vi.* 横切る. *n.* 裁断線.

treat *vt., vi.* とり扱う.

treatment *n.* 処理.
～ contrast 処理対比. ～ effect 処理効果.

tree *n.* 木, 樹木, 樹形; 樹木[形]曲線.
binary ～ 二分木. directed ～ 有向木. spanning ～ 全域木. ～ diagram 樹形図. ～ of game ゲームの樹木. ～ representation 樹木表現.

trend *n.* 傾向. *vi.* 向かう, 傾く.

tri- *pref.*「三, 三重, 三倍」などの意.

triad *n.* 三面組, 三つ組, トライアド.

triadic *n.*, *a.* トリアディック.

trial *n.* 試行.
Bernoulli ~[s] ベルヌーイ試行[試行(の)列]. independent ~s 独立試行. sequence of ~s, ~ sequence 試行(の)列. ~ function 試験関数.

triangle *n.* 三角形; 三角定規【米】.
acute ~ 鋭角三角形. circular ~ 円弧三角形. curvilinear ~ 曲線三角形. defect of ~ 角不足. elements of ~ 三角形の要素. equilateral ~ 等辺[正]三角形. five centroids of ~ 三角形の五心. isosceles ~ 二等辺三角形. obtuse ~ 鈍角三角形. Pascal's ~ パスカルの三角形. pedal ~ 垂足三角形. polar ~ 極三角形. rectangular ~ 直角三角形. regular ~ 正三角形. right(-angled) ~ 直角三角形. scalene ~ 不等辺三角形. solution of ~ 三角形の解法. spherical ~ 球面三角形. ~ inequality 三角不等式. ~ law 三角形法則. ~ test 三点比較法.

triangular *a.* 三角形の; 三角の.
~ inequality 三角不等式. ~ matrix 三角行列. ~ net 三角網. ~ number 三角数. ~ prism 三角柱. ~ pyramid 三角錐. upper ~ matrix 上三角行列.

triangulation *n.* 三角形に分けること, 三角形分割.

trichotomy *n.* 三分法.

triclinic *a.* 三斜の.
~ system 三斜晶系.

Tricomi, Francesco G. (1897-1978) トリコミ.

trident *n.* 三叉曲線. *a.* 三叉の.

tridiagonal *a.* 三重対角の.
~ matrix 三重対角行列.

trifocal *a.* 三焦点の.

trifurcate *a.* 三叉の. *vt.* 三叉に分かれる.

trigamma *n.* トリガンマ.
~ function トリガンマ関数.

trigon *n.* 三角形.

trigonal *a.* 三角(形)の.

trigonometric *a.* 三角法の, 三角-.
inverse ~ function 逆三角関数. ~ equation 三角方程式. ~ function 三角関数. ~ integral 三角積分. ~ moment problem 三角モーメント問題. ~ polynomial 三角多項式. ~ ratio 三角比. ~ series 三角級数. ~ sum 三角和.

trigonometrical =trigonometric.
~ identity 三角恒等式. ~ measurement 三角測量. ~ series 三角級数. ~ surveying 三角測量.

trigonometry *n.* 三角法.
plane ~ 平面三角法. solid ~ 立体三角法. spherical ~ 球面三角法.

trihedral *a.*, *n.* 三面の; 三面体(の), 三面角(の).
coordinate ~ 座標三面角. moving ~ 動三面角. ~ angle 三面角. trirectangular ~ angle 三直角三面角.

trihedron (*pl.* -rons, -dra) *n.* 三面体.

trilateral *a.* 三辺の. *n.* 三辺形, 三角形.

trilinear *a.* 三線の.
~ coordinates 三線座標.

trillion *n.*, *a.* 【米】百万の二乗(の), 一兆(の); 【英】百万の三乗(の).

trimetric *a.* 斜方晶系の; 斜方の.
~ projection 斜方投象[射影].

trinomial *n.* 三項式. *a.* 三項の, 三項式の.
~ equation 三項方程式. ~ expression 三項式.

triple *a.* 三倍の; 三重の. *n.* 三倍の数; 三つ組.
~ integral 三重積分. ~ interaction 三因子交互作用. ~ point 三重点. ~ product 三重積. ~ ratio 三乗比. ~ root 三重根. ~ series 三重級数.

triplet *n.* 三つ組, 三幅対.

triplex *n.* 三つ組.

triplicate *a.* 三重の; 三倍の.
~ ratio 三乗比; 立方根比.

triply *ad.* 三重に.

tripod *n.* 三面角; 三脚台.

tripolar *a.* 三極の.
~ coordinates 三極座標.

trirectangular *a.* 三直角の.
~ spherical triangle 三直角球面三角形. ~ trihedral angle 三直角三面角.

trisecant *n.* トライセカント, 三等分曲線.

trisect *vt.* 三分する，三等分する．

trisection *n.* 三等分．
~ of angle 角の三等分．

trisector *n.* 三等分線．

trisectrix *n.* 三等分線; 蝸牛形．

trivial *a.* 自明な; つまらない，平凡な; トリビアルな．
~ solution 自明な解． ~ subgroup 自明な部分群． ~ valuation 自明な付値．

triviality *n.* 自明なこと．

trochoid *n.* トロコイド，余擺線．

trochoidal *a.* トロコイドの．
~ wave トロコイド波．

true *n.* 真; 真理． *a.* 真の; 厳密な．
~ anomaly 真近点離角． ~ error 真の誤差． ~ length 実長． ~ solution 真の解． ~ value 真の値．

truncate *a.* (斜)截頭の，切頭の． *vt.* 切頭する，先を切る．
~(d) cone 截[切]頭錐，欠錐． ~(d) distribution 切断分布． ~(d) prism 截[切]頭角錐．

truncation *n.* 切ること．
~ error 打切り誤差．

truth *n.* 真; 真理．
~ and falsity 真偽． ~ table 真理(値)表． ~ value 真理値． ~(-value) function 真理(値)関数．

TSCHEBYSCHEFF ＝CHEBYCHEV.

TSCHIRNHAUSEN, Ehrenfried Walter von (1651–1708) チルンハウゼン．

tube *n.* 管．
stream ~ 流管． ~ domain [region] 管状領域． vector ~ ベクトル管． vortex ~ 渦管．

tubular *a.* 管状の．
~ domain [region] 管状領域． ~ mapping 管状写像． ~ neighbo(u)rhood 管状近傍． ~ surface 管面．

-tuple *suf.*「倍の; 重の」の意．
n~ root *n* 重根． *n*~ vector *n* 項(数)ベクトル．

turbulence *n.* 乱れ; 乱流．
homogeneous ~ 一様乱流． isotropic ~ 等方乱流．

turbulent *a.* 荒れた，乱れた．
~ flow 乱流．

TURING, Alan Mathison (1912–1954) テューリング．

turn *vt.* 曲がる; 回す; 向ける． *vi.* 回る; 回転する; ころがる． *n.* 回転; 転換; 順番．
capital ~ 資本回転率． in ~ 順次に． ~ing point 変わり点．

turn-over *n.* 一期の総売上高．
~ of inventories 在庫回転率．

turnpike *n.* ターンパイク．
~ theorem ターンパイク定理．

twice *ad.* 二度; 二倍に．

twin *a.* 双子の、*n.* 双子．
~ prime numbers 双子素数．

twist *vt.* ねじる． *vi.* ねじれる． *n.* ねじり．

twisted *a.* ねじれた．
~ cone ねじれ錐． ~ curve 空間曲線，ねじれた曲線． ~ quadrangle [quadrilateral] ねじれ四角[辺]形．

two *a.* 二の． *n.* 二，二個．
~-bin system 二棚法． ~(-)body problem 二体問題． ~-by-~ contingency table 2×2 分割表． ~-dimensional 二次元(的)の． ~ pairs 二対．

two-factor *a.* 二因子の．
~ interaction 二因子交互作用．

twofold *a.* 二倍の，二重の．

two-phase *a.* 二相の．
~ method 二段階法． ~ sampling 二相抽出(法)．

two-point *a.* 二点の．
~ boundary value problem 二点境界値問題．

two-sample *n.* 二標本．
~ test 二標本検定．

two-sided *a.* 両側の．
~ ideal 両側イデアル． ~ surface 両側曲面． ~ test 両側検定．

two-stage *a.* 二段の．
~ sampling 二段抽出法．

two-valued *a.* 二価の; 二値の．
~ function 二価関数， ~ logic 二値論理．

two-way *a.* 二重の，二方向の，両変の．
~ classification 二重分類，二元配置法． ~

control chart 複式管理図. ~ layout 二元配置.

-ty *suf.* ラテン系の形容詞から「性質，状態」を表す名詞をつくる【-ity, -ety ともなる】: affinity, variety.

TYCHONOFF, Andrey Nikolayevich (1906–1993) チホノフ.

type *n.* 型; タイプ.
elliptic ~ 楕円型. hyperbolic ~ 双曲型. maximal [minimal] ~ 極大[小]型. (m, n)-~ (m, n) 型. ordinary ~ 通常型. parabolic ~ 放物型. positive ~ 正型. ~ number 特性数. ~ problem 型問題.

U

UFD, U.F.D. Unique Factorization Domain. 一意分解整域.

ultimate *a.* 究極の，根本的な.
～ definite language 究極的限定言語.

ultra- *pref.*「極端な，超」の意.

ultrabornologique *a.* (*F.*) 超有界型の.
espace ～ 超有界型空間.

ultraconvergence *n.* 超収束. ⇨ overconvergence.

ultraconvergent *a.* 超収束の.
～ power series 超収束ベキ級数.

ultrafilter *n.* 極大フィルター.

ultrainfinite *a.* 超無限の.
～ point 超無限遠点.

ultraproduct *n.* 超積.

ultrasonic *a.* 超音波の. *n.* 超音波.
～ wave 超音波.

ultraspherical *a.* 超球の.
～ polynomial 超球多項式.

umbilic *n.* 臍点. *a.* 臍の.
～-free 臍点のない. ～ surface 臍点曲面.

umbilical *a.* =umbilic.
totally ～ 全臍的な. ～ hypersurface 臍点超曲面. ～ point 臍点. ～ surface 臍点曲面.

umbilicus (*pl.* -es, -lici) *n.* 臍点.

Umlaufszahl *n.* (*G.* =rotation(-)number; winding number.) 回転数.

U.M.V. Uniformly Minimum Variance. 一様最小分散.

un- *pref.*「否定」の意.

unbiased *a.* 偏りのない，不偏な.
～ confidence region 不偏信頼域. ～ estimate 不偏推定値. ～ estimator 不偏推定量. ～ regression estimator 不偏回帰推定量. ～ variance 不偏分散.

unbounded *a.* 非有界な，有界でない; 無限の.
～ function 非有界関数. ～ set 非有界集合.

uncertainty *n.* 不確定性.

unconditional *a.* 無条件の，絶対的な.
～ convergence 無条件収束. ～ inequality 絶対不等式.

unconditionally *ad.* 無条件に.
～ convergent 無条件[絶対]収束の.

uncorrelated *a.* 無相関の.

uncountable *a.* 非可算の; 不可算の; 数えることのできない.
～ set 非可算集合.

undecagon *n.* 十一角形，十一辺形.

undefined *a.* 無定義の.
～ concept [notion] 無定義概念. ～ term 無定義用[術]語.

undemonstrable *a.* 論証[証明]できない.

undemonstrated *a.* 論証[証明]され(てい)ない.

undenary *n.* 十一進法.

under *prep.* の下に; のもとに. *ad.* 下に. *a.* 下の.
a little ～ 10 percent 一割弱. ～ this condition この条件のもとに.

undercross *vi.* 下交差[叉]する，下で交わる.
～ing point 下交差点.

underdetermined *a.* 劣決定の. ⇨ overdetermined.
～ system 劣決定系.

underflow *n.* アンダーフロー.

undergraduate *n.* 学部学生. *a.* 学部学生

の.

underlying *a.* 基礎の.
~ field 基礎体. ~ group 基礎群. ~ space 基底空間. ~ topological space 基礎[基底]位相空間.

undeterminate *a.* 不定の.

undetermined *a.* 未定の. ⇨ indeterminate.
~ coefficient(s) 未定係数.

undirected *a.* 無向の.
~ graph 無向グラフ.

undulation *n.* 波動.

undulatory *a.* 波動[状]の.
~ theory 波動説.

uneliminated *a.* 消去され(てい)ない.

unequal *a.* 等しくない, 不等の.
of ~ sides 不等辺の.

uneven *a.* 平らでない, 一様でない; 奇数の.

unexpanded *a.* 展開され(てい)ない.

unfold *vi., vt.* 開く; 展開する. *vt.* 明らかにする, 表明する.

ungula (*pl.* -lae) *n.* 蹄状体[形].

unharmonic *a.* 調和でない.

uni- *pref.*「一, 単」の意.

unicity *n.* 一致性; 一意性, 単一性.
~ principle 一致性(の)原理. ~ theorem 一致の定理.

unicursal *a.* 一筆書きできる; 有理的に一意化可能な.
~ curve 有理(化可能)曲線. ~ figure 一筆書き可能な図形.

unified *a.* 統一された.
~ field theory 統一場理論.

uniform *a.* 一様な; 一意な; 一価の.
~ boundedness 一様有界性. ~ continuity 一様連続(性). ~ convergence 一様収束. ~ density 一様密度. ~ distribution 一様分布. ~ function 一意[一価]関数. ~ motion 等速運動. ~ norm 一様ノルム. ~ probability distribution 一様確率分布. ~ scale 普通目盛[尺]. ~ space 一様空間. ~ topology 一様位相. ~ velocity 等速度.

uniformity *n.* 一様性.

uniformizable *a.* 一意化可能な; 一様性を許容する.

uniformization *n.* 一意化.
local ~ 局所一意化.

uniformize *vt.* 一意化する.

uniformizer *n.* 一意化(媒介)変数.

uniformizing *a.* 一意化する.
~ parameter 一意化(媒介)変数.

uniformly *ad.* 一様に.
converge ~ 一様(に)収束する. ~ better 一様に良い. ~ bounded 一様(に)有界な. ~ consistent test 一様一致検定. ~ continuous 一様(に)連続な. ~ convergent 一様収束の. ~ convex 一様凸な. ~ equivalent 一様同相な. ~ integrable 一様に積分可能な, 一様可積な. ~ in the wider sense 広義(の)一様に. ~ minimal variance 一様最小分散. ~ most powerful 一様最強力な.

unify *vt.* 統一する. 一様にする.

unilateral *a.* 単側の; 片側の.
~ shift 片側シフト. ~ surface 単側曲面.

unimodal *a.* 単峰形の.
~ distribution 単峰(形)分布.

unimodular *a.* ユニモジュラーな.
~ equivalence ユニモジュラー対等. ~ group ユニモジュラー群. ~ matrix ユニモジュラー行列.

union *n.* 合併, 和集合; 連合.
disjoint ~ 直和. ~ of sets 和集合. ~ of surface elements 面素連合.

unipotency *n.* ベキ単(性).

unipotent *a.* ベキ単の.
~ component ベキ単成分. ~ group ベキ単群. ~ matrix ベキ単行列.

unique *a.* 一つしかない, 唯一の; 一意(的)な. *n.* 唯一(のもの).
~ continuation 一意接続. ~ decomposition 一意分解. ~ existence theorem 一意存在定理. ~ factorization domain (UFD) 一意分解整域. ~ factorization ring 一意[素元]分解環. ~ solution 唯一つの解.

uniquely *ad.* 単独に; 一意に; 唯一に.

uniqueness *n.* 単独性; 一意性.
~ condition 単独[一意]性の条件. ~ theorem 一意性の定理.

uniserial *a.* 単列の.

～ algebra 単列多元環.

unit *n.* 単位; 単元; 装置.
group of ～s, ～ group 単数群, 単元群. imaginary ～ 虚数単位. matrix ～ 行列単位. ～ area 単位面積. ～ ball 単位球(体). ～ cell 単位胞体. ～ circle 単位円. ～ cube 単位立方体. ～ disk [disc] 単位円板. ～ dyadic 単位ダイアディック. ～ element 単位元. ～ fraction 単位分数. ～ function 単位関数. ～ ideal 単位イデアル. ～ mass 単位質量. ～ matrix 単位行列. ～ normal 単位法線. ～ of a ring 環の単元. ～ operator 単位作用素. ～ point 単位点. ～ price 単価. ～ representation 単位表現. ～ sphere 単位球. ～ square 単位正方形. ～ vector 単位ベクトル. ～ volume 単位体積.

unital *a.* 単位的な.
～ module 単位(的)加群.

unitarity *n.* ユニタリ性.

unitary *a.* 単位的な; ユニタリな; 一元の.
～ algebra 単位的多元環. ～ group ユニタリ群. ～ matrix ユニタリ行列. ～ module 単位(的)加群. ～ polynomial 単多項式. ～ representation ユニタリ表現. ～ ring [semigroup] 単位的環[半群]. ～ space ユニタリ空間. ～ symplectic group ユニタリ・シンプレクティック群. ～ transformation ユニタリ変換.

unity *n.* 一; 単位; 単一(性).
partition of ～ 単位の分割[分解], 一の分割[分解]. primitive *n*-th root of ～ 1の原始 *n* 乗根. root of ～ 1のベキ根[累乗根].

univalence *n.* 単葉性.

univalency ＝univalence.
radius of ～, ～ radius 単葉(性)半径.

univalent *a.* 単葉な. ⇨ (*G.*) schlicht.
～ correspondence 一意対応. ～ function 単葉関数. ～ mapping 単葉写像.

univariate *a.* 一変量の.

universal *a.* 普遍的な.
～ arithmetic 一般算術, 代数. ～ bundle 普遍(ファイバー)束. ～ constant 普遍[宇宙]定数. ～ covering group 普遍被覆群. ～ covering surface 普遍被覆面. ～ curve 万有曲線. ～ domain 万有体. ～ gravitation 万有引力. ～ mapping 普遍[万有]写像. ～ property 普遍的性質. ～ proposition 全称命題. ～ quantifier 全称記号【∀】. ～ set 普遍集合, 全体集合.

universe *n.* 宇宙; 母集団.

unknotted *a.* 結ばれていない.

unknown *a.* 未知の. *n.* 未知数; 未知量.
equation with *n* ～s *n* 元方程式. ～ function 未知関数. ～ quantity 未知量.

unless *conj.* もし…でなければ, …でない限り.

unlimited *a.* 無限の; 境界のない.

unmixed *a.* 非混合の, 純粋な.
～ ideal 純イデアル.

unmixedness *n.* 純性.

unnecessary *a.* 不必要な.

unnumbered *a.* 数えられない, 無数の.

unordered *a.* 順序のついていない.
～ pair 非順序対.

unoriented *a.* 無向の, 向きづけられていない.

unprovable *a.* 証明できない, 証明不可能な.

unramified *a.* 非分岐な, 不分岐な.
～ covering surface 非[不]分岐(な)被覆面. ～ extension 非[不]分岐拡大.

unsolvability *n.* 非可解性, 決定不可能性; 解決不可能性.

unsolvable *a.* 非可解な; 解けない.
～ equation 非可解方程式. ～ problem 解けない問題.

unsolved *a.* 未解決の; 解かれていない.
～ problem 未解決(の)問題.

unstable *a.* 不安定な.
～ solution 不安定解. ～ state 不安定状態.

up *ad., prep.* 上へ, 上方へ. *a.* 上に向かう.
carry ～ 繰上げる.

up-ladder *n.* 上昇演算子.

upper *a.* 上の, 上方の; 上位の. ⇨ lower.
～ base 上底. ～ bound 上界. ～ central series 昇中心列. ～ derivative 上微分係数. ～ half-plane 上半(平)面. ～ integral 上積分. ～ limit 上限; 上端. ～ limit function 上限関数. ～ quartile 上側四分位数. ～

semicontinuous 上(に)半連続な. ~ sum 過剰和. ~ tail probability 上側確率. ~ triangular matrix 上三角行列.

upsilon *n.* ユプシロン, ウプシロン. ギリシア語アルファベットの 20 番目の文字【Υ, υ】.

upward *ad.* 上に. ⇨ downward.

upwards *ad.* 上に. =upward.
bounded ~ 上(方)に有界な. ~ concave [convex] 上に凹[凸]な. ~ to the right 右上りに.

URYSO(H)N, Pavel Samuilovič (1898-1924) ウリソーン.

use *n.* 使用(法). *vt.* 用いる.

user *n.* 使用者, ユーザー.

utility *n.* 効用.
expected ~ 期待効用. marginal ~ 限界効用. ~ function 効用関数.

utilization *n.* 利用.
~ factor [rate] 利用率.

V

V, v （ローマ数字の）5.
vague *a.* 漠然とした; 漠-.
～ convergence 漠収束. ～ topology 漠位相.
vaguely *ad.* 漠位相的に.
～ bounded 漠有界な. ～ continuous 漠連続な.
valence *n.* 価.
uni～ 単葉性.
valency ＝valence.
-valent *a.* 価の, 葉の.
p-～ function *p* 葉関数. uni～ 単葉な.
valid *a.* 妥当な; 有効な.
validity *n.* 妥当性.
general ～ 一般妥当性.
VALIRON, Georges（1884–1955）バリロン.
valuation *n.* 付値, 賦値.
additive ～ 加法(的)付値. Archimedean ～ アルキメデス付値. discrete ～ 離散付値. exponential ～ 指数付値. generalized ～ 一般付値. multiplicative ～ 乗法(的)付値. normal ～ 正規付値. normalized ～ 正規化された付値. trivial ～ 自明な付値. ～ field 付値体. ～ ideal 付値イデアル. ～ ring 付値環. ～ vector 付値ベクトル.
value *n.* 値.
absolute ～ 絶対値. approximate ～ 近似値. boundary ～ 境界値. characteristic ～ 特性値. constant ～（一)定値. critical ～ 臨界値. extremal ～ 極値. limit ～ 極限値. maximal [maximum] ～ 最大値. mean ～ 平均値. minimal [minimum] ～ 最小値. negative ～ 負(の)値. numerical ～ 数値; 絶対値. positive ～ 正(の)値. principal ～ 主値. singular ～ 特異値. stationary ～ 定常[停留]値. ～ distribution 値分布. ～ group 値群.
valued *a.* 値[価]をもつ.
many-～ 多価の, 多値の. multiple-～ 多値の. *n*-～ *n* 価の, *n* 値の. single-～, one-～ 一価の.
VANDERMONDE, Alexandre Théophile（1735–1796）ヴァンデルモンド.
VAN DER WAERDEN, Bartel Leedert（1903–1996）ファン・デル・ヴェルデン.
vanish *vi.* 消滅する; 零になる.
vanishing *a.* 消える. *n.* 消えること.
～ line 消(尽)線. ～ point 消(尽)点. ～ theorem 消滅定理.
Var. Variance. 分散.
variability *n.* 変化性, 可変性.
～ region 変化範囲.
variable *a.* 変わる, 可変の; 不定の. *n.* 変数; 変量.
change of ～(s) 変数変換. complex ～ 複素変数. dependent ～ 従属変数. dual ～ 双対変数. function of one ～ 一変数の関数. independent ～ 独立変数. primal ～ 主変数. random ～ 確率変数. real ～ 実変数. separation of ～s 変数分離. several ～s 多変数. ～ component 変化成分. ～ point 変点, 動点; 可変小数点.
variance *n.* 分散, 平方偏差.
analysis of ～ 分散分析. population ～ 母分散. sample ～ 標本分散. unbiased ～ 不偏分散. ～ matrix 分散行列. ～ ratio 分散比.
variate *n.* 変量.
multi～ 多変量. normalized ～ 正規化変量.

variation *n.* 変分; 変動; 振幅.
among-class ~ 級間変動. calculus of ~s 変分学[法]. coefficient of ~ 変動係数. first ~ 第一変分. negative ~ 負変動. of bounded ~ 有界変動の. positive ~ 正変動. second ~ 第二変分. strong ~ 強変分. total ~ 全変動. ~ of constant(s) [parameter(s)] 定数変化(法). ~ of sign 符号変化. weak ~ 弱変分. within-class ~ 級内変動.

variational *a.* 変分の.
~ derivative 変分導関数. ~ equation 変分方程式. ~inequality 変分不等式. ~ method 変分的方法. ~ principle 変分原理. ~ problem 変分問題.

variety *n.* 多様体. ⇨ manifold.
abelian ~ アーベル多様体. affine (algebraic) ~ アフィン(代数)多様体. algebraic ~ 代数多様体. differentiable ~ 可微分多様体. modulus ~, ~ of moduli モジュラス多様体. support ~ 台多様体. toric ~ トーリック多様体.

vary *vi.* 変じる, 変化する. *vt.* 変える.
~ directly [inversely] as に正比例[逆比例]して変化する.

Veblen, Oswald (1880-1960) ヴェブレン.

vector *n.* ベクトル, 方向量.
basic [basis] ~ 基底ベクトル. characteristic ~ 特性ベクトル. column ~ 列ベクトル. component ~ 成分ベクトル. constant ~ 定ベクトル. constrained ~ 束縛ベクトル. contravariant ~ 反変ベクトル. covariant ~ 共変ベクトル. free ~ 自由ベクトル. geometric ~ 幾何ベクトル. *n*-~ *n* 次元ベクトル. normal ~ 法ベクトル. normalized ~ 正規化されたベクトル. null ~ 零ベクトル. number [numerical] ~ 数ベクトル. position ~ 位置ベクトル. pricing ~ 評価ベクトル. pseudo-~ 擬ベクトル. radius ~ 動径ベクトル. row ~ 行ベクトル. tangent ~ 接ベクトル. unit ~ 単位ベクトル. valuation ~ 付値ベクトル. ~ analysis ベクトル解析. ~ bundle ベクトル束. ~ equation ベクトル方程式. ~ field ベクトル場. ~ flux ベクトル流. ~ function ベクトル関数. ~ lattice ベクトル束. ~ line ベクトル線. ~ potential ベクトル・ポテンシャル. ~ product ベクトル積. ~ quantity ベクトル量. ~ space ベクトル空間. ~ triple product ベクトル三重積. ~ tube ベクトル管. velocity ~ 速度ベクトル. zero ~ 零ベクトル.

vectorial *a.* ベクトルの.
~ angle (極座標の)角座標. ~ equation ベクトル方程式. ~ representation ベクトル表現.

vectormotive *a.* 超ベクトルの.

vector-valued *a.* ベクトル値の.
~ function ベクトル値関数.

velocity *n.* 速度, 速さ.
absolute ~ 絶対速度. angular ~ 角速度. component ~ 分速度. constant [uniform] ~ 等速度. phase ~ (位)相速度. relative ~ 相対速度. ~ potential 速度ポテンシャル. ~ vector 速度ベクトル.

Venn, John (1834-1923) ベン.

verification *n.* 確認; 検算.

verify *vt.* 証明する; 確かめる; 検孔する.

vernier *n.* 副尺, バーニア.

Veronese, Giuseppe (1854-1917) ベロネーゼ.

versed *a.* 反対の.
co~ sine 余矢【covers】. ~ sine 正矢【vers】.

versiera *n.* =witch.

versine *n.* versed sine. 正矢【vers】.

version *n.* 説明; 視点.

versor *n.* ベルソル.

vertex (*pl.* -tices, -texes) *n.* 頂点; 角頂.
corresponding ~tices 対応頂点. four ~ theorem, theorem of four ~tices 四頂点定理. ~ angle 頂角. ~ of angle 角の頂点.

vertical *a.* 水平面に直角な, 垂直な, 鉛直な; 頂点の. *n.* 垂直線[面].
~ angle 頂角. ~ angle(s) 対頂角. ~ component 垂直成分. ~ line 垂直[鉛直]線. ~ motion 上下運動. ~ plane 垂直[鉛直]面; 立画面. ~ trace 直立跡. ~ly opposite angle(s) 対頂角.

very *ad.* 極めて; 非常に. *a.*【the などと共

に使用】ちょうどその.
～ ample system 非常に豊富な系.

vibrate *vi.* 振動する.
～ting membrane [string] 振動膜[弦].

vibration *n.* 振動. ⇨ oscillation.
amplitude of ～ 振幅. damped ～ 減衰振動. forced ～ 強制振動. free ～ 自由振動. normal ～ 規準振動. self-excited ～ 自励振動.

vicenary *a.* 二十進法の.
～ notation 二十進法.

vice versa *ad.* (*L.*) 逆もまた同様.

vicinity *n.* 近傍.

vicious *a.* 不備な; 誤った; 悪い.
～ circle 循環論法.

vid. ＝vide.

vide *vi.* (*L.* ＝see.) 見よ.
～ ante 前を見よ. ～ infra 下を見よ.

videlicet *ad.* (*L.* ＝namely.) すなわち, 換言すれば.

VIÈTE, François (1540–1603) ヴィエト.

view *n.* 視野; 考慮; 見解. *vt.* 見る; 調べる; 考察する.
in ～ of を考慮して, のゆえに. point of ～, ～point 見地, 観点.

vigesimal ＝vicenary.

vinculum (*pl.* -la) *n.* 括線.

VINOGRADOV, Ivan Matveevič (1891–1983) ビノグラドフ.

virgule *n.* 斜線.

virtual *a.* (名目上ではなく)実質上の; 仮の; 仮想の, 仮設の; 虚の【物理用語】.
～ degree 仮想次数. ～ displacement 仮想変位. ～ genus 仮想種数. ～ image 虚像. ～ work 仮想[設]仕事.

virtually *ad.* (名目は異なるが)実質的に.

viscosity *n.* 粘性.
～ solution 粘性解.

viscous *a.* 粘性の.
～ fluid 粘性流体.

visual *a.* 視覚の.
～ angle 視角. ～ center 視心.

VITALI Giuseppe (1875–1932) ヴィタリ.

viz. ＝videlicet.

void *a.* 空の. ⇨ empty.
～ set 空集合.

Vol., vol. ＝volume. 巻.

VOLTERRA, Vito (1860–1940) ヴォルテラ.

volume *n.* 体積, 容積; 巻.
～ element 体積要素.

VON NEUMANN, John [Janos] (1903–1957) フォン・ノイマン.

vortex (*pl.* -tices, -texes) *n.* 渦; 渦心.
～-free 渦なしの. ～ motion 渦(運)動. ～ tube 渦管.

vorticity *n.* 渦度, 渦量.

v.p. *n.* (*F.*) 主値【valeur principale の略称】.

vulgar *a.* 一般の.
～ fraction (通)常分数.

v.v. ＝vice versa.

W

Wald, Abraham (1902–1950) ワルド.

walk *n.* 歩行.
random ~ 酔歩, 乱歩.

Wallis, John (1616–1703) ウォリス.

wandering *a.* 遊走的な.

Waring, Edward (1734–1798) ワーリング.

warn *vt.* 注意する.

Watson, George Neville (1886–1965) ワトソン.

wave *n.* 波, 波動.
dispersive ~ 分散波. electric ~ 電波. longitudinal ~ 縦波. plane ~ 平面波. shock ~ 衝撃波. sine ~ 正弦波. spherical ~ 球面波. standing [stationary; steady] ~ 定常波. surface ~ 表面波. transversal ~ 横波. trochoidal ~ トロコイド波. ~ equation 波動方程式. ~ front 波面. ~ function 波動関数. ~ length 波長. ~ mechanics 波動力学. ~ number 波数. ~ optics 波動光学.

wavelet *n.* ウェーブレット.

wavy *a.* 波状の.
~ line 波線.

weak *a.* 弱い. ⇨ strong.
~ convergence 弱収束. ~ limit 弱極限. ~ solution 弱解. ~ summation [summability] 弱総和法. ~ topology 弱位相. ~ variation 弱変分.

weakly *ad.* 弱く. ⇨ strongly.
~ continuous representation 弱連続表現. ~ converge 弱収束する. ~ integrable 弱可積な. ~ stationary 弱定常な. ~ symmetric 弱対称な.

web *n.* 織.
hexagonal ~ 六角形織.

Weber, Heinrich (1842–1913) ウェーバー.

Wedderburn, Joseph Henry Maclagan (1882–1948) ウェダーバーン.

wedge *n.* 楔; 楔形, 直三角柱.
~ product 楔積【記号 ∧】.

Weierstrass, Karl Theodor Wilhelm (1815–1897) ワイエルシュトラス.

Weierstrassian *a.* ワイエルシュトラスの.
~ theory ワイエルシュトラスの理論.

weight *n.* 重み, 重さ; 荷重(値), 加重(値); 評量値, 秤量値; 重要度; ウェート. *vt.* 重みをつける; 加重する.
~ enumerator 重み母関数.

weighted *a.* 重みのついた.
~ mean 荷[加]重平均.

Weil, André (1906–1998) ヴェイユ.

Weingarten, Julius (1936–1910) ワインガルテン.

well-chained *a.* 鎖状連結の.

well-defined *a.* はっきり定義された; 明確な; 定義可能な; 矛盾なく定義される(た); ウェルディファインド.

well-formed *a.* 整合的な.
~ formula 整合論理式.

well-order *n., vt.* 整列順序(をつける).
~ing theorem 整列(可能)定理.

well-ordered *a.* 整列された.
inversely ~ set 逆整列集合. ~ set 整列集合.

well-posed *a.* 適切な.

Wessel, Caspar (1745–1818) ウェッセル.

Weyl, Claude Hugo Hermann (1885–1955) ワイル.

white *a.* 白い.
~ noise 白色雑音.

WHITEHEAD, Alfred North（1861-1947） ホワイトヘッド.

WHITTAKER, Sir Edmond Taylor（1873-1956） ホイッテ[タ]カー.

whole *a.* 全体の; 完全な; 整数の.
～ event 全事象. ～ number 整数(＝integer). ～ set 全体集合. ～ space 全空間.

wide *a.* 広い, 幅広い.
in the ～ sense 広義に. 3 feet ～ 幅3フィートの.

wider *a.* より広い.
in the ～ sense 広義に.

width *n.* 幅; 広さ; 横.

WIENER, Norbert（1894-1964） ウィーナー.

wild *a.* 野性[生]的な.
～ space 野性[生]的空間. wild ramification 野性[生]分岐.

WILES, Andrew（1953- ） ワイルス.

WILSON, John（1741-1793） ウィルソン.

winding *a.* 回転の; 巻きつく.
～ index 回転指数. ～ number 回転数, 巻き数.

-wise *suf.*「様式, 方法」の意.
clock～ 時計の針の方向の[に]. coefficient～ 係数ごとの[に]. component～ 成分ごとの[に]. element-～ 元ごとの[に]. piece-～ 区分ごとの[に]. point～ 各点ごとの[に].

within-class *a.* 級内の.
～ variation 級内変動.

WITT, Ernst（1911-1991） ヴィット.

W.K.B. method Wentzel-Kramers-Brillouin method. W.K.B.法.

word *n.* 語, 単語.
～ length 語長. ～ problem 語の問題.

work *n.* 仕事, 作業; 業績.
select(ed) ～s 選集. virtual ～ 仮想仕事.

working *a.* 動く; 作業用の; 実用上の.
～ mean 仮の平均.

world *n.* 世界.
～ line [point] 世界線[点].

write *vt., vi.* 書く; 書き込む.

WRONSKI, Höéné Joseph Maria（1778-1853） ロンスキ.

Wronskian *n.* ロンスキアン, ロンスキの行列式.

***W*-surface** *n.* W曲面.

X

X, x (ローマ数字の)10.

xi *n*. クシー，グザイ．ギリシア語アルファベットの 14 番目の文字【Ξ, ξ】.

-xion *suf*. ＝-ction. ⇨ -tion.

Y

yard *n.* ヤード.
　～-pound system ヤードポンド法.

year *n.* 年; 年度, 学年.
　calendar ～ 暦年. duration ～s 耐用年数. light-～ 光年. school [scholatic, academic] ～ 学年.

yield *vt., vi.* 生じる; 産する. *vt.* 与える. *n.* 利回り; 収量.
　～ing interest 利回り.

Young, Alfred (1873-1940) ヤング.

Young, William Henry (1862-1942) ヤング.

Z

z *n.* ゼット; ズィー.
z-plane ゼット平面. ~-transformation ゼット変換.

Zaremba, Stanisław (1863–1942) ザレンバ.

Zariski, Oscar (1899–1986) ザリスキ.

Zassenhaus, Hans Julius (1912–1991) ツァッセンハウス.

zenith *n.* 天頂.
~ angle 天頂角. ~ distance 天頂距離. ~ projection 天頂投影.

zenithal *a.* 天頂の.
~ angle [distance] 天頂角[距離].

Zeno(n) of Elea (490?–429? B.C.) エレアのツェノン, ゼノン.

Zermelo, Ernest (1871–1953) ツェルメロ.

zero (*pl.* -(e)s) *n.* ゼロ, 零; 零点. *a.* 零の.
aleph ~ アレフゼロ. characteristic ~ 標数零. of measure ~ 測度ゼロの. simple ~ 一位の零点. ~ chain 零鎖. ~ correlation 零相関. ~ divisor 零因子. ~ element 零元. ~-free 零点のない. ~ matrix 零行列. ~-one law 0–1 法則, 零または一の法則, 悉無律. ~ point 零点. ~ (point) of *k*th order [order *k*] *k* 位の零点. ~ system 零系. ~ vector 零ベクトル.

zero-sum *n.* 零和. *a.* 零和の.
~ game 零和ゲーム.

zeta *n.* ゼータ, ジータ. ギリシア語アルファベットの 6 番目の文字【Z, ζ】.
elliptic ~ function 楕円ゼータ[ツェータ]関数. ~ function ゼータ関数.

ZF Zermelo-Fraenkel set theory.

zonal *a.* 帯の; 帯状の.
~ harmonics 帯球調和関数. ~ spherical function 帯球関数.

zone *n.* (球面・円錐などの)帯; 域; ゾーン.
spherical ~ 球帯. ~ of indifference 検査続行域. ~ of preference for acceptance 合格域. ~ of preference for rejection 不合格域. ~ of sphere 球帯.

Zorn, Max (1906–1993) ツォルン.

Zygmund, Antoni (1900–1992) ジグムント.

Zyklographie *n.* (*G.*) 円点投象法.

II

和英の部

JAPANESE-ENGLISH

あ・ア

アイ・アイ・ディー i. i. d; independent and identically distributed.

あいことなる 相異なる pairwise distinct. ⇨異なる.

アイコナール (*G.*) Eikonal.

アイゼンシュタイン F. G. M. Eisenstein (1823–1852).
～既約判定法 Eisenstein criterion.

アイソスピン isospin.

あいだ 間 between; among; during.
(二者の)～ between. (三者以上の)～ among. (時間的な)～ during.

あいまい 曖昧(性) ambiguity; vagueness.
～な ambiguous; vague; equivocal; uncertain; fuzzy. ～な文法 ambiguous grammar.

アウトプット output. ⇨出力.

あおんそく 亜音速(の) subsonic.
～波 subsonic wave.

あかいけ・ひろつぐ 赤池弘次 Akaike Hirotugu (1927–2009).
赤池情報量基準 Akaike information criterion; AIC.

あきらか 明らか(な) evident; obvious; clear.
～に evidently; obviously; clearly.

アキレス Achilles.
～と亀 Achilles and tortoise.

アークサイン arcsine. ⇨逆正弦.

あぐん 亜群 groupoid.

あし 足; 脚 foot (*pl.* feet); pedestal; leg. 垂線の～ foot of perpendicular.

あずまや・ごろう 東屋五郎 Azumaya Goro (1920–2010).

あたい 値 value.
～[価]をもつ valued. ～をとる attain; assume. ～分布 value distribution. ～群 value group. ～の範囲 range of values.

あたえる 与える give; yield.

アダマール J. S. Hadamard (1865–1963).
～空隙 Hadamard's gap.

あたり 当り hit; success.
当たる hit; strike. まぐれ～ chance hit.

アーチ arch.
～形の arched; arch-shaped.

アーチモデル ARCH model; autoregressive conditional heteroskedasticity model.

あつかい 扱い treatment; management.
扱者 server.

あつかう 扱う treat; deal with; manage; address.

あつさ 厚さ thickness.
3センチの～ three centimeters thick [in thickness].

あっしゅく 圧縮 compression.
～性; ～率 compressibility. ～性の compressible. ～する compress. ～(性)流体 compressible fluid. ～限界 compressed limit.

あつめる 集める collect; gather; group.

あつりょく 圧力 pressure; stress.
～方程式 pressure equation.

あてはめ 当てはめ fitting.
～る fit. 曲線の～ curve fitting.

アデール (*F.*) adèle.
主～ principal adèle. ～群 adèle group. ～環 adèle ring.

あと 跡 ⇨跡(せき).

あといれさきだし 後入れ先出し last-in first-out; LIFO.

アドミッタンス admittance.
～行列 admittance matrix.

アトラス ATLAS.
有限群の～ ATLAS of Finite Groups.
アトラス atlas.
アドレス address.
アトレッド atled.【記号 ∇, delta の逆読み.】⇨ナブラ.
あな 穴; 孔 hole; puncture.
～をあける puncture. ～をあけた平面[円板] punctured plane [disc; disk].
アナログ analog【米】; analogue【英】.
～・データ analog(ue) data. ～・シミュレーション analog(ue) simulation.
アニェージ M. G. Agnesi (1718–1799).
～の魔女 witch of Agnesi, Agnesi's witch.
アーネシー，アグネシ ⇨アニェージ.
アフィン(的) affine.
～的に affinely. ～合同な affinely congruent. ～群 affine group. ～環 affine ring. ～同型[形] affine isomorphism. ～変換 affine transformation; affinity. ～写像 affine mapping. ～的性質 affine property. ～幾何学 affine geometry. ～微分幾何学 affine differential geometry. ～空間 affine space. ～座標 affine coordinates. ～法線 affine normal. ～弧長 affine arc length. ～枠 affine frame. ～接続 affine connection. ～対応 affine correspondence. ～スキーム affine scheme. ～束 affine bundle. ～(代数)多様体 affine (algebraic) variety.
アフィノール affinor.
アファイン ⇨アフィン.
アプリオリ (*L.*) a priori.
あふれ 溢れ overflow.
～る overflow. 行～ line overflow.
アーベル N. H. Abel (1802–1829).
～的な Abelian; abelian. 非～的な non-abelian. ～群 Abelian group; commutative group. ～拡大 Abelian extension. ～関数体 Abelian function field. ～微分 Abelian differential. ～積分 Abelian integral. ～多様体 Abelian variety. ～圏 Abelian category.
アポステリオリ (*L.*) a posteriori.
アポロニウス Apollonius [Apollonios] (262–200? B.C.).
～の Apollonian; Apollonius'. ～の円 Apollonius' [Apollonian] circle.
あまり 余り remainder; rest; residue; residuum (*pl.* -dua); surplus. ⇨剰余.
～の residual.
あみ 網 net.
三角形の～ triangular net. ～目 mesh; network. 被覆の～目 mesh of covering. ～目構造 network structure. ～目の幾何学 geometry of nets.
あやまった 誤った wrong; mistaken; incorrect; false; vicious.
～計算 incorrect calculation; miscalculation.
あやまり 誤り error; erratum (*pl.* -ta); mistake.
～の訂正 error-correction. 第一[二]種の～ error of the first [second] kind. ～を立証する disprove. ～訂正符号 error correcting code.
あゆみ 歩み step.
～幅 step-size.
あらい 粗い coarse; rough.
～評価 rough estimate.
アラビア Arabia.
～(人)の Arabian; Arabic. ～数字 Arabian cypher; Arabic figure [numeral]. ～(の)数学 Arabian mathematics. ～記数法 Arabian notation of numbers.
あらわす 表す denote; express; represent; write.
あらわれる 現れる appear; occur.
ありそう ありそうな likely; possible; probable; plausible.
アリマモデル ARIMA model; autoregressive integrated moving average model.
ある 或る a [an]; one; certain; some.
アルキメデス Archimedes (287–212 B.C.).
～の; ～的な Archimedean. ～順序体 Archimedean ordered field. ～螺線 Archimedean spiral. ～的付値 Archimedean valuation.
アルゴリズム algorithm; algorism.
～の algorithmic. 多項式時間～ polynomial time algorithm.

アルティン E. Artin (1898-1962).
～的な Artinian. ～(的)加群 Artinian module. ～環 Artinian ring. ～・シュライアー拡大 Aritin-Schreier extension. ～・リースの補題 Artin-Rees lemma.

アルファ alpha; 【A, α】.

アルファベット alphabet.
～の; ～順の alphabetical. ～順 alphabetical order. ～順に. alphabetically; in alphabetical order.

アレフ aleph; 【ℵ】.
～・ゼロ aleph zero; 【$\aleph_0$】.

あれんぞく 亜連続(な) hypocontinuous.
～性 hypocontinuity.

あん 鞍 saddle; col. ⇨鞍点.
～点 saddle point. ～(部)点法 saddle point method; method of steepest descent. ～状領域 saddle region.

あんごう 暗号 cipher.
～(法) cryptography.

あんごうか 暗号化 encryption.
～する encrypt.

あんごうかいどく 暗号解読 decryption.

あんごうけい 暗号系 cryptosystem.
公開鍵～ public key cryptosystem.

あんざん 暗算 mental calculation; mental arithmetic.
～の mental. ～する calculate mentally.

あんじょう- 鞍状
～点[領域] saddle point [region].

あんぜん 安全 safety; security.
～余裕 safety allowance.

アンダーフロー underflow.

あんてい 安定(性) stability.
～な; ～した stable. ～して stably. ～状態 stable state. ～域 stable range. ～平衡 stable equilibrium. ～解 stable solution. ～分布 stable distribution. ～過程 stable process. ～人口 stable population. 漸近～性 asymptotic stability. 軌道～性 orbital stability. ～平行性をもつ stably parallelizable. ～同値な stably equivalent. 両側～な stable in both directions. 絶対～な absolutely stable. ～化 stabilization. ～化する stabilize. ～化子 stabilizer. ～化部分群 stabilizer; stabilizing subgroup. ～性解析 stability analysis. ～同値 stable equivalence.

あんてん 鞍点 saddle point; col.
～法 saddle point method; method of steepest descent; (*F.*) méthode de col.

あんないにん 案内人 pilot; guide.

アンビグ ambig.
～類 ambig class.

あんぶ 鞍部 ⇨鞍(あん).

あんぶん 按分(する) divide in proportion.
～比例 proportional distribution.

あんぶんせん 按分線 split.

い・イ

い　位　order; rank; place.
一～の極 simple pole; pole of the first order. 高～の of high(er) order. 末～ last location [place]. (十進)小数～ decimal place.

いえん　緯円　latitude circle.

イオタ　iota; 【I, ι】.

いか　位価　local value.

いか　以下　less than or equal to; not greater than; at most.

いかん　移換　transvection.

-いき　域　domain; region; zone; range. 定義～ domain [region] of definition. 作用～ operator domain. 変～ domain [region] of variability. 整～ integral domain; domain of integrity. 収束～ convergence region [domain]. 点～ point range. 合格[不合格]～ zone of preference for acceptance [rejection]. 検査続行～ zone of indifference. 採択～ acceptance region. 棄却～ range of rejection; critical region. 信頼～ confidence region. 数～ numerical range. 安定～ stable range. 係数～ range of parameter.

いぎ　意義　⇨意味.

いきょ　緯距　latitude distance.

いご　以後　following; (*L.*) sequentia; sequentes; s(e)q.
23ページ～ p. 23 (et) seq.

いこう　移行　transference; shift.

いこう　移項　transposition.
～する transpose.

いしつ　異質(性)　heterogeneity.
～の; ～的な heterogeneous. ～性の検定 heterogeneity test; test of heterogeneity.

いしゅ　異種(の)　exotic.
～球面 exotic sphere.

いじょう　以上　greater [more] than or equal to; not less than; at least.

いじょう　異常　abnormality; anomaly; extraordinariness; singularity.
～な abnormal; anomalous; extraordinary; singular; improper. ～点 singular point; singularity. ～閾 anomalous threshold. ～球面 exotic sphere.

いすう　位数　order.
接触～ order of contact. ～条件 order condition. 有限[無限]～ finite [infinite] order. 無限大[無限小]の～ order of infinity [infinitesimal]. 群の～ order of group. 元の～ order of element.

いずれにせよ　in any case.

いせん　緯線　(parallel of) latitude; latitude circle.

いそう　位相　topology.
～的な topological. ～づける; ～化する topologize. ～数学; ～幾何学 topology. 相対～ relative topology. 離散～ discrete topology. 強[弱]～ strong [weak] topology. 積[商]～ product [quotient] topology. 直積～ direct product topology. 直和～ direct sum topology. 誘導～ induced topology. ～空間 topological space. ～(的)構造 topological structure. ～写像 topological mapping. ～多様体 topological manifold. ～群 topological group. ～積 topological product. ～不変量 topological invariant. ～胞体 (topological) cell. ～(的)モデル topological model. ～同値な topologically equivalent. ～的に不変な

topologically invariant. 位相～ topological module. クルル～ Krull topology.

いそう 位相 phase.
～角 phase angle. 初期～ initial phase. ～のずれ phase shift. ～速度 phase velocity.

いそう 移送 transfer; transport(ation).

いそうきかがく 位相幾何学 topology.
一般～ general topology. 組合せ[代数的]～ combinatorial [algebraic] topology. 集合論的[微分]～ set-theoretic [differential] topology.

いそうすうがく 位相数学 topology. ⇨トポロジー.

いそうどうけい 位相同型[形] homeomorphy; homeomorphism.
～な homeomorphic. ～写像 homeomorphism; homeomorphic mapping.

イソトピー isotopy.
～型 isotopy type.

イソトープ(な) isotopic.

いぞん 依存 dependence; dependency.
に～する depend on [upon]. ～領域 domain [region] of dependence.

イーター eta; 【H, η】.

いたるところ everywhere; throughout; (*F.*) partout; (*L.*) passim.
～ない nowhere. 殆んど～ almost everywhere; a.e.; (*F.*) presque partout; p.p. ～疎 nowhere dense.

いち 一 one; unity. ⇨第一(の).
～価の single [one]-valued. ～の分解[分割] partition of unity. ～パラメータ族[群] one-parameter family [group]. 1の原始 n 乗根 primitive n-th root of unity.

いち 位置 position; situation; location; place; site; locus.
～の positional. 一般[標準]の～ general [standard] position. 対称[相似]の～ position of symmetry [similarity]. ～ベクトル position vector. ～幾何学 geometry of position. ～母数 location parameter. ～の問題 placement problem.

いち 位置(の) potential.
～エネルギー potential energy.

いち- 一 uni-; mono-; one. ⇨単(たん-).

いちい 一意(性) uniqueness; unicity; univalency.
～(的)な unique; univalent. ～に uniquely. ～性の定理 uniqueness theorem. ～分解 unique decomposition. ～分解整域 unique factorization domain; UFD. ～対応 univalent correspondence.

いちい 一意(な) uniform.
～関数 uniform function.

いちいか 一意化 uniformization.
～する uniformize. ～可能な uniformizable. 局所～ local uniformization. ～(媒介)変数 uniformizer; uniformizing parameter. 有理的に～可能な unicursal.

いちいでんしせい 一遺伝子性(の) monogenic.

いちげん 一元 one unknown; one-dimension.
～の of one unknown; one dimensional; monogenic; unitary. ～性 monogeneity. ～方程式 equation with one unknown. ～計画(法) design of one dimension.

いちじ 一次(の) linear. ⇨線形(の).
～的に linearly. ～式 linear expression. ～方程式 linear equation; equation of first degree. ～関数 linear function. ～形式 linear form. ～分数式 linear fractional expression. ～変換 linear transformation. ～結合 linear combination. ～独立な linearly independent. ～従属な linearly dependent.

いちじ 一次(の) primary; first.
～方程式 equation of the first degree; linear equation. ～抽出単位 primary sampling unit.

いちじ 一時(の) temporary; temporal; transient.
～(的に) temporarily. ～解 transient solution.

いちじへんかん 一次変換 linear transformation; Möbius transformation.

いちそうついさ 一双対鎖 crossed homomorphism; 1-cochain.

いちたいいち 一対一[1 対 1](の) one-to-one; biunivoque.

～に one-to-one; biunivoquely. ～写像 one-to-one mapping; injection. ～対応 one-to-one correspondence; bijection.

いちぶ 一部 part; portion.
～の; ～分の partial. ～釣合 partial balance. ～釣合型計画(法) partially balanced design.

いちぶじっし 一部実施(法) fractional replication.

いちへんりょう 一変量(の) univariate.

いちよう 一様(な) uniform; equal; even.
～に uniformly; equally; even. 広義(の)～に uniformly in the wider sense. ～にする make uniform; unify. ～性 uniformity. ～収束 uniform convergence. ～収束の uniformly convergent. ～(に)有界[連続; 凸]な uniformly bounded [continuous; convex]. ～密度 uniform density. ～分布 uniform distribution; (*G.*) Gleichverteilung. ～位相 uniform topology. ～ノルム uniform norm. ～一致検定 uniformly consistent test. ～最小分散 uniformly minimal variance.

いちようらんりゅう 一様乱流 homogeneous turbulence.

いちらんひょう 一覧表 list; table.

いちれつ 一列
～にする align; aline. ～にすること alignment; alinement. ～の tandem.

いっか 一価(の) single-valued; one-valued; uniform.
～関数 single [one]-valued function; uniform function.

いっかい 一階 first order.
～微分方程式 differential equation of first order; first order differential equation. ～導関数 derivative of first order; first order derivative. ～差分 difference of first order; first difference. ～述語論理 first order predicate logic.

いっかい 一回 one time; once.
～抜取検査 single sampling inspection.

いっかせい 一価性 monodromy; single-valuedness.
～の定理 monodromy theorem.

いっかつ- 一括
～処理 batch processing. ～して (*F.*) en bloc.

いっき 一期 term; period.
～の総売上高 turn-over.

いっそうつよいりゆうで いっそう強い理由で (*L.*) a fortiori.

いつだつど 逸脱度 deviance.

いっち 一致 coincidence; consistency; consistence; concordance; unicity.
～する coincide; consist. ～した coincident; consistent. ～検定 consistent test. ～推定量 consistent estimator. ～統計量 consistent statistic. ～性のない統計量 inconsistent statistic. ～条件 consistency condition. ～係数 coefficient of concordance. ～の定理 unicity theorem; theorem of identity. ～性の原理 unicity principle.

いっつい 一対 pair; couple.
～比較(法) paired comparison.

いってい 一定な constant; fixed; invariable; definite; determinate.

いっぱん 一般(な) general; generic; ordinary; universal.
～に in general; generally. ～性 generality. ～化 generalization. ～化する generalize. ～項 general term. ～な位置 general position. ～解 general solution. ～微分係数 general derivative. ～螺旋 general helix. ～線形群 general linear group. ～ディリクレ級数 general Dirichlet series. ～(理)論 general theory. ～集合論 general set theory; general theory of sets. ～関数論 general theory of functions. ～トポロジー general topology. ～相対性原理 general principle of relativity. ～伝染理論 general theory of epidemics. ～妥当性 general validity. ～性を失うことなく without loss of generality. ～座標 generalized coordinates. ～関数 generalized function. ～(化された)ゼータ関数 generalized zeta function. ～距離 generalized distance. ～付値 generalized valuation. ～等周問題 generalized isoperimetric problem. ～分散 generalized variance. ～算術 universal arith-

metic.

いっぱんか 一般化 generalization.
~された generalized. ~加法モデル generalized additive model; GAM. ~逆行列 generalized inverse (matrix). ~線形モデル generalized linear model; GLIM.

イデアル ideal.
~群 ideal group. ~類 ideal class. 左[右]~ left [right] ideal. 両側~ two-sided ideal. 素~ prime ideal. 主[単項]~ principal ideal. 主[単項]~整域 principal ideal domain. 極大~ maximal ideal. ~環 ideal ring. 縮約~ contracted ideal. 許容~ admissible ideal. 単位~ unit ideal. 同[斉]次~ homogeneous ideal. 無縁~ irrelevant ideal.

イデール (*F.*) idèle.
~群 idèle group. 主~ principal idèle.

いでん 遺伝 heredity; inheritance.
~的に受け継ぐ inherit. ~的な hereditary; inherited. 左[右]~的な left [right] hereditary. ~誤差 inherited error. ~形質 inherited character [quality].

いでんがく 遺伝学 genetics.
集団~ population genetics.

いど 緯度 latitude; parallel.
~の latitudinal.

いどう 移動 movement; displacement; translation; transfer; transference; shift.
~する move; translate; shift. ~させる displace, translate; transfer; shift. 平行~ translation; parallel displacement. 左[右]~ left [right] translation. ~可能な translatable. ~演算子 translation operator. ~平均 moving average; running mean.

いにゅう 移入 injection.
~的な injective. ~分解 injective resolution. ~加群 injective module.

いぬきょくせん 犬曲線 ⇨犬曲線(けんきょくせん).

イプシロン epsilon; 【E, ε, ϵ】.
~・デルタ論法 epsilon-delta technique; epsilontics.

いぶんさんせい 異分散性の heteroskedastic.

いみ 意味 sense; meaning; significance.
~する mean; implicate; imply.

いみろん 意味論 semantics.

いらい 以来 since.

いれこ 入れ子(構造) nest.
~にする nest. ~型計画 nested design.

いん 陰(の) implicit.
~に implicitly. ~関数 implicit function.

いんがかんけい 因果関係 causality; causal relation.

いんがりつ 因果律 law of causality; law of cause and effect.

いんかんすう 陰関数 implicit function.

いんし 因子 factor; divisor.
共通~; 公~ common factor. 積分~ integrating factor. ~群 factor group. ~団 factor set. 直和~ direct summand. ~分析(法) factor analysis. ~荷重 factor loading. ~スコア factor score. ~に含む contain (as factor). 素~ prime divisor; prime factor. ~類 divisor class. 零~ zero divisor. 非退化~ non-degenerate divisor. 単~ elementary divisor. ~表現 factor representation. ~分解 factorization. ~分解可能性 factorability. ~分解定理 factorization theorem. 乗法~ multiplicator; multiplier. ~加群 factor module. 正~ positive divisor. 組成~ composition factor. 有効~ effective divisor.

いんしょう 印象 impression.
~的な impressive. ~主義 impressionism.

いんすう 因数 factor; divisor. ⇨因子.
共通~ common factor. ~定理 factor theorem. ~分解 factorization. ~に分解する factor; factorize. ~分解公式 factorization formula; formula of factorization. ~分解可能性 factorability.

インタフェース interface.

インデックス index (*pl.* -dexes, -dices).

インド India.
~の Indian. ~数学 Indian mathematics.

インピーダンス impedance.
~行列 impedance matrix.

インプット input. ⇨入力.

インフレーション inflation.

インボリュート involute.

～歯車 involute gear.

いんよう 引用 quotation; reference. ⇨参照. ～する quote; cite; refer【to】. ～符 quotation mark. ～文献[書目] reference(s); bibliography. 上記～文中[箇所]に (*L.*) loco citato; loc. cit.

いんりょく 引力 attraction; attractive force; gravitation, gravity. 万有～ universal gravitation. 地球の～ terrestrial gravitation; gravity.

う・ウ

うえ　上　⇨下(した).
　〜の upper; upward.　〜に upwards; above.　〜に凹[凸]な upwards concave [convex].　〜に有界な bounded (from) above; bounded upwards.　〜(に)半連続な upper semicontinuous.　〜への写像 onto-mapping.　〜三角行列 upper trianqular matrix.
うえがわ　上側(の)　upper.
　〜確率 upper tail probability.　〜四分位数 upper quartile.
ウェート　weight.　⇨重み.
ウェーブレット　wavelet.
ウェルディファインド　well-defined.
うかい　迂回　detour.
　〜行列 detour matrix.
うかぶ　浮かぶ　float.　⇨浮動(する).
うけいれ　受入　acceptance.
　〜れる accept.　〜検査 acceptance inspection.
うけいれかのう　受け入れ可能な　acceptable.
うけとり　受取(り)　receipt.
　〜る receive; accept.　〜人 receiver; recipient.
うごく　動(く)　move.
　〜かない fixed; stationary.　〜く[かない]特異点 movable [fixed] singular point.
うしろむき　後ろ向き(の)　retrospective.
　〜研究 retrospective study.
うず　渦　vortex (*pl.* -texes, -tices); swirl.
　〜のある rotational.　〜なしの irrotational; rotationless; vortex-free; lamellar.　〜運動 vortex motion.　〜度 vorticity.
うすい　薄い　thin.
うずまき-　渦巻　spiro-; spiral; volute.
　〜線 spiral.
うたんどうしゅつ　右端導出　rightmost derivation.
うち　内　inside; interior.　⇨外(そと).
　〜の inner; interior.　〜向き(の)法線 inward normal; inner [interior] normal.
うちきり　打切り　truncation; censoring.
　〜誤差 truncate(d) [truncation] error.
うちせん　迂地線　⇨アニェージ.
うちのり　内法　inside measurement.
うちゅう　宇宙　universe; cosmos.
　〜の universal; cosmic.　〜定数 universal constant.　〜論 cosmology.
うつす　移す　transfer; remove; map; carry; shift; translate.
　〜こと transference.
うつす　写す　map.
うへん　右辺　right(-hand) side [member].
うめこみ　埋込(み)　embedding; imbedding; immersion.　⇨埋蔵.
　〜む embed; imbed.　〜定理 embedding [imbedding] theorem.　正則(な)〜 regular embedding [imbedding].　〜原理 embedding [imbedding] principle.
うら　裏　reverse.
　(含意命題の)〜 converse of contrapositive.
うらがえし　裏返し　reflection; reflexion; inversion.
　〜する turn over.
うんえい　運営　operation; management.
　〜研究 operational research; operations research; OR.
うんざん　運算　calculation; operation.
うんてん　運転　running; working; operation.
うんどう　運動　motion; movement.
　等速〜 uniform motion.　相対〜 relative

motion. 自由[平均]～ free [mean] motion. 円[楕円]～ circular [elliptic] motion. ～の三法則 three laws of motion. ～の方程式 equation of motion. ブラウン～ Brownian motion. 無限小～ infinitesimal motion. ～群 group of motions.

うんどう 運動(の) kinetic.

～エネルギー kinetic energy. ～学 kinetics; kinematics. ～学の kinematic.

うんどうりょう 運動量 momentum (*pl.* -ta -tums).

角～ angular momentum. 一般～ generalized momentum. ～の保存 conservation of momentum.

え・エ

え　絵　picture.
～グラフ pictogram; pictograph.
エイ・アイ・シー　AIC; Akaike information criterion.
えいかく　鋭角　acute angle.
～の acute(-angled). ～三角形 acute triangle.
えいきょう　影響　influence; effect.
～する affect; influence; act. ～領域 domain [region] of influence.
えいしゃ　映写　projection. ⇨投射.
～する project.
えいせい　衛星　satellite.
～関手 satellite functor.
えいねん　永年(の)　secular.
～方程式 secular equation. ～摂動 secular perturbation.
えいびん　鋭敏(な)　sharp; acute.
えがく　描く，画く　draw; plot.
エス　S.
～字形[状]の S-shaped; sigmoid; serpentine. ～字形[状]曲線 sigmoid curve; serpentine curve.
エス・エヌ　sn.
～関数 sn-function.
えだ　枝　branch. ⇨分枝(ぶんし).
エヌ・ピー　NP; non-deterministic polynomial time.
～完全 NP complete. ～困難 NP hard.
エネルギー　energy.
運動～ kinetic energy. 位置～ potential energy. 相互～ mutual energy. 共鳴～ resonance energy. ～積分 energy integral. ～原理 energy principle. ～方程式[不等式] energy equation [inequality]. ～論 energetics.
エピグラフ　epigraph.
えびら　箙　quiver.
エフ　F.
～分布 F-distribution. ～検定(法) F-test. ～・シグマ集合 F_σ set.
エム　M.
～判定法 M-test; majorant test.
えらぶ　選ぶ　choose; select. ⇨選択.
えり　襟　collar.
～をもつ collared. ～(つき)近傍 collared neighbo(u)rhood.
エル　L.
～字形の L-shaped.
エルゴード(性)　ergodicity.
～の; ～的な ergodic. ～仮説 ergodic hypothesis. ～理論 ergodic theory. ～条件 ergodic condition. ～定理 ergodic theorem. ～面 ergodic surface. ～伝送容量 ergodic transmission capacity.
エル・ピー　LP. ⇨線形計画(法).
エルミート　Ch. Hermite (1822-1901).
～の[的な] Hermitian. ～形式 Hermitian form. ～行列 Hermitian matrix. ～変換 Hermitian transformation. ～拡大 Hermitian extension. ～計量 Hermitian metric. ～積 Hermitian product. 反～的な anti-Hermitian.
エレクトロニクス　electronics.
エレクトロン　electron.
エレメント　element. ⇨要素.
えん　円　circle.
～の circular. 大～ great circle. 小～ small circle. 反転～ circle of inversion. 曲率～ circle of curvature. 測地～ geodesic

circle. オイラー～ Euler('s) circle. 単位～ unit circle. 収束～ convergence circle. ～の求積 quadrature of circle; cyclometry. ～の切り口 cyclic [circular] section. ～順列 circular permutation. ～弧 circular arc. ～部分 circular parts. ～錐[柱] circular cone [cylinder]. ～領域 circular domain. ～グラフ circle [circular] graph. ～運動 circular motion. ～振子 circular pendulum.

えんえき 演繹 deduction.
～的な deductive; syllogistic. ～する deduce. ～(法) deductive method; syllogism.

えんえんたいおう 円々対応 circle-to-circle correspondence; (*G.*) Kreisverwandtschaft.

えんかくへいこう 遠隔平行(性) teleparallelism; absolute parallelism.
～の teleparallel.

えんかん 円環 annulus (*pl.* -li, -luses); circular ring.
～の annular. 同心～ annulus; concentric circular ring. ～領域 annular domain; annulus.

えんかんすう 円関数 circular function.

えんかんたい 円環体 torus (*pl.* -ri); solid torus.
～の toric.

えんかんめん 円環面 torus (*pl.* -ri); anchor ring.
～の toric. ～関数 torus function. デュパンの～ Dupin's cyclide.

えんきかがく 円幾何学 circle geometry.

えんきんほう 遠近(画)法 perspective (representation).

えんけい 円形の circular, circled.

えんけいとつ 円形凸な absolutely convex.

えんこ 円弧 circular arc.
四分～ quadrant. 六分～ sextant. 八分～ octant. ～二角形 lune; circular digon. ～三角形 circular triangle. ～多角形 circular polygon. ～截線 circular slit.

えんこき 円弧規 arcograph.

えんざん 演算 operation; calculation.
～の operational. 基本～ elementary operation. 四則～ four (arithmetic) operations [rules]. ～数 operand. 論理～ calculus of logic; logical operation. ピボット～ pivoting; pivot operation. 区間～ interval arithmetic.

えんざんし 演算子 operator. ⇨作用素.
～法 operational calculus. 微分[積分]～ differential [integral] operator. ラプラス～ Laplace operator; Laplacian. ハミルトン～ Hamiltonian. 移動～ translation operator. 上昇～ step-up operator; up-ladder. 下降～ step-down operator; down-ladder. 昇降～ ladder (operator).

えんじつてん 遠日点 ophelion (*pl.* -lia).

えんしゅう 円周 circle; circumference; periphery of circle.
～の circular. ～率 circle ratio; circular constant; Ludolph number; pi [π]. ～角 angle of circumference; inscribed angle. ～等分多項式 cyclotomic polynomial. ～法 circle method.

えんしゅう 演習 exercise; ex.; practice.

えんしゅうりつ 円周率 circle ratio; circular constant; Ludolph number; pi [π].

えんすい 円錐 (circular) cone.
～面 (surface of) (circular) cone. 直～ right (circular) cone. 斜～ oblique (circular) cone. 直～台 frustum of right (circular) cone. ～条件 cone condition. ～図法 conic projection. 正角～法 conformal conic projection.

えんすいきょくせん 円錐曲線 conics; conic section.
～の conic; conical. 有心～ central conics; conics with center. 無心～ noncentral conics; conics without center. 固有な～ proper conics. 退化～ degenerate conics. 共焦～ confocal conics. 極～ polar conics. 焦～ focal conics. ～回転面[体] toroid.

えんせき 円積 quadrature of circle. ⇨求積.

えんちゅう 円柱 (circular) cylinder.
～(状)の cylindric; cylindrical. ～形の cylindroid. 直～ right (circular) cylinder. ～関数 cylindrical function. ～座標 cylindrical coordinates.

えんちょう 延長 prolongation; continuation;

extension; production.
～する prolong; extend; produce. ～線 prolongation, prolonged line; production. 線形～ linear extension.

えんちょく 鉛直な[鉛直線] vertical; perpendicular.

えんてんとうしょうほう 円点投象法 (*G.*) Zyklographie.

えんとう 円筒; 円壔 cylinder; solenoid.
～形の cylindrical; solenoidal.

えんとうぶん 円等分 ⇨円分.

エントロピー entropy.
平均～ mean entropy. 条件つき～ conditional entropy. 等～の homentropic.

えんばん 円板 circular disc [disk].
単位～ unit disc [disk]. 閉～ closed disc [disk]. 開～ open disc [disk]. 多重～ polydisc; polydisk.

えんぶん 円分 cyclotomy.
～の cyclotomic. ～体 cyclotomic field. ～多項式 cyclotomic polynomial.

お・オ

オー・アール　OR. ⇨オペレーション.

おいだす　追い出す　eliminate. ⇨消去.

オイラー　L. Euler (1707-1783).
　～の Eulerian; Euler's. ～線 Euler line. ～の数[角] Eulerian number [angle]. ～の関数[定数] Euler's function [constant]. ～のトーシェント関数 Euler's totient function.

おう　凹(性)　concavity. ⇨凸(性).
　～な concave. ～角(の) reentrant. ～多角形 concave polygon; reentering polygon. ～関数 concave function. ～計画問題 concave programming problem. 狭義(に)～な strictly concave. 下に～な downward(s) concave.

おうかく　凹角　reentrant; concave angle.
　～の reentrant.; reentering.

おうぎがた　扇形　sector. ⇨扇形(せんけい).

おうごん　黄金(の)　golden. ⇨中外比.
　～分割 golden section; golden cut. ～比 golden ratio.

おうだんせい　横断性　transversality.
　横断(性)条件 transversality condition.

おうだんせん　横断線　transversal (line); cross cut.
　～の transversal.

おうだんてき　横断的な　transverse; transversal.

おうだんめん　横断面　transverse section.
　～の場 transverse field.

おうとう　応答　response; acknowledgment.
　～する respond 【to】; acknowledge. ～時間 response time.

おうは　横波　transversal [transverse] wave.

おうふく　往復　return trip; round trip.

おうよう　応用　application.
　～の applied. ～する apply. ～できる applicable. ～数学 applied mathematics. ～問題 application problem; excercise; practical exercise.

おうりょく　応力　stress.
　接[法]線～ tangential [normal] stress. ずれ～ shearing stress. ～テンソル stress tensor.

おおう　覆[被]う　cover. ⇨被覆.

おおきい　大きい　big; large; great; major.
　十分～ sufficiently large.

おおきさ　大きさ　size; magnitude; dimension(s); bulk.
　母集団[標本]の～ size of population [sample].

おおざっぱ　大ざっぱな　rough; gross.

おおもじ　大文字　capital (letter).
　～の capital.

おきかえ　置き替[換]える　replace; substitute; displace; shift; reset.

おきまり　おきまりの　routine; usual; customary.
　～手順 routine; usual procedure.

おく　億　hundred millions.

おく　置く　put; place; set.

おくれ　遅れ　retardation; delay; lag; retard.
　～る retard; delay. ～型 retarded type. ～型差分微分方程式 difference-differential equation of retardation type. ～相関 lag correlation. 時刻の～ time lag.

おこる　起こる　occur.

オーシー　OC.
　～曲線 OC curve; operating characteristic curve.

おしだし 押し出し pushout.
オッズ odds.
〜比 odds ratio.
オートマトン automaton (*pl.* -ta).
プッシュダウン・〜 pushdown automaton. 有限〜 finite automaton. セル〜 cellular automaton.
オートメーション automation.
おなじ 同じ same; identical.
〜ような similar; like; analogous. 〜強さの equally strong.
おのおの 各々(の) each; respective.
オーバーフロー overflow. ⇨溢れ.
おび 帯 band; strip. ⇨帯(たい).
〜グラフ band graph. メービウスの〜 Möbius' band. 〜行列 band matrix.
おびはば 帯幅 bandwidth.
おびる 帯びる carry.
オペランド operand.
オペレーショナル operational.
〜・リサーチ operational research; OR.
〜・ゲーミング operational gaming.
オペレーション operation. ⇨演算.
〜ズ・リサーチ operations research; OR.
オペレータ operator. ⇨作用素.
オペレーティング operating.
オミクロン omicron; 【O, o】.
オメガ omega; 【Ω, ω】.
〜同型[形] Ω-isomorphism. 〜準同型[形] Ω-homomorphism. 〜群 Ω-group.
おもさ 重さ weight. ⇨重み.
おもな 主な principal; chief; main; leading.
おもみ 重み weight.
〜のついた weighted. 最高の〜 highest weight. 〜つき平均 weighted mean. 〜母関数 weight enumerator.
おやひょうほん 親標本 master sample.
およそ about; nearly; (*L.*) circa; ca. 【circa の略】; approximately.
〜2 メートル about [nearly] 2 meters. 〜の数 approximate number; round number.
および and; as well as.
おりかえし 折返し reflection; reflexion; regression.
おれせん 折(れ)線 polygonal line; broken line.
〜グラフ line graph.
おわり 終(り) end; termination.
〜りの terminal; last. 〜る end; terminate.
おんきょうがく 音響学 acoustics.
おんそく 音速 sound velocity.
〜の sonic. 超〜の supersonic. 亜〜の subsonic.
おんちょう 音調 tone; tune.
〜学 tonetics.
おんど 温度 temperature.
〜計 thermometer.

か・カ

か 価 value; price; cost.
一～の single-valued; one-valued. 多～の many-valued; multi(-)valued. n～の n-valued. 一～関数 single [one]-valued function; uniform function. 多～関数 many-valued [multi(-)valued] function. 単～ unit price. 正～ net price. 売[買]～ selling [buying] price. 原～ prime cost.

か 科 department; dept; dpt.
～の departmental. 数学～ department of mathematics; mathematics department.

かい 回 time(s).
一～ one time; once. 二～ two times; twice. 三～ three times, thrice. 数～ several times. n[無限]～微分可能な n-times [infinitely] differentiable. ～数 number of times.

カイ chi; 【X, χ】.
～二[自]乗分布 chi-square distribution.
～二[自]乗判定法 chi-square [χ^2] test.

かい 解 solution.
方程式の～ solution [root] of equation. 自明な～ trivial solution. 数値～(法) numerical solution. 近似～ approximate solution. ～曲線[曲面] solution curve [surface]. ～空間 solution space. 一般[特殊; 特異]～ general [particular; singular] solution. 完全～ complete solution. 基本～ fundamental solution; elementary solution. 素～ elementary solution. 原始[固有]～ primitive [proper] solution. 形式～ formal solution. 級数～ series solution. 安定～ stable solution. 周期～ periodic solution. 最大[小]～ maximal [minimal] solution. 最適～ optimal solution. 唯一の～ unique solution. 強[弱]～ strong [weak] solution. ～法 (method of) solution.

かい 階 order; rank. ⇨階数.
(第)二～[次]導関数 derivative of second order; second order derivative. 二～偏微分方程式 partial differential equation of second order; second order partial differential equation.

かい 可移(的な) transitive.
～群 transitive group. ～置換群 transitive permutation group.

かい- 開 open. ⇨閉(へい-).
～曲線 open curve. ～曲面 open surface. ～区間 open interval. ～集合 open set. ～弧 open arc. ～核 open kernel. ～被覆 open covering. ～基 open base. ～近傍 open neighbo(u)rhood. ～写像 open mapping. ～多様体 open manifold. ～円板 open disc [disk]. ～球 open sphere [ball].

がい- 外 external; exterior; extra-; outer. ⇨外(そと); 内(ない-).

がい- 概 approximately; almost.
～微分可能な approximately derivable. ～微分係数 approximate derivative. ～連続な approximately continuous. ～解析的な approximately analytic. ～周期関数 almost periodic function. ～周期 translation number. ～複素構造[多様体] almost complex structure [manifold]. ～計量構造 almost metric structure. ～加法的な almost additive. ～収束 almost (everywhere) convergence. ～体 near fild.

かいが 絵画 picture; painting.

〜図表 pictogram; pictograph.

かいかく 解核 resolvent (kernel).

がいかく 外角 external angle; exterior angle.

がいかくど 外拡度 outer [exterior] content.

かいき 回帰 recurrence; regression.

〜的な recurrent; recurring; regressive. 〜関係 recurrence relation; recurring relation. 〜数列 recurrent sequence. 〜公式 recurrence formula. 〜係数 regression coefficient. 〜曲線 regression curve. 〜(直)線 regression line. 〜超平面 regression hyperplane. 〜関数 regression function. 〜推定値 regression estimate. 〜解析[分析] regression analysis. 自己〜 autoregression. 重〜 multiple regression.

かいぎ 会議 meeting; conference; congress.

かいきゅう 階級 class; rank; grade.

〜値 class mark.

がいけい 外形 external [outward] form; shape.

がいけい 外径 outside diameter.

がいけい 概型 scheme.

代数的〜 algebraic scheme. ログ〜 logarithmic scheme.

かいけつ 解決 solution.

〜する solve. 〜可能な solvable. 〜可能性 solvability. 〜不能性 unsolvability.

がいこう 外項 external term(s); extreme.

かいこてき 回顧的(な) retrospective.

かいさ 階差 difference. ⇨差分.

〜方程式 difference equation. 高階〜 difference of higher order. 〜表 difference table. 〜数列 progression of differences.

がいさん 概算 rough estimation.

かいし 開始 start; beginning; commencement.

〜する start; bigin; commence.

かいしゃく 解釈 interpretation; explanation; exposition.

〜する interpret; explain. 〜の interpre(ta)tive; explanatory.

がいしゅうき 概周期 translation number.

〜関数[系] almost periodic function [system].

かいしゅうきかん 回収期間 pay-back period.

⇨返済.

かいしょう 解消 resolution.

特異点の〜 resolution of singularities.

かいじょう 階乗 factorial.

〜の factorial. n の〜 n factorial; factorial n; factorial of n. 〜関数 factorial function. 〜級数 factorial series. 〜モーメント factorial moment.

がいしん 外心 circumcenter; center of circumcircle.

かいすう 回数 number of times; frequency.

かいすう 階数 rank; order. ⇨階.

行列の〜 rank of matrix. 最大〜 full rank. 微分の〜 order of differentiation. 〜低下 depression [reduction] of order.

がいすう 概数 round number.

がいせい 外生(の) exogen(e)ous.

〜変数 exogenous [external] variable.

かいせき 解析 analysis (*pl.* -ses).

〜する analyze. 代数〜 algebraical analysis. フーリエ〜 Fourier analysis. ベクトル[テンソル]〜 vector [tensor] analysis. 関数〜 functional analysis. 調和〜 harmonic analysis. 留数〜 residue analysis. 次元〜 dimensional analysis. 多変量〜 multivariate analysis. 数値〜 numerical analysis. 数学〜 mathematical analysis. 安定性〜 stability analysis. 感度〜 sensitivity analysis. 区間〜 interval analysis. 漸近〜 asymptotic analysis. 凸〜 convex analysis.

かいせき- 解析 analytic; analytical. ⇨解析的(な).

〜幾何学 analytic(al) geometry. 〜力学 analytical dynamics; analytical mechanics. 〜(的)整数論 analytic theory of numbers. 〜関数 analytic function; monogenic analytic function. 実〜関数 real analytic function. 〜接続 analytic prolongation; analytic continuation. 〜接続可能な analytically prolongable [continuable]. 〜曲線 analytic curve. 〜空間 analytic space. 〜構造 analytic structure. 〜(的)集合 analytic set. 〜写像 analytic mapping. 〜微分 analytic differential.

がいせき 外積 outer product; exterior prod-

uct.
～代数 exterior algebra. ～ベキ(作用素) exterior power.
かいせきがく 解析学 analysis (*pl.* -ses).
数学～ mathematical analysis.
かいせきき 解析機 analyzer.
微分[調和]～ differential [harmonic] analyzer.
かいせきせい 解析性 analyticity.
実[複素]～ real [complex] analyticity.
かいせきてき 解析的(な) analytic; analytical. ⇨解析(かいせき-).
～に analytically. ～方法 analytic(al) method. ～整数論 analytic theory of numbers; analytic number theory. ～集合 analytic set. ～近傍 analytic neighbo(u)rhood. ～部分空間 analytic subspace. ～完備化 analytic completion. ～中心 analytic center. ～容量 analytic capacity. ～層 analytic sheaf. 実[複素]～な real [complex] analytic. ～に独立な analytically independent. ～(に)完備な analytically complete. ～(に)非分岐の analytically unramified. ～に細い analytically thin.
かいせつ 解説 exposition.
～的な expository.
かいせつ 回折 diffraction.
がいせつ 概説 survey.
がいせつ 外接 circumscription.
～させる circumscribe. ～円 circumcircle; circumscribed circle. ～多角形 circumscribed polygon. ～球 circumsphere; circumscribed sphere.
がいせつ- 外接 circum-.
～円 circumcircle; circumscribed circle. ～球 circumsphere; circumscribed sphere.
かいせん 界線 horocycle; limiting curve.
がいぜん 蓋然(的) probable.
～性 probability. ～誤差 probable error.
かいそう 階層 hierarchy.
解析[算術]的～ analytic [arithmetic] hierarchy. ～(の)定理 hierarchy theorem. ～(の)理論 theory of hierarchies. ～モデル hierarchical model.
がいそう 外挿 extrapolation. ⇨補外(法).
～する extrapolate.
がいそくど 外測度 outer measure.
がいたいせき 外体積 outer volume; exterior volume.
かいだん 階段 echelon.
(行)～行列 (row) echelon matrix [form]. 簡約行～行列 reduced row echelon matrix [form].
かいだん 階段 step.
～関数 step function. ～状の step-formed. ～状グラフ step-formed graph. ～的分類 hierarchial subdivision.
がいちゅうひ 外中比 extreme and mean ratio.
かいてい 改訂 revision.
～する revise. ～増補版 revised and enlarged edition.
かいてん 回転 rotation; revolution; roll; turn.
～する rotate; revolve; turn. ～軸 axis of rotation [revolution]. ～の中心 center of rotation. ～半径 radius of rotation. 固有[非固有]～ proper [improper] rotation. 無限小～ infinitesimal rotation. 瞬間～ instantaneous rotation. ～面 surface of revolution. ～体 solid of revolution; body of rotation. ～群 rotation group. ～(指)数 rotation [winding] number; winding index. ～定理 rotation theorem. ～放物面[双曲面] paraboloid [hyperboloid] of revolution. ～楕円面[体] ellipsoid of revolution; ellipsoidal surface [body] of revolution; spheroid. 在庫～率 turn-over of inventories.
かいてん 回転 rotation; curl.
ベクトル場の～ rotation [curl] of vector field. ～(演算)子 rot; curl.
がいてん 外点 exterior point; outer point.
かいてんぎ 回転儀 gyroscope.
～の gyroscopic. ～項 gyroscopic term.
かいてんすう 回転数 winding number [index]; rotation number.
かいてんだえんたい 回転楕円体 ellipsoid of revolution; spheroid.
～の spheroidal. 扁平[扁長]～ oblate

[prolate] spheroid. ～波動関数 spheroidal wave function.

かいてんだえんたいちょうわかんすう 回転楕円体調和関数 spheroidal harmonics.

かいとう 解答 solution; answer.

かいどく 解読 decoding.
～する decode; decipher; decrypt. ～器 decoder.

がいねん 概念 concept; notion.
～的な conceptional, conceptual; notional. 無定義～ undefined concept; primitive notion.

がいはいせん 外擺線 epicycloid.
～の epicycloidal.

がいびぶん 外微分 exterior derivative; exterior differentation.
～形式 exterior differential form.

がいぶ 外部 exterior; outside. ⇨内部.
～の exterior; external; outer; outside. ～自己同型[形] outer automorphism. ～問題 exterior problem. ～積 exterior [external] product. ～言語 external language.

がいぶん 外分 external division.
～する divide externally. ～点 externally dividing point. ～比 external ratio.

がいぶんさん 外分散 external variance.

かいへい 開平 extraction (of square root).
～する extract square root. ～剰余 square remainder.

かいへん 改変 modification; change.
固有[解析的]～ proper [analytic] modification. 球面～ spherical modification.

かいほう 解法 (method of) solution.
三角形の～ solution of triangle. 代数的～ algebraic solution. 数値～ numerical method [solution].

かいほう 開方 extraction (of root); evolution.

かいほう 会報 report; bulletin; transactions; proceedings.

かいほう-ゲーム 開放ゲーム open game.

がいめん 外面 outside; exterior (face).

がいめんせき 外面積 outer area; exterior area.

がいようりょう 外容量 outer capacity.

がいよはいせん 外余擺線 epitrochoid.
～の epitrochoidal.

がいりゃく 概略 outline; summary.
～の outlined; summarized; summary; rough.

かいりゅう 開立(法) extraction of cubic root. ～する extract cubic root. ～商 cubic root. ～剰余 cubic remainder.

かいろ 回路 circuit; network.
論理～ logical circuit. 電気～ electric circuit [network]. 相反～ reciprocal [bilateral] network. 接点～ contact network.

かいわ 会話 conversation.
～の conversational.

ガウス C. F. Gauss (1777-1855).
～の Gaussian; Gauss'. ～分布 Gaussian distribution. ～平面 Gaussian plane. ～曲線[正規曲線] Gaussian curve. ～(の)曲率 Gaussian curvature. ～型確率過程 Gaussian stochastic process. ～の整数 Gaussian integer. ～の和 Gauss' sum. ～の公式 Gauss' formula. ～の記号 Gauss' symbol; symbol of Gauss. ～の消去法 Gaussian elimination.

かえす 返す return; give back.

かえる 変える change; alter; shift.
符号を～ change sign. 向きを～ turn; shift.

かえる 帰る return; come [go] back.

カオス chaos.
～的な chaotic.

かかい 下界 lower bound. ⇨上界.
最大～ greatest lower bound; (*L.*) infimum (*pl.* -fima); inf.

かかい 可解(な) solvable; resolutive.
～群 solvable group. ～性 solvability; resolvability. ～代数 solvable algebra. ～コンパクト化 resolutive compactification.

かかく 価格 price; value.

かがく 科学 science.
自然[社会]～ natural [social] science. 数理～ mathematical science. 総合～ integrated science. 精密～ exact science. コンピューター～ computer science. 情報～ information science. 生命～ life science.

かかん 可換(な) commutative; commutable; permutable.
～性 commutativity; commutability. ～でない, 非～な non-abelian. ～律[法則] commutative law. ～群 commutative group; Abelian group. ～環[体] commutative ring [field]. ～多元環 commutative algebra. ～子(環) commutor. ～収束 commutative convergence. ～な図式 commutative diagram. ～関数 permutable function. ～環のスペクトル spectrum of a commutative ring.

かかんすう 下関数 hypofunction.

かかんでない 可換でない non-abelian, non-commutative.

かぎゃく 可逆(性) invertibility; reversibility.
～(的)な invertible; inversible; reversible. ～元 invertible element. ～変換 invertible transformation. ～行列 invertible matrix. ～過程 reversible process. ～振子 reversible pendulum. ～層 invertible sheaf.

かぎゅうけい[せん] 蝸牛形[線] (*F.*) limaçon.

かきょくげん 下極限 inferior limit; (*L.*) limes inferior (*pl.* limites inferiores).
～事象 inferior limit event.

かく 各 each; every.
～点で at each point. ～個に separately; individually.

かく 核 kernel.
対称～ symmetric kernel. エルミート～ Hermitian kernel. 差～ difference kernel. 反復～ iterated kernel. 随伴[共役]～ adjoint [conjugate] kernel. 積分～ integral kernel. 退化～ degenerate kernel. 多項式型の～ kernel of polynomial type; degenerate kernel. 解～; 逆～ resolvent (kernel). 特異～ singular kernel. 再生～ reproducing kernel. ～関数 kernel function. ～超関数 kernel distribution. ～微分式 kernel differential. 開～ open kernel. 等測～ measurable kernel. 素～ elementary [fundamental] kernel. 準同型[形]の～ kernel of homomorphism.

かく 核 nucleus (*pl.* -lei, -leuses).
～の nuclear. ～(型)作用素 nuclear operator. ～形空間 nuclear space.

かく 書く; 描く write; draw; describe.

かく 角 angle.
～の angular. 直～ right angle. 平～ straight angle. 周～ perigon. 鋭～ acute angle. 鈍～ obtuse angle. 余～ complementary angle. 補～ supplementary angle. 同位～ corresponding angle. 頂角～ vertical angle. 対頂～ vertical angle(s); vertically opposite angle(s). 錯～ alternate angle. 内対～ interior opposite angle. 内～ internal angle. 外～ external angle. 中心～ central angle. 接～ adjacent [contiguous] angle. 仰～ angle of elevation. 俯[伏]～ angle of depression [inclination]. 方位～ direction angle. 天頂～ zenith(al) angle. 入射～ angle of incidence. 立体～ solid angle. 離心～ eccentric angle. 非ユークリッド～ non-Euclidean angle. オイラーの～ Eulerian angle. 偏～ argument. 傾～ inclination. 位相～ phase angle.

かく- 角 angular.
～速度 angular velocity. ～運動量 angular momentum. ～振動数 angular [circular] frequency. ～距離 angular distance. ～測度 angular measure. ～変換 angular transformation. ～領域 angular domain [region]. ～微係数 angular derivative. ～極限 angular limit. ～錐 pyramid. ～柱 prism.

-がく 学 -logy.
生物～ biology. 言語～ philology. 生理～ physiology.

がくい 学位 (doctor's) degree; doctorate.
～論文 (doctoral; doctorate) thesis; dissertation.

かくこ 各個(に) separately; individually. ⇨ 各個(の)(かっこ).
～連続な separately continuous.

がくさいてき 学際的 interdisciplinary.

かくさん 拡散 diffusion.
～係数 diffusion coefficient. ～過程 diffusion process. ～方程式 diffusion equation.

がくしゅう 学習 learning; study.
～する learn; study. ～模型 learning

model. ～者 learner. ～指導要領 course of study. 機械～ machine learning.

かくしんどうすう 角振動数 angular [circular] frequency.

かくすい 角錐 pyramid.
～の pyramidal. 正～ regular pyramid. 三～ triangular pyramid. ～台 frustum of pyramid; prismoid. ～面 pyramidal surface.

かくだい 拡大 extension; dilatation; magnification; augmentation; enlargement.
～する extend; dilate; magnify; augment; enlarge. ～可能な extendable; extensible. 相似～ similar extension. 中心～ central extension. 代数[超越](的)～ algebraic [transcendental] extension. ガロア～ Galois extension. アーベル～ Abelian extension. 単純～ simple extension. 有限[無限](次)～ finite [infinite] extension. 正規～ normal extension. 群～ group extension. 係数～ scalar extension. 解析(的)～ analytic extension. ～環 extended ring; overring; extension of ring. ～体 extension field; extended field; overfield. ～図 extended figure; magnification [magnified] figure. ～率 dilatation ratio; magnification ratio. ～係数行列 enlarged coefficient matrix. ～次数 extension degree. クンマー～ Kummer extension. 巡回～ cyclic extension.

かくだい 角台 prismoid.

かくだいかん 拡大環 overring; extended ring; extension of ring.

かくだいたい 拡大体 overfield; extension field; extended field.

かくちゅう 角柱 prism.
～の prismatic. 正～ regular prism. 直[斜]～ right [oblique] prism. 四～ quadrangular prism. 切[截]頭～ truncate(d) prism. ～面 prismatic surface. 類似～ symmedian prism.

かくちょう 拡張 extension; generalization.
～する extend; generalize. ～可能な extendable; extensible. 写像の～ extension of mapping. 強[弱]～ strong [weak] extension. 定理の～ generalization of theorem. ～定理 extension theorem. ～された平面 extended plane.

かくちょう 角頂 vertex (*pl.* -texes, -tices).

かくてい 確定 determination; decision; confirmation.
～する determine. ～的な definite; deterministic; definitely decided; strictly determined. ～的過程 deterministic process. 面積～の of definite area.

かくていした 確定した certain, determined.

かくていとくいてん 確定特異点 regular singularity; regular singular point.

かくていねんきん 確定年金 annuity certain.

かくてん 各点(ごと) pointwise ⇨点別の[に].
～収束 pointwise convergence.

かくど 角度 angle.
～計 goniometer.

かくど 拡度 content.
内[外]～ inner [outer] content.

かくど 確度 likelihood ratio.

がくは 学派 school.

がくぶ 学部 faculty; school; college.
理～ faculty of science.

かくふそく 角不足 defect [deficiency] of triangle.

かくべつ 格別の particular; specific; distinguished; noticeable.

かくりつ 確率 probability.
～の; ～(論)的な probabilistic; stochastic; random. ～的に probabilistically; stochastically; at random. ～論 theory of probability; probability theory. 数学的～ mathematical probability. 統計[経験]的～ statistical [experimental; empirical] probability. 幾何(学的)～ geometric probability. 主観～ subjective probability. ～事象 probability event. ～標本 probability sample; random sample. ～空間 probability space. ～分布 probability distribution. ～測度 probability measure. ～素分[エレメント] probability element. ～密度 probability density; random measure. ～誤差 probability error; random error. ～限界 probability limit. 原因の～ probability of causes.

事後～ posteriori probability. 事前～ priori probability. ～(方眼)紙 probability paper. ～ベクトル probability vecter. ～変数 stochastic variable; random variable; chance variable. ～母関数 probability generating function. ～過程 stochastic process; random process. ～(論的)模型 stochastic model. ～収束 stochastic convergence; convergence in probability. ～行列 stochastic matrix. ～積分 stochastic integral. ～微分方程式 stochastic differential equation. ～フィルタリング stochastic filtering. ～抽出 random sampling. ～的[無作為]配置 random arrangement. ～化 randomization. ～(化)操作 random(ization) operation. ～化された検定 randomized test. ～的に大き[小さ]い stochastically large [small]. 上側～ upper tail probability.

かくりつろんてき 確率論的 stochastic; probabilistic; probability-theoretic.
～モデル[模型] stochastic model. ～処理 stochastic [probabilistic; probability-theoretic] treatment.

かぐん 加群 additive group. ⇨加法群.

かぐん 加群 module.
左[右]*A*～ left [right] *A*-module. 既約～ irreducible module. 自由～ free module. 双対～ dual module. 直既約～ indecomposable module. 誘導～ induced module. 輪体[境界]～ module of cycles [boundaries]. 位相～ topological module. 移入[単射, 入射]～ injective module. 射影～ projective module. 剰余～, 因子～ factor module. 相補～ complementary module. ネーター～ Noetherian module. 単純～ simple module. 補～ complementary module.

かげ 陰 shade.
日～ shade. ～をつける shade.

かげ 影 shadow.
～の価格 shadow price.

かけざん 掛け算 multiplication. ⇨乗法.
～の multiplicative. ～九々表 multiplication table; Pythagorean table.

かける 掛ける multiply.
5に3を～ multiply 5 by [into] 3.

かげん 仮言 hypothesis (*pl.* -ses).
～の hypothetical; hypothetic. ～三段論法 hypothetical syllogism.

かげん 下限 greatest lower bound; g. l. b.; (*L.*) infimum (*pl.* -fima); inf.; (*L.*) finis inferior; lower limit.
～関数 lower limit function.

かげんじょうじょ 加減乗除 addition, subtraction, multiplication and division; four operations; four rules; arithmetical operations.

かこう 下降 descent. ⇨降下.
～する descend; drop; step down. ～枝 descending branch. ～演算子 down-ladder; step-down operator. ～定理 going down theorem.

かごう 加号 addition symbol; plus sign.

かこむ 囲む enclose; inclose; contain; encircle.

かさねあわせ 重ね合わせ superposition.
～る superpose. ～の原理 principle of superposition.

かさん 加算 addition.
～する add.

かさん 可算(性) countability; enumerability; denumerability.
～な countable; enumerable; denumerable. ～的に countably; enumerably; denumerably. ～集合 countable set; enumerable set; denumerable set. ～順序数 countable ordinal number. ～基 countable base. ～公理 countability axiom. ～被覆 countable covering. ～モデル enumerable model. ～加法的[コンパクト]な countably additive [compact]. ～ノルム空間 countably normed space. 局所～な locally countable.

かし 華氏 Fahrenheit.
～百度 hundred degrees Fahrenheit; 100°F.

カージオイド cardioid.

カージナル cardinal. ⇨基数.
～数 cardinal number.

かじゅう 荷[加]重 weight; load. ⇨重み.
～平均 weighted mean.

かしゅく 可縮(性) contractibility.
～な contractible. 局所～な locally contractible.
かじゅつ 可述(的) predicative.
かじょ 可除(の) divisible.
～部分群 divisible subgroup.
かしょう 仮商 temporary quotient.
かしょう 過小(な) too low [little].
～近似値 lower approximate value. ～評価 underestimation. ～緩和 under-relaxation.
かじょう 過剰 excess; redundance; abundance.
球面～ spherical excess. ～近似 approximation in excess. ～数 abundant number. ～和 upper sum.
かじょうけってい 過剰決定 overdetermination.
～の overdetermined. ～系 overdetermined system.
かじょうてん 渦状点 focal point; focus (*pl.* -cuses, -ci).
流れの～ focus of flow.
かしらもじ 頭文字 capital letter; initial (letter).
～の capital; initial.
かしんてん 渦心点 central point; center.
渦状～ focus-center.
かすう 加数 addend; summand.
かすう 仮数 mantissa.
カスケード cascade.
～法 cascade method.
カスプ cusp. ⇨尖点.
かせい 仮性(の) improper.
～[広義]積分 improper integral. ～特異点 improper singularity [singular point].
かせき 可積(性) integrability. ⇨積分可能(性).
～な integrable. ～関数 integrable function; summable function. ～条件 condition of integrability; integrability condition. 絶対～な absolutely integrable. 局所～な locally integrable. 弱～な weakly integrable.
かせきぶん 可積分 ⇨可積(性).
かせつ 仮説 hypothesis (*pl.* -ses).
～の hypothetical. 連続体～ continuum hypothesis. 統計(的)～ statistical hypothesis. 帰無～ null hypothesis. 対立～ alternative hypothesis. 複合～ composite hypothesis. ～(の)検定 test of statistical hypothesis.
かせつ 仮設 ⇨仮定; 仮想(の).
かせん 火線 caustic; caustic curve.
～の caustic. 反射～ catacaustic (curve). 屈折～ diacaustic (curve).
かせん 可遷(性) transitivity. ⇨推移.
～的な transitive. n重～な n-ply [n-fold] transitive. ～群 transitive group. ～置換群 transitive permutation group. 測度～的な metrically transitive.
かそう 仮想(の) virtual.
～次[指]数 virtual degree. ～種数 virtual genus. ～変位 virtual displacement. ～仕事 virtual work. ～点 ideal point.
がぞう 画像 picture.
かぞえる 数える count; number; enumerate.
かそく 加速 acceleration.
～する accelerate. ～度 acceleration. ～緩和 over-relaxation. ～モデル, ～故障モデル accelarated failure model.
かそく 可測(性) measurability.
～な measurable. ～集合[事象] measurable set [event]. ～空間 measurable space. ～関数 measurable function. ルベーグ～な Lebesgue [L-]measurable. ボレル～な Borel [B-]measurable.
かそくど 加速度 acceleration.
角～ angular acceleration. 等～ constant [uniform] acceleration. 負の～ negative acceleration; deceleration. ～ベクトル acceleration vector.
かた 過多 superabundance.
～の superabundant.
かた 型 type; shape.
～問題 type problem. ～作用素 shape operator.
-かた -型(の) ⇨型(の)(-がた).
-がた 型(の) -like; -shaped; -type.
星～領域 star-like [-shaped] domain. 楕円[双曲; 放物]～の of elliptic [hyperbolic;

parabolic] type. フックス～の of Fuchsian type.

かたい 硬い stiff.
～系 stiff system.

かだい 課題 subject; theme; problem; exercise.

かだい 過大(な) too high [big]; excessive.
～近似値 upper approximate value. ～評価 overestimation. ～緩和 over-relaxation.

かたがわ 片側 one side. ⇨単側(たんそく).
～の one-sided; unilateral. ～極限 one-sided limit. ～イデアル one-sided ideal. ～安定過程 one-sided stable process. ～検定 one-sided test. ～カイ二[自]乗検定 one-sided chi-square test. ～シフト unilateral shift.

カタストロフィー catastrophe.
～理論[写像] catastrophe theory [map].

かたち 形 form; shape.
～づくる form; shape. の～の of the form; -shaped; -formed. ～作用素 shape operator.

かたむき 傾き slant; slope; inclination; gradient.
傾く incline; lean; slant; trend.

かたより 偏り bias; polarization.
偏らせる bias. 偏った推定量 biased estimator. ～のない unbiased.

カタログ catalog(ue).

かたん 下端 lower limit.

がっか 学科 department; dept.; dpt.; discipline.
数～ department of mathematics; mathematics department; dept. of math.

かっこ 括弧 parenthesis (*pl.* -ses), ( ); brace, { }; bracket, [ ].
～に入れる put in parentheses [braces; brackets]; parenthesize. ～積 bracket product. ～演算 bracket operation. ～式 bracket expression.

かっこ 各個(の) each; every; individual; respective; separate. ⇨個別(の).
～連続な separately continuous. ～逐次検査 item-by-item sequential inspection.

カッシーニ J. D. Cassini (1625–1712).
～の Cassinian. ～曲線 Cassinian (curve).

かっしゃく 滑尺 sliding scale.

かっせん 括線 vinculum (*pl.* -la).

かっせん 割線 secant.

かって 勝手(な) arbitrary. ⇨任意(性).
～に arbitrarily.

カット cut.
最小～ minimum cut.

かつどう 活動 activity.
～的な active. ～力 activity. ～分析 activity analysis.

カッパ kappa; 【K, κ】.

カップ cup; 【記号 ∪】.
～積 cup product.

がっぺい 合併 union; collection; aggregation. ⇨和集合.
～する unite; combine; collect; aggregate. ～集合 union.

かてい 仮定 assumption; supposition; hypothesis (*pl.* -ses); presumption; postulate; postulation; antecedent.
～の assumptive; hypothetical. ～する assume; suppose. ～法 rule of supposition.

かてい 過程 process.
確率～ stochastic process. マルコフ～ Markov process. 決定～ deterministic process; decision process. 確定的～ deterministic process. 逆進～ reversed process. 推[遷]移～ transition process. 非可逆～ irreversible process. 加法～ additive process. 定常～ stationary process. 安定～ stable process. 拡散～ diffusion process. 出生[死亡]～ birth [death] process. 計数～ counting process.

カテゴリー category. ⇨圏; 類.
～の categorical. 加法的～ additive category. ～代数 categorical algebra.

カテナリー catenary. ⇨懸垂(けんすい-).
～環 catenary ring.

カテノイド catenoid. ⇨懸垂(けんすい-).

かでん 荷電 (electric) charge.
～不変性; ～独立 charge independence.

かてんめん 可展面 applicable surface; developable surface.

かと 過渡(的な) transient.

～現象 transient phenomenon. ～(現象の)問題 transient problem. ～的成分 transient component.

かど 渦度 vorticity.

かどう 渦動 vortex (*pl.* -texes, -tices). ⇨渦(うず).

かどう 可動(な) movable; mobile.
～性 movability; mobility. ～分岐点 movable branch point. ～特異点 movable singularity.

かどう 稼動 operation.
～の operating; busy. ～時間 operating time. ～日数 total of workdays. 全～期間 busy period.

かなめせん 要線 pivotal line.

かならず 必ず necessarily; certainly; surely.
～しも…ない not necessarily; not always.

かね 鐘 bell.
～形曲線 bell(-shaped) curve.

かのう 可能(な) possible; feasible; eventual; -able; -ible.
～にする enable. ～性 possibility; feasibility. ～解 feasible solution. 微分～な differentiable. 接続～な prolongeable. 分割～な divisible. 分離～な separable; distinguishable. 作図～な of possible construction.

かはん 下半 lower half.
～(平)面 lower half-plane. ～連続な lower semi-continuous.

かひのり 加比の理 componendo.

かびぶん 可微分(な) differentiable. ⇨微分可能(な).
～性 differentiability. ～写像 differentiable mapping. ～構造 differentiable structure. ～多様体 differentiable manifold.

かびぶんけいすう 下微分係数 lower derivat(iv)e.

がひょう 画表 (*F.*) tableau (*pl.* -leaux).

カーブ curve.
サイン・～ sine curve. ～・トレーシング curve tracing.

かふごう 可符号(の) orientable.
～多様体 orientable manifold. ～曲面 orientable surface.

かふばん 可付番(性) countability; enumerability; denumerability. ⇨可算(性).
～な countable; enumerable; denumerable.

かぶん 可分(性) separability.
～な separable. ～距離空間 separable metric space. 完全～な perfectly separable.

かぶんさん 過分散 overdispersion.

かぶんすう 仮分数 improper fraction.

かへん 可変(な) variable.
～小数点 variable point.

かほ 可補(的な) complemented.
～束 complemented lattice.

かほう 加法 addition.
～の; ～的な additive. ～九々 addition table. 省略～ approximate addition. ～公式 addition formula. ～定理 addition theorem. ～族 additive class [family]. ～群 additive group. ～付値 additive valuation. ～過程 additive process. ～性 additivity. 有限[完全]～性 finite [complete] additivity. ～的整数論 additive number theory. ～的汎関数 additive functional. ～的作用素 additive operator. ～圏[的カテゴリ] additive category. 可算～的な countably additive. 概～的な almost additive. ～的関手 additive functor. ～モデル additive model.

かほう 下包 lower envelope.
～原理 lower envelope principle.

かほう 下方(の) lower; downward; inferior.
～に有界な bounded downwards. [(from) below].

がほう 画法 description; drawing.
～幾何学 descriptive geometry. (中心)投象～ perspective drawing.

かほうぐん 加法群 additive group.

かみ 紙 paper.

かめん 火面 caustic; caustic surface.
～の caustic.

がめん 画面 picture (plane); scene.
平～ plan; horizontal plane. 立～ elevation; vertical plane. 側～ side plane.

かやく 可約(性) reducibility; reductivity. ～な reducible; decomposable. ～分数 reducible fraction. ～元 reducible element.

〜多項式 reducible polynomial. 完全〜な completely reducible.

かよう 可容(な) capacitable.
〜集合 capacitable set. 〜性 capacitability.

かよう 可容(な) admissible.
〜制御 admissible control.

かり 仮(の) temporary; provisional; tentative; transient; virtual; assumed.
〜規格 tentative standard. 〜平均 working mean; temporary average.

かりこみへいきん 刈り込み平均 trimmed mean.

カルテシアン Cartesian; cartesian. ⇨デカルト.
〜座標 Cartesian coordinates. 〜空間 Cartesian space. 〜積 Cartesian product.

カルデロン A. P. Calderón (1920–1998).

カレント current; (*F.*) courant.
彷徨〜 random current.

ガロア, ガロワ É. Galois (1811–1832).
〜群[拡大] Galois group [extension]. 〜理論 Galois theory.

がわ 側 side.
一方〜 one side. 両(方)〜 both sides. 向い[反対]〜 opposite side.

かわり 代り(に) in place of; instead of.

かわりてん 変(わ)り点 turning point.

かわる 変わる change; vary; alter; shift.

かん 巻 volume; Vol.; vol.

かん 管 tube. ⇨管状(の).
〜状の tubular. ベクトル〜 vector tube. 流〜 stream tube. 〜状近傍 tubular neighbo(u)rhood. 〜状領域 tube domain. 〜面 tubular surface.

かん 環 ring; algebra. ⇨多元環.
〜つきの ringed. 同心円〜 concentric circular ring; annulus. 〜面 ring surface. 球〜 spherical ring. 部分〜 subring. 単純〜 simple ring. 単位的〜 unitary ring. 可換[非可換]〜 commutative [non-commutative] ring. ノルム〜 normed ring. 商〜 quotient ring; ring of quotients. 全商〜 total ring of quotients; ring of total quotients; total quotient ring. 群〜 group ring. 群(多元)〜 group algebra. 集合〜 ring of sets. 多項式〜 polynomial ring. 局所〜 local ring. 半局所〜 semilocal ring. 係数〜 ring of scalars. 係数添加〜 supplemented algebra; augumented algebra. 付値〜 valuation ring. 基礎〜 ground ring; basic ring. 作用素〜 ring of operators; operator algebra. 主[単項]イデアル〜 principal ideal ring. 被約な〜 reduced ring. 〜同型[形] ring isomorphism. 〜準同型[形] ring homomorphism. 〜つき空間 ringed space. バナッハ〜 Banach algebra. リー〜 Lie algebra. C^*〜 C^*-algebra. (可換)〜のスペクトル spectrum of a (commutative) ring. カテナリー〜 catenary ring. 〜の単元 unit of a ring. 形式的ベキ級数〜 formal power series ring. コーエン・マコーレー〜 Cohen-Macaulay ring. 乗法子〜 multiplier algebra. 正則局所〜 regular local ring. W^*〜 W^*-algebra.

がんい 含意 implication.
〜の implicational. 〜する imply. 〜[条件]命題 implicational proposition.

かんかく 間隔 space; interval.
〜をおく space. 等〜の equally spaced. 抽出〜 sampling interval.

かんかく 感覚 sense; sensibility; feeling.
〜器官 receptor; sense organ; sensorium (*pl.* -s, -ria).

がんきん 元金 principal; capital.

かんけい 関係 relation; relationship; reference; connection; respect; concern.
〜する relate to; be concerned with. 〜のある related; relational; referring. 〜式 relation. 対称な〜 symmetric relation. 対称〜 symmetry relation. 逆〜 inverse relation. 同値〜 equivalence relation. 二項〜 binary relation. n項〜 n-ary relation. 関数〜 functional relation. 順序〜 order relation. 射影[配景]的〜 projective [perspective] relation. 射影〜 projectivity. 極限〜 limit relation. 〜環 relationship [relation] algebra.

かんけい 環形 annulus (*pl.* -li, -luses).

かんげき 間隙 gap. ⇨空隙.

かんけつ 簡潔な brief; concise; short.

〜証明 brief proof.

かんげん 還元 reduction.
〜する reduce. 〜できる reducible. 〜公式 reduction formula.

かんげん 換言すれば in other words; that is (to say); (*L.*) id est; i.e.; namely; (*L.*) videlicet; viz.

がんけんせい 頑健性 robustness.

かんさ 監査 inspection; check.
〜検査[限界] check inspection limit.

かんさつ 観察 observation; survey.
〜する observe. 〜者 observer.

かんさん 換算 conversion; exchange. ⇨換算(かんざん).
〜する convert. 〜表 conversion table. 〜率 exchange rate.

かんざん 換算 reduction. ⇨換算(かんさん).
〜する reduce. 〜公式 reduction formula. 〜[通約]できる reducible.

かんじ 漢字 Chinese character.

かんしつ 換質 obversion.
〜する obvert. 〜命題 obverse.

かんして 関して about; concerning; regarding; with regard to; in relation to; in connection with; in respect of; referring to; relative to.

かんしゅ 関手 functor.
〜の functorial. 加法的〜 additive functor. 完全〜 exact functor. 導来〜 derived functor. 反[共]変〜 contravariant [covariant] functor. 〜性 functoriality. 〜(間)の同型[形] functorial isomorphism.

かんしょう 干渉 interference; intervention.

かんしょう 緩衝 buffer.
〜器; 〜回路 buffer. 〜効果 buffer effect. 〜用在庫 buffer inventory.

かんじょう 勘定 calculation; computation; account; cast.

かんじょう 管状(の) solenoidal; tubular; tubulous.
〜近傍 tubular neighbo(u)rhood. 〜流 solenoidal stream [flow].

かんじょう 環状(の) annular; ring-formed.
〜領域 ring domain.

かんしん 関心 concern.
…に〜がある be concerned with ….

かんすう 関数, 函数 function.
〜の; 〜的な functional. 〜的に functionally. 一[多]変数(の)〜 functon of one [several] variable[s]. 実[複素]〜 real [complex] function. 実[複素]変数(の)〜 function of real [complex] variable(s). 〜論 theory of functions; function theory. 定数〜 constant function. 一価〜 single [one]-valued function. 多価〜 many-valued [multi(-)valued] function. 陰〜 implicit function. 一次〜 linear function. 分数〜 fractional function. 有理〜 rational function. 無理〜 irrational function. 代数〜 algebraic function. 合成〜 composite [compound] function. 初等〜 elementary function. 超越〜 transcendental function. 三角[指数; 対数]〜 trigonometric [exponential; logarithmic] function. 逆三角〜 inverse trigonometric function. 双曲線〜 hyperbolic function. 楕円〜 elliptic function. 球〜 spherical function. 周期〜 periodic function. モジュラー〜 modular function. 円柱[筒]〜 cylindrical function. 特殊〜 special function. 奇[偶]〜 odd [even] function. 逆〜 inverse function. 有界〜 bounded function. 連続〜 continuous function. 不連続〜 discontinuous function. 微分可能な〜 differentiable function. 滑らかな〜 smooth function. 単調〜 monotone function. 有界変動(の)〜 function of bounded variation. 凸[凹]〜 convex [concave] function. 劣[優]調和〜 subharmonic [superharmonic] function. 導〜 derived function; derivative. 原始〜 primitive function. 被積分〜 integrand. 解析〜 analytic function. 正則〜 regular [holomorphic] function. 単葉〜 univalent [(*G.*) schlicht, simple] function. ベール〜 Baire function. ベッセル〜 Bessel function. メビウス〜 Möbius function. 母〜 generating function. 命題〜 propositional function. 〜関係 functional relation. 〜記号 functional symbol. 〜列 sequence of functions. 〜行列式 function-

al determinant; Jacobian. ～方程式 functional equation. ～空間 function space. ～要素 function element. ～群 function group. ～解析(学) functional analysis. ～目盛 functional scale. ～的に独立[従属]な functionally independent [dependent]. ～論的零集合 function-theoretic null set. 優～ superior function. リスク[危険]～ risk function. ハザード[危険]～ hazard function. メンバーシップ～ membership function. 距離～ metric; distance function. 伝達～ transfer function. 復号化～ decoding function.

かんすうぎょうれつしき 関数行列式 functional determinant; Jacobian.

かんせい 慣性 inertia. ⇨惰性.
～の inertial. ～法則 law of inertia. ～能率[モーメント] moment of inertia. ～楕円体 ellipsoid of inertia. ～主軸 principal axis of inertia. ～指数 index of inertia. ～場 inertial field. ～系 inertial system.

かんせい 完成 completion; perfection.
～する complete; accomplish; finish. ～品 finished product.

かんせつ 間接(の) indirect; mediate; implicit.
～に indirectly. ～法 indirect method. ～証明(法) indirect proof. ～接続 indirect continuation [prolongation]. ～超越特異点 indirect transcendental singularity. ～測定 indirect measurement. ～的定義 implicit [indirect] definition.

かんぜん 完全(性) completeness; perfection; absoluteness; integrity.
～な complete; perfect; whole; full. ～に completely; perfectly; wholly; fully. ～帰納法 complete induction. ～順列 complete permutation. ～剰余系 complete system of residues. ～四辺[角]形 complete quadrilateral [quadrangle]. ～六点列 complete range of six points. ～系 complete system. ～正規直交系 complete orthonormal system. ～加法性 complete additivity. ～加法的な completely additive; totally additive. ～数 complete number; perfect number. ～解 complete solution. ～可約な completely reducible. ～積分可能な completely integrable. ～楕円積分 complete elliptic integral. ～ベクトル場 complete vector field. ～センサス complete census. ～性定理 completeness theorem. ～平方[立方] perfect square [cube]. ～集合 perfect set. ～可分な perfectly separable. ～正規空間 perfectly normal space. ～流体 perfect fluid. ～マッチング perfect matching. ～グラフ complete graph. ～体 perfect field. ～代表系 complete set of representative. ～分岐 totally ramified

かんぜん 完全(性) exactness.
～な exact. ～微分 exact differential. ～微分形[方程]式 exact differential form [equation]. ～(系)列 exact sequence. ホモロジー[ホモトピー]～系列 homology [homotopy] exact sequence. ～関手 exact functor. 半～な semiexact; half-exact. 左[右]～な left [right] exact.

かんぜんれつ 完全(系)列 exact sequence.
短～ short exact sequence. 長～ long exact sequence.

かんぞうか 緩増加 slow increase.
～関数 slowly increasing function.

かんそく 観測 observation.
～する observe. ～値 observed value. ～ベクトル observation vector. ～方程式 observation equation. 直接[間接]～ direct [indirect] observation. ～者 observer.

かんたん 簡単 brevity; simplicity.
～な brief; simple. ～にする simplify; abbreviate. ～のため for the sake of brevity [simplicity]. ～な証明 simple [brief, short] proof.

かんつう 貫通 penetration.
～する penetrate.

かんど 感度 sensitivity.
～解析 sensitivity analysis.

かんとう 冠頭 prenex.
～標準形 prenex normal form.

かんにゅう 陥入 inden(ta)tion.
～させる indent.

かんねん 観念 notion; concept; conception;

idea.

かんのう 官能 sense.

～検査 sensory test.

かんび 完備(性) completeness.

～な complete. ～計画(法) complete block. ～束 complete lattice. ～体 complete field. ～空間 complete space. 測地的～な geodesically complete. 条件[制限]～な conditionally complete. ～化 completion. ～拡大 completion. 解析的～化 analytic completion. ～線形系 complete linear system.

ガンマ gamma; 【Γ, γ】.

～関数 gamma function; Γ(-)function. ～分布 gamma distribution. ～構造 gamma structure. ディ～ digamma. トリ～ trigamma. テトラ～ tetragamma. ポリ～ polygamma.

かんまん 緩慢 slowness.

～な slow. ～に slowly.

かんやく 簡約 simplification; abbreviation; cancellation; reduction.

～する abbreviate; cancel; reduce. ～法則 cancellation law. ～可能な reducible; reductive. ～行階段行列 reduced row echelon matrix [form]. ～勾配 reduced gradient.

かんよ 関与 concern; participation.

～する take part in; be concerned in; participate in.

かんり 管理 control; administration; management.

～する control; manage; administrate. ～検査 control inspection. ～水準 control level. ～限界 control limit. ～図 control chart. ～状態 state of control. 統計的～ statistical control. 品質～ quality control. 在庫～ inventory control [management]. ～欠如 lack of control. ～はずれの out of control. ～者 administrator.

がんり 元利 principal and interest.

～合計 amount with interest added.

かんりゃく 簡略(な) simple; brief.

～にする simplify; abbreviate. ～化 simplification; abbreviation. ～化(した)記法[記号] abbreviated notation.

かんれん 関連 relation; connection; connexion; regard; respect.

～した relevant; related; allied; connected; associated. に～して in [with] relation to; in connection with; with regard to. ～方程式 related equation. ～収束半径 associated convergence radii.

かんわ 緩和 relaxation.

～振動 relaxation oscillation. ～法 relaxation method. ～表 relaxation table. 線状～ line relaxation. 加速～ over-relaxation. 過大～ over-relaxation. 過小～ under-relaxation.

かんわー 緩和 slack.

～変数 slack variable. ～ベクトル slack vector.

かんわぬきとり 緩和抜取り reduced sampling.

～検査 reduced sampling inspection.

き・キ

き 基 base; basis (*pl.* -ses); radix (*pl.* -dices).
直交～ orthogonal basis. 正規直交～ orthonormal basis. 双対～ dual basis. 可算～ countable basis. 自由～ free base. 標準～ canonical basis [base]. 整数～ integral basis. 開～ open base. 部分～ subbase; subbasis. ～変換行列 base change matrix.

き 木 tree. ⇨樹木.
全域～ spanning tree. 二分～ binary tree. 有向～ directed tree.

き- 奇 odd. ⇨偶(ぐう-).
～数 odd number. ～順列[置換] odd permutation. ～関数 odd function. ～点 odd point. ～偶性 parity.

キー key.
～ワード[語] key word; keyword.

ぎ 偽 falsehood; fallacy; falsity.
～の false. ～の命題 false proposition. 真～ truth and falsity.

ぎ- 擬 quasi-. ⇨準(じゅん-).
～逆元 quasi-inverse; quasi-inverse element. ～可逆元 quasi-invertible element. ～正則元 quasi-regular element. ～順序 quasi-order. ～正則イデアル quasi-regular ideal. ～線形方程式 quasi-linear equation. ～等角写像 quasi-conformal mapping. ～解析関数 quasi-analytic function. ～一様収束 quasi-uniform convergence.

ぎ- 擬 pseudo-.
～ベクトル pseudo-vector. ～付値 pseudo-valuation. ～順序 pseudo-order. ～群 pseudo-group. ～加群 pseudo-module. ～微分作用素 pseudo-differential operator. ～距離 pseudo-distance; pseudo-metric. ～距離空間 pseudo-metric space. ～距離づけ可能な pseudo-metrizable. ～コンパクトな pseudo-compact. ～線形な pseudo-linear. ～解析関数 pseudo-analytic function. ～多項式 pseudo-polynomial. ～凸な pseudo-convex. ～勾配 pseudo-gradient. ～球面 pseudo-sphere. ～多様体 pseudo-manifold. ～幾何(学)的な pseudo-geometric. ～有向な pseudo-directed.

ぎえんすい 擬円錐 conoid.
～形の conoid. ～体 conoid.

きおく 記憶 memory.
無～性 lack of memory.

きおん 基音 tonic.

きか- 幾何 geometric(al); geometrico-.
～平均 geometric mean. ～調和平均 geometrico-harmonic mean. 算術～平均 arithmetico-geometric mean. ～数列 geometric progression. ～級数 geometric series. ～図形 geometrical figure. ～ベクトル geometric vector. ～確率 geometric probability. ～分布 geometric distribution. ～種数 geometric genus. ～光学 geometric optics. 超～ hypergeometric. ～的重複度 geometric multiplicity.

きかい 機会 opportunity; occasion; chance.
～費用 opportunity cost.

きかい 機械 machine; machinery.
～語 machine language. 決定性チューリング～ deterministic Turing machine. ～学習 machine learning.

きかいてき 器械的 mechanical.
～作図 mechanical construction. ～求積[積分] mechanical quadrature [integration].

きかがく　幾何学　geometry. ⇨幾何的.
～の; ～的な; ～図形の geometric; geometrical. ～的に geometrically. ユークリッド～ Euclidean geometry. 非ユークリッド～ non-Euclidean geometry. 楕円[双曲; 放物](的)～ elliptic [hyperbolic; parabolic] geometry. リーマン～ Riemannian geometry. 平面[立体]～ plane [solid] geometry. 球面～ spherical geometry; spherics. 初等～ elementary geometry. 解析～ analytic(al) geometry. 座標～ coordinate geometry. 総合～ synthetic geometry. 自然～ natural geometry. 球面～ spherical geometry. 画法～ descriptive geometry. 射影[アフィン; 共形]～ projective [affine; conformal] geometry. 位相～ topology. 微分～ differential geometry. 積分～ integral geometry. 代数～ algebraic geometry. ～基礎論 foundations of geometry. ～的公理 geometric axiom. ～的量 geometric(al) quantity [magnitude]. ～的図形 geometrical figure. ～的次元 geometric dimension. 計算～ computational geometry. 情報～ information geometry.

きかく　規格　standard; norm.
暫定～ temporary standard. ～化 standardization; normalization. ～化する standardize; normalize.

ぎかくちゅう　擬角柱　prismatoid

きかてき　幾何的　geometric; geometrical.
～図形 geometrical figure. ～次元 geometric dimension. ～胞複体 geometric cell complex. ～重複度 geometric multiplicity. ～差分方程式 geometric difference equation.

きかん　帰還; 饋還　feedback.

ぎかんすう　擬関数　improper function; pseudo-function.

ききゃく　棄却　rejection.
～する reject. ～検定 rejection test. ～限界 rejection limit. ～(個)数 rejection number. ～域 range of rejection; critical [rejection] region. ～点, ～限界値 critical point.

きぐう　奇偶(性)　parity. ⇨パリティ.
～検査 parity check.

きけつ　帰結　consequence; consequent; conclusion; result.

きけん　危険　risk.
～関数 risk function; hazard function. ～率 level of significance; significant level. 相対～度 relative risk.

きこう　機構　mechanism; structure; device.

きごう　記号　symbol; notation; sign; signal; signature.
～の symbolic; signal. 代数[関数]～ algebraic [functional] symbol. 総和～ summation symbol. 三添字～ three index symbol. 対象～ individual symbol. ～言語 symbolic language. ～化された symbolized. ～化されたパターン symbolized pattern. ～的解法 symbolic (method of) solution. ～論理(学) symbolic logic. クロネッカーの～ Kronecker('s) symbol. ～法 system of notation. 略～ abridged notation. 階乗～ factorial notation. ～代数 signal algebra. 述語～ predicate symbol. ～ベキ symbolic power.

ぎこう　技巧　technique; technical skill.
～的な ingenious, artful; skillful.

ぎこうばい　擬勾配　pseudo-gradient.
～ベクトル場 pseudo-gradient vector field.

きざみ　刻み　pitch; step.
～幅 step-size.

ぎじ　擬似(の)　dummy.

ぎじ-　擬似　pseudo-.
～コード pseudo-code. ～乱数の pseudo-random. ～乱数 pseudo-random numbers. ～逆行列 pseudo-inverse.

きじゅつ　記述　description.
～の; ～的な descriptive. ～する describe. ～集合論 descriptive set theory. ～統計学 descriptive statistics.

ぎじゅつ　技術　technique.
～上の[的な] technical. ～上[的に] technically. 計算～ calculation technique.

きじゅん　基準　standard; criterion (*pl.* -ria, -rions). ⇨標準(的).
～の standard; canonical; basic; original; fiducial. ～順列 standard [original] per-

mutation. ～ベクトル空間 standard vector space. ～化 standardization. 単数～ regulator. 単体～ simplex criterion. ～領域 canonical domain. ～積 canonical product. ～基底 canonical base. ～点 fiducial point. ～線 base line. ～方程式 canonical equation.

きじゅん 規準(の) normal.
～振動 normal vibration. ～ゲーム rigid game.

ぎじゅんじょ 擬順序 pseudo-order; quasi-order.

きじゅんめん 基準面 indicatrix.

きすう 奇数 odd number.
～の odd; odd-numbered.

きすう 基数 cardinal; cardinal number; cardinality; potency.
超限～ transfinite cardinal number. 到達不能～ inaccessible cardinal.

きすう 基数 base; radix (*pl.* -xes, -dices).
～法(による分類) radix sorting; sorting by radix.

きすうほう 記数法 numeration system; scale of notation.

ぎせいそく 擬正則(性) quasi-regularity.
～な quasi-regular. ～イデアル quasi-regular ideal. ～元 quasi-regular element.

きせき 軌跡 locus (*pl.* loci); geometric locus.
基本～ fundamental locus [loci]. ～の限界 limit of locus. ～交会法 intersection of loci. 結節点の～ node-locus. 接触点の～ path of contact; tac-locus.

きせつへんどう 季節変動 seasonality.

きせん 基線 ground line; base [basic] line.

きせん 軌線 trajectory; orbital path.

ぎせんけい 擬線形(性) quasi(-)linearity; pseudo-linearity.
～な quasi(-)linear; pseudo-linear. ～方程式 quasi(-)linear equation; pseudo-linear equation.

きそ 基礎 foundation; base; basis; elements.
～的な basic; fundamental; elementary. ～にある underlying. 数学～論 foundations of mathematics. ～面 basic surface. ～三角形 fundamental triangle. ～体 ground field; basic field; underlying field. ～群 underlying group. ～空間 base space; underlying space. ～位相空間 underlying topological space. ～概念 basic concept [notion]. ～知識 elementary knowledge; background. 三角法の～ elements of trigonometry. ～科目 primary subject.

きそうする 起草する draft.

きそく 規則 rule; law.
推論(～) rule of inference.

きた 北 north.
～の north. ～半球 north(ern) hemisphere.

きたい 期待 expectation.
～する expect. ～値 expectation; expected value. 条件つき～値 conditional expectation. 品質水準～ expected quality level.

ぎたこうしき 擬多項式 pseudo-polynomial.
特殊～ distinguished pseudo-polynomial.

ぎたようたい 擬多様体 pseudo-manifold; quasi-manifold.

きち 既知(の) known; determinate.
～数 known quantity; known number; datum (*pl.* -ta).

きちゃく 帰着 reduction.
～する reduce; result (in). ～させる reduce.

きてい 基底 base; basis (*pl.* -ses).
～の basic; underlying. 非～の nonbasic. ～の変換 change of bases. 順序～ ordered basis. 超越～ transcendence basis. 準[部分]～ subbase; subbasis. ～部分 basic component. ～変数 basic variable. 非～変数 nonbasic variable. ～ベクトル basic [basis] vector. ～方程式 basic equation. ～解 basic solution. ～成分 basic component. ～形式 basic form; canonical form. ～空間 underlying space. グレブナー～ Gröbner basis. ～変換行列 base change matrix.

きてん 起点 origin; starting point; initial point; fiducial point.

きてん 基点 base point; cardinal point.

きてん 危点 critical point.

きどう 軌道 orbit; trajectory; orbital path;

locus (*pl.* loci); path.
〜の orbital. 〜曲線 orbit; orbital curve. 〜面 plane of orbit; orbital plane. 〜空間 orbit space. 〜閉包 orbit closure. 〜要素 orbital element. 〜安定性 orbital stability. 最適〜 optimal locus. 〜多角形 path-polygon. 直交〜 orthogonal trajectory. 極限周期〜 limit cycle. 周期〜 periodic orbit.

ぎとうかく 擬等角(性) quasi-conformality.
〜な quasi(-)conformal. 〜写像 quasi(-)conformal mapping.

ぎとつ 擬凸(な) pseudo(-)convex.
〜領域 pseudo-convex domain. 強〜な strongly pseudo-convex.

ぎねん 疑念 doubt.

きのう 帰納 induction.
〜的な inductive. 〜的に inductively. 〜する induce. 〜法 induction; inductive method. 数学的〜(法) mathematical induction. 完全〜(法) complete induction. 二重[多重]〜(法) double [multiple] induction. 〜法の仮定 inductive assumption [hypothesis]; assumption [hypothesis] of induction. 〜的定義 inductive definition. 〜的推理 inductive analogy. 〜系 inductive system; direct system. 〜的極限 inductive limit; direct limit. 〜的次元 inductive dimension. 〜的順序集合 inductively ordered set.

きのう 機能 function.

きのうてき 帰納的(な) recursive.【以下の用例では"recursive"を「再帰的」と訳すことが多くなってきている】
〜に recursively. 〜定義 recursive definition. 〜関数 recursive function. 〜集合 recursive set. 〜に可算な recursively enumerable.

きのうほう 帰納法 induction; inductive method; recursion.
数学的〜 mathematical induction. 完全〜 complete induction. 二重[多重]〜 double [multiple] induction. 超限〜 transfinite induction.

きびゅうほう 帰謬法 ⇨背理法.

きほう 記法 notation.
略〜 abridged notation.

きぼう 希望 ⇨期待.

ぎほう 技法 technique. ⇨技術.

ぎほうせん 擬法線 pseudo-normal.

きほん 基本(的) elementary; rudimentary; rudimental; fundamental; basic.
〜演算 elementary [rudimentary; fundamental] operation(s). 〜変形 elementary transformation. 〜対称式 elementary symmetric expression. 〜対称多項式 elementary symmetric polynomial. 〜交代式 elementary alternating expression. 〜交代(多項)式 elementary alternating polynomial. 〜作図 fundamental construction(s). 〜軌跡 fundamental locus [loci]. 〜量 fundamental quantity. 〜点 fundamental point. 〜ベクトル(場) fundamental vector (field). 〜近傍系 fundamental system of neighbo(u)rhoods. 〜空間[関数] fundamental space [function]. 〜図形[曲線] fundamental figure [curve]. 〜列 fundamental sequence. 〜解 fundamental solution; elementary solution. 〜系 fundamental system. 〜形式 fundamental form; ground form. 〜周期 fundamental period; primitive period. 〜群 fundamental group. 〜領域 fundamental region. 〜定理[補助定理] fundamental theorem [lemma]. 〜多元環 basic algebra.

きほんしゅうき 基本周期 primitive period; fundamental period.
〜帯 primitive period strip. 〜平行四辺形 primitive period(-)parallelogram.

きまつ 期末 end of a term; term-end.
〜の terminal; term-end.

ぎむ 義務 duty; obligation; liability.

きむかせつ 帰無仮説 null hypothesis.

きむぶんぷ 帰無分布 null distribution.

ぎもん 疑問 question; interrogation.
〜文 interrogative sentence. 〜符 question mark; interrogation mark.

きゃく 客 customer.

きやく 規約 agreement; convention.
総和〜 summation convention.

きやく 既約(性) irreducibility.

〜な irreducible; reduced. 〜分数 irreducible fraction. 〜剰余系 irreducible system of residues. 〜剰余類 reduced residue class. 〜多項式 irreducible polynomial. 〜成分 irreducible component. 〜元 irreducible element. 〜指標 irreducible character. 〜表現 irreducible representation. 〜加群 irreducible module. 絶対〜な absolutely irreducible. 直〜な directly irreducible; indecomposable. 〜分解 irreducible decomposition.

ぎゃく 逆 converse; opposite.
〜の converse; opposite; reverse. 〜に conversely; oppositely. 〜[転換]命題 converse proposition. 〜の向き opposite direction [sense; orientation]. 〜進過程 reversed process.

ぎゃく 逆(の) inverse; reciprocal; reverse.
〜数 reciprocal (number). 〜比例 inverse proportion; reciprocal proportion. 〜比 reciprocal ratio. 〜元 inverse element; inverse. 〜行列 inverse matrix. 〜関数 inverse function. 〜三角関数 inverse trigonometric function. 〜写像 inverse mapping. 〜像 inverse image; preimage. 全〜像 complete inverse image. 〜系 inverse system. 〜極限 inverse limit. 〜変換 inverse transformation. 〜作用素 inverse operator. 〜射 inverse morphism. 〜補間 inverse interpolation. 〜関係 inverse relation. 〜回転 inverse rotation. 〜の道[路] inverse path. 〜整列集合 inversely well-ordered set. 〜差分 reciprocal difference. 〜匝線 reciprocal spiral. 〜向きの不等式 reverse inequality. 〜の向き opposite sense [orientation].

ぎゃく- 逆 anti-.
〜同値 anti-equivalence. 〜同型[形] anti-isomorphism. 〜自己同型[形] anti-automorphism. 〜準同型[形] anti-homomorphism. 〜自己準同型[形] anti-endomorphism. 〜エルミート的な anti-Hermitian. 〜エルミート行列 anti-Hermitian matrix. 〜平行な anti-parallel. 〜正則な anti-regular. 〜対称な anti-symmetric; skew-symmetric. 〜等角な anti-conformal; indirectly conformal.

ぎゃくかく 逆核 resolvent.

ぎゃくかんすう 逆関数 inverse function.

ぎゃくぎょうれつ 逆行列 inverse matrix.
一般化〜 generalized inverse (matrix). 疑似〜 pseudo-inverse.

ぎゃくさん 逆算する count back.

ぎゃくすう 逆数 reciprocal (number).
〜目盛 reciprocal scale.

ぎゃくせいげん 逆正弦 arcsine.
〜関数 arcsine function. 〜変換 arcsine transformation. 〜法則 arcsine law.

ぎゃくせつ 逆説 paradox.
〜的な paradoxical.

ぎゃくぞう 逆像 inverse image; preimage.
全〜 complete inverse image.

ぎゃくたいしょう 逆対称 skew-symmetry; anti-symmetry.
〜行列 skew-symmetric matrix; anti-symmetric matrix.

きゃくちゅう 脚注 footnote.

ぎゃくとうかく 逆等角な anti-conformal; indirectly conformal.

ぎゃくへいこう 逆平行 anti-parallelism.
〜な anti-parallel.

ぎゃくもまたどうよう 逆もまた同様 (*L.*) vice versa; v.v.

ぎゃくり 逆理 paradox. ⇨背理.
〜的な paradoxical.

きゃっか 却下 rejection. ⇨棄却.
〜する reject.

きゃっかんせい 客観性 objectivity.

キャップ cap; 【記号 ∩】.
〜積 cap product.

キャビネット cabinet.
〜投象(法) cabinet projection.

キャラクタ character.

キャリヤ carrier. ⇨台.

キュー・イー・ディ Q.E.D.; q.e.d.; (*L.*) quod erat demonstrandum (= which was to be proved).

きゅう 球 sphere; ball.
〜の spherical. 単位〜 unit sphere [ball]. 〜面 spherical surface; sphere. 〜錐 spher-

ical cone. ～台[冠; 欠] spherical segment. ～扇形 spherical sector. ～帯 zone of sphere. ～体 solid sphere. ～殻 spherical shell. ～座標 spherical coordinates. ～表現 spherical representation. ～関数 spherical function. リーマン～(面) Riemann sphere. 開～ open sphere [ball]. 半～ hemisphere. 超～ hypersphere.

きゅう 級 class.
～代表値 class value. ～の幅; ～間隔 class interval. ～下[上]限 class lower [upper] limit. ～間 among-class. ～内 interclass; within-class. 二～曲線 curve of second class.

きゅう- 九- nona-; nine.
～去法 excess of nines; casting out nines. ～点円 nine-point circle. ～元数 nonion. ～進法(の) nonary.

きゅうかく 球殻 spherical shell.

きゅうかん 球冠 spherical segment (of one base).

きゅうかん 球環 spherical ring.

きゅうかん 級間(の) among-class; between--class.
～変動 among [between]-class variation.

きゅうきょく 究極(の) ultimate.
～的限定言語 ultimate definite language.

きゅうけい 弓形 crescent; segment; lune; bow shape.
～の segmental; segmentary. ～の角[弧] angle [arc] of segment.

きゅうけい 球形 globular [spherical] form.

きゅうけつ 球欠 spherical segment.

きゅうげつけい 弓月形 lune.

きゅうげんしょう 急減少 rapid decrease.
～関数 rapidly decreasing function. ～超関数 rapidly decreasing distribution.

きゅうげんすう 九元数 nonion.

きゅうし 休止 rest; repose.
～点 rest point.

きゅうしゅう 吸収 absorption.
～的 absorptive; absorbing. ～する absorb. ～法則[律] absorption law. ～壁 absorbing [absorptive] barrier. ～連鎖[状態] absorbing chain [state]. ～断面(積) absorption cross section.

きゅうしんほう 九進法(の) nonary.

きゅうすう 級数 series (*pl.* -).
算術[等差]～ arithmetic series. 幾何[等比]～ geometric series. 調和[汎調和]～ harmonic [panharmonic] series. 有限[無限]～ finite [infinite] series. 収束[発散]～ convergent [divergent] series. 定[不定]発散～ definitely [indefinitely] divergent series; properly [oscillatingly] divergent series. 正項～ series with [of] positive terms; positive series. 交代[交項]～ alternating series. 優～ majorant (series). ベキ[整]～ power series. 二[多]項～ binomial [multinomial] series. 超幾何～ hypergeometric series. テイラー[ローラン]～ Taylor [Laurent] series. 三角～ trigonometric series. フーリエ～ Fourier series. 共役～ conjugate series; allied series. 二[多]重～ double [multiple] series. 階乗～ factorial series. 指数[対数]～ exponential [logarithmic] series. 多項式～ polynomial series. 漸近～ asymptotic series. 空隙～ gap [lacunary] series. 超収束～ overconvergent [ultraconvergent] series. ～解(法) series solution.

きゅうせき 求積 quadrature; mensuration.
～可能な quadrable. 数値～ numerical quadrature. 器械的～ mechanical quadrature. 区分～ quadrature [mensuration] by parts. ～法 (method of) quadrature; stereometry. 立体～ cubature. ～器 integrator; instrument of quadrature.

きゅうせんけい 球扇形 spherical secter.

きゅうそう 球叢 sphere congruence.

きゅうそく 急速 rapidity.
～な quick; rapid. ～に quickly; rapidly.

きゅうたい 球体 solid sphere; ball.

きゅうたい 球帯 zone of sphere.

きゅうだい 球台 spherical segment.

きゅうちょう 求長 rectification.
～する rectify. ～可能な rectifiable.

きゅうちょうわかんすう 球調和関数 solid harmonic function; solid harmonics.
体～ solid harmonic function; solid har-

monics. 帯［縞］～ zonal ［tesseral］ harmonics.

きゅうない 級内(の) interclass; within-class.
～相関 interclass correlation. ～変動 within-class variation.

きゅうめん 球面 spherical surface; sphere.
～の spherical. 単位～ unit sphere. 異種～ exotic sphere. 擬～ pseudosphere. ～幾何学 spherical geometry; spherics. ～三角法 spherical trigonometry; spherics. ～天文学 spherical astronomy. ～三角形 spherical triangle. ～過剰 spherical excess. ～表示 spherical representation. 複素～ complex sphere. リーマン～ Riemann sphere. ホモトピー～ homotopy sphere. ～距離 spherical distance. ～座標 spherical coordinates. ～曲線 spherical curve. ～定理 sphere theorem. ～(調和)関数 spherical harmonic function; spherical harmonics; surface harmonics. ～導関数 spherical derivative. ～改変 spherical modification. ～波 spherical wave. ～振子 spherical pendulum. ～投影(法) globular projection. ～束 sphere bundle.

きゅうめんかんすう 球面関数 spherical harmonic function; spherical harmonics; surface harmonics.

きゅうめんきかがく 球面幾何学 spherical geometry; spherics.

きゅうめんさんかくけい 球面三角形 spherical triangle.
三直角～ trirectangular spherical triangle. 象限弧～ quadrantal spherical triangle.

きゅうめんさんかくほう 球面三角法 spherical trigonometry; spherics.

きゅうめんちょうわかんすう 球面調和関数 spherical harmonic function; spherical harmonics; surface harmonics.

きゅうりょう 穹稜 groin.

キュムラント cumulant.

きょ- 虚 imaginary; virtual.
～数 imaginary number. ～数単位 imaginary unit. ～(数)部(分) imaginary part. 純～数 purely imaginary number. ～軸 imaginary axis. ～根 imaginary root. ～円［楕円］imaginary circle ［ellipse］. ～幾何学 imaginary geometry. ～円点 imaginary circular point. ～変換 imaginary transformation. ～像 virtual image.

きよ 寄与 contribution.
～する contribute. ～率 coefficient of determination.

きょう 莢 hull.
～核 hull-kernel. ～核位相 hull-kernel topology.

きょう 強
一割～ a little over 10 percent.

きょう- 強 strong. ⇨弱(じゃく-).
～位相 strong topology. ～収束 strong convergence. ～極限 strong limit. ～解 strong solution. ～混合性 strong mixing property. 大数の～法則 strong law of large numbers. ～可測な strongly measurable. ～連続表現 strongly continuous representation. ～楕円的な［型の］ strongly elliptic. ～擬凸な strongly pseudo-convex. ～定常過程 strongly stationary process. ～多項式の strongly polynomial.

きょう- 共 co-; con- ［b, m, p の前では com-, l の前では col-, r の前では cor-］.
～線の collinear. ～点の concurrent; copunctal. ～面の coplanar. ～円の concircular. ～役な conjugate. ～形の conformal. ～分散 covariance.

ぎょう 行 row.
～ベクトル row vector. ～階数 row rank. ～空間 row space. ～有限行列 row-finite matrix. ～階段行列 row echelon matrix. ～基本変形 elementary row operation.

きょうあつてき 強圧的(な) coercive.
～条件 coercive condition.

きょういく 教育 education; instruction.
～のある educated. ～する educate; instruct. ～制度 educational system. ～課程 curriculum (*pl.* -la). ～学 pedagogy; pedagogics.

きょうえい 鏡映 reflection; reflexion; (orthogonal) symmetry.
～面 reflection surface.

きょうえん 共円(性) concircularity.

～の concircular; concyclic. ～的に concircularly. ～点 concircular points. ～変換 concircular transformation. ～曲率テンソル concircular curvature tensor. ～的に平坦な concircularly flat.

きょうかい 境界 boundary; border; frontier.
～点 boundary point; frontier point. ～(曲)線 boundary curve; border line; circumference. ～値 boundary value. ～値問題 boundary value problem. ～条件 boundary condition. ～作用素 boundary operator. ～輪体 boundary cycle; boundary. ～成分[帯] boundary component [strip]. ～要素 boundary element. ～性状 boundary behavio(u)r. 自然～ natural boundary. 理想～ ideal boundary. 相対～ relative boundary. 零～ null boundary. ～のある with border [boundary]. ～のない unlimited; without border [boundary].

きょうかく 夾角 included angle.

ぎょうかく 仰角 angle of elevation.

きょうぎ 狭義(の) strict.
～に strictly; in the strict [restricted] sense. ～(の)帰納的極限 strict inductive limit. ～の[等号を含まない]不等式 strict inequality. ～の包含(関係) strict inclusion. ～(に)単調な strictly monotone. ～に増加[減少]の strictly increasing [decreasing]. ～(の)凸[凹]関数 strictly convex [concave] function.

きょうぎ 競技 game; play.

きょうけい 共形(性) conformality.
～な conformal. ～的に conformally. ～変換 conformal transformation. ～接続 conformal connection. ～(微分)幾何学 conformal (differential) geometry. ～空間 conformal space. ～曲率テンソル conformal curvature tensor. ～的対応[変形] conformal correspondence [deformation]. ～的線素[捩率] conformal arc element [torsion]. ～的に平坦な conformally flat.

きょうけい 共傾(の) cogradient.

きょうごう 競合 competition.

きょうし 教師 teacher; instructor; schoolmaster.
数学の～ teacher of mathematics.

きょうじく 共軸(の) coaxial; coaxal.
～円 coaxial circles. ～円錐曲線 coaxial conics. ～円系 pencil of circles.

きょうしつ 教室 school [class] room; department; dept.; dpt.
～の departmental. 数学～ mathematics department; department of mathematics. 階段～ lecture theater.

きょうじゅ 教授 professor.
准～ associate professor. 名誉～ emeritus professor; professor emeritus. ～会 faculty meeting.

きょうしゅう 共終(の) cofinal.
～部分集合 cofinal subset. ～度 cofinality.

ぎょうしゅう 凝集 condensation.
～点 point of condensation; condensation point. ～判定法 condensation test.

きょうしょう 共焦(の) confocal.
～楕円 confocal ellipses. ～双曲線 confocal hyperbolas. ～放物線 confocal parabolas. ～二次曲面 confocal quadrics [conicoids]. ～円錐曲線 confocal conics.

きょうしん 共振 resonance. ⇨共鳴.

きょうしん 共心(の) ⇨同心(の).

きょうせい 強制(された) forced.
～振動 forced oscillation [vibration].

きょうせん 共線(性) collinearity.
～の collinear; collineatory. ～的に collinearly. ～写像 collineation in the wider sence. 射影的～変換 projective collineation. ～変換 collineation. ～平面 collinear planes. ～点 collinear points. ～図表 alignment chart; collinear diagram [nomogram].

きょうせんへんかん 共線変換 collineation.
アフィン～ affine collineation. 射影的～ projective collineation.

きょうそう 競争 competition; contest; contention.
～の competitive. ～する compete; contend. ～者 competitor. ～経済 competitive economy. 完全～均衡 (complete) competitive equilibrium. ～関数 competing function.

きょうぞう 鏡像 mirror image, reflected image; reflection; reflexion; inversion; symmetry.
～の原理 principle of reflection [inversion]; reflection [inversion] principle; symmetry principle.

きょうぞん[そん]せい 共存性 consistency.

きょうだ 強打 bang; bang-bang; heavy blow.

きょうちょ 共著 collaboration; joint work [paper]; collaborated work.
～者 coauthor; joint author; collaborating author.

きょうつう 共通(の) common.
～点 common point. ～因数[子] common factor. ～根 common root. ～内[外]接線 internal [external] common tangent. ～弦 common chord.

きょうつうぶぶん 共通部分 intersection; common part.

きょうてん 共点(性) concurrence.
～の concurrent; copunctal. ～図表 intersection chart. ～平面 copunctal planes.

きょうど 強度 strength; intensity.

きょうどう 共同(の) joint; common; bipartite.
～作業 joint work.

きょうふく 狭幅(の) leptokurtic.
～[幅ほそ; 急尖]分布 leptokurtic distribution.

きょうぶんさん 共分散 covariance.
～分析 analysis of covariance. ～行列 covariance matrix. 母集団[標本]～ population [sample] covariance.

きょうへん 共変(性) covariance.
～的な covariant. ～性 convariance ～量 covariate. ～ベクトル covariant vector. ～テンソル[スピノル] covariant tensor [spinor]. ～関手 covariant functor. ～微分 covariant differentiation. ～成分 covariant component. ～指標[添数] covariant index. ～接ベクトル束 cotangent bundle. 絶対[多重]～式 absolute [multiple] covariant.

きょうめい 共鳴 resonance.
～エネルギー resonance energy. ～定理 resonance theorem. ～散乱 resonance scattering.

きょうめん 共面(性) coplanarity.
～の coplanar. ～的に coplanarly. ～点 coplanar points.

きょうやく 共訳 joint translation.

きょうやく 共役[軛](な) conjugate.
～弧 conjugate arc. ～複素数 conjugate complex (number); complex conjugate. ～根 conjugate root. ～点 conjugate point. 調和～点 harmonic conjugate(s). 等線分～点 isometric [isotomic] conjugate point. ～双曲線[双曲面] conjugate hyperbola [hyperboloid]. ～軸 conjugate axis. ～直径 conjugate diameter. ～元 conjugate (element). ～ベクトル conjugate vector. ～調和関数 conjugate harmonic function; harmonic conjugate. ～微分 conjugate differential. ～類 conjugate class. ～作用素 conjugate operator. ～核 conjugate kernel. ～空間 conjugate space; dual space. ～表現 conjugate representation. ～勾配[傾斜]法 conjugate gradient method. 複素～化 complex conjugation. ～(化)写像 conjugation (mapping). ～転置行列 adjoint matrix.

きょうやくきゅうすう 共役級数 conjugate series; allied series.

きょうやくさせき 共役差積 different; (*G.*) Differente.

きょうゆう 共有(の) common. ⇨共通(の).
～点 common point. ～根 common root. 両辺～の角 coterminal angles.

きょうりょく 協力 cooperation; collaboration.
～の cooperative. ～者 cooperator; collaborator. ～ゲーム cooperative game.

ぎょうれつ 行列 matrix (*pl.* -rices, -rixes).
単位～ unit matrix; identity matrix. 転置～ transposed matrix. 対称[交代]～ symmetric [alternating] matrix. 歪[逆]対称～ skew-symmetric matrix; anti-symmetric matrix. エルミート～ Hermitian matrix. ヤコビ～ Jacobian matrix. 対角～ diagonal matrix. 正方[三角]～ square [triangular]

matrix. ラムダ～ lambda [λ] matrix. 逆～ inverse matrix. 擬似逆～ pseudo--inverse. 正則～ regular matrix. 正規～ normal matrix. 直交[ユニタリ]～ orthogonal [unitary] matrix. 正定値[負定値]～ positive-[negative-] definite matrix. 無限～ infinite matrix. 拡大係数～ enlarged coefficient matrix. 余因子～ cofactor matrix. 変換～ transformation matrix. ～単位 matrix unit. ～要素 matrix element. ～表現 matrix representation. ～群 matrix group. 全～環 total matrix ring. ～方程式 matrix equation. 一般化逆～ generalized inverse matrix. 簡約行階段～ reduced row echelon matrix [form]. 行階段～ row echelon matrix. スパース～ sparse matrix. 生起～, 接続～ incidence matrix. 生成～ generator matrix. 疎～ sparse matrix. 置換～ permutation matrix. 表現～ representation matrix. 密～ dense matrix. 隣接～ adjacency matrix.

ぎょうれつしき 行列式 determinant.
～の determinantal. 小～ minor (determinant). 巡回～ cyclic determinant. 随伴～ adjoint determinant. 関数～ functional determinant; Jacobian. ヤコビ(の)～ Jacobian. ロンスキの～ Wronskian. ～因子 determinant divisor. ～方程式 determinantal equation.

きょぎ 虚偽 fallacy; falsehood; falsity.

きょく 極 pole; polar singularity.
～の polar. 北[南]～ north [south] pole. ～に凸[凹]な convex [concave] toward pole. 単～; 一位の～ simple pole. n 位の～ pole of order n. ～因子 pole divisor. 対数的～ logarithmic pole. 双～ dipole.

きょく- 極 polar.
～性 polarity. ～三角形 polar triangle. ～四面体 polar tetrahedron. ～円 polar circle. ～円錐曲線 polar conics. ～(曲)線 polar line; polar. ～平面 polar plane. ～(曲)面 polar surface; polar. ～系 polar system. ～関係 polarity. ～(相)反 polarity. ～距離 polar distance. ～座標 polar coordinates. 接線～座標 polar tangential coordinates. ～形式 polar form. ～表示 polar representation [expression]. ～射[投]影 polar projection. ～要素 polar element. ～集合 polar set. ～錐 polar cone.

きょくげん 極限 limit; (*L.*) limes (*pl.* -mites).
～値 limit value; limiting value. 上～ superior limit; (*L.*) limes superior. 下～ inferior limit; (*L.*) limes inferior. 主～ principal limits; (*L.*) limites principales. 順序～ order limit. ～点 limit point. ～関数[分布] limit function [distribution]. ～関係 limit relation. ～順序数 limit ordinal number. 片側～ one-sided limit. 強[弱]～ strong [weak] limit. 帰納的～ inductive limit; direct limit. 直[順]～ direct limit. 射影的～ projective limit. 逆～ inverse limit. ～事象 limit event. ～定理 limit theorem. 中心～定理 central limit theorem. ～に近づく tend to limit; approach limit. ～周期軌道 limit cycle.

きょくしょ- 局所 local; locally; in the small; (*G.*) im Kleinen.
～的な local. ～的に locally; in the small; (*G.*) im Kleinen. ～(的)性(質) local property. ～座標 local coordinates. ～(媒介)変数 local parameter. ～環[体] local ring [field]. ～化 localization. ～有界な locally bounded. ～有限な locally finite; scattered. ～可積な locally integrable. ～閉の locally closed. ～コンパクトな locally compact. ～連結の locally connected. ～ユークリッド的な locally Euclidean. ～(的に)平坦な locally flat. ～凸な locally convex. ～対称な locally symmetric. ～環つき空間 local-ringed space. ～絶対 p 葉な locally absolute p-valent. ～準同型[形](写像) local homomorphism. ～探索 local search. ～閉集合 locally closed set.

きょくしょう 極小 minimum (*pl.* -ma, -mums); min.; relative minimum; local minimum. ⇨極小(きょくしょう-).
～の minimal. ～に minimally.

きょくしょう- 極小 minimal.
～値 minimal value; minimum. ～条件

minimal condition. ～曲面[曲線] minimal surface [curve]. ～イデアル minimal ideal. ～概周期群 minimally almost periodic group.

きょくしょてき 局所的(な) local.
～に locally; in the small; (*G.*) im Kleinen. ～理論 local theory. ～性質 local property. ～最小解 local minimizer. ～最大解 local maximizer. ～収束性 local convergence.

きょくせん 曲線 curve; curved line.
～の curvilinear. ～の追跡[あてはめ] curve tracing [fitting]. 連続～ continuous curve. 単純[一]～ simple curve. ジョルダン～ Jordan curve. 漸近～ asymptotic curve. 円錐～ conics; conic section. 二[三]次～ quadratic [cubic] curve. 二級～ curve of second class. 楕円～ elliptic curve. 正弦～ sine curve. 指数[対数]～ exponential [logarithmic] curve. 歯形～ tooth curve. 有理～ rational curve. 代数[超越]～ algebraic [transcendental] curve. 正則～ regular curve. 解析～ analytic curve. 有理型～ meromorphic curve. 極小～ minimal curve. 万有～ universal curve. 解～ solution curve. 座標～ coordinate curve. 定傾[定幅]～ curve of constant inclination [bredth]. ～定理 curve theorem. ～座標 curvilinear coordinates. ～積分 curvilinear integral. モジュラー～ modular curve.

きょくせん 極線 polar; polar line.

きょくせんちゅう 曲線柱 cylindroid.

きょくそうはん 極相反 polarity.

きょくだい 極大 maximum (*pl.* -ma; -mums); max.; relative maximum; local maximum. ⇨極大(きょくだい-).
～の maximal. ～に maximally. ～の原理 principle of maximum; maximum principle.

きょくだい- 極大 maximal.
～値 maximal value; maximum. ～条件 maximal condition. ～整環[イデアル] maximal order [ideal]. ～フィルター maximal filter; ultrafilter. ～独立系 maximal independent system. ～概周期群 maximally almost periodic group. ～マッチング maximal matching.

きょくたん 極端な extreme; excessive; ultra-. 極端に～ extremely.

きょくたんたい 曲単体 continuous simplex.

きょくち 極値 extremal value; extreme value; (*L.*) extremum (*pl.* -ma, -mums). (局所的)～ local extremum. ～点 extreme [extremal] point. ～関数 extremal (function). ～問題 extremal problem. 条件つき～ constrained extremum. ～曲線[曲面] extremal (curve [surface]). ～計量 extremal metric.

きょくちてき 極値的 extremal.
～に extremally. ～方法 extremal method. ～長さ extremal length. ～距離 extremal distance. ～計量 extremal metric.

きょくはん 極反 polarity.

きょくびな 極微な minute; atomic.

きょくめん 曲面 surface; curved surface.
～論 theory of surfaces. 凸～ convex surface. 回転～ surface of revolution. 閉～ closed surface. 斜～ skew surface. 代数[超越]～ algebraic [transcendental] surface. 展直～ rectifying surface. 複素～ complex surface. 特殊～ special surface. 極小～ minimal surface. 接線～ tangent surface. 定曲率～ surface of constant curvature. 線織～ ruled surface. 中心～ center surface. 単[両]側～ unilateral [bilateral] surface. 平行～ parallel surface. (微分方程式の)解～ integral surface. ～片 surface patch. 超～ hypersurface.

きょくめん 極面 polar surface; polar.

きょくめんせき 曲面積 surface area.

きょくりつ 曲率 curvature.
～半径 radius of curvature. ～中心 center of curvature. ～円 circle of curvature. ～テンソル[形式] curvature tensor [form]. 主～ principal curvature. 平均[全]～ mean [total] curvature. 絶対～ absolute curvature. 法～ normal curvature. 断面～ sectional curvature. 定～空間 space of constant curvature. ～線 line of curvature.

きょこう 虚構 fiction.

～の fictitious; fictional.

きょこん 虚根 imaginary root.

きょじく 虚軸 imaginary axis.

きょすう 虚数 imaginary number. ⇨虚(きょ-).

～単位 imaginary unit. 純～ purely imaginary number. ～部分 imaginary part. ～乗法 complex multiplication.

きょすうじょうほう 虚数乗法 complex multiplication.

きょせん 距線 diagonal.

きょっか 極化 polarization.

～恒等式 polarization identity.

きょどう 挙動 behavio(u)r.

の～ behavio(u)ral. ～する behave. 漸近的～ asymptotic behavio(u)r.

きょよう 許容 tolerance.

～差 tolerance. ～限界 tolerance limit(s). ～域 tolerance region. ～区間 tolerance interval.

きょよう 許容(的な) admissible; allowed.

～性 admissibility. ～する admit. ～同型[形] admissible isomorphism. ～準同型[形] admissible [allowed] homomorphism. ～関数 admissible function. ～列 admissible sequence. ～イデアル admissible ideal.

きょようりょういき 許容領域 feasible region.

きょり 距離 distance; metric.

～の metric. 最短～ shortest distance. 球面～ spherical distance. 角～ angular distance. 弦～ chordal distance. 天頂～ zenith distance; coaltitude. 擬～ pseudo-distance. 非ユークリッド～ non-Euclidean distance. ～円[点] distance circle [point]. フレシェの(擬)～ Fréchet's écart. ～関数 metric; distance function. ～空間 metric space. 誘導～ induced distance [metric]. 焦点～ focal distance [length]. 有向～ oriented distance. 極値的～ extremal distance. 有効～ effective range. ハミング～ Hamming distance. 設計～ designed distance.

きょりづけ 距離づけ metrization.

～可能な metrizable.

きりあげ 切上げ round-up; raising.

～る raise. 端数の～ raising of fraction. ～誤差 round-up error.

きりくち 切り口 ⇨断面.

ギリシア Greece.

～の Greek; Grecian; Greco-. ～文字 Greek letter.

きりすて 切捨て round-off; cut-off; omission; discarding.

～る omit; discard. 端数の～ omission [discarding] of fraction. ～誤差 round-off error.

きりつめる 切り詰める curtail.

キリング W. K. J. Killing (1847–1923).

～形式 Killing form.

きる 切る cut; cross; truncate.

キロ kilo-.

～メートル kilometer.

きろく 記録 record; document.

～する record; register.

ぎろん 議論 discussion; argument.

～する discuss; argue.

きわめて 極めて very; extremely; exceedingly.

きんこう 均衡 balance; equilibrium (*pl.* -ria, -riums).

～点[状態] equilibrium point [state]. ～凸集合 balanced convex set. 完全競争～ complete competitive equilibrium. 力の～ balance of power.

きんこうけいすう 近交係数 inbreeding coefficient.

きんし 禁止 inhibition; prohibition.

～する inhibit; forbid; prohibit.

きんじ 近似 approximation.

～の; ～的な approximate. ～的に approximately. ～する approximate. ～値 approximate value. ～作図 approximate construction. ～式[多項式; 関数] approximate expression [polynomial; function]. ～関数等式 approximate functional equation. ～定理 approximation theorem. ～法 approximation method; method of approximation. 逐次～ successive approximation. 多項式～ polynomial approximation. 単体～ simplicial approximation. ～

の飽和 saturation of approximation. 最良～ best (fit) approximation. 線形～ linear approximation. 最小二［自］乗～ least square approximation. 過剰［不足］～ approximation in excess [in deficiency]. ～可能な approximable. 差分～ difference approximation.

きんじち 近似値 approximate value.
過大［小］～ upper [lower] approximate value.

きんじつてん 近日点 perihelion (*pl.* -lia).

きんじぶんすう （連分数の）近似分数 convergent.
第 n～ n-th convergent. 主［中間］～ principal [intermediate] convergent.

きんしんこうはい 近親交配 inbreeding.

きんせつ 近接（の） adjacent; neighbo(u)ring; immediate; near.
～関数 proximity function; (*G.*) Schmiegungsfunktion.

きんてんりかく 近点離角 anomaly.
離心～ eccentric anomaly. 真～ true anomaly.

きんとう 均等 equality.
～の equal; even. ～割 per capita rate.

きんぼう 近傍 neighbo(u)rhood; vicinity.
～系 neighbo(u)rhood system. ε～ ε-neighbo(u)rhood. 開～ open neighbo(u)rhood. 凸～ convex neighbo(u)rhood. 管状～ tubular neighbo(u)rhood. 除外～ deleted neighbo(u)rhood. 基本～（系） fundamental (system of) neighbo(u)rhood(s). 座標～ coordinate neighbo(u)rhood.

ぎんみ 吟味 examination; investigation; scrutinization; discussion.
～する examine; investigate; scrutinize (【英】scrutinise); sift; discuss; try; test.

きんみつ 緊密 tightness; closeness.
～な tight; close.

きんりん 近隣 adjacency.

く・ク

く- 九- nona-; nine. ⇨九(きゅう-).
クイバー quiver.
くう 空 empty; void; null.
～の empty; void. ～でない non-empty; non-void. ～集合 empty [void; null] set. ～事象 empty [void] event. ～語 empty word. ～理 empty theory.
ぐう- 偶 even. ⇨奇(き-).
～数 even number. ～順列[置換] even permutation. ～関数 even function. ～点 even point.
くうかん 空間 space.
～の; ～的な spatial [spacial]; solid. ～的に spatially. ～で (*L.*) in solido. ～曲線 space curve. ～幾何学 solid geometry; space geometry. 射影[アフィン]～ projective [affine] space. 無限遠～ space at infinity. 半～ half-space. 定曲率～ space of constant curvature. 抽象～ abstract space. 位相[距離]～ topological [metric] space. ノルム～ normed space. 線形～ linear space. ベクトル～ vector space. 関数～ function space. 部分～ subspace. 共役～ conjugate [dual] space. 双対～ dual space. 商～ quotient space. 接[法]～ tangent [normal] space. 被覆～ covering space. 固有～ eigenspace. 直積～ product space. 確率～ probability space. 標本～ sample space. 相～ phase space. レンズ～ lens space. ユークリッド～ Euclidean space. リーマン～ Riemannian space. ヒルベルト[バナッハ]～ Hilbert [Banach] space. シュタイン～ Stein space. n次元～ n-dimensional space; n-space. 実[複素]～ real [complex] space. 基礎～ basic [base] space; space of elementary events. 母～ generating space. ～形 space form. ～格子 space lattice. ～的同型[形]な spatially isomorphic. ネーター～ Noetherian space. 補間～ interpolation space. 零～ null space.
ぐうき 偶奇(性) parity. ⇨奇偶(性); パリティ.
くうげき 空隙 gap.
～級数 gap series; lacunary series. ～定理 gap theorem. アダマール(の)～ Hadamard's gap. ～値 gap value.
くうしょ 空所 blank; space.
ぐうすう 偶数 even number.
～の even; even-numbered.
ぐうぜん 偶然 chance; accident.
～な accidental. ～に accidentally; by chance; contingently. ～誤差 accidental error. ～性 contingency. ～表 contingency table.
くうはく 空白 blank; null.
～の blank.
ぐうりょく 偶力 couple (of forces).
くかく 区画 block. ⇨ブロック.
くかっけい 九角形 nonagon; enneagon.
くかん 区間 interval.
開[閉; 半開]～ open [closed; semi-open] interval. 有限[無限]～ finite [infinite] interval. ～関数 interval function. ～推定 interval estimate. 許容～ tolerance interval; admissible interval. 信頼～ confidence interval. ～縮小法 method of diminishing [nested] intervals; (*G.*) (Methode der) Intervallschachtelung. ～演算 interval arithmetic. ～解析 interval analysis.
くき 茎 stalk.

前層の～ stalk of presheaf.
くぎり 区切り break; end; stop; punctuation.
～点 break-point.
くく 九々
掛け算～表 multiplication table; Pythagorean table. 加法～ addition table.
くけい 矩形 rectangle. ⇨長方形.
～の rectangular; oblong.
くさび 楔 wedge.
～形 wedge(-shape). ～積 wedge product.
～形文字 cuneiform (character).
クシー Xi; 【Ξ, ξ】.
くじ 籤 lottery.
富～ lottery. 複合～ compound lottery.
ぐたいてき 具体的(な) concrete.
～に concretely.
くっきょく 屈曲 bending; winding; flection; flexion; flexure.
くっせつ 屈折 refraction.
～させる refract. ～率 refractive index.
～光線 refracted ray.
くっせつかせん 屈折火線 diacaustic.
～の diacaustic.
くっせつしょうせん 屈折焦線 ⇨屈折火線.
くっせつせん 屈折線 polygonal line. ⇨折(れ)線.
くつひも 靴ひも bootstrap.
グーデルマン C. Gudermann (1798-1851).
～関数 Gudermannian.
くねった tortuous.
くびかざり 首飾り necklace. ⇨数珠(じゅず).
くぶん 区分 section; division; classification; partition.
～求積(法) quadrature [mensuration] by parts.
くぶんごと 区分ごとの[に] piecewise.
くぶんてき 区分的(に) piecewise.
～線形の piecewise linear. ～連続関数 piecewise continuous function. ～滑らかな piecewise smooth.
くべつ 区別 distinction; difference; discrimination.
～する distinguish; discriminate. ～された distinguished.
くへんけい 九辺形 nonagon; enneagon.
くみ 組 class; group; set; pair.
二～ two classes [pairs]. 上[下]～ upper [lower] class.
-ぐみ 組 ⇨組.
くみあわせ 組合せ combination.
～的; ～論の combinatorial; combinatory. ～る combine. 重複～ repeated combination. ～(論的)位相数[幾何]学 combinatorial topology. ～論 combinatorial analysis; combinatorics. ～多様体 combinatorial manifold. ～最適化 combinatorial optimization
くみいと 組み糸 braid.
～群 braid group.
くみこみ 組込み(の) built-in.
くみたて 組立
～単位 derived unit. ～除法 synthetic division.
くみつくし 汲み尽し exhaustion.
～法 exhaustion method; method of exhaustion.
くみひも 組み紐 braid.
くみわけ 組分け grouping; division into classes.
くめんたい 九面体 nonahedron (*pl.* -rons, -dra); enneahedron (*pl.* -rons, -dra).
くら 鞍 ⇨鞍(あん).
くらい 位 location; place; position; grade; rank; order.
～取り radix (*pl.* -dices). ～取り記数法 positional notation.
クライン F. Klein (1849-1925).
～の Kleinian; Klein's. ～の管[壷] Klein's bottle. ～の四元群 Klein four group. ～群 Kleinian group.
グラスマン H. G. Grassmann (1809-1877).
～代数[座標] Grassmann algebra [coordinates].
グラフ graph; chart.
～(式)の graphic(al). 棒～ bar [rectangular] graph. 帯～ band graph. 円～ circle [circular] graph. 面積～ area graph. 有限～ finite (linear) graph. ～理論 graph theory. 閉～定理 closed graph theorem.

関数の～ graph of function. ～[図]表現 graphic representation. 正方形～ square diagram. 完全～ complete graph. 双対～ dual graph. 単純～ simple graph. 二部～ bipartite graph. 部分～ subgraph. 平面～ plane graph. 平面的～ planar graph. 補～ complement graph. 無向～ undirected graph. 有限～ finite graph. 有向～ directed graph; digraph; quiver. 連結～ connected graph.

グラム J. P. Gram (1850–1916).

グラム・シュミット Gram-Schmidt.
～の正規直交化法 Gram-Schmidt orthonormalization.

クラメル G. Cramer (1704–1752).
～の公式 Cramer's formula.

くりあがり (足し算の)繰上り carrying.

くりあげ 繰上げる carry; carry up.

くりかえす 繰返す repeat; iterate.
～もの repeater.

クリーク clique.
～数 clique number.

くりさがり (引き算の)繰下り borrowing.

くりさげ 繰下げる carry; carry down.

クリスタルコホモロジー crystalline cohomology.

グリム GLIM; generalized linear model.

グリーン G. Green (1793–1841).
～関数[空間] Green function [space].

クルル W. Krull (1899–1971).
～位相 Krull topology. ～次元 Krull dimension. ～の標高定理 Krull's height [altitude] theorem.

グレコラテン Greco-Latin.
～方格[方陣] Greco-Latin square.

グレブナー W. Gröbner (1899–1980).
～基底 Gröbner basis.

クロス cross.
～(乗)積 cross product.

グロス gross. (*pl.*-)
大[十二]～ great gross. 小～ small gross.

クロソイド clothoid.

クロネッカー L. Kronecker (1823–1891).
～のデルタ[記号] Kronecker's delta [symbol].

クローバー clover.
～[三つ葉]型のもの clover-leaf. 四つ葉～型のもの four-leaf clover.

くわえざん 加え算 addition.

くわえる 加える add (to); sum (up).
aにbを～ add b to a. 2に3を～と5になる Two and three make [are] five.

くわけ 区分け partition.

ぐん 群 group.
有限[無限]～ finite [infinite] group. アーベル～ abelian group. 可換～ commutative group. 対称[交代]～ symmetric [alternating] group. 単純～ simple group. 部分～ subgroup. 置換～ permutation group. 商～ quotient group. 因子～ factor group. 古典～ classical group. 巡回～ cyclic group. ガロア～ Galois group. (正)多面体～ polyhedral group. 四元数～ quaternion group. ベッチ～ Betti group. ホロノミー～ holonomy group; group of holonomy. 変換～ transformation group. 運動～ group of motions. 被覆～ covering group. 基本～ fundamental group. モジュラー～ modular group. 連続[不連続]～ continuous [discontinuous] group. リー～ Lie group. 位相～ topological group. 加～ additive group; module. 乗法～ multiplicative group. 自由～ free group. 代数～ algebraic group. 構造～ structural group. 固定～, 不変～ fixed group. 行列～ matrix group. 混～, 混合(アーベル)～ mixed group. 亜～ groupoid. ～(多元)環 group algebra. ～環 group ring. ～多様体 group manifold. ～指標 group character. ～芽 group germ. ～拡大 group extension. ～符号 group code. ～逐次抜取り検査 grouped sequential inspection. ～対象 group object. ～の位数 order of group. 散在型単純～ sporadic simple group. 剰余～ factor group. 単数～, 単元～ group of units; unit group. 非可換～ non-abelian group. 例外～ exceptional group.

クンマー E. E. Kummer (1810–1893).
～拡大 Kummer extension.

け・ケ

けい (定理の)系 corollary; cor.

けい 系 porism.
～論[不定命題] porism.

けい 系 system.
公理～ system of axioms. 座標～ coordinate system. 方程式～ system of equations. 直交～ orthogonal system. 有理数～ system of rational numbers. 生成～ system of generators. 指標～ character system. 絶対単位～ system of absolute units. 硬い～ stiff system. 完備線形～ complete linear system. 帰納～ inductive system; direct system. 逆～ inverse system. 射影(的)～ projective system. 線形～ linear system. 直～, 順～ direct system. パラメータ～ system of parameters. 複雑～ complex system.

-けい 形 form; shape.
J字～ J-shape. J字～の J-shaped. 樽～ barrel shape.

けいえい 経営 management.
～科学 management science. ～費 operating cost; running expenses.

けいかく 計画 plan; scheme; plot; design; project; schedule; program.
～する plan; plot; design. ～行列 design matrix. ブロック～ block design. 実験～ design of experiments. 線形～(法) linear programming; LP. 凸～(法) convex programming. ～数学 planning mathematics. ～(手順) planning. 日程～ scheduling. ～表 schedule; program. 入れ子型～ nested design. 数理～(法) mathematical programming. 大域的～法 global programming. 半無限～(法) semi-infinite programming.

けいかく 傾角 inclination; angle of inclination; oblique angle; dip; obliqueness.

けいけってん 軽欠点 minor defect.

けいけん 経験(的な) empirical; experimental; experienced.
～式 experimental [empirical] formula. ～的確率 experimental [empirical] probability. ～分布関数 empirical distribution function. ～論 empiricism.

けいこう 傾向 trend; tendency.
中心～ central tendency.

けいざい 経済 economy.
競争～ competitive economy. ～学 economics. 計量～学 econometrics.

けいさん 計算 calculation; computation.
～する calculate; compute; cipher. ～的な; ～上の computational. ～法 method of computation; calculus. ～者 computer; numerator. 命題～ propositional calculus. 述語～ predicate calculus. 数値～ numerical calculation [computation]. 図式～ graphical calculation. ～違い miscalculation. ～可能な computable. 高精度～ high precision computation. ～量 computational complexity [quantity]. ～言語学 computational linguistics. ～幾何学 computational geometry. ～複雑性 computational complexity. 最悪～量 worst-case complexity. 非決定性～ non-deterministic computation. 平均～量 average-case complexity. 並列～ parallel computation.

けいさんき 計算機 computer; calculator.
デジタル[計数型]～ digital computer. 電子～ electronic computer. ～代数 compu-

ter algebra.

けいさんじゃく 計算尺 slide rule.
特殊～ particular slide rule.

けいさんずひょう 計算図表 nomogram; nomograph.
～学 nomography. ～学(の) nomographic.

けいさんりょう 計算量 computational complexity [quantity].
最悪～ worst-case complexity. 平均～ average-case complexity.

けいしき 形式 form.
～の; ～的な formal. ～的に formally. 一次[双一次; 二次]～ linear [bilinear; quadratic] form. 定値[半定値]～ definite [semi-definite] form. 整～ integral form. 極～ polar form. 保型～ automorphic form. 微分～ differential form. 積分二次～ quadratic integral form. 基本～ fundamental form; ground form. 同伴～ associated form. テンソル～ tensorial form. 曲率[接続]～ curvature [connection] form. エルミート～ Hermitian form. ヘッセ～ Hessian (form). パッフ～ Pfaffian (form). ～的ベキ級数 formal power series. ～的ベキ級数環 formal power series ring. ～的展開 formal expansion. ～解 formal solution. ～的次数 formal degree. ～群 formal group. ～環 formal ring; associated graded ring. ～的自己随伴な formally selfadjoint. ～主義 formalism. ～主義者 formalist.

けいじじょうがく 形而上学 metaphysics.
～的な metaphysical.

けいじてき 経時的(な) longitudinal; chronological, chronologic.
～データ longitudinal data.

けいしゃ 傾斜 inclination; slope; slant; gradient; obliqueness; tilting.
～した inclined; slant; gradient; oblique. ～的な tilting. ～角 angle of inclination; dip. ～面 slope; slant. ～度 gradient; pitch. ～法 gradient method. 共役～[勾配]法 conjugate gradient method. 上り～ acclivity. 下り～ declivity. ～複体 tilting complex. ～理論 tilting theory.

けいすう 係数 coefficient.
～ごとの[に] coefficientwise. 二項[多項]～ binomial [multinomial] coefficient. 微分～ differential coefficient. ～域 range of parameter. ～群[環; 体] coefficient group [ring; field]. ～行列 coefficient matrix. 変動～ coefficient of variation. 変位～ coefficient of displacement. 一致～ coefficient of concordance. 相関～ correlation coefficient. 主～ leading coefficient. 最高次の～ coefficient of highest degree. 最大～ maximal coefficient. ～問題 coefficient problem. 資本～ capital coefficient. 粘性～[率] coefficient of viscosity. 近交～ inbreeding coefficient. 結合～ incidence number. ～添加環 augmented [supplemented] algebra. 決定～ coefficient of determination.

けいすう 係数 scalar.
～環 ring of scalars. ～拡大 scalar extension. ～制限[変更] scalar restriction [change].

けいすう 径[係]数 parameter. ⇨パラメータ.
接続(の)～ parameter of connection. 正則～ parameter of regularity.

けいすう 計数 cardinal number; numeration; enumeration; counting.
～論的な cardinal. ～データ[値] enumeration [enumerated] data; discrete value. ～過程 counting process.

けいすうがた 計数型(の) digital. ⇨デジタル.
～計算機 digital computer.

けいせい 形成 formation; building.
～する form; build up.

けいせん 経線 longitude; meridian; circle of longitude.

けいぞく 継続 continuation.
符号の～(性) continuation [permanence] of sign.

けいど 傾度 inclination; slope; slant.

けいど 経度 longitude.
～の longitudinal.

けいとう 系統 system.
～的な systematic. 神経～ nervous system.

～(的)誤差 systematic error. ～(的)抽出法 systematic sampling. ～化 systematization.

けいめん 径面 diametral plane.

けいりょう 計量 metric.
～の metric; metrical. ～的な metrical. ～的に metrically. ～化 metrization. ～化可能な metrizable. ～ベクトル[線形]空間 metric vector space. ～テンソル metric tensor. ～的性質 metric property. ユークリッド[非ユークリッド]～ Euclidean [non-Euclidean] metric. リーマン～ Riemannian metric. ケーラー[ポアンカレ]～ Kählerian [Poincaré] metric. 極値的～ extremal metric. 支持～ supporting metric. 誘導～ induced metric.

けいりょう 計量 measuring; mensuration. ⇨測定.
～する measure.

けいりょう-…-がく 計量…学 -metrics.
～経済～ econometrics. ～生物～ biometrics. ～心理～ psychometrics.

けい-りろん *K* 理論 *K*-theory.

けいれつ 系列 sequence; series.
～の sequential; serial. 完全～ exact sequence. 補～ complementary series. 連続～ continuous series. 疎～ discrete series. ～相関 serial correlation.

けいろ 経路 course; route; path.
～積分 path-integral.

げか 外科 surgery.
～的な surgical

ゲージ guage.
～変換 gauge transformation.

けす 消す cancel; annihilate; eliminate.

けた 桁 place; figure.
三～の数 three place number; number of three figures; number of three ciphers. ～を誤まる misplace the figure.

けた- 桁
～落ち cancelling. ～溢れ overflow. 循環～上げ end-around carry.

けっか 結果 result; consequence; effect; event.
～の resultant. ～表 result table.

けつごう 結合 combination; composition; connection; association; incidence; joint.
～的な composite; associative; joint; conjoint. ～する combine; compose; connect; join. 一次[線形]～ linear combination. ～の原理 principle of combination. ～定理 composition theorem. ～法則[律] associative law. ～的代数 associative algebra. ～環 associative ring; association algebra. ～性 associativity. ～集合 associated set. ～の公理 incidence axiom. ～係数 incidence number. ～分布 joint distribution. ～確率変数 joint random variable.

けつじょ 欠如 lack; lacking.
管理～ lack of control.

けつじょ 欠除 deficiency.
～次数 deficiency. ～値 deficient value.

けっしょう 結晶 crystal.
～学 crystallography. ～の; ～学上の crystallographic. ～系 crystal system. ～類 crystal class. ～群 crystallographic group.

けっしょうせい 結晶性(の) crystalline.

けっすい 欠錐 truncate(d) cone.

けっせつじょう 結節状(の) nodal.
～領域 nodal region.

けっせつてん 結節点 node; nodal point; ordinary double point.
～の軌跡 node-locus.

けっそくち 欠測値 missing data; missing plot; missing value.

けったく 結託 coalition; collusion.

けってい 決定 determination; decision.
～する determine; decide. ～(論)的な deterministic. ～可能な determinable; decidable. ～条件 determinating condition. ～集合 determinating set. ～系 determined system. ～的ゲーム strictly determined game. ～関数 decision function. ～問題 decision problem; (*G.*) Entscheidungsproblem. ～空間 decision space. ～過程 decision process. ～版 final edition; definitive edition. ～方程式 determining equation. (指数)～方程式 indicial equation. ～係数 coefficient of determination. ～性チューリング機械 deterministic Tu-

ring machine.

けっていろんてき 決定論的 deterministic.
～モデル[過程] deterministic model [process].

けってん 欠点 defect; fault.
軽[重]～ minor [major] defect.

けつろん 結論 conclusion; consequence; consequent.
～する conclude.

ゲート gate. 【論理回路の一種】.

ゲーム game.
～をする play game. ～(の)理論 theory of games; game theory. ～の樹形図[樹木] game tree. 凹[凸]～ concave [convex] game. 零和～ zero-sum game. 協力～ cooperative game. 逐次～ sequential game. 双行列～ bimatrix game.

ケーラー E. Kähler (1906-2000).
～の Kählerian. ～計量[多様体] Kählerian metric [manifold].

ケーリー・ハミルトン Cayley-Hamilton.
～の公式 Cayley-Hamilton formula.

けん 圏 category.
～の categorical. 集合[群](の)～ category of sets [groups]. 双対[加法]～ dual [additive] category. 商～ quotient category. アーベル～ Abelian category. 部分～ subcategory. ～代数 categorical algebra. 導来～ derived category.

げん 元 element.
集合の～ element of set. 零～ zero element. 単位～ unit element; identity element. 逆～ inverse element. 可逆～ invertible element. 共役～ conjugate (element). 生成～ generating element. 代数的～ algebraic element. 超越的～ transcendental element. ～ごとの[に] elementwise. ～の位数 order of element. ねじれ～ torsion element.

げん 元 unknown.
一[二]～方程式 equation with one [two] unknown[s].

げん 弦 chord; subtense.
～の chordal. 補～ complement chord. 共通～ common chord. 焦[双焦]～ focal [bifocal] chord. ～距離 chordal distance.

げん 絃 string.
振動～ vibrating string.

げん 源 source.
熱[光]～ heat [light] source.

けんいん 牽引 traction.
～曲線 tractrix.

げんいん 原因 cause.
～の確率 probability of causes.

げんか 原価 cost price; prime cost.
～計算 cost accounting. ～管理 cost control.

げんか 現価 present worth.
正味～ net present worth.

げんかい 限界 bound; (*L.*) finis; limit; margin.
～を設ける bound. 検査～ check bound. 処置～ action bound. 誤差(の)～ limit of error. ～法線 limit normal. ～超球 limiting hypersphere. ～の marginal. ～検査 marginal checking. ～効用 marginal utility.

げんかく 厳格 strictness; rigor; tightness.
～な strict; rigorous; tight; sharp. ～抜取り検査 tightened sampling inspection.

けんきゅう 研究 research; study; investigation.
～する research; study; investigate. ～費 research expenses. ～員 research fellow. 運営～ operational research; operations research; OR. ～報告 (research) report; memoir; transactions. ～方法 approach.

けんきゅうしつ 研究室 laboratory; seminar.

けんきょくせん 犬曲線 curve of pursuit; tractrix.

げんきょり 弦距離 chordal distance.

げんけい 原型 prototype; model.

げんご 言語 language.
プログラム～ programming language. 計算機～ computer language. 目的～ target language. 正規～ regular language. 外部～ external language. 原始～ source language. 限定～ definite language. 文脈依存～ context-sensitive language. 文脈自由～ context-free language. ～の linguistic.

～(学) linguistics. 計算[数理]～学 computational [mathematical] linguistics. ～代数 linguistic algebra. ～工学 linguistic [language] engineering.

げんご 原語 source language; original language [word].

げんこう 原稿 manuscript; draft.

けんさ 検査 inspection; examination; test; check.

抜取り～ sampling inspection. つまみとり～ spot inspection. 受入[監査]～ acceptance [check] inspection. 管理[破壊]～ control [destructive] inspection. 全数～ 100% inspection; total inspection. 選別～ screening inspection. 平均～量 expected amount of inspection. ～特性曲線 operating characteristic curve; OC curve. ～ルーチン test routine; checking routine. ～ビット check bit. パリティ[奇偶]～ parity check. 冗長度～ redundancy check.

けんさく 検索 retrieval; search; reference.

情報～ information retrieval; IR.

けんさぞっこういき 検査続行域 zone of indifference.

けんざん 検算 proof; check; verification.

～する prove; check; verify. ～法 (method of) casting out.

げんし 原子 atom.

～の atomic. ～元 atomic element.

げんし 原始(的) primitive.

～関数 primitive; primitive function. ～根 primitive root. ～元 primitive element. ～多項式[方程式] primitive polynomial [equation]. ～解 primitive solution. ～イデアル primitive ideal. ～環 primitive ring. 1の～n乗根 primitive n-th root of unity. ～ベキ等元 primitive idempotent.

げんし- 原始 source.

～言語 source language. ～資料 source material.

けんしゅく 牽縮 retract. ⇨レトラクト.

けんしゅつ 検出 detection.

妨害～ interference detection.

けんしゅつりょく 検出力 power of test.

～関数 power function.

げんしょう 減少 decrease; decrement. ⇨増加.

～の decreasing. ～する decrease. ～関数 decreasing function. ～列 decreasing sequence. 単調～の monotone decreasing. 急～の rapidly decreasing. ～の状態 decreasing state. ～量 decrement.

げんしょう 現象 phenomenon (*pl.* -mena).

ギブズ～ Gibbs phenomenon. 自然[社会]～ natural [social] phenomenon.

げんじる 減じる subtract; diminish; decrease.

けんすい 懸垂 suspension.

～定理 suspension theorem. ～同型[形](写像) suspension isomorphism. 約～ reduced suspension.

けんすい- 懸垂

～線 catenary. ～面 catenoid.

げんすい 減衰 damp; damping; attenuation.

～させる damp. ～比 damp ratio; damping ratio. ～振動 damped oscillation [vibration]. ～歪み attenuation distortion. 対数～率 logarithmic decrement.

げんすう 減数 subtracter; subtrahend.

げんせん 原線 initial line; polar axis; primitive line.

げんせん 源泉 source.

げんぞう 原像 inverse image; preimage.

げんそく-ニュートン-ほう 減速ニュートン法 damped Newton method.

げんだい 現代(の) modern; present day; contemporary.

～数学 modern mathematics.

げんち 現地 field.

～作業 field work. ～指導員 field supervisor.

げんちょ 原著 original work [paper].

～者 author.

けんてい 検定 test; testing.

～関数 test function. ～統計量 test statistic. カイ二[自]乗～ chi-square test. 適合(度)～ test of goodness of fit. 符号～ sign test. 確率化された～ randomized test. スチューデント～ Student test. エフ～ F-test. 片[両]側～ one [two]-sided test. ミ

ニマックス〜 minimax test. 仮説(の)〜 test of statistical hypothesis. 不変[不偏]水準〜 invariant [unbiased] level test. 有意性(の)〜 significance test. 棄却〜 rejection test. 一致〜 consistent test. 逐次〜 sequential test. 正確〜 exact test. 二標本〜 two-sample test.

げんてい 限定 restriction; limitation; definition.

〜された restricted, limited; definite. 〜する restrict; limit; specify. 〜型 restricted type. 〜言語 definite language. 〜記号[作用素] quantifier. 〜版 limited edition.

げんてい 限定 bound.

分枝〜 branch and bound.

げんていさようそ 限定作用素 quantifier.

げんてん 原点 origin.

げんてん 限点 limiting point(s).

げんぴょう 原表 original table.

げんぶん 減分 decrement. ⇨増分.

げんぽう 減法 subtraction.

げんみつ 厳密(性) strictness; exactness, exactitude.

〜な strict; close; exact; precise; true. 〜に strictly; in the strict sense; exactly. 〜な含意 strict implication. 〜でない inexact.

げんり 原理 principle.

形式[算法]不易の〜 principle of conservation of form [calculation]. 双対〜 duality principle; principle of duality. エネルギー〜 energy principle. 相対性〜 principle of relativity. 鏡像の〜 principle of reflection; reflection principle. 最大(絶対)値の〜 maximum (modulus) principle. 偏角の〜 argument principle. 連続性〜 continuity principle. 包除〜 principle of inclusion and exclusion. 重ね合わせの〜, 重畳〜 principle of superposition. 最適性の〜 principle of optimality. 代入の〜 substitution principle.

げんりょうきごう 限量記号 quantifier.

げんりょうし 限量子 quantifier.

げんろん 原論 Elements.

こ・コ

こ 弧 arc.
優[劣]～ major [minor] arc. 円弧 circular arc. 弓形の～ arc of segment. 共役～ conjugate arc. 開(いた)[閉(じた)]～ open [closed] arc. 連続～ continuous arc. 正則[解析(的)]～ regular [analytic] arc. 単純～ simple arc. ジョルダン～ Jordan arc. 測地～ geodesic arc. ～長 arc length. ～(長)要素 arc element. ～状に arcwise. ～状連結の arcwise connected.

ご 五 five; V; v; pentad.
～- penta-; quinqu(e)-. ～の quinary. ～倍(の) five times; quintuple. ～心 five centroids. ～角形 pentagon. ～角形の pentagonal; quinquangular. ～角数 pentagonal number. ～辺形 pentagon. ～辺形の pentagonal. ～面の pentahedral. ～面体 pentahedron (*pl.* -rons, -dra). ～面体の pentahedral. ～球座標 penta(-)spherical coordinates. ～点形 quincunx. 5数要約 five number summary. 5補題, 5項補題 five lemma.

ご 語 word; term.
～の問題 word problem. ～長 word length. 符号～ code word.

こう 項 term.
定数～ constant term. 同類～ like [similar] terms. 初[末]～ initial [last] term. 剰余～ remainder term. 誤差～ error term. 高位の～ term of higher order. ～別の[に] termwise. ～別微[積]分 termwise differentiation [integration]. 二～関係 binary relation. 二[三; 多]～式 binomial [trinomial; polynomial] (expression). 二[多]～定理 binomial [multinomial] theorem.

こう- 公 common.
～約数[元] common divisor; common measure. ～倍数[元] common multiple. ～分母 common denominator. ～差 common difference. ～比 common ratio.

-こう 項 -ary; -tuple.
二～関係 binary relation. n～関係 n-ary relation. n～ベクトル n-tuple vector.

ごう 号 number; (*L.*) numero; No; no. (*pl.* Nos., nos.).

こうい 高位の higher; of high(er) order.
～無限大 higher order infinity; infinity of higher order. ～無限小 higher order infinitesimal; infinitesimal of higher order.

こうえつ 校閲 ⇨校訂.

こうえん 講演 address; lecture.

こうか 効果 effect; result.
～的な effective. ～的に effectively. 乗数～ multiplier effect. 処理～ treatment effect. ブロック～ block effect. 変量～ random effect. 緩衝～ buffer effect. ～的に作用する act effectively.

こうか 降下 descent; fall; drop.
最急～線 line [curve] of steepest descent. 最速～線 line of swiftest descent. 最短～線 brachistochrone. 最急～法 method of steepest descent; steepest descent method. 最急～ steepest descent. 最速～ swiftest descent. ～方向 descent direction.

こうかい 高階 high(er) order.
～導関数 derivative of higher order; higher order derivative. ～差分 difference of higher order; higher order difference. ～微分方程式 differential equation of higher order; higher order differential equation.

〜述語論理 predicate logic of higher order.

こうがい 梗概 outline; summary; synopsis (*pl.* -ses).

〜を述べる outline; summarize.

こうかいかぎあんごうけい 公開鍵暗号系 public key cryptosystem.

こうかいせん 航海線 loxodrome; loxodrome curve; loxodromic curve; rhumb line.

こうかく 交角 angle of intersection; intersectional angle.

こうがく 工学 engineering; technology.

人間〜 human engineering.

こうがく 光学 optics.

〜の optical. 〜器械 optical instrument. 〜的距離 optical distance. 幾何〜 geometric optics. 波動〜 wave optics.

ごうかく 合格 pass, passing; acceptance; success; eligibility.

〜する be accepted; succeed in; pass; be found eligible [qualified]. 〜判定個数 acceptance number. 〜品質水準 acceptable quality level; AQL. 〜線 acceptance line. 〜域 zone of preferance for acceptance.

こうかん 交換 exchange; interchange.

〜する exchange; interchange. 〜法 sorting by exchanging.

こうかん 交換 commutation; permutation. ⇨可換(な).

〜的な commutative. 〜可能な commutable; permutable. 〜する commute; permute. 〜法則 commutative law. 〜子 commutator. 〜関係 commutation relation.

こうかんし 交換子 commutator.

〜群 commutator subgroup.

こうぎ 講義 lecture.

〜ノート lecture note.

こうぎ 広義(に) in the wider [generalized] sense.

〜(の)一様収束 uniform convergence in the wider sense; generalized uniform convergence. 〜積分 improper integral.

こうきてん 交軌点 node. ⇨ノード.

〜の nodal.

こうきゅうかんすう 縞球(調和)関数 tesseral harmonics.

ごうけい 合計 sum total; total (amount).

〜する sum [add] up; total. 〜して in amount; aggregately; in the aggregate.

こうげん 光源 light source; source of light.

こうご 交互(の) alternate; alternating; mutual.

〜に alternately. 〜近似 alternative approximation. 〜方向法 alternating direction iteration; ADI.

こうこう 公項 general term.

こうこう 後項 consequent.

こうこうさ 光行差 aberration.

日周〜 diurnal aberration.

こうごさよう 交互作用 interaction.

二因子〜 simple [two factor] interaction.

三因子〜 three-factor interaction.

こうさ 公差 common difference; tolerance.

こうさ 交差[叉] intersection; crossing.

〜する intersect; cross; secant. 有限〜性 finite intersection property. 完全〜 complete intersection. 〜積 cross product; intersection product. 〜比 cross ratio. 〜三円筒のなす面 groin. ⇨穹稜.

こうさ 降鎖 descending chain.

〜条件 descending chain condition; d.c.c.

こうさつ 考察 consideration; examination; discussion.

〜する consider; examine; discuss.

こうさん 公算 probability; likelihood. ⇨確率; 尤度.

〜比 likelihood ratio. 〜関数 likelihood function.

こうし 格子 lattice.

〜点 lattice point. 〜点公式 lattice-point formula. 単[複]〜 simple [double] lattice. 四重〜 quadruple lattice. 空間〜 space lattice. 立方〜 cubic lattice. 双対〜 dual lattice. 臨界〜 critical lattice. 〜群 lattice group. 〜分布 lattice distribution. 〜方格(法) lattice square (method). 〜計画 lattice design. 底心[体心; 面心]〜 base [body; face]-centred lattice.

こうし 講師 lecturer; instructor.

専任～ full-time lecturer; instructor. 非常勤～ part-time lecturer.

こうじ　高次 higher degree; higher order. ⇨高階.
～方程式 equation of higher degree. ～導関数 derivative of higher order.

こうしき　公式 formula (*pl.* -las, -lae).
加法[乗法]～ addition [multiplication] formula. 正弦～ sine formula. 展開～ expansion formula. 補間～ interpolation formula. 漸化～ recurrence [recursion; recurring; recurrent; recursive] formula. 積分～ integral formula. X氏の～ formula of X; X's formula. ～化 formulation. クラメルの～ Cramer's formula. 明示～ explicit formula.

こうじく　交軸 transverse axis.

こうじく　光軸 optical axis.

こうしじま　格子縞 plaid; cross stripes.

こうしゃ　後者 successor; the latter.

こうじゅん　公準 postulate.
第五～ fifth postulate.

ごうじょひのり　合除比の理 componendo and dividendo.

こうしん　更新 renewal.

こうしん　後進(の) backward.
～差分 backward difference. ～方程式 backward equation.

こうず　構図 composition.
～する compose.

こうすう　恒数 constant.

こうせい　構成 construction; formation; composition; constitution.
～的な constructive; compositive. ～する construct; compose; constitute. ～可能な constructible. ～的方法 constructive method. ～的順序数 constructive ordinal. ～規則 formation rule. ～要素 component; constituent (element).

こうせい　校正 proofreading; revisal.
～する read [correct] proofs; revise. ～刷 proof sheet. ～済みの corrected. ～者 reviser.

ごうせい　合成 composition; convolution.
～の composite; compound. ～数 composite number. ～命題 compound statement [proposition]. ～体 composite field. ～関数 composite function; composed function. ～写像 composed mapping; composite mapping. 二次～ secondary composition. ～変換 convolution (transformation); composite transformation. ～積 convolution; (*G.*) Faltung.

ごうせい　剛性 rigidity; rigidness.

ごうせいせき　合成積 convolution; (*G.*) Faltung.

こうせいてき　構成的 constructive.
～順序数[方法] constructive ordinal [method].

こうせいど　高精度 high precision.
～計算 high precision computation.

こうせいようそ　構成要素 component; constituent (element).

こうせき　航跡 wake; furrow; track.

こうせつ　交切 intersection.

ごうせつ　合接 conjunction.
～命題 conjunction [conjunctive] proposition.

こうせん　交線 line of intersection.

こうせん　光線 ray; light.
反射～ reflected ray. 屈折～ refracted ray. 近軸～ paraxial ray.

こうぞう　構造 structure.
～(上)の structural. 網目～ network structure. 可微分[微分可能]～ differentiable structure. ～定理 structure theorem. ～群 structure group. ～空間 structure space. ～分析 structure analysis. ～安定性 structural stability. ～方程式 structural equation; structure equation. ～式 equation of structure. ～定数 structure constant; structural constant. ～射 structure morphism. 複素[概複素]～ complex [almost complex] structure. 解析～ analytic structure. 可微分[平滑]～ differential [smooth] structure. 接触～ contact structure. 葉層～ foliated structure; foliation. ログ～ logarithmic structure.

こうそく　拘束 restraint; restriction; constraint. ⇨束縛.

こうたい 交代 alternation; interchange.
～の alternate; alternating; alternant; skew-symmetric. ～式 alternating expression; alternate expression; alternant; alternating polynomial. ～多項式 alternating polynomial. 基本～式 elementary alternating expression. 基本～(多項)式 elementary alternating polynomial. ～級数 alternating series. ～演算 alternating operation. ～微分 alternating derivation. ～形式 alternating form. ～行列 skew-symmetric matrix; alternating matrix; alternant. ～群 alternating group. ～体 alternative field. ～代数 alternative algebra. ～テンソル skew-symmetric tensor; alternating tensor. ～結び目 alternating knot. ～化作用素 alternizer.

こうたい 後退(の) backward; backspace.
～差分 backward difference. ～移動平均 backward moving average. ～方程式 backward equation. ～段階 backward step. ～代入 backward substitution.

ごうたい 剛体 rigid body.
～運動 rigid motion.

こうちゅうしんれつ 降中心列 descending central series; lower central series.

こうつう 交通 traffic; transportation.
～量 (volume of) traffic. ～(量)調査 traffic survey.

こうてい 工程 process; amount [rate] of work.
生産～ manufacturing process.

こうてい 肯定 affirmation.
～的な affirmative. ～的に affirmatively. ～する affirm. ～命題 affirmative proposition.

こうてい 校訂 revision. ⇨校正.
～する revise. ～版 revised edition.

こうていしき 肯定式 (*L.*) modus ponens.
否定～ modus tollendo ponens.

こうていせん 航程線 rhumb.

こうてき 好適な suitable; nice.
～関数 nice function.

こうてつのり 更迭の理 alternendo.

こうてん 交点 point of intersection; intersection point.
～数 intersection number. ～定理 intersection theorem.

こうてんてき 後天的(に) (*L.*) a posteriori.

こうど 高度 height; altitude; level.

こうとう- 恒等-
～(的)な identical. ～的に identically. ～式 identity; identical equation. ～変換 identical transformation; identity transformation. ～写像[関数] identity mapping [function]. ～置換 identity permutation. ～元 identity element. ～作用素 identity operator. ～的に等しい identically equal.

こうとう 高等(な) higher; advanced.
～数学 higher mathematics; advanced mathematics. ～微(分)積分学 advanced calculus. ～超越関数 higher transcendental function.

こうとう 口頭(の) oral; verbal.
～で orally; verbally.

こうどう 行動 action; behavio(u)r; movement.
～空間 action space. ～科学 behavio(u)ral science.

こうどう 黄道 ecliptic.

ごうどう 合同 congruence.
～な congruent; congruous. ～性 congruity. ～(の)公理 axiom(s) of congruence. ～な図形 congruent figures. ～変換 congruent transformation. ～部分群 congruence subgroup. ～方程式 congruence equation. ～式 congruence expression. ～式ゼータ関数 congruence zeta [ζ-] function. アフィン～な affinely congruent. 乗法～な multiplicatively congruent.

こうはい 交配 crossover; crossbreeding.
～法 crossover design. 近親～ inbreeding.

こうばい 勾配 slope; gradient.
～関数 slope function; gradient function. ～ベクトル gradient vector. 簡約～ reduced gradient. 共役～[傾斜]法 conjugate gradient method. ～法 gradient method. 縮約～ reduced gradient. 劣～ subgradient.

こうばいすう[げん] 公倍数[元] common mul-

tiple.
最小～ least [lowest] common multiple; L.C.M.
こうはん 広範な extensive.
こうばん- 交番 ⇨交代.
～級数 alternating series. ～二進コード reflected binary code.
こうひ 公比 common ratio.
ごうひのり 合比の理 componendo.
こうふく 広幅(の) platykurtic.
～[幅ひろ; 緩尖]分布 platykurtic distribution.
こうふん 興奮 excitement; stimulation.
～させる excite; stimulate.
こうぶんぼ 公分母 common denominator.
こうへい 公平な fair.
～ゲーム fair game.
こうべき 降ベキ descending powers.
～の順 descending order of powers.
こうべつ 項別
～の; ～に termwise. ～微分 termwise differentiation. ～積分 termwise integration.
こうほう 航法 navigation.
天測～ celestial navigation.
こうほう 後方(へ)の backward.
こうもく 項目 item; head; heading.
～反応理論 item response theory.
こうやくすう[げん] 公約数[元] common divisor; common measure.
最大～ greatest common measure [divisor]; G.C.M. [G.C.D]. ～のない incommensurable; relatively prime.
こうよう 効用 utility; effect.
限界～ marginal utility. 期待～ expected utility. ～関数 utility function.
こうらく 交絡 confounding.
～させる confound. 部分～した partially confounded.
こうり 公理 axiom.
～の; ～的な axiomatic. 普通[普遍]～ common axiom. ユークリッドの～ Euclidean axiom(s). 平行線(の)～ axiom of parallels; axiom of parallel lines. 合同[連続性](の)～ axiom(s) of congruence [continuity]. 順序～ axiom(s) of order. 選択[出]～ axiom of choice. 可算～ countability axiom. ～系 system of axioms. ～的集合論 axiomatic set theory. ～的方法 axiomatic method. ～化する axiomatize. ～化 axiomatization. ～主義 axiomatism.
こうりつ 効率 efficiency.
～因子 efficiency factor.
こうりつ 後率 consequent.
ごうりてき 合理的な rational; reasonable.
～群 rational subgroup.
ごうりゅう 合流 confluence.
～した[の] confluent. ～型の of confluent type. ～超幾何関数 confluent hypergeometric function.
ごうりょく 合力 resultant; resultant force.
こえる 越[超]える exceed.
越[超]えて beyond; over.
コーエン I. S. Cohen (1917-1955).
～・マコーレー環 Cohen-Macaulay ring.
コーエン P. J. Cohen (1934-2007)
ごがく 語学 language study; linguistics.
～の linguistic. ～者 linguist; philologist.
ごかん 互換(性) transposition; interchange; interchangeability.
～的な interchangeable.
こきょくせん 弧曲線 tractrix.
こくさい 国際(的な) international.
～会議 international conference; international congress. ～記号 international notation.
ここ 個々(の) individual; separate.
～に individually; separately.
こころみ 試み trial; test; experiment; attempt.
～る try; make trial; test; attempt.
ごさ 誤差 error; residual.
～(の)伝播 propagation of errors; error propagation. ～(の)限界 limit of error. ～平方和 error sum of squares. 平均二乗[平方]～ mean square error. 相対～ relative error. 絶対～ absolute error. 実効～ effective error. 離散化～ discretization error. 打切り～ truncation error. 丸め～ round-off error; rounding error. 計算～ computational error. 器械～ instrumental

error. (原因)不明～ residual error. 偶然～ accidental error. 確率～ probability error. 累積～ cumulative error; accumulated error. 抽出～ sampling error. ～論 theory of errors. ～関数 error function. ～空間 error space. ～項 error [residual] term. ～定数 error constant. 平均絶対～ mean absolute error.

コサイン cosine; cos. ⇨余弦.

ごさん 誤算 miscalculation; miscount.
～する miscalculate; miscount.

コーシー A. L. Cauchy (1789-1857).
～・リーマンの方程式 Cauchy-Riemann equation. ～[基本]列 Cauchy [fundamental] sequence.

ごじ 五次(の) quintic.
～関数 quintic (function). ～(代数)曲線 quintic curve. ～方程式 quintic equation.

コシグマ cosigma.
～関数 cosigma function.

ごじゅう 50, fifty; L.; l.

ごじょ 互除 mutual division.
～法 method of mutual division; Euclidean algorithm.

こしょう 故障 hindrance; obstacle; trouble; damage; fault.
～時間 downtime; fault time.

こじょう 弧状(の) arcwise.
～に arcwise. ～連結の arcwise connected.

ごしん 五心 five centroids.
三角形の～ five centroids of triangle.

ごしんほう 五進法 quinary (scale).

こすう 個数 number (of articles).
～関数 counting function; (*G.*) Anzahlfunktion. ～測度 counting measure.

コスツル, コジュール, コシュール, コスール J-L. Koszul (1921-).
～複体 Koszul complex.

コセカント cosecant; cosec. ⇨余割.

こそんりつ 呼損率 loss probability.

こたい 個体 individual.
～の individual.

こだい 古代(の) ancient.
～数学 ancient mathematics; mathematics of antiquity.

こたえ 答 answer; reply; result; solution. ～る answer; reply.

コタンジェント cotangent; cot. ⇨余接.

こちょう 弧長 arc length.
アフィン(的)～ affine arc length. ～媒介変数 arc length parameter. ～要素 arc element.

こっかく 骨格 skeleton.
単体の～ skeleton of simplex.

こてい 固定(の) fixed; rigid.
～する fix. ～体 fixed field. ～群 fixed group. ～効果 fixed effect. ～点 fixed point; fix point. ～小数点(表示) fixed point (representation.) ～尺 fixed scale.

こていぶぶんぐん 固定部分群 stabilizer.

こてん 古典 classics.
～的な classic; classical ～解析学 classical analysis. ～群 classical group. ～リー代数 classical Lie algebra. ～的理論 classical theory. ～論理 classical logic.

こど 弧度 circular measure; radian; rad.
～法 (method of) circular [radian] measure; radian system.

コード code.
二進～ binary code. 三増し～ excess-three code.

ことなる 異なる differ from; different; distinct.
ゼロと～ different from zero. (相)～根 distinct roots. 相～ pairwise distinct.

コノイド conoid. ⇨擬円錐.
正～ right conoid.

ごはさん 御破算 clearance; resetting.
～する clear; reset.

コピュラ copula.

ごぶごぶ 五分五分 even break; fifty-fifty.
～の equally [even-] matched.

コベクトル covector.
～場 covector field.

こべつ 個別(の) individual; item-by-item; particular. ⇨各個(の)(かっこ).
～エルゴード定理 individual ergodic theorem. ～的危険論 individual risk theory. ～逐次検査 item-by-item sequential inspection.

コホモトピー cohomotopy.
〜群 cohomotopy group.

コホモロジー cohomology.
〜の cohomological. 〜類 cohomology class. 〜群 cohomology group. 〜環 cohomology ring. 〜代数 cohomology algebra. 〜完全系列 exact cohomology sequence. 〜次元 cohomological dimension. 〜関手 cohomological functor. エタール・〜 étale cohomology. クリスタル〜 crystalline cohomology. チェックの〜 Čech cohomology.

コボルダント cobordant. ⇨同境.

コボルディズム cobordism. ⇨同境.

こまか 細か(い) fine.
〜な類別 fine classification.

こゆう- 固有(な) proper; characteristic; eigen-.
〜円錐曲線 proper conics. 〜二次曲面 proper quadrics; proper surface of second degree. 〜回転 proper rotation. 〜n次元空間 properly n-dimensional space. 〜値 eigenvalue; eigen value; proper value; characteristic value; latent root. 〜根 characteristic root; primitive root. 〜ベクトル eigenvector; proper vector. 〜元 eigenelement. 〜関数 eigenfunction; characteristic function. 〜多項式 eigenpolynomial; characteristic polynomial. 〜(値)方程式 characteristic equation; proper equation; secular equation. 〜空間 eigenspace. 〜振動 eigenoscillation. 〜解 proper solution; primitive solution. 〜射 proper morphism. 〜改変 proper modification.

こゆうち 固有値 eigenvalue; eigen value; characteristic value; proper value; latent root.
〜問題 eigenvalue problem. 重複〜 multiple eigenvalue.

こゆうわ 固有和 trace. ⇨跡(せき).

コラム column.

こりつ 孤立 isolation.
〜の isolated. 〜させる isolate. 〜順序数 isolated ordinal number. 〜点 isolated [isolate] point; acnode. 〜集合 isolated set. 〜特異点 isolated singularity [singular point].

こりつした 孤立した sporadic; isolated.

こりょう 呼量 traffic intensity.

コルニュ M. A. Cornu (1841–1902).
〜の螺線 Cornu's spiral; clothoid.

ころがる 転がる roll; turn.

こわす 壊す break; destroy. ⇨破壊.
壊れる break; be broken.

こん 根 root; radical.
〜号 root [radical] sign. 平方〜 square root. 立方〜 cubic root; cube root. ベキ〜 power root; radical (root). 累乗〜 radical root. 方程式の〜 root of equation. 単〜 simple root. 二[三]重〜 double [triple] root. 重(複)〜 multiple root. 実[虚]〜 real [imaginary] root. 正[負](の)〜 positive [negative] root. 共役〜 conjugate root. 原始〜 primitive root. 特性〜 characteristic root. 〜系 root system. 〜の分離 separation of roots. 〜軸 radical axis. 〜数 radical number. 1の原始n乗〜 primitive n-th root of unity. 共通〜 common root.

こん- 混 mixed.
〜循環小数 mixed recurring decimal. 〜群 mixed group.

こんき 根基 radical.
ベキ零〜 nilpotent radical. ベキ零元〜 nilradical.

こんきょ 根拠 basis; reason; ground; foundation; cause.

こんぐん 混群 mixed group; (G.) Mischgruppe.

こんげんじしょう 根元事象 elementary event; fundamental event.

コンコイド conchoid.
〜の conchoidal. 〜曲線 conchoidal curve; generalized conchoid.

こんごう 根号 radical sign; root sign.

こんごう 混合 mixing; mixture; amalgamation; blending; merging.
〜の mixed. 〜する mix; blend. 〜型 mixed type. 〜偏導関数 mixed partial derivative. 〜境界値問題 mixed boundary

value problem. ～スピノル mixed spinor. ～テンソル mixed [hybrid] tensor. ～模型 mixed model. ～優[劣]ゲーム mixed dominant [recessive] game. ～戦略[術策] mixed strategy. ～性 mixing property. ～問題 mixing [mixed] problem; blending problem. ～(アーベル)群 mixed group. ～分布 mixture distribution.

こんごう 混合(の) hybrid.
～テンソル hybrid tensor.

こんごうけい 混合形(の) hybrid.

こんごうほう 混合法 alligation.

こんせい 混成(の) hybrid.

コンテンション contention.

こんどう 混同 confusion.
～する confuse; confound.

こんとん 混沌 chaos.

コンパクト compact.
～な compact. ～開位相 compact-open topology. ～集合 compact set; compactum (*pl.* -ta). ～空間 compact space. ～型 compact type. ～化 compactification. 一点～化 one-point compactification. 可解～化 resolutive compactification. 局所[可算]～な locally [countably] conpact. 点列～な sequentially compact.

コンパス compass (通例 *pl.* -es).

コンピュータ computer. ⇨計算機.
～科学 computer science.

こんぽんてき 根本的 basic; fundamental; radical; ultimate.

こんらん 混乱 confusion; disorder.
～する be confused [disordered].

さ・サ

さ 差 difference.
～群 difference group. ～積[商] difference product [quotient]. ～集合 difference set. 対称～ symmetric difference. 階～[～分] difference. 公～ common difference.

さ 鎖 chain.
～複体 chain complex. ～写像 chain mapping. ～変換 chain transformation. ～ホモトピー chain homotopy. 有限[無限]～ finite [infinite] chain. 零～ zero chain. 昇[降]～ ascending [descending] chain. ～状連結の well-chained. ～線 chained line.

ざ 座 place.

さい 賽 die (*pl.* dice).
乱数～ random die.

さい 差異 difference; distinction.
有意な～ significant difference [distinction].

さい- 細 fine.
～位相 fine topology. ～収束 fine convergence.

さいあくけいさんりょう 最悪計算量 worst-case complexity.

さいいそう 細位相 fine topology.

さいかくち 最確値 most probable value.

さいかん 最簡(の) simplest.
～交代式 simplest alternating polynomial [expression]; difference product. ～直交多項式 simplest orthogonal polynomial.

さいき 再帰 recurrence; recursion. ⇨漸化.
～の[的な] recursive; recurrent; recurring; reflexive. ～的に recursively. ～的定義 recursive definition. ～式 recursive formula. ～的関数 recursive function. ～的集合 recursive set. ～定理 recurrence theorem; recursion theorem. ～的に可算な recursively enumerable. ～連鎖 recurrent chain. ～事象 recurrent event. ～関係[公式] recurrence relation [formula]. 正[零]～の positive [null] recurrent. 非～の transient.

さいきほう 再帰法 recursion.

さいきゅうこうかせん 最急降下線 line [curve] of steepest descent.

さいきゅうこうかほう 最急降下法 method of steepest descent; steepest descent method.

さいきんぼうほう 最近傍法 nearest neighbor method.

サイクライド cyclide.
共焦～ confocal cyclides.

サイクル cycle.
～・タイム cycle time.

サイクロイド cycloid.
～の cycloidal. 内～ hypocycloid. 外～ epicycloid. ～振子 cycloid(al) pendulum.

さいげん 際限のない endless; limitless; unlimited; boundless.

ざいこ 在庫 stock.
～品 inventory; stocks. ～管理 inventory control; inventory management; stock management. 安全～ safety stock. 緩衝用～ buffer inventory. ～回転率 turn-over of inventories.

さいこう 最高(の) highest.
～次[階]の項 term of highest degree [order]. ～次(の)係数 leading coefficient. ～の重み highest weight.

さいさ 歳差 precession.

〜運動 precession (of equinoxes).

さいさいてきか 再最適化 postoptimization.

さいしゅう 最終(の) last; final; terminal.

〜定理 last theorem. 〜乗式 last multiplier. 〜決定 final [terminal] decision.

さいしゅうそく 細収束 fine convergence.

さいしょ 最初(の) first; primal.

さいしょう 最小 minimum (*pl.* -ma, -mums); min.; absolute minimum.

〜な least; minimal; smallest. 〜にする minimize. 〜値 minimum value. 逐次〜 successive minima. 〜公倍数 least common multiple; L.C.M. 〜上界 least upper bound; sup.; l.u.b. 〜解 minimal solution; minimum solution. 〜多項式 minimal polynomial. 〜二[自]乗法 method of least squares. 〜二[自]乗推定量 least square estimator. 〜カイ二[自]乗推定量 minimum chi-square estimator. 〜化列 minimizing sequence. 〜化 minimization. 〜カット minimum cut. 〜分解体 minimal splitting field.

さいしょうかい 最小解 minimizer.

局所的〜 local minimizer. 大域的〜 global minimizer.

さいしょうきょくげん 最小極限 ⇨下極限.

さいしょうこうばいすう 最小公倍数 least common multiple; L.C.M.

さいしょうさいだい 最小最大 minimax. ⇨ミニマックス.

〜原理 minimax principle.

さいしょうじじょう 最小自乗 ⇨最小二乗.

さいしょうにじょう 最小二乗 least square.

〜法 method of least squares. 間接〜法 method of indirect least squares. 二段階〜法 method of two-stage least squares. 〜近似 least square approximation. 〜推定量 least square estimator. 反復重み付き〜法 iteratively reweighted least squares.

さいしょく 彩色(の) chromatic.

〜数 chromatic number.

さいせい 再生 renewal.

〜過程 renewal process.

さいせい 再生 reproduction.

〜の[的な] reproducing. 〜する reproduce. 〜核 reproducing kernel. 〜性 reproducibility; reproductivity; reproducing property. 〜精度 exactness [precision] of reproduction.

さいぜん 最善の optimal.

さいそくこうかせん 最速降下線 line of swiftest descent. ⇨最短降下線.

さいだい 最大 maximum (*pl.* -ma, -mums); max.; absolute maximum.

〜な greatest; maximal; largest. 〜にする maximize. 〜値 maximum value; maximal value. 〜公約数 greatest common measure [divisor]; G.C.M.; G.C.D. 〜下界 greatest lower bound; inf.; g.l.b. 〜値(の)原理 maximum (modulus) principle. 〜解 maximal solution; maximum solution. 〜作用素 maximal operator. 〜項 maximal term; maximum term. 〜変形率 maximal dilatation. 〜不足数 maximal deficiency. 〜尤度 maximum likelihood. 〜階数 maximum rank; full rank. 〜化列 maximizing sequence. 〜化 maximization. 〜マッチング maximum matching. 〜流, 〜フロー maximum flow.

さいだいかい 最大解 maximizer.

局所的〜 local maximizer. 大域的〜 global maximizer.

さいだいきょくげん 最大極限 ⇨上極限.

さいだいこうやくすう 最大公約数 greatest common measure [divisor]; G.C.M.; G.C.D.

さいだいさいしょう 最大最小 maximin; max-min. ⇨マクシミン.

〜原理 maximin [max-min] principle.

さいたく 採択 acceptance.

〜する accept. 〜できる acceptable. 〜域 acceptance region.

さいたん 最短(の) shortest; minimal.

〜距離 shortest distance. 〜路 shortest path. 〜表現 shortest representation.

さいたんこうかせん 最短降下線 brachistochrone. ⇨最速降下線.

さいだんせん 裁断線 traverse. ⇨横断線.

さいちょうさ 再調査 re-examination; reinvestigation; follow-up; call back.

さいてき 最適 optimum (*pl.* -ma, -mums).

〜な optimal. 〜性 optimality. 〜化 optimization. 〜解 optimal [optimum] solution. 〜軌道 optimal locus. 〜計画(法) optimum programming. 〜問題 problem of optimization. 〜停止問題 optimal stopping problem. 〜停止時刻問題 optimal stopping-time problem. 〜推定量 optimum estimate. 〜在庫 optimal inventory. 〜制御 optimal control. 〜政策 optimal policy. 〜戦略 optimal strategy. 〜配分[割当] optimal allocation. 〜線形予報 optimal linear prediction. 〜性の原理 principle of optimality. 条件つき〜化 constrained optimization. 組合せ〜化 combinatorial optimization. 〜化理論 optimization theory. 〜性条件 optimality condition. 制約付き〜化 constrained optimization. 大域的〜化 global optimization. 離散〜化 discrete optimization.

さいてきか 最適化 optimization.
再〜 postoptimization. 〜理論 optimization theory. 制約付き〜 constrained optimization. 組合せ〜 combinatorial optimization. 大域的〜 global optimization. 離散〜 discrete optimization.

さいてきせい 最適性 optimality.
〜条件 optimality condition. 〜の原理 principle of optimality.

さいてん 臍点 ⇨臍点(せいてん).

さいにゅう 再入 reentering.

さいはいれつ 再配列 rearrangement.
〜する rearrange.

サイバネティックス cybernetics.

さいひんち 最頻値 mode. ⇨モード.

さいぶ 細部 detail(s).
〜測量 detail survey.

さいぶん 細分 refinement; subdivision.
〜する refine; subdivide. 〜定理 refinement theorem. 重心〜 barycentric subdivision. 星型[状]〜 star-refinement.

さいへん 細片 patch; small piece.

さいぼう 細胞 cell. ⇨胞体.
〜の cellular. ジョルダン〜 Jordan cell [block].

さいもく 細目 detail(s).

さいゆう 最尤
〜の maximum likelihood. 〜推定法 maximum likelihood estimation. 〜推定量 maximum likelihood estimator. 〜復号法 maximum likelihood decoding.

さいりょう 最良 best.
〜な best (possible). 〜近似 best approximation. 〜評価 best (possible) estimation. 〜不変推定量 best invariant estimator. 〜棄却域 best critical region. 〜漸近正規 best asymptotically normal; BAN. 〜線形不偏推定量 best linear unbiased estimator; BLUE.

さいりょう 最良 optimum. ⇨最適.
〜な optimal. 〜線形予報 optimal linear prediction. 〜線形予報値 optimal linear predictor.

サイン sine; sin. ⇨正弦.

さきいれさきだし 先入れ先出し first-in first-out; FIFO.

さぎょう 作業 work; operation; job.
〜計画 schedule. 職種機種別〜計画 job-shop scheduling.

さく 朔 crescent; new moon.

さくいん 索引 index (*pl.* -dexes, -dices).
〜をつける index. 著者〜 author index. 事項〜 subject index.

さくかんすう 柵関数 barrier.

さくじょ 削除 deletion; elimination; excision.
〜する delete; eliminate; excise.

さくず 作図 construction; drawing.
〜の constructive. 〜する construct; draw. 〜公準[の公法] postulate for construction. 〜(問)題 construction problem; problem for [of] construction. 〜不(可)能な of impossible construction. 〜可能な of possible construction; constructible. 近似〜 approximate construction. 定規とコンパスによる〜 ruler and compass construction.

さけめ 裂け目 slit; break.

ささえ 支え support; carrier.
〜る support.

さしょう 差商 difference quotient.

させき 差積 difference product. ⇨最簡(の).

させん 鎖線 chained line.

さだまった 定まった definite; constant; fixed.

ざつおん 雑音 noise.
～のある noisy. ～のない noiseless. ～のある情報路 noisy channel. 白色～ white noise.

さっかく 錯角 alternate interior angle(s).

ざっしゅ 雑種 hybrid.
～の hybrid. ～培殖 hybridization.

さどく 査読 referee; review.
～する review. ～者 referee; reviewer.

サービス service.
集団～ bulk service.

ざひょう 座標 coordinate.
～系 coordinate system; system of coordinates. ～軸[平面] coordinate axis [plane]. デカルト～ Cartesian coordinates. 第 i ～ ith coordinate. 直角～ rectilinear coordinates. 平行～ parallel coordinates. 斜交～ oblique [skew] coordinates. 極～ polar coordinates. 多極～ multipolar coordinates. 曲線～ curvilinear coordinates. 平面～ plane coordinates. 円柱～ cylindrical coordinates. 楕円[楕円面]～ elliptic [ellipsoidal] coordinates. 放物線[面]～ parabolic [paraboloidal] coordinates. 同[斉]次～ homogeneous coordinates. 正規～ normal coordinates. 直交～ orthogonal coordinates. 重心～ barycentric coordinates. 等温～ isothermal coordinates. 測地～ geodesic coordinates. 三線～ trilinear coordinates. 四円[五球]～ tetracyclic [pentaspherical] coordinates. 一般～ generalized coordinates. 局所～ local coordinate(s). ～変換 coordinate transformation. ～表現 coordinate representation. ～近傍系 system of coordinate neighbo(u)rhoods.

サブ sub-. ⇨部分(ぶぶん-); 劣(れつ-).
～オーディネーション subordination.

サフィックス suffix (*pl.* -fices, -fixes). ⇨添数.

さふくたい 鎖複体 chain complex.
商[相対; 二重]～ quotient [relative; double] chain complex. 有向[順序]～ oriented [ordered] chain complex. 部分～ chain subcomplex. 双対～ cochain complex.

さぶん 差分 difference. ⇨階差.
～法 calculus of (finite) differences; difference calculus. ～表 difference table. (第)二階～ difference of second order; second difference. ～方程式 difference equation. ～微分方程式 difference-differential equation. ～商[係数] difference quotient. 中心～ central difference. 前進[後退]～ forward [backward] difference. ～近似 difference analogue [approximation]. 有限～ finite difference.

さぶんしょう 差分商 difference quotient; divided difference.

さへん 左辺 left(-hand) side member.

サーボ servo.
～理論 servo theory. ～機構 servo-mechanism.

さもないと otherwise; (or) else.

さよう 作用 operation; action.
～の operational. ～する operate; act. ～素 operator. ～域 operator domain. ～同型[形] operator isomorphism. ～反～の法則 law of action and reaction. 最小～ least action. ～積分 action integral. 群の～ action of group. 簡約可能(な)～ reductive action. 相互～ interaction.

さようそ 作用素 operator. ⇨演算子.
恒等[逆]～ identity [inverse] operator. 線形～ linear operator. 加法的～ additive operator. 共役[双対]～ conjugate [dual] operator. 微分[積分]～ differential [integral] operator. 楕円[双曲]型～ elliptic [hyperbolic] operator; operator of elliptic [hyperbolic] type. ラプラス～ Laplace [Laplacian] operator. 射影～ projection; projection operator; projector. ホモトピー～ homotopy operator. 面～ face operator. 境界～ boundary operator. ～環 ring of operators; operator algebra. ～解析 operational calculus. 限定～ quantifier.

対称化～ symmetrizer. 交代化～ alternizer. 型(かた)～ shape operator. 形(かたち)～ shape operator. シェイプ～ shape operator. 散逸～ dispersive operator. ～代数 operator algebra.

サラス Sarrus.
～の規則[法則] Sarrus' rule [law].

さん 三 three.
～の three. ～次元の three-dimensional. ～円定理 three circle theorem. ～体問題 three(-)body problem; problem of three(-) bodies. 四分の～ three fourths; three quarters. ～シグマ法 three sigma method. ～増しコード excess-three code.

さん- 三 tri-; ter-.
～角形 triangle. ～焦点の trifocal. ～元の ternary.

さんいつ 散逸(的な) dispersive.
～作用素 dispersive operator.

さんいんし 三因子 three factors.
～の three-factor. ～交互作用 three-factor interaction; triple interaction.

さんか 参加 participation; entry.
～する participate【in】; take part in; join. ～者 participant; entrant.

さんかく 三角(の) triangular.
～数 triangular number. ～錐[柱] triangular pyramid [prism]. ～行列 triangular matrix. 上[下]～行列 upper [lower] triangular matrix. ～定規【米】triangle,【英】set square. ～不等式 triangle inequality.

さんかく- 三角 trigonometric(al).
～関数 trigonometric function. ～比 trigonometric ratio. ～和 trigonometric sum. ～恒等式 trigonometrical identity. ～方程式 trigonometric(al) equation. ～多項式 trigonometric polynomial. ～級数 trigonometric(al) series. ～積分 trigonometric integral. ～モーメント問題 trigonometric moment problem. ～測量 trigonometrical measurement [surveying]; triangulation. ～点 point of triangulation. ～網 triangular net. ～組 triad.

さんかくけい 三角形 triangle; trigon.
～の triangular. 正[等辺]～ regular [equilateral] triangle. 二等辺～ isosceles triangle. 直角～ right(-angled) triangle. 垂足～ pedal triangle. 中点～ medial triangle. 極～ polar triangle. 自極～ self-polar triangle. 球面～ spherical triangle. 両直角～ birectangular triangle. 曲線～ curvilinear triangle. 円弧～ triangle bounded by circular arcs; circular triangle. ～分割 triangulation. ～法則 triangle law. ～に内[外]接する円 circle inscribed in [circumscribed about] triangle; incircle [circumcircle] of triangle. ～の五心 five centroids of triangle.

さんかくほう 三角法 trigonometry.
～の trigonometric; trigonometrical. 平面～ plane trigonometry. 立体～ solid trigonometry. 球面～ spherical trigonometry; spherics.

さんぎょう 産業 industry.
～連関 interindustry relation. ～連関分析 analysis of interindustry relations; input-output analysis.

さんきょく 三極(の) tripolar.
～座標 tripolar coordinates.

さんげん 三元(の) ternary.
～方程式 equation with three unknowns.

さんこう 参考 reference.
～書(目録) reference(s); bibliography.

さんこうしき 三項式 trinomial (expression).
～の trinomial.

さんさ 三叉(の) trident; trifurcate.
～する trifurcate. ～曲線 trident.

ざんさ 残差 residual.
～分散 residual variance. ～二乗和 residual sum of squares.

さんざい 散在 scatter.
～させる scatter.

さんじ 三次(の) cubic; cubical; of degree three.
～式 cubic expression. ～方程式 cubic equation. ～分解(方程)式 cubic resolvent. ～関数 cubic function [polynomial]. ～曲線 cubic (curve).

さんしゃしょうけい 三斜晶系 triclinic system.

さんじゅう 三重(の) triple; three(-)fold;

triplicate.
～に triply. ～根 triple root. ～積 triple product. ～積分 triple integral. ～対角行列 tridiagonal matrix. ～連結の triply connected. ～点 triple point.

さんしゅつ 算出 computation; calculation.
～する compute; calculate.

さんじゅつ 算術 arithmetic.
～の arithmetic; arithmetical. ～的に arithmetically. 初等～ elementary arithmetic. ～的関数 arithmetical function. ～平均 arithmetic mean. ～数列 arithmetic progression. ～級数 arithmetic series. ～種数 arithmetic genus. ～化 arithmetization. ～家 arithmetician.

さんじゅつ- 算術 arithmetico-.
～幾何平均 arithmetico-geometric mean.
～調和平均 arithmetico-harmonic mean.

さんしゅつかんすう 算出関数 enumeration function.

さんしょう 参照 reference. ⇨見よ.
～させる; ～する refer to. ～せよ see; cf. ((*L.*)＝confer); vid. ((*L.*)＝vide); compare. 前後～ cross reference. ～符 reference mark【* (asterisk, star), † (obelisk, dagger), ‡ (double dagger), § (section), ¶ (paragraph); → (arrow), など】.

さんじょう 三乗 cube. ⇨立方.
～の cubic. ～する cube. ～根 cubic root.
～比 cubic ratio; triplicate ratio.

さんしん 三進(の) ternary.
～集合 ternary set. ～法 ternary system.

さんすう 算数 (elementary) arithmetic.

さんせん 三線(の) trilinear.
～座標 trilinear coordinates. ～定理 three line theorem.

ざんそん 残存 survival.
～者[物] survival. ～率 survival rate.

さんたいもんだい 三体問題 three(-)body problem; problem of three(-)bodies.

さんだんろんぽう 三段論法 syllogism.
～(的)な syllogistic. 仮言～ hypothetical syllogism. 選言～ disjunctive syllogism. 定言～ categorical syllogism.

さんち 三値(の)～ three-valued.
～論理 three-valued logic; ternary logic.

ざんてい 暫定(の) provisional; tentative; temporary.
～規格 temporary standard.

さんていきょくせん 算定曲線 logistic curve.

さんてん 三点 three points.
～問題 three point problem. ～比較法 triangle test.

さんとうぶん 三等分 trisection.
～する trisect. 角の～ trisection of angle.
～曲線 trisecant; trisectrix. ～線 trisector; trisectrix.

さんばい 三倍 three times; thrice; three-fold.
～の triple; threefold. ～にする triplicate.
～角の公式 formula for [of] triple angle.

さんぱつてき 散発的な sporadic.

さんぷ 散布 scatter; dispersion; dissemination.
～の scedastic. ～線 scedastic curve. ～図 scatter diagram. ～的層 scattered sheaf.
～指数 index of dispersion. ～度 degree of scattering [dissemination].

さんぷくつい 三幅対 triplet; triad.

サンプリング sampling. ⇨抜取(り).

サンプル sample. ⇨標本.
マスター・～ master sample.

さんぶん 三分 trisection; trichotomy.
～する trisect. ～法 trichotomy.

さんへいほうのていり 三平方の定理 Pythagorean theorem; theorem of three squares.

さんぺんけい 三辺形 trilateral.

さんぽう 算法 calculation; algorithm; arithmetic; (law of) composition; operation.
四則～ four arithmetical operations. 二項～ binary operation. 無限小～ infinitesimal calculus. ユークリッドの～[互除法] Euclid's [Euclidean] algorithm. ～言語 algorithmic language.

さんめん 三面(の) trihedral.
～体 trihedron (*pl.* -rons, -dra). ～体の trihedral. ～角 trihedral angle; tripod.

さんめんかく 三面角 trihedral; trihedral angle; tripod.
座標[動]～ coordinate [moving] trihedral.

ざんよ 残余 remainder; residue; rest.

さんようすうじ 算用数字 Arabic numeral [figure].

さんようむすびめ 三葉結び目 clover-leaf knot.

さんらん 散乱 scattering; scatter.
～の長さ scattering length. ～の幅 width of scattering. 共鳴～ resonance scattering. ポテンシャル～ potential scattering. 弾性[非弾性]～ elastic [inelastic] scattering. ～作用素 scattering operator. ～された scattered.

ざんりゅう 残留(の) residual.
～誤差 residuum (*pl.* -dua); residual error.

ざんりょう 残量 residual quantity.

し・シ

し　四　four.
～の four. ～則 four rules. ～則算法 four arithmetic operations. ～元ベクトル four vector. ～色問題 four colo(u)r problem; problem of four colo(u)rs. ～頂点定理 four-vertex theorem; theorem of four vertices. ～分円 quadrant. ～分の一 quarter. ～分位数 quartile. ～角[辺]形 quadrangle [quadrilateral]. ～元数 quaternion.

し　歯　tooth (*pl.* teeth).
～形の toothed. ～形曲線 tooth(ed) curve. ～面 tooth(ed) surface. ～筋 tooth trace.

し-　四　quadri-; quadru-; tetra-. ⇨四(し), 四(よ-), 四(よん).

じ　字　letter; character.
太[細]～ heavy [slender] character.

じ　時　hour.
～速 speed per hour. 毎～40キロ(メートル) speed of fourty kilometers an hour; 40km/h.

じ　次　degree.
～数 degree. n～の多項式 polynomial of degree n.

ジー　G.
～・デルタ集合 G_δ set.

ジェー　J.
～型分布 J-shaped distribution.

ジー・エイ・エム　GAM; generalized additive model.

シェイプ　shape.
～作用素 shape operator.

ジェット　jet.
～空間 jet space.

シー・エヌ　cn.
～関数 cn-function.

シェーマ　(*G.*) schema.

しえん　四円　four circles.
～の tetracyclic. ～座標 tetracyclic coordinates.

しかい　視界　visual field.
～半径[俯角] radius [depression] of visual field.

じかい　磁界　magnetic field.

しかく　資格　qualification.

しかく　視角　visual angle.

しかく　視覚　vision; sight.
～の visual.

しかく-　四角　quadrangular.
～数 quadrangular number. ～柱[錐] quadrangular prism [pyramid]. ～柱六点[六点図形] quadrangular set of six points.

しかくけい　四角形　quadrangle; tetragon. ⇨四辺形.
～の quadrangular; tetragonal. 正～ regular quadrangle [tetragon]; square. 完全～ complete quadrangle.

じかん　時間　time; period.
～的な time-like; temporal. ～反転 time reversal. ～割 time table; (teaching) schedule; time-sharing. ～最適問題 time-optimal problem. ～的に一様な temporally homogeneous.

しき　式　expression.
分数[無理]～ fractional [irrational] expression. 整[有理]～ integral [rational] expression. 代数～ algebraic expression. 近似～ approximate expression. 展開～ expanded expression; expansion. 微分～ differential expression. 発散～ divergence

expression.

しき 式 formula (*pl.* -las, -lae).
～で示す formulate. 漸化～ recurrence [recursion; recurring; recurrent] formula.

しき (論理)式 formula; logical formula; well-formed formula; logical expression. ⇨論理式.

しき (推論様)式, (推論形)式 (*L.*) modus.
肯定～ modus ponens. 否定～ modus tollens. 否定肯定～ modus tollendo ponens. 肯定否定～ modus ponendo tollens.

じき 磁気 magnetism.
～の magnetic. ～誘導 magnetic induction.

しきい 閾 threshold.
異常～ anomalous threshold. ～値 threshold value. ～値ヤコビ法 threshold Jacobi method.

しきべつ 識別 discrimination; discernment.
～する distinguish; discriminate; discern.

じきょうやく 自共役な self-conjugate. ⇨自己共役.

じきょく 自極 self-polarity.
～の self-polar. ～三角形[四面体] self-polar triangle [tetrahedron].

じきょく 磁極 magnetic pole.

しきり 仕切 partition; division; screen.

しきり 仕切 lot.
～番号 lot number. ～検査 lot inspection. ～品質保護 lot quality protection. ～許容不良率 lot tolerance percentage defective; LTPD.

じく 軸 axis (*pl.* axes).
～の axial. 座標～ coordinate axes. 横[縦](座標)～ axis of abscissas [ordinates]. 直[斜]交～ orthogonal [oblique] axes. 実[虚]～ real [imaginary] axis. 長[短]～ major [minor] axis. 共役[副]～ conjugate axis. 主～ principal axis. 回転[対称]～ axis of revolution [symmetry]. 光～ optical axis. ～対称 axial symmetry; axisymmetry.

じくう 時空 time and space; space-time.
～系座標 frame of space and time.

ジグザグ zigzag.

じくせい 軸性(の) axial.
～ベクトル axial vector.

じくそく 軸測(の) axonometric.
～投影[象](法) axonometric projection. 等[斜]～投影[象] isometric [oblique] axonometric projection. ～線 axonometric line.

じくたいしょう 軸対称 axial symmetry; axisymmetry.
～な axisymmetric; axially symmetric.

シグマ sigma; 【Σ, σ, ς】.
～関数 sigma function; σ-function. ～・コンパクト sigma compact; σ-compact. ～有限測度 σ-finite measure. 三～法 three sigma [σ] method. コ～ cosigma.

しけいきょくせん 歯形曲線 tooth(ed) curve.

じけいれつ 時系列 time series.

しげき 刺激 stimulus (*pl.* -li); excitation. ～する stimulate; excite.

しけん 試験 examination; experiment; test; trial.
～する examine; experiment; test. ～関数 trial function. ～調査 pilot test; pilot survey. ～採集 sampling. 選抜～ selective [competitive] examination.

じげん 次元 dimension.
～の dimensional. 有限～の finite(-)dimensional. 無限～の infinite(-)dimensional. 二～(的)の two-dimensional. ～論 dimension theory. ～公理 dimension axiom. ～解析 dimensional analysis. 被覆～ covering dimension. 局所～ local dimension. ホモロジー～ homological dimension. 帰納的～ inductive dimension. 余～ codimension. クルル～ Krull dimension. ～の呪い curse of dimensionality.

しげんぐん (クラインの)四元群 (Klein) four group.

しげんすう 四元数 quaternion.
～の quaternionic. ～算法 quaternion calculus. ～環 quaternion algebra. ～群 quaternion group. ～双曲型空間 quaternion hyperbolic space. ～斜体 skew field of quaternions. ～体 quaternion field. ～ベクトル束 quaternion vector bundle. ～構

造 quaternionic structure.
しげんほう　四元法　quaternions.
じこ-　自己　auto-; self-.
～同型[形] automorphism. ～双対な self-dual. ～移入[入射]的な self-injective. ～まつわり[纏絡]数 self-linking coefficient.
じご　事後(の)　posterior.
～分布 posterior distribution. ～確率 posterior probability; a posteriori probability. ～危険 posterior risk.
じこいにゅうてき　自己移入的な　self-injective.
しこう　試行　trial.
～列 sequence of trials; trial sequence. ベルヌーイ～ Bernoulli trial. 独立～ independent trials.
じこう　事項　subject; matter; item.
～索引 subject index.
じこうめこみ　自己埋め込み　self-embedding.
じこかいき　自己回帰　autoregression.
～の autoregressive. ～過程 autoregressive process. ～条件付き異分散モデル autoregressive conditional heteroskedasticity model; ARCH model. ～和分移動平均モデル autoregressive integrated moving average model; ARIMA model.
じこきょうぶんさん　自己共分散　autocovariance.
～関数 autocovariance function.
じこきょうやく　自己共役　self-adjointness; self-conjugacy.
～な self-adjoint; self-conjugate. ～[自己随伴]作用素 self-adjoint operator. ～点 self-conjugate point.
じこく　時刻　time.
～変数 time variable. ～助変数 time parameter. ～のずれ[遅れ] time lag.
じここうさ　自己交叉　self-intersection.
じここうてんすう　自己交点数　self-intersection number.
じこしゃぞう　自己写像　self-mapping.
じこじゅうそくてき　自己充足的な　self-contained.
じこじゅんどうけい　自己準同型[形]　endomorphism.
～環 endomorphism ring. 反～ anti-endomorphism.
じこじょうほう　自己情報(量)　self-adjoint information.
じこずいはん　自己随伴(な)　self-adjoint.
～[自己共役]作用素 self-adjoint operator.
～方程式 self-adjoint equation.
しごせん　子午線　meridian (circle).
～の meridian.
じこそうかん　自己相関　autocorrelation.
～関数 autocorrelation function.
じこそうつい　自己双対(な)　self-dual.
じこそうはん　自己相反　self-reciprocity.
～な self-reciprocal. ～関数 self-reciprocal function.
じこたいしょう　自己対称　self-symmetry.
～な self-symmetric. ～図形 self-symmetric figure.
じこちゅうみつ　自己稠密(な)　self-dense; dense in itself; (*G.*) in sich dicht.
じこちょうみつ　自己稠密　⇨自己稠密(な)(じこちゅうみつ).
じこてんらく　自己纏絡(の)　self-linking.
～数 self-linking coefficient.
じこどうけい　自己同型[形]　automorphism.
内部～ inner automorphism. 対合的～ involutive automorphism. ～群 automorphism group. ～写像 automorphism. 反～(写像) anti-automorphism.
じこにゅうしゃてき　自己入射的な　self-injective.
じこまいぞう　自己埋蔵　self-embedding; self-imbedding.
じこむじゅんのない　自己矛盾のない　self-consistent.
しさ　視差　parallax.
～の parallactic(al). 年周～ annual parallax. 地心～ geocentric parallax.
しじ　支持　support.
～する support. ～線 supporting line. ～点 support point; supporting point. ～平面[超平面] supporting plane [hyperplane]. ～(線)関数 supporting (line) function. ～汎関数 support functional. ～計量 supporting metric.
しじ　指示　indication; instruction.

〜する indicate; instruct. 〜(曲)線 index line; indicatrix (*pl.* -trices, -trixes).

しじしん 指時針【日時計の】 gnomon.

じじつ 事実 fact; case. ⇨実際(の).
〜上 as a matter of fact; in fact. 〜に反して contrary to the fact.

ししゃごにゅう 四捨五入 rounding.
〜する round; drop decimals after a certain significant place; count fractions of 5 and over as a unit and disregard the rest.

しじゅう 四重(の) fourfold; quadruple; quadruplicate.
〜に quadruply. 〜格子 quadruple lattice.

ししゅうごう 始集合 initial set.

しじゅんひょう 視準標 target.

じしょ 辞書 dictionary; lexicon.
〜の lexical. 〜式の lexicographic(al). 〜式順序 lexicographic(al) order.

しじょう 市場 market.
〜分析 market research. 〜調査 market survey [research].

じしょう 事象 event.
素[根元]〜 elementary event. 基本[根元]〜 fundamental event. 全[空]〜 whole [empty] event. 単(純)[複(合)]〜 simple [compound] event. 再帰〜 recurrent event. 和[積]〜 sum [product] event. 独立[従属]〜 independent [dependent] event(s). 排反〜 exclusive event(s). 余〜 complementary event. 可測[正則]〜 measurable [regular] event. 確率〜 probability event. 極限〜 limit event.

じじょう 自乗 ⇨二乗.

ししん 視心 visual center.

しすい 死水 dead water. ⇨静水.

シズィジ syzygy.
〜理論 syzygy theory.

しすう 指数 exponent; index (*pl.* -dices, -dexes).
〜の exponential; indicial. 〜法則 law(s) of exponents; exponential law(s). 〜関数[曲線] exponential function [curve]. 〜方程式[不等式] exponential equation [inequality]. 〜多項式 exponential polynomial. 〜級数 exponential series. 〜写像 exponential mapping. 〜付値 exponential valuation. ベキ〜 power exponent. 収束〜 convergence exponent; exponent of convergence. 分岐〜 ramification exponent [index]. 不足〜 deficiency index; defect. 回転〜 winding index; rotation number. 不動点〜 fixed point index. 〜分布 exponential distribution. 〜乱数 exponential random numbers. 〜平滑法 exponential smoothing. 〜のベキにあげる exponentiate. 〜定理 index theorem. (〜)決定方程式 indicial equation. 散布〜 index of dispersion. 慣性〜 index of inertia. 物価〜 index number of prices. 〜型分布族 exponential family. 二重〜関数型公式 double exponential formula.

じすう 次数 degree.
全[補充]〜 total [complementary] degree. 仮想[有効]〜 virtual [effective] degree. 形式的〜 formal degree. 相対〜 relative degree. 超越〜 transcendence degree; degree of transcendence. 最高〜 highest degree. 拡大〜 extension degree. 非分離〜 inseparable degree. 分離〜 separable degree.

じすうつき 次数つき(の) graded.
〜環[加群] graded ring [module]. 〜リー環 graded Lie algebra.

システム system.
〜工学 system engineering. 〜・シミュレーション system simulation.

しずめこみ 沈め込み submersion.

しせいすう 示性数 genus (*pl.* -nera, -nuses).

じせいすう 示性数 ⇨示性数(しせいすう).

しせん 始線 initial line.

しせん 矢線 ⇨矢線(やせん).

しぜん 自然 nature.
〜の; 〜な natural. 〜性 naturality. 〜数 natural number. 〜尺度 natural scale. 〜対数 natural logarithm. 〜幾何学 natural geometry. 〜方程式 natural equation; intrinsic equation. 〜標構 natural frame. 〜境界 natural boundary. 〜変換 natural transformation. 〜な(準)同型[形] natural homomorphism. 〜射影 natural projec-

tion.

じぜん 事前(の) prior; advance.
~分布 prior distribution. ~確率 prior probability; a priori probability. ~推定値 advance estimate.

しぜんすう 自然数 natural number; positive integer.

しぜんたいすう 自然対数 Napierian logarithm; natural logarithm; ln.

しぜんほうていしき 自然方程式 intrinsic equation; natural equation.

しそく 四則(演算) four operations; four rules; (four) arithmetical operations. ~混合算 computation involving [containing] four rules.

じぞく 持続 continuation.
~的 durable. ~期間 duration.

した 下 ⇨上(うえ).
~の lower; downward. ~に downwards; below. ~組 lower class. ~に凸[凹]の downwards convex [concave]. ~に有界な bounded (from) below; bounded downwards. ~(に)半連続な lower semicontinuous. ~三角行列 lower trianglar matrix.

シータ theta; 【Θ, θ, ϑ】.
~関数 theta function; ϑ-function. ~級数 theta series.

じだい 時代 time(s); period; epoch; age; era.
同~の contemporary.

しだいしょう 始対象 initial object; cofinal object.

したがわ 下側(の) lower.
~四分位数 lower quartile.

しち 七 seven.
~の septimal. ~倍; ~倍(の) septuple. ~倍の septimal. ~倍する septuple.

しち- 七 hepta-; sept-.

しちかくけい 七角形 heptagon; septangle.
~の heptagonal; septangular. 正~ regular heptagon.

じつ 実(の) real.
~数 real number. ~根 real root. ~軸 real axis. ~(数)部(分) real part. ~変数[関数] real variable [function]. ~数値関数 real-valued function. ~線形[ベクトル]空間 real linear [vector] space. ~射影空間 real projective space. ~二次体 real quadratic field. ~解析関数 real-analytic function. ~表現 real representation. ~閉体 real closed field.

じつ- 実 true. ⇨実(の)(じつ).
~長 true length.

じっかくけい 十角形 ⇨十角形(じゅっかくけい).

じっけん 実験 experiment.
~の: ~的な experimental; empirical. ~式 experimental [empirical] formula. ~的解析(法) experimental analysis. ~計[企]画(法) design of experiment; experimental design. ~心理学 experimental psychology. ~定数 empirical constant. ~室 laboratory.

じつげん 実現 realization.
~する realize. ~可能な realizable.

じっこう 実行 execution.
~する execute; carry out.

じっこう 実効(の) effective; actual.
~誤差 effective error.

じっこうかのう 実行可能(な) feasible.
~でない infeasible. ~解 feasible solution. ~領域 feasible region.

じっこん 実根 real root.

じっさい 実際(の) actual; practical.
~に actually; in practice; in fact; really; indeed; effectively; as a matter of fact. ~に計算可能な effectively calculable.

じつざい 実在 (actual) existence; entity; reality.
~の existent. ~論 realism.

じっさん 実算 practice; practical calculation.

じつじかん 実時間の real-time.

じっしつ 実質 substance; reality.
~的な substantial, virtual, practical. ~的に substantially, virtually, practically.

しつじん 悉尽 exhaustion. ⇨汲み尽し.
~法 method of exhaustion.

じっすう 実数 real number.
~の real. ~値の real-valued. ~値関数 real-valued function. ~部分 real part. ~体 real number field; field of real numbers.

じっせん 実線 solid line; actual line.
シッソイド cissoid.
〜の; 〜的な cissoidal. 一般〜曲線 cissoidal curve; generalized cissoid.
しっそうせん 疾走線 cissoid. ⇨シッソイド.
じっそく 実測 actual measurement.
〜値 actually measured value.
じっちょう 実長 true length.
しつてき 質的(な) qualitative.
〜変数 qualitative variable.
しっぱい 失敗 failure.
〜する fail.
しつむりつ 悉無律 zero-one law; 0–1 law; law of naught or one.
しつもん 質問 question; inquiry.
じつよう 実用 practical use.
〜の practical; applied. 〜数学 practical mathematics. 〜関数 applied function.
しつりょう 質量 mass.
〜中心 center of mass. 点〜 point mass. 単位〜 unit mass. 〜分布 mass distribution.
してい 指定 specification; designation; assignment.
してん 始点 origin; initial point; starting point.
してん 視点 point of view; visual point; point of sight.
してん 支点 fulcrum (*pl.* -crums, -cra); supporting point.
してん 示点 supporting point.
じてん 辞典; 事典 dictionary. ⇨辞書.
百科〜 encyclop(a)edia.
じてん 時点 instant.
〜表示 instantaneous description.
じてん 示点 ⇨示点(してん).
しどう 指導 guidance; leadership.
〜物; 〜者 leader; director.
じどう 自動(の) automatic; servo-; self-.
〜制御 automatic control. 〜調整 self-adjustment. 〜加重標本 self-weighting sample. 〜微分 automatic differentiation.
じどうけんさ 自動検査 automatic check; built-in check.
ジーナス genus (*pl.* -nera, -nuses). ⇨種数.
じなぞ 字謎 anagram.
〜(的)の anagrammatic.
シヌソイド sinusoid.
じば 磁場 ⇨磁界.
しはい 支配 domination. ⇨優越.
〜する dominate. 〜的な dominant; dominating.
しはらい 支払 payment; pay-off.
〜行列 pay-off matrix. 〜高 the amount paid.
しばる 縛る tie.
シー・ピー・エム CPM; critical path method.
しひょう 指標 characteristic; character.
対数の〜 characteristic of logarithm. 群の〜 character of group. 〜群 character group. 単純〜 simple character. 剰余〜 residue character. 既[被]約〜 irreducible [reduced] character. 誘導〜 induced character. 積分〜 integral character.
しひょう 指標 index (*pl.* -dexes, -dices).
シフト shift.
片側〜 unilateral shift.
しぶんい 四分位
〜の quartile. 〜偏差 quartile deviation. 〜間の interquartile. 〜範囲 interquartile range.
しぶんいすう 四分位数 quartile.
上側〜 upper quartile. 下側〜 lower quartile.
しぶんえん 四分円 quadrant.
〜の quadrantal.
じぶんかつ 時分割 time-sharing.
しぶんのいち 四分の一 quarter; quart.; fourth.
しぶんのさん 四分の三 three quarters; three fourths.
〜の three-quarter.
じへいせん 自閉線 loop; looped curve; self-closed curve.
しへんけい 四辺形 quadrilateral. ⇨四角形.
〜の quadrilateral. 円に内接する〜 quadrilateral inscribed in circle; inscribed quadrilateral. 捩れ〜 twisted [skew] quadrilateral. 完全〜 complete quadrilateral.
じぼ 字母 letter; alphabet; matrix.

しぼう 死亡 death.
~過程 death process. ~率 death rate; mortality (rate). ~数 mortality number. ~生残表 life table; mortality table.

しぼうりつ 死亡率 mortality; mortality rate; death rate.
~(統計)表 mortality table.

しほん 資本 capital.
~の capital. ~金 capital. ~回転(率) capital turn; capital circulation. ~係数 capital coefficient.

しま 島 island.

シミュレーション simulation.
シミュレートする simulate. アナログ[システム]・~ analog(ue) [system] simulation. 数値~ numerical simulation.

シミュレータ simulator.

じめい 自明(な) trivial; obvious; self-evident.
~なこと triviality. ~な付値 trivial valuation. ~な[でない]解 trivial [non-trivial] solution. ~な部分群 trivial subgroup.

しめん 歯面 tooth(ed) surface.

しめんたい 四面体 tetrahedron (*pl.* -rons, -dra).
~の tetrahedral. ~幾何学 geometry of tetrahedrons. 正~ regular tetrahedron. 垂心[直稜]~ orthocentric [orthogonal] tetrahedron. 極[自極]~ polar [self-polar] tetrahedron. ~群 tetrahedral group.

しゃ 射 morphism; arrow.
単~ monomorphism; injection. 全~ epimorphism; surjection. 双[全単]~ bijection. 逆~ inverse morphism. 固有~ proper morphism. 対角~ diagonal morphism. 自己同型[形]~ automorphism. 同型[形]~ isomorphism; equivalence. 構造~ structure morphism.

しゃ- 斜 oblique; skew; slant.
~円錐 oblique (circular) cone. ~角柱 oblique prism. ~対称(変換) oblique symmetry. ~角 oblique angle. ~投影[象](法) oblique projection. ~曲面 skew surface. ~体 skew field; skew-field; skewfield. ~高 slant height. ~積 slant product.

ジャイロスコープ gyroscope.

しゃえい 射影 projection; projection map.
~の; ~的な projective. ~的に projectively. ~する project. 正~; 直交~ orthogonal projection. 斜~ oblique projection. 中心~ central projection. 自然~ natural projection. 立体~ stereographic projection. 斜方~ trimetric projection. ~面 plane of projection. ~幾何学 projective geometry. ~微分幾何学 projective differential geometry. ~平面[空間] projective plane [space]. ~(的)変換[対応; 写像] projective transformation [correspondence; mapping]. ~作用素 projection; projection operator; projector. ~的性質 projective property. ~的共線変換 projective collineation. ~不変量[不変性] projective invariant [invariance]. ~(的)分解 projective resolution. ~加群 projective module. ~(的)系 projective system. ~的極限 projective limit. ~的に平坦な projectively flat. ~被覆 projective cover. ~閉包 projective closure.

じゃく 弱
一割~ a little under 10 percent.

じゃく- 弱 weak. ⇨強(きょう-).
~位相 weak topology. ~収束 weak convergence. ~極限 weak limit. ~解 weak solution. ~連続表現 weakly continuous representation. ~収束の weakly convergent. ~可測[可積]な weakly measurable [integrable].

しゃくど 尺度 scale.
自然[標準]~ natural [canonical] scale. 比~ ratio scale. ~単位 unit of scale. ~係数 scale factor. ~母数 scale parameter. ~ベクトル measure vector. ~構成 scaling. 名義~ nominal scale.

しゃこう 斜高 slant height.

しゃこう 斜交 oblique intersection.
~座標 oblique coordinates. ~軸 oblique axis [axes]. ~群 symplectic group.

しゃこう 斜航(的) loxodromic.
~(的)変換 loxodromic transformation.

しゃこうざひょう 斜交座標 oblique (Cartesian) coordinates; skew coordinates.

しゃこうじく 斜交軸 oblique axis (*pl.* axes).

しゃじく 斜軸の scalene.

しゃせん 斜線 oblique line; solidus; virgule; 【 / 】. ～の oblique. ～[陰影]の部分 shaded portion.

しゃせん 射線 ray.
放～ radial ray. ～錐体 emission.

しゃぞう 写像 mapping; map; (*F.*) application.
恒等～ identity [identical] mapping. ～の制限 restriction of mapping. 逆～ inverse mapping. 転置～ transposed mapping. 単射～ injective mapping; injection. 全射～ surjective mapping; surjection. 配景～ perspective mapping. 線形～ linear mapping. 位相～ topological mapping. 定値～ constant mapping. 連続～ continuous mapping. 開[閉]～ open [closed] mapping. 有理～ rational mapping. 可微分～ differentiable mapping. 解析～ analytic [holomorphic] mapping. 重心～ barycentric mapping. 胞体～ cellular mapping. 特性～ characteristic mapping. アフィン～ affine mapping. ～空間 mapping space. ～度 degree of mapping; mapping degree. 等角～ conformal mapping; conformal representation. 単葉～ univalent [(*G.*) schlicht] mapping. 星型[凸]～ star(-)like [convex] mapping. 内[外]～半径 inner [outer] mapping radius. ～錐[柱] mapping cone [cylinder]. 約～錐, 被約～錐 reduced mapping cone. 縮小～ contraction mapping. 制限～ restriction map. 標準(的)～ canonical map.

しゃたい 斜体 skewfield; skew-field; non-commutative field; (*G.*) Schiefkörper; division ring.

しゃたいしょう 斜対称(変換) oblique symmetry.

しゃてい 射程 range.
有効～ effective range.

しゃとうえい 斜投影 oblique projection.

しゃへん 斜辺 hypotenuse; oblique side.

しゃほう 斜方(の) rhombic; trimetric.
～形 rhombus (*pl.* -buses, -bi). ～晶系 rhombic system. ～投象 trimetric projection.

しゃめん 斜面 slant; slanting surface; inclined plane.
～の slant; oblique.

しゅ 種 sort; kind; class; genus (*pl.* -nera, -nuses); species (*pl.* -).

しゅ- 主 principal; main; chief; leading.
～軸 principal axis. ～方向 principal direction. ～法線 principal normal. ～平面 principal plane. ～曲率 principal curvature. ～曲率半径[方向] radius [direction] of principal curvature. ～接線曲線 asymptotic curve. ～行列 principal matrix. ～小行列式 principal minor. ～イデアル principal ideal. ～イデアル整域 principal ideal domain; PID. ～アデール principal adèle. ～束 principal bundle. ～種 principal genus. ～値 principal value. ～近似分数 principal convergent. ～極限値 principal limits; (*L.*) limites principales. ～係数 principal [leading] coefficient. ～定理 main [principal] theorem. ～効果 main effect. ～変数 primal variable. ～問題 primal problem.

シューア I. Schur (1875-1941).
～分解 Schur decomposition.

しゅう 周 periphery; circumference; perimeter.
～の peripheral; circumferential. ～上の特異点 peripheral singularity.

じゅう 十 ten; X; x. ⇨千.
～ダース ten dozen(s); small gross. ～万 hundred thousands. ～億 billion. ～の百乗 googol. ～を底とした～の百乗ベキ googolplex.

じゅう- 十 deca-. ⇨十角形; 十辺形; 十進(の).
～年間 decade. ～面体 decahedron. ～面体の decahedral.

じゅう 重 multiple; poly-; bi-.
～根 multiple root. ～積分 multiple integral. ～調和な biharmonic. ～回帰 multiple regression.

じゅう 重(の) -fold; -ple; -tuple.
～に -fold; -ply. 二～の[に] two(-)fold; double [doubly]. 三～積[積分] triple product [integral]. n～根 n-tuple root; root of multiplicity n. n～連結の n-ply connected. p～対称な p-ply symmetric. k～推移的な k-ply transitive.

じゆう 自由 freedom; liberty.
～な free. ～に freely. ～運[可]動(性) free mobility. ～ベクトル[変数] free vector [variable]. ～群 free group. ～加群 free module. ～積 free product. ～境界 free boundary. ～振動 free oscillation. ～度 degree of freedom. ～に作用する act freely. ～分解 free resolution.

しゅうい 周囲 circumference; periphery; perimeter. ⇨周.

しゅういき 終域 codomain.

じゅういち 十一 eleven.
～角[辺]形 undecagon. ～角[辺]形の undecagonal. ～進法 undenary (scale).

しゅうえき 収益 profit; gain; revenue.

しゅうえん 周縁 margin. ⇨周辺.
～の marginal. ～度数 marginal frequency.

じゅうかいき 重回帰 multiple regression.

しゅうかく 周角 round angle; full circle; periphery of circle; perigon.

しゅうき 周期 period; cycle.
～的な periodic; cyclic. ～的に periodically. ～性 periodicity. 基本～ primitive period. ～帯 period strip. ～平行四辺形 period(-)parallelogram. ～関数 periodic function. 単[二重]～関数 simply [doubly] periodic function. 概～関数 almost periodic function. ～解 periodic solution. ～群[系] periodic group [system]. ～母数 periodicity modulus. ～関係式[行列] period relation [matrix]. ～不等式 period inequality. ～軌道 periodic orbit. 極限～軌道 limit cycle.

じゅうきょうやく 重共役な bidual.

しゅうけつ 終結 conclusion. ⇨結論.

しゅうけつしき 終結式 resultant.

じゅうけってん 重欠点 major defect.

じゅうご 十五 fifteen.
～分 fifteen minutes; quarter. ～角[辺]形 quindecagon.

しゅうごう 集合【物質的な】 aggregation; aggregate; collection; concurrence.
～する aggregate; gather; meet. ～的な collective. ～的危険論 collective risk theory. (線・面の)～点 concurrence.

しゅうごう 集合【数学的な】 set; (*G.*) Menge; (*F.*) ensemble.
～論 set theory; theory of sets. ～論的な set-theoretic. ～(演)算 set operation. 和～ union; sum. ベキ～ power set. 差～ difference set. 空～ empty [void; null] set. ～系[族] system [family] of sets. ～束[体] lattice [field] of sets. 点～ point set. 可算～ countable [enumerable; denumerable] set. 開[閉]～ open [closed] set. エフ・シグマ[ジー・デルタ]～ F_σ [G_δ] set. カントル[ボレル]～ Cantor [Borel] set. ルベーグ可測～ Lebesgue measurable set. 普遍～ universal set. 全体～ whole [universal] set. 特異～ singular set. 特異点～ singularity set; set of singularities [singular points]. ～列 sequence of sets. ～関数 set function. ～値の set-valued. 超～ superset. 順序～ ordered set. 乗法的閉～, 積閉～ multiplicatively closed set; multiplicative set. 真凸～ proper convex set. ファジィ～ fuzzy set.

しゅうごうさん 集合算 set operation(s); class calculus; (*G.*) Klassenkalkül.

しゅうごうち 集合値(の) set-valued.
～関数 set-valued function.

しゅうごうろん 集合論 set theory; theory of sets.
～的な set-theoretic. 記述～ descriptive set theory. ～的位相幾何学 set-theoretic topology.

じゅうこん 重根 multiple root; equal root.
二[三]～ double [triple] root. n～ n-triple root; root of multiplicity n.

しゅうさ 収差 aberration.
色[球面]～ chromatic [spherical] aberration.

じゅうさんこ　十三個　thirteen; baker's dozen.

しゅうし　終止　stop; termination; end.
～符 full stop; period.

しゅうし　修士　master.
理学～ master of science [M.S.; M.Sc.].

じゅうじひ　十字比　cross ratio; anharmonic ratio.

しゅうしゅう　収集　collection.
～する collect; gather.

しゅうしゅうごう　終集合　final set.

しゅうしゅく　収縮　shrinking; contraction.
～する shrink; contract. ～列 shrinking sequence. ～写像 contraction (map); deflation.

しゅうしん　終身
～年金 life annuity [interest].

じゅうしん　重心　barycentre, -ter; center of mass; center of gravity; centroid.
～の barycentric. ～細分 barycentric subdivision. ～写像 barycentric mapping. ～座標 barycentric coordinates.

しゅうせい　修正　correction; modification; amendment; revision.
～する correct; modify; amend; revise. ～子 corrector. 連続～ continuity correction.

しゅうせき　集積　accumulation; clustering.
～する accumulate; cluster. ～点 accumulation point; cluster point; limit point. ～値 accumulation value; cluster value. ～値集合 cluster set.

しゅうせきちしゅうごう　集積値集合　cluster set. 内～ inner cluster set. 境界～ boundary cluster set.

じゅうせきぶん　重積分　multiple integral.

じゅうそうかん　重相関　multiple correlation.
～係数 multiple correlation coefficient.

じゅうそうつい　重双対な　bidual.

しゅうそく　収束　convergence; convergency.
～の convergent. ～する converge; tend to. ～級数[(数)列] convergent series [sequence]. ～半径 convergence radius; radius of convergence. 連関[同伴]～半径 associated convergence radii. ～円[区間] convergence circle [interval]; circle [interval] of convergence. ～域 domain of convergence. ～(横)座標[軸] abscissa [axis] of convergence. ～指数 convergence index. 各点[単純]～ pointwise [simple] convergence. 絶対[条件]～ absolute [conditional] convergence. 無条件[可換]～ unconditional [commutative] convergence. 一様[平等]～ uniform convergence. 広義の一様に～の uniformly convergent in the wider sense. 強[弱]～ strong [weak] convergence. 漸近～ asymptotic convergence. 平均～ convergence in the mean; mean convergence. 概～ almost everywhere convergence; almost convergence. 漠～ vague convergence. 確率[法則]～ convergence in probability [law]. 有界～の boundedly convergent. 部分～ partial convergence. 星型～ star convergence. 超～ overconvergence; ultraconvergence. ～判定法 convergence criterion [test]. 局所的～性 local convergence. ～率, ～速度 rate of convergence. 大域的～性 global convergence. 超一次～ superlinear convergence.

じゅうぞく　従属(性)　dependence; dependency.
～する depend【on; upon】. ～な dependent. 一次[線形]～な linearly dependent. 関数[代数]的～(性) functional [algebraic] dependence. ～変数 dependent variable. ～事象 dependent event(s). 試行列に～な path-dependent. 整～な integrally dependent.

じゅうぞく　従属　subordination.
～な; ～する subordinate. ～関数 subordinate function. ～(性); ～操作 subordination. ～原理 subordination principle. ～(化)過程 subordinator.

じゅうそくかのうせい　充足可能性　satisfiability.

しゅうたいしょう　終対象　final object.

しゅうだん　集団　collective; group; mass; body; population.
～的な collective. ～現象 collective [mass] phenomenon. ～遺伝学 population genetics.

しゅうだん-　集団　bulk; batch.

～待ち行列 bulk queue. ～サービス bulk service. ～到着 batch arrival.

じゅうちょうわ 重調和(性) biharmonicity.
～な biharmonic. ～関数 biharmonic function.

しゅうてん 終点 end point; terminal (point); terminus (*pl.* -ni, -nuses).

じゅうてん 充填 packing.
～する pack. 最密～ closest packing. ～密度 packing density.

じゅうでん 充電 charge; electrification.
～する charge; electrify.

しゅうてんえん 周転円 epicycle.

じゅうとくせい 従特性(の) bicharacteristic.
～曲線 bicharacteristic curve.

じゅうに 十二 twelve; dozen.
～角[辺]形 dodecagon. ～角[辺]形の dodecagonal. ～面体 dodecahedron. ～面体の dodecahedral. 正～角形 regular dodecagon. 正～面体 regular dodecahedron. ～進(法) duodenary [duodecimal] (scale). ～ダース twelve dozen(s); gross.

しゅうはい 周髀 gnomon. ⇨ノーモン.

しゅうはすう 周波数 frequency.
～変調 frequency modulation.

じゅうふく 重複 duplication; repetition; overlap; redundancy. ⇨重複(ちょうふく).
～した multiple; duplicate; repeated. ～度 multiplicity. ～根 multiple root. ～点 multiple point. ～積分 multiple integral. ～順列[組合せ] repeated permutation [combination]. ～対数 iterated logarithm. ～連結の multiply-connected.

じゅうふくど 重複度 multiplicity.
代数的～ algebraic multiplicity. 幾何的～ geometric multiplicity.

しゅうぶん 秋分(点) autumnal equinox.

じゅうぶん 十分(性) sufficiency.
～な sufficient. ～に sufficiently. ～大きい[小さ; 多]い sufficiently large [small; many]. ～条件 sufficient condition. ～統計量 sufficient statistic.

しゅうへん 周辺 margin; periphery.
～の marginal; peripheral; surrounding. ～度数[分布] marginal frequency [distribution]. ～の長さ perimeter.

じゅうほうせん 従法線 binormal. ⇨陪法線.
～標形 binormal indicatrix.

しゅうまつ 終末 termination; end.

じゅうまん 充満(な) full.
～忠実な fully faithful. ～部分圏 full subcategory.

しゅうらく 集落 colony; cluster.
～の colonial. ～化 colonization. ～抜取り cluster sampling.

じゅうりょう 重量 weight.

じゅうりょく 重力 gravity; gravitation.
～の gravitational. ～加速度 gravity [gravitational] acceleration. ～の場 field of gravitation. ～単位系 system of gravitational units.

しゅうれん 収斂 ⇨収束.

じゅうろくげんすう 十六元数 sedenion.

じゅうろくしん 十六進(の) hexadecimal; sexadecimal.
～法 hexadecimal [sexadecimal] notation.
～数字 hexadecimal [sexadecimal] digit.

しゅかん 主観(的な) subjective.
～確率 subjective probability.

しゅきょくげんち 主極限値 principal limits; (*L.*) limites principales.

しゅくごう 縮合 contraction.

しゅくしゃ 縮写 reduction; miniature.
～図 reduced drawing; miniature copy.

しゅくしゃく 縮尺 reduced scale.

しゅくじゅう 縮重 degeneracy.

しゅくしょう 縮小 contraction; diminution.
～する contract. ～可能な contractible. ～作用素 contraction operator. 区間～法 method of diminishing [nested] intervals. ～写像 contraction mapping.

しゅくず 縮図 reduced drawing; miniature.
～器 pantograph.

しゅくたい 縮退 degeneracy; reduction. ⇨退化.

しゅくぶん 縮分 reduction.

しゅくへいせん 縮閉線 evolute.

しゅくやく 縮約 contraction.
～する contract. ～テンソル contracted tensor. ～イデアル contracted ideal.

しゅくやくこうばい 縮約勾配 reduced gradient.

じゅけい 樹形 tree.
~曲線 tree (curve). ~図 tree diagram.

しゅこうか 主効果 main effect.

しゅざ- 主[首]座 principal.
~小行列式 principal minor.

しゅざん 珠算 calculation on [by] abacus.

しゅじく 主軸 principal axis (*pl.* axes).
~変換 principal axis transformation. 慣性~ principal axis of inertia.

しゅじゅつ 手術 surgery.

じゅしんしゃ 受信者 receiver.

じゅず 数珠 rosary; chaplet.
~順列 necklace permutation.

しゅすう 種数 genus (*pl.* -nera, -nuses).
算術[幾何]~ arithmetic [geometric] genus. 有効~ effective genus. 無限~ infinite genus. 仮想~ virtual genus.

しゅせん 首線 initial line.

しゅだい 主題 (main) subject; theme.

しゅだん 手段 method; means; way.

しゅち 主値 principal value; p.v.; (*F.*) valeur principale; v.p.

しゅちょう 主張 assertion; claim; contention; argument.
~する assert; claim; contend.

じゅっかくけい 十角形 decagon.
~の decagonal.

しゅつげん 出現 occurrence; appearance.
~頻度 frequency of occurrence.

じゅつご 術語 technical term; terminology.

じゅつご 述語 predicate.
~の predicate. 多項式~ polynomial predicate. ~記号 predicate symbol. ~論理 predicate logic. ~変数 predicate variable. ~計算 predicate calculus. 一階~論理 first order predicate logic.

じゅっさく 術策 strategy.

しゅっしょ 出所 origin; source.

しゅっしょう 出生 ⇨出生(しゅっせい).

じゅっしん 十進(の) decimal; denary.
~(法)の decimal. ~(法)で decimally. ~(法)化する decimalize. ~(法)化 decimalization. ~法 decimal system; denary scale. ~記数法 decimal notation; decimal numeration system. ~小数 decimal fraction.

しゅっせい 出生 birth.
~率 birth ratio [rate]. ~過程 birth process. ~関数 birth function. ~死亡連鎖 birth and death chain.

しゅっぱつ 出発 start; departure.
~する start. ~点 starting point. ~値 starting value.

しゅっぱん 出版 publication; issue.
~する publish; put (bring) out; issue. 限定~ limited publication. ~権 publication right. ~社 publisher. ~部数 the issue; circulation. ~目録 catalog(ue) of publications.

じゅっぺんけい 十辺形 decagon.
~の decagonal.

しゅつりょく 出力 output. ⇨入力.
~域 output area.

じゅどう 受動(的) passive.
~境界点 passive boundary point.

しゅびいっかん 首尾一貫した self-consistent.

シュプール (*G.*) Spur; trace. ⇨跡(せき); トレース.

じゅほうしゃ 受報者 destinator; receiver.

じゅみょう 寿命 life; life expectancy.
~曲線 life curve.

じゅもく 樹木 tree. ⇨木.
~曲線 tree (curve). ~表現 tree representation. ゲームの~ tree of game. ~状 arborescence.

しゅよう 主要(な) principal; primary; main; leading; primal.
~部 principal part. ~根 principal root. ~解 principal solution; primary solution. ~定理 main [principal] theorem. ~項 leading term. ~管理点 milestone.

じゅよう 需要 demand.
~予測 demand forecasting. ~関数 demand function.

しゅるい 種類 kind; sort; genus (*pl.* -nera, -nuses); type.

シュレーディンガー E. R. J. A. Schrödinger (1887–1961).

シュワルツ H. A. Schwarz (1843–1921).

〜の Schwarzian; Schwarz('). 〜の導関数 Schwarzian derivative. 〜の補助定理 Schwarz(') lemma.

じゅん 純(な) pure; unmixed.
〜に purely. 〜虚数 purely imaginary number. 〜小数 pure decimal. 〜循環小数 pure recurring decimal. 〜二次方程式 pure quadratic equation. 〜イデアル unmixed ideal. 〜無限な purely infinite. 〜不連続な purely discontinuous. 〜非分離的な purely inseparable.

じゅん 順(の) direct.
〜に directly. 〜像 direct image. 〜系 direct system. 〜極限 direct limit. 〜結び tame knot. 〜分岐 tame ramification.

じゅん- 準 quasi-; pseudo-. ⇨擬(ぎ-).
〜凹の quasiconcave. 〜凸の quasiconvex. 〜凹関数 quasiconcave function. 〜凸関数 quasiconvex function. 〜群 quasi-group. 〜ノルム quasi-norm. 〜ノルム空間 quasi-normed space. 〜解析関数 quasi-analytic function. 〜アフィンな quasi-affine. 〜連続な quasi-continuous. 〜p葉な quasi-p-valent. 〜フロベニウス多元環 quasi-Frobenius algebra. 〜ラテン方格 quasi-Latin square.

じゅんい 順位 rank (order); grade.
〜づけ ranking; grading. 〜法 ranking method. 〜相関係数 rank correlation coefficient. 同〜 tied rank. 符号付き〜 signed rank.

じゅんえき 純益 net profit [gain].

じゅんえん 準円 director circle.

じゅんおう 準凹(の) quasiconcave.
〜関数 quasiconcave function.

じゅんか 純化 refinement; purification.
〜する refine; purify.

じゅんかい 巡回 cycle; cycling.
〜的な cyclic. 〜方程式 cyclic equation. 〜置換[行列] cyclic permutation [matrix]. 〜群[多元環] cyclic group [algebra]. 〜ベクトル cyclic vector. 〜表現[拡大] cyclic representation [extension]. 〜部分 cyclic part. 〜符号 cyclic code. 〜ホモロジー cyclic homology.

じゅんかいせき 準解析(性) quasi-analyticity.
〜的な quasi-analytic. 〜関数 quasi-analytic function.

じゅんかい-セールスマン-もんだい 巡回セールスマン問題 traveling salesman problem.

じゅんかぎゃく 準可逆(な) quasi-invertible.
〜元 quasi-invertible element.

じゅんかほてき 準可補的 quasi-complemented.

しゅんかん 瞬間 moment; instant.
〜の instantaneous; momentary. 〜速度 instantaneous velocity. 〜回転 instantaneous rotation. 〜状態 instantaneous state. 〜力 impulsive force.

じゅんかん 循環 recurrence.
〜の recurrent; recurring; repeated; cyclic. 〜する recur; repeat. 〜小数 recurring decimal; repeated [repeating] decimal; circulating decimal. 〜連分数 recurring continued fraction. 〜節 recurring period. 〜数列 recurring sequence. 〜座標 cyclic coordinates.

じゅんかん 循環 circulation.
〜の circulating; circulate; cyclic. 〜する circulate. 〜小数 circulating decimal; recurring decimal.

じゅんかんしょうすう 循環小数 recurring decimal; repeated [repeating] decimal; circulating decimal.
純[混]〜 pure [mixed] recurring decimal.

じゅんかんせつ 循環節 recurring period.

じゅんかんび 準完備(性) quasi-completeness.
〜な quasi-complete. 〜空間 quasi-complete space.

じゅんかんろんぽう 循環論法 vicious circle; (*L.*) circulus viciosus; circular argument [reasoning].

じゅんぎゃく 準逆(の) quasi-inverse.
〜元 quasi-inverse element.

じゅんきょうじゅ 准教授 associate professor.

じゅんきょくげん 順極限 direct limit.

じゅんきょくしょ 準局所(的) quasi-local.

じゅんぐん 準群 quasi-group; semigroup.

じゅんざんか 純残価 disposal value.

じゅんしゃえい 準射影(的) quasi-projective.

～代数的多様体 quasi-projective algebraic variety.

じゅんじょ 順序 order.

～づける order. ～の ordinal. ～関係 order relation. ～公理 axiom(s) of order. 線形[全]～ linear [total] order. 半～ semiorder; partial order. 前～ preorder. 擬～ pseudo-order; quasi-order. 辞書式～ lexicographical order. 整列～ well order. ～基底 ordered basis. ～和 ordered sum. ～積 ordered product. ～対 ordered pair. ～集合 ordered set. ～複体 ordered complex. ～位相 order topology. ～同型(写像) order isomorphism. ～型 order type. ～統計量 order statistic. 束～の lattice-ordered. ～尺度 ordinal scale.

じゅんじょしゃくど 順序尺度 ordinal scale.

じゅんじょすう 順序数 ordinal; ordinal number. ⇨序数.

構成的～ constructive ordinal. 孤立[極限]～ isolated [limit] ordinal number. 有限[超限]～ finite [transfinite] ordinal number.

じゅんすい 純粋 purity.

～な pure. ～数学 pure mathematics. ～幾何学 pure geometry. ～戦略 pure strategy.

じゅんせいき 準正規(な) quasi-normal.

～族 quasi-normal family.

じゅんせいそく 準正則(な) quasi-regular.

～列 quasi-regular sequence.

じゅんせん 準線 directrix (*pl.* -trixes, -trices).

じゅんせんけい 準線形(性) quasi-linearity. ⇨擬線形(性).

～な quasi-linear. ～方程式 quasi-linear equation.

じゅんそ 準素(な) primary.

～イデアル primary ideal. ～群 primary group. ～成分 primary component.

じゅんそう 準層 presheaf (*pl.* -eaves). ⇨前層.

じゅんそうつい 準双対(性) quasi-duality.

～な quasi-dual.

じゅんだえん 準楕円(的) hypoelliptic.

～型の hypoelliptic. ～型作用素 hypoelliptic operator.

じゅんたようたい 準多様体 quasi-manifold; pseudo-manifold.

じゅんたるがた 準樽型(の) (*F.*) quasi-tonnelé.

～空間 quasi-tonnelé space; (*F.*) espace quasi-tonnelé.

じゅんたんじゅん 準単純(な) quasi-simple.

～環 quasi-simple ring.

じゅんどうけい 準同型[形] homomorphism.

～な homomorphic. ～写像 homomorphism; homomorphic mapping. ～定理 homomorphism theorem. ～像 homomorphic image. 環[作用]～ ring [operator] homomorphism. 辺[束]～ edge [lattice] homomorphism. 双対[連結]～ dual [connecting] homomorphism. 局所～ local homomorphism. 有理[許容]～ rational [allowed] homomorphism. 自己～ endomorphism. 反～ anti-homomorphism. 標準的～(写像) canonical homomorphism.

じゅんどうち 準同値(な) quasi-equivalent.

～表現 quasi-equivalent representation.

じゅんとつ 準凸(の) quasiconvex.

～関数 quasiconvex function.

じゅんはんきょくしょ 準半局所(的) quasi-semilocal.

じゅんび 準備 preparation.

～の preparatory; preliminary. ～する prepare.

じゅん-ピッチ 準ピッチ reduced pitch.

じゅんふへん 準不変(な) quasi-invariant.

～測度 quasi-invariant measure.

しゅんぶん 春分(点) vernal equinox.

じゅんゆうかい 準有界(性) quasi-boundedness.

～な quasi-bounded. ～関数 quasi-bounded function.

じゅんよういん 準要因 quasi-factor.

～の quasi-factorial.

じゅんれつ 順列 permutation.

全～ all permutation. 基準～ original permutation. 奇[偶]～ odd [even] permutation. 円[数珠]～ circular [necklace] permutation. 重複～ repeated permutation.

じゅんれんせつ 準連接な quasi-coherent.

じゅんれんぞく 準連続(性) quasi-continuity.
～な quasi-continuous. ～関数 quasi-continuous function.

しょい 初位 initial location [place].
～の数字 most significant digit; M.S.D.

しょう 商 quotient.
～法則 quotient law. ～位相[集合] quotient topology [set]. ～空間[対象] quotient space [object]. ～群 quotient group. ～環 quotient ring; ring of quotients. ～体 quotient field; field of quotients. ～表現 quotient representation. ～測度 quotient measure. 差[差分]～ difference quotient. 微分～ differential quotient. 整～ integral quotient. 仮～ temporary quotient.

しょう 章 chapter.
第1[一]～ chapter [Chap.] 1. ～末 end of chapter.

しょう- 焦 focal.
共～の confocal. ～点 focus; focal point. ～曲面 focal surface. ～円錐曲線 focal conics. ～線 focal line; caustic curve. ～面 focal plane [surface]; caustic (surface).

しょう- 昇 ascending.
～ベキ ascending powers. ～ベキの順 ascending order of powers. ～鎖 ascending chain. ～鎖条件 ascending chain condition; a.c.c.

しよう 使用(法) use.
～者 user.

しよう 仕様(書) specifications.

じょう- 常 ordinary; vulgar; regular; normal; proper.
～螺旋 ordinary helix. ～分数 vulgar fraction. ～超球 proper hypersphere.

じょう- 上 upper; super-; hyper-.

しょういきてき 小域的(に) in the small; locally; (*G.*) im Kleinen. ⇨局所的(な).

しょうがい 障害[碍] obstruction; obstacle.
～理論 theory of obstructions. 第一～(類) primary obstruction. 第二～(類) secondary obstruction. 拡張への～ obstruction to extension.

じょうかい 上界 upper bound. ⇨下界.
最小～ least upper bound; (*L.*) supremum (*pl.* -rema); sup.

しようかのう 使用可能(性) availability.
～な available. ～時間 available time.

じょうぎ 定規[木] rule; ruler; square.
直角～ square. 三角～【米】triangle,【英】set square. ティ～ T-rule; T-square. ～で線を引く rule.

じょうきゅう 上級(の) superior.

しょうきょ 消去 elimination.
～する eliminate. ～できる eliminable. ～法 method of elimination; elimination method. ガウスの～法 Gaussian elimination. ～式 eliminant. 前進～ forward elimination.

じょうきょう 状況 situation.

しょうぎょうれつ 小行列 submatrix.
余[補]～ complementary submatrix.

しょうぎょうれつしき 小行列式 minor; minor determinant.
主(座)[首座]～ principal minor.

じょうきょくげん 上極限 superior limit; (*L.*) limes superior (*pl.* limites superiores).
～事象 superior limit event.

しょうきょくてき 消極的 passive; negative.

しょうきょくめん 焦曲面 focal surface.

しようきんし 使用禁止 disability.
～にする disable.

しょうげき 衝撃 impulse; shock.
～の; ～的な impulsive. ～関数 impulse function. ～波 shock wave. ～形 bang-bang type.

しょうげん 象限 quadrant.
～の quadrantal. 第一[二]～ the first [second] quadrant. 正の～ positive quadrant. ～儀 quadrant. ～弧球面三角形 quadrantal spherical triangle.

じょうけん 条件 condition.
～つきの conditional. 必要[十分]～ necessary [sufficient] condition. 必要十分[完全]～ necessary and sufficient condition; complete condition. 付帯～ subsidiary [accessory; incidental] condition; side condition. 初期[境界]～ initial [boundary]

condition. ～収束 conditional convergence. ～安定性 conditional stability. ～文 conditional. ～完備な conditionally complete. 最適性～ optimality condition. ～数 condition number. 追加された～ additional condition.

じょうけん 条件 qualification.

じょうげん 上限 least upper bound; l.u.b.; (*L.*) supremum (*pl.* -rema); sup.; (*L.*) finis superior; upper limit.
～関数 upper limit function. 本質(的)～ essential supremum.

じょうけんつき 条件付(の) conditional; constrained.
～不等式 conditional inequality. ～極値 conditional extremum. ～極値問題 constrained extremal problem. ～確率 conditional probability; relative probability. ～期待値 conditional expectation. ～平均 conditional mean. ～測定 conditional measurement.

しょうこ 証拠 evidence; proof.

しょうごう 照合 comparison; check; collation.

じょうこう 条項 article.

しょうこうえんざん 昇降演算 ladder operation.
～子 ladder (operator). ～子法 ladder method.

じょうこん 乗根 power root; radical (root).

しょうさ 昇鎖 ascending chain.
～条件 ascending chain condition; a.c.c.

しょうさい 詳細 details; particulars.
～に in detail.

しょうさん 消散 dissipation.
～的な dissipative. ～する dissipate. ～部分 dissipative part. ～系 dissipative system. ～作用素 dissipative operator.

じょうさんさようそ 乗算作用素 multiplier.
フーリエ～ Fourier multiplier.

じょうしき 乗式 multiplier.
最終～ last multiplier.

じょうしょう 上昇 ascent; rise.
～の ascending. ～する ascend; rise; step up. ～枝 ascending branch. ～演算子 up-ladder; step-up operator.

しょうじる 生じる yield; produce; cause; give rise to; occur.

じょうじる 乗じる multiply.

しょうすう 小数 decimal.
～の decimal. ～で; ～の形で decimally. ～点 decimal point. ～部 decimal part. ～位 decimal place. 有限[無限]～ terminating [infinite] decimal. 循環～ recurring [repeated; circulating] decimal. 十進～ decimal fraction.

じょうすう 乗数 multiplier; multiplicator.
～法 multiplier method. ～効果 multiplier effect. 特性～ characteristic multiplier. ラグランジュ～ Lagrange multiplier.

じょうすう 常[定]数 constant; const. ⇨定数(ていすう).

しょうすうてん 小数点 decimal point.
固定[浮動]～ fixed [floating] point.

しょうすうのほうそく 小数の法則 law of small numbers.

しょうずる 生ずる occur. ⇨生じる.

じょうせき 乗積 product.
～記号 product notation; Π. 連～ continued product. 無限～ infinite product.

しょうせん 焦線 focal line; caustic; caustic curve. ⇨火線.
～の caustic.

しょうせん 消線 vanishing line.

しょうぜんてい 小前提 minor premise; reason.

じょうたい 状態 state.
定常～ stationary state. 安定～ stable state. 管理～ state of control. ～空間 state space. ～和 sum over states. 増加[減少]の～ increasing [decreasing] state.

じょうたん 上端 upper limit.

しょうちゅうしんれつ 昇中心列 ascending central series; upper central series.

じょうちょう 冗長 redundance; redundancy.
～な redundant. ～度 redundancy. ～(度)検査 redundancy check. ～符号 redundant code.

しょうてん 焦点 focus (*pl.* -cuses, -ci); focal point.

～の focal. ～が合う; ～を合わす focus. ～距離 focal length; focus; focal distance. ～半径 focal radius. 楕円の～ focus [foci] of ellipse. 三～の trifocal.

しょうてん 消点 vanishing point.

じょうはん 上半 upper half.
～(平)面 upper half-plane. ～連続な upper semi-continuous.

しょうひ 消費 consumption.
～する consume. ～者; ～財 consumer.

じょうびぶんけいすう 上微分係数 upper derivat(iv)e.

じょうびぶんほうていしき 常微分方程式 ordinary differential equation.

しょうひょうほん 小標本 small sample.

しょうぶごと 勝負事 game; match; play.

しょうぶぶんかぐん 小部分加群 small submodule.

じょうぶんすう 常分数 vulgar fraction.

しょうへき 障壁 barrier.

しょうべき 昇ベキ ascending powers.
～の順 ascending order of powers.

じょうべき 乗ベキ power.

じょうほう 乗法 multiplication.
～の; ～的な multiplicative. ～的に multiplicatively. ～表 multiplication table. スカラー～ scalar multiplication. 虚数～ complex multiplication. ～群 multiplicative group. ～的関数 multiplicative function. ～定理 multiplication theorem. ～付値 multiplicative valuation. ～合同 multiplicative congruence. ～的閉集合 multiplicatively closed set; multiplicative set. ～因子 multiplicative factor; multiplicator; multiplier. ～子 multiplicator. ～子環 multiplier algebra.

じょうほう 情報 information.
～理論 information theory. ～行列 information matrix. ～多項式 information polynomial. ～科学 information science; informatics. ～量 amount of information. ～源[路] information source [channel]. ～処理 information processing. ～検索 information retrieval; IR. ～安定性 information stability. ～交換用米国標準コード American national standard code for information interchange; ANSCII. ～のある, 有～の, ～に富んだ informative. ～のない, 無～の noninformative. ～幾何学 information geometry. 無～事前分布 noninformative prior distribution. 有～事前分布 informative prior distribution.

じょうほう 上方(の) upper; upward; superior.
～に有界な bounded upwards [(from) above].

じょうほういんし 乗法因子 multiplicator; multiplier; multiplicative factor.

しょうみ 正味(の) net.
～現価 net present worth.

しょうめい 証明 proof; demonstration; verification; certification.
～する prove; demonstrate; verify; certificate. ～可能な provable. ～され(てい)ない unproved; undemonstrated. ～できない unprovable; undemonstrable. 偽であることを～する; 反証する disprove. ～法 method of proof. 直接[間接]～ direct [indirect] proof. 背理法による～ proof by reductio ad absurdum [reduction to absurdity]. ～図 diagram of proof. ～終り this concludes [completes] the proof; which was to be proved; (*L.*) quod erat demonstrandum; Q.E.D.; q.e.d.

しょうめつ 消滅 extinction; killing; vanishing.
～確率 extinction probability. ～時刻[時間] extinction time. ～法 killing method. ～測度 killing measure. ～作用素 killing operator. ～定理 vanishing theorem.

しょうめつする 消滅する collapse; vanish.

しょうめん 焦面 focal plane [surface]; caustic; caustic surface. ⇨火面.
～の causric.

しょうめん 正面 front.
～の front. ～図 (front) elevation.

じょうよ 剰余 remainder; residue.
～の residual. ～項 remainder term. ～定理 remainder theorem. 開平[開立]～ square [cubic] remainder. 平方～ quad-

ratic residue. n ベキ〜 residue of nth power. 〜類 residue [remainder] class; coset. 左[右; 両側]〜類 left [right; double] coset. 〜(類)群[空間] residue class group [space]. 〜指標 residue character. 〜スペクトル residual spectrum. 〜群 factor group. 中国式〜定理 chinese remainder theorem.

じょうようたいすう 常用対数 common logarithm.

じょうよるい 剰余類 residue [remainder] class; coset.
〜群 residue class group. 左[右]〜 left [right] coset. 両側〜 double coset.

しょうらい 将来の prospective.

じょうらせん 常螺旋 ordinary helix.
〜面 ordinary helicoid.

しょうりゃく 省略 abbreviation; abridg(e)ment; omission.
〜する abbreviate; abridge; omit. 〜算 abridged calculation; contracted calculation. 〜符号 ellipsis (*pl.* -ses).

しょうりゃく- 省略 approximate.
〜代数 approximate algebra. 〜加法[乗法] approximate addition [multiplication].

じょがい 除外 exception; exclusion.
〜の exceptional; excluded. 〜する except; exclude. 〜値 exceptional value; deficient value. 〜点[曲線; 集合] exceptional point [curve; set]. 〜近傍 deleted neighbo(u)rhood.

じょがいしすう 除外指数 defect.

しょき 初期(の) initial.
〜位相 initial phase. 〜条件 initial condition. 〜値 initial value. 〜値問題 initial value problem. 〜データ initial data. 〜分布 initial distribution.

じょきょ 除去 removal; discharge; exclusion.
〜する remove; exclude. 〜可能な removable. 〜可能(な)特異点 removable singularity. 二重否定の〜 discharge of double negation.

じょきょう (大学の)助教 assistant.

しょくしゅうごう 触集合 closure; (*F.*) adhérence; (*G.*) Berührungsmenge.

しょくせん 触線 supporting line. ⇨支持.

しょこう 初項 first term; initial term.

しょざい 所在 location.

しょしき 書式 form; format.

しょしょ 諸所(に) at several places; (*L.*) passim.

じょすう 序数 ordinal; ordinal number.
〜の ordinal.

じょすう 除数 divisor.

じょする 除する divide.

しょそく 初速 initial velocity.

しょぞく 所属する belong [be attached] to.

しょち 処置 dealing; treatment; action.
〜する deal with; treat.

しょとう 初等(の) elementary.
〜数学 elementary mathematics. 〜幾何[代数](学) elementary geometry [algebra]. 〜整数論 elementary theory of numbers. 〜解法 elementary solution. 〜関数 elementary function. 〜超越関数 elementary transcendental function.

しょとうすう 諸等数 compound number.
〜の除法 compound division. 〜の減法 compound subtraction.

じょひのり 除比の理 dividendo.

ジョブ job.

しょぶん 処分 disposal.
〜価値 disposal value.

じょへんすう 助変数 parameter. ⇨媒介変数.
〜の parametric. 標準[局所]〜 canonical [local] parameter. 等温〜 isothermal parameter. 〜表示 parametric representation.

しょほ 初歩 element; the first step; the ABC.
〜の elementary; rudimentary.

じょほう 除法 division.
〜の divisional. 〜定理 division algorithm [theorem]. 連〜 division algorithm. 組立〜 synthetic division.

しょめい 署名 signature.
デジタル〜 digital signature.

しょゆうする 所有する carry; possess.

しょようじかん 所要時間 duration; the time required.

しょり 処理 treatment; management; pro-

cess; processing; disposal.
～する deal with; treat; manage; process. ～効果 treatment effect. ～対比 treatment contrast. ～過程(の)誤差 processing error. 情報～ information processing. 数式～ formula manipulation.
ジョルダン C. Jordan (1838-1922).
～曲線[弧] Jordan curve [arc]. ～領域 Jordan domain [region]. ～細胞 Jordan cell [block].
じょろん 序論 introduction; introductory remark(s).
ジョン F. John (1910-1994).
じりつ 自律 autonomy.
～的な autonomous; autonomic.
しりょう 資料 data (*sing.* -tum).
しりょう 試料 sample. ⇨標本.
シリンダー cylinder. ⇨柱(ちゅう).
シール seal.
しるし 印 sign; mark; symbol; tag.
しれい 指令 command.
～する command.
じれい 自励(の) autonomous; self-excited; self-exciting.
～振動 autonomous oscillation; self-excited vibration. ～系 autonomous system; self-excited system.
しん 真 truth; reality.
～の true; real; genuine. ～の値 true value. ～の解 true solution; genuine solution. ～近点離角 true anomaly. ～偽 truth and falsity.
しん 真(の) proper.
～に properly. ～分数 proper fraction. ～部分集合 proper subset. ～部分群 proper subgroup. ～部分空間 proper subspace. ～凸関数 proper convex function. ～凸集合 proper convex set.
-しん 進 -adic.
α～の α-adic. p～整数 p-adic integer. p～体 p-adic number field.
じんい 人為(的) artificial.
～変数 artificial variable.
シンガー I. M. Singer (1924-).
しんかいせん 伸開線 involute.
しんぎ 真偽 truth and falsity. ⇨真理.
～関数 truth function.
じんきょくせん 腎曲線 nephroid.
しんけい 神経 nerve.
～の nervous.
しんげつずけい 新月図形 meniscus (*pl.* -cuses, -ci); crescent figure.
しんこう 進行 advance; progress.
～性の progressive. ～[進み]型 advanced type.
しんごう 信号 sign; signal.
じんこう 人口 population; number of inhabitants.
～調査 census (of population). ～統計(学) population statistics; demography. 過剰～ surplus population. ～増加率[法則] rate [law] of progression.
じんこう 人工(の) artificial; mechanical.
～知能 artificial intelligence.
じんこうとうけいがく 人口統計学 population statistics; demography.
～の demographic.
しんさりつ 心差率 eccentricity. ⇨離心.
しんし 振子 pendulum.
単～ simple pendulum. 円[球面]～ circular [spherical] pendulum. サイクロイド～ cycloid(al) pendulum. 可逆～ reversible pendulum.
しんしゃずほう 心射図法 gnomonic projection.
しんじゅ 真珠 pearl.
～曲線 pearl curve.
しんすう 真数 anti-logarithm.
じんすう 尽数 commensurable number.
不～ incommensurable number.
しんせい 真性(の) essential; proper.
～特異点 essential singularity. ～不連続な properly discontinuous.
じんせん 尽線 vanishing line.
しんぞうけい 心臓形 cardioid.
しんだん 診断 diagnosis (*pl.* -noses).
～の diagnostic.
じんてん 尽点 vanishing point.
しんど 深度 depth.
しんどう 振動 oscillation; vibration.

〜的な oscillatory. 〜する oscillate; vibrate. 〜数列 oscillating sequence. 〜定理 oscillation theorem. 〜子 oscillator; vibrator. 単〜 simple harmonic motion; SHM. 調和〜 harmonic oscillation. 非線形〜 nonlinear oscillation. 固有〜 eigen(-)oscillation. 自由[強制]〜 free [forced] oscillation. 減衰〜 damped oscillation. 緩和〜 relaxation oscillation. 自励〜 autonomous oscillation; self-excited vibration. 規準〜 normal [canonical] vibration. 〜絃[膜; 板] vibrating string [membrane; plate].

しんどうすう 振動数 frequency.
角〜 angular [circular] frequency.

しんぷく 振幅 amplitude (of oscillation [vibration]); oscillation; saltus; fluctuation.
〜関数 amplitude function. 確率〜 probability amplitude. 極限〜 limit oscillation.

シンプレクティック symplectic.
〜群 symplectic group. 射影[ユニタリ]〜群 projective [unitary] symplectic group. 〜変換 symplectic transformation.

シンプレックス simplex (*pl.* -lexes, -lices). ⇨単体.
〜法 simplex method. 〜・タブロー simplex tableau.

シンポジウム symposium (*pl.* -ums, -sia).

シンボル symbol. ⇨記号.

しんらい 信頼 confidence; reliance.
〜限界 confidence limit; fiducial limit. 〜係数 confidence coefficient. 〜水準 confidence level. 〜区間 confidence interval; fiducial interval. 〜帯[域] confidence belt [region]. 〜性 reliability.

しんり 真理 truth.
〜関数 truth(-value) function. 〜値 truth value. 〜値表 truth table.

じんりょう 尽量 commensurable quantity.

しんろ 進路 course; track.

しんわ 親和 amicability.
〜的な amicable. 〜数 amicable number.

しんわすう 親和数 amicable number.

す・ス

ず 図 figure; fig.; picture; diagram; chart. ベン～ Venn diagram. 矢線～ arrow diagram. 相関～ correlation diagram. 証明～ proof diagram; diagram of proof. 付～ attached figure [chart]. 投影～ projection chart. 流れ～ flow chart; flowchart.

すい 錐 cone.
～の conic; conical. 円[楕円]～ circular [elliptic] cone. 直[斜]円～ right [oblique] circular cone. 二次～ quadratic cone. 凸～ convex cone. 漸近～ asymptotic cone. 角～ pyramid. 三角～ triangular pyramid. 角～台 frustum of pyramid. 多面～ polyhedral cone. ～台 frustum of cone. ～面[体] cone; conical surface [body]. 欠[截頭]～ truncate(d) cone. 約～, 被約～ reduced cone. 写像～ mapping cone. 約写像～, 被約写像～ reduced mapping cone. 極～ polar cone. 正則～ regular cone. 接～ tangent cone. 法～ normal cone.

すいい 推移 transition. ⇨可遷(性).
～的な transitive. ～的に transitively. ～性 transitivity. ～法則 transitive law. ～的関係 transitive relation. ～群 transitive group. ～的置換群 transitive permutation group. ～過程 transition process. ～確率 transition probability ～行列 transition matrix. ～図 transition diagram. 単純[測度]～的な simply [metrically] transitive.

ずいい 随意(の) optional; free; voluntary; arbitrary.

すいけい 推計
～の stochastic(al). ～的に stochastically. ～学 theory of inference; inductive statistics; stochastics. ～紙 stochastic paper.

すいこう 遂行 execution; accomplishment.
～する execute; accomplish; perform.

すいこみ 吸込み sink.

すいさん 推算 calculation; estimation; estimate.

すいじゅん 水準 level; (*F.*) niveau.
～線[面] level curve [surface]; niveau curve [surface]. ～測量 level(l)ing. 管理～ level of control. 有意～ level of significance; significant level. 信頼～ confidence level.

すいしん 垂心 orthocenter.
～の orthocentric. ～四面体 orthocentric tetrahedron.

すいしん 推進 propulsion.
～力 propulsion.

すいせん 垂線 perpendicular.
～の足[長さ] foot [length] of perpendicular.

すいそく 垂足(の) pedal.
～面; ～線; ～線の pedal. ～三角形 pedal [orthic] triangle. ～曲線 pedal curve.

すいそく 推測 conjecture; guess; inference; presumption.

すいそくとうけいがく 推測統計学 theory of inference; inductive statistics; stochastics. ⇨推計.

すいたい 錐体 cone; conical body.

すいちょく 垂直 perpendicularity; verticality.
～な perpendicular; vertical; normal. ～変位 perpendicular [normal] displacement. ～条件 condition of perpendicularity. ～二等分線 perpendicular bisector; perpendicular at midpoint; midperpendicular;

bisecting normal. ～線[面] perpendicular; vertical; normal (line [surface]). ～成分 vertical component.

すいてい 推定 estimation.
～する estimate. ～可能な estimable. ～値 estimate. ～量 estimator. 最尤～法 maximum likelihood estimation. 統計的～ statistical estimation. 区間～ interval estimation. 点～ point estimation. ～空間 estimation space. ノンパラメトリック～ non-parametric estimation. こみにした～ pooled estimation.

すいていち 推定値 estimate.

すいていりょう 推定量 estimator.
不変～ invariant estimator. 不偏～ unbiased estimator. 最尤～ maximum likelihood estimator. 一致～ consistent estimator. 有効～ efficient estimator. 最小二[自]乗～ least square estimator.

ずいはん 随伴(の) adjoint; ajoint; associated; concomitant.
～(性) adjointness; concomitance. ～行列 adjoint matrix; associated matrix. ～行列式 adjoint determinant. ～式 adjoint (expression); concomitant. ～系 adjoint system. ～群[関手] adjoint group [functor]. ～表現 adjoint representation. ～微分方程式 adjoint differential equation. ～核[作用素] adjoint kernel [operator]. ～曲面 associated surface. ～素イデアル associated prime ideal. ～変数 concomitant variable.

ずいはんへんすう 随伴変数 concomitant variable.

すいへい 水平(な) horizontal.
～線 horizon; horizontal line. ～面 horizontal plane. ～成分 horizontal component. ～座標 horizontal coordinates. ～投象 horizontal projection. ～跡 horizontal trace.

すいほ 酔歩 random walk.

すいめん 錐面 cone; conical surface.
二次～ quadratic cone. 漸近～ asymptotic cone. 包絡～ enveloping cone.

すいり 推理 inference; reasoning.
統計的～ statistical inference. 論証的～ demonstrative reasoning.

すいろん 推論 deduction, inference; reasoning; argument.
～(規則) rule of inference. 統計的～ statistical inference.

すう 数 number; numeric.
～の numeral; numerary; numeric; numerical. 正[負](の)～ positive [negative] number. 自然～ natural number. 整～ integer; integral number. 分～ fraction; fractional number. 有理[無理]～ rational [irrational] number. 実[虚; 複素]～ real [imaginary; complex] number. 四元～ quaternion. 代数的[超越]～ algebraic [transcendental] number. 素[合成]～ prime [composite] number. 基～ cardinal number. 順序～ ordinal (number). 場合の～ number of cases. ルドルフ～ Ludolph number. ベルヌーイ(の)～ Bernoulli's [Bernoullian] number. オイラー(の)～ Euler's [Eulerian] number. ルベーグ～ Lebesgue number. 三[五]角～ triangular [pentagonal] number. 不足～, 輪～ deficient number. ～直線 real line; number line. ～体 number field. ～値 numerical value. ～係数 numerical coefficient. ～ベクトル numerical vector. ～的表現 numerical representation [evaluation]. クリーク～ clique number. 条件～ condition number. 染色～, 彩色～ chromatic number.

すうがく 数学 mathematics; math.
～の; 的な mathematical. ～的に mathematically. 初等[高等]～ elementary [advanced; higher] mathematics. 応用[純粋]～ applied [pure] mathematics. 古代(の)～ ancient mathematics; mathematics of antiquity. 現代～ modern mathematics. 物理～[数理物理学] mathematical physics. 計画～ mathematics for [of] programming. ～解析(学) mathematical analysis. ～基礎論 foundations of mathematics. ～的帰納法 mathematical induction. ～的確率 mathematical probability. ～的構造

[体系] mathematical structure [system]. ～的モデル mathematical model. ～遊戯[パズル] mathematical game [puzzle]. ～会 mathematical society. ～物理学会 physico-mathematical society. ～者 mathematician. ～原論 Élements de mathématique【ブルバキの著作】. 離散～ discrete mathematics.

すうけい 数系 number system; system of numbers.
有理[実; 複素]～ system of rational [real; complex] numbers.

すうし 数詞 numeral.

すうじ 数字 figure; digit; number; numeral.
～の digital; numeral; numerical; numerary. 二進～ binary digit. ～方程式 numerical equation. アラビア[ローマ]～ Arabic [Roman] numeral.

すうしき 数式 mathematical expression [formula].
～処理 formula manipulation.

すうじく 数軸 number axis.

すうじく 枢軸 pivot.
～の pivotal. ～演算 pivot operation; pivoting.

すうち 数値 numerical value.
～の numerical. ～的に numerically. ～方程式 numerical equation. ～解析 numerical analysis. ～計算 numerical calculation [computation]; numerical calculus. ～解法 numerical method [solution]. ～解(法) numerical solution. ～微分 numerical differentiation. ～積分 numerical integration. ～管理[制御] numerical control. ～的に同値[不安定]な numerically equivalent [unstable].

すうちょくせん 数直線 real line; number line.

すうり- 数理 mathematical.
～物理学 mathematical physics. ～論理学 mathematical logic. ～生物学 mathematical biology. ～経済学 mathematical economics. ～言語学 mathematical linguistics. ～科学 mathematical science. ～計画(法) mathematical programming. ～ファイナンス mathematical finance.

すうりょう 数量 quantity; scalar.

すうりょうか 数量化 quantification.
～する quantify. ～法 quantification method.

すうれつ 数列 sequence (of numbers); progression. ⇨列.
～の progressional. 等比[幾何]～ geometric progression; G.P. 等差[算術]～ arithmetic progression; A.P. 調和～ harmonic progression; H.P. 単調～ monotone sequence. 増加[減少]～ increasing [decreasing] sequence. 収束[発散; 振動]～ convergent [divergent; oscillating] sequence. 二[多]重～ double [multiple] sequence. 回帰～ recurrent sequence. 循環～ recurring sequence.

すうろん 数論 theory of numbers; number theory. ⇨整数論.
～的な number-theoretic; arithmetic.

ずかい 図解 illustration; explanation by figure(s); schema (*pl.* -mata).

ずがく 図学 graphics.

スカラー scalar.
～の scalar. ～量 scalar quantity. ～倍 scalar multiple. ～積 scalar product. ～乗法 scalar multiplication. ～三重積 scalar triple product. ～行列 scalar matrix. ～曲率 scalar curvature. ～場 scalar field. ～・ポテンシャル scalar potential. 曲率～ curvature scalar.

スキーム scheme.
アフィン～ affine scheme. 正規～ normal scheme.

すくない 少ない few; little.
少なくとも at least.

スクレロノーム(な) scleronomic; scleronomous.

ずけい 図形 figure; fig.
幾何～ geometrical figure. 平面[立体]～ plane [solid] figure. 対称～ symmetric figure. ディンキン～ Dynkin diagram. ヤング～ Young diagram.

スケーリング scaling.

スコア score.
ティ・～ T-score. 因子～ factor score. ～

検定 score test.

ずしき 図式 diagram; schema (*pl.* -mata).
～の graphic; graphical. ～算法 graphics. ～計算(法) graphic(al) calculation. ～解(法) graphic solution. ～微分[積分] graphic differentiation [integration]. ～補間 graphic interpolation. ～推計法 graphic(al) method of statistical inference. ～力学 graphic mechanics. ベン(の)～ Venn diagram. 可換な～ commutative diagram.

ずしん 図心 centroid.

ステラジアン steradian.

ストロフォイド strophoid.

すなわち すなわち namely; (*L.*) videlicet; viz.; that is; (*L.*) id est; i.e.; or.

スパイラル spiral. ⇨螺線.

スパン span.

スピノル spinor.
混合～ mixed spinor. 付点[無点]～ dotted [undotted] spinor. 共変[反変]～ covariant [contravariant] spinor. ～群 spinor group.

ずひょう 図表 graph; chart; diagram.

スピン spin.
～の spinorial. ～・ノルム spinorial norm. ～束 spin bundle. ～表現 spin representation. ～写像 spin mapping.

スプライン spline.
～関数 spline function.

スペクトル spectrum (*pl.* -tra, -trums).
～の spectral. 連続～ continuous spectrum. 離散～ discrete spectrum. 点～ point spectrum. (可換)環の～ spectrum of a (commutative) ring. プライム～ prime spectrum. ～分解 spectral decomposition [resolution]. ～表示[表現] spectral representation. ～解析 spectral analysis. ～半径 spectral radius. ～系列 spectral sequence. ～の重複度 spectral multiplicity. ～測度 spectral measure. ～積分 spectral integral. ～総合 spectral synthesis. ～同型[形]な spectrally isomorphic.

すべて すべての all; whole; every.
殆んど～ almost all; a.a.

ずほう 図法 projection.
透視～ perspective projection. 正角円錐～ conformal conic projection. 心射～ gnomonic projection.

スポット spot.

ずめん 図面 drawing; plan; sketch; plot.
～をかく draw; sketch; plot.

ずらしのへんかん ずらしの変換 shift transformation.

スラント-せき スラント積 slant product.

すり 刷り impression.
第3版の2～ the second impression of the third edition.

するどい 鋭い acute; sharp.

ずれ ずれ shear; lag; discrepancy; difference; drift.
～応力 shearing stress. ～の検定 location test.

すんぽう 寸法 size; measure; dimension.

せ・セ

せ　背　back; ridge.

せい　整　integrity.
～の integral; entire.　～に integrally.　～数 integral number; integer.　～式 integral expression; polynomial.　～商 integral quotient.　～因子 integral divisor.　～形式 integral form.　～元 integral element.　～閉包 integral closure.　～域 integral [integrity] domain.　～拡大 integral extension.　～関数 integral function; entire function.　～従属の integrally dependent.　～閉の integrally closed.　～級数 power series; series of positive powers.　～式級数 polynomial series.

せい　正(の)　positive; plus.　⇨負(の).
～に positively.　～(記)号 positive sign; plus sign.　～(の)数 positive number.　～(の)値 positive value.　～値の positive-valued; positive definite.　～の向き positive direction; positive orientation.　～値関数 positive(-valued) function.　～(の)根 positive root.　～の無限大 positive infinity.　～領域 positive region [domain].　半～(定)値の positive semi-definite.　～値形式 positive definite form.　～項級数 positive term series; series with [of] positive terms.　～の象限 positive quadrant [orthant].　～変動 positive variation.　総[全]～の totally positive.　～因子 positive divisor.

せい-　正　regular; direct; orthogonal; right.
～多角形[多面体] regular polygon [polyhedron].　～多面角 regular polyhedral angle.　～比例 direct proportion.　～射影[投影, 投象(法)] orthogonal projection.　～角錐[角柱] regular pyramid [prism].　～コノイド right conoid.

せいいき　整域　integral domain; integrity domain; domain of integrity.
ユークリッド～ Euclidean domain.　デデキント～ Dedekind domain.　素～ prime integral domain.　一意分解～ unique factorization domain; UFD.　主[単項]イデアル～ principal ideal domain; PID.

せいか　正価　net price.

せいかい　正解　correct answer [solution].

せいかく　正[精]確　correctness; accuracy; precision; exactness; sharpness.
～な accurate; precise; exact; correct.　～な評価 exact [precise, sharp] estimation.　～検定 exact test.

せいかくすい　正角錐　regular pyramid.

せいかくちゅう　正角柱　regular prism.

せいかつ　正割　secant; sec.

せいかん　整環　order.
極大～ maximal order.　左[右]～ left [right] order.

せいかんすう　整関数　entire function; integral function.
超越～ transcendental entire [integral] function.　有理～ rational integral function; polynomial.

せいき　世紀　century.
(第)二十～ twentieth century.　一～の centennial.　一～半の sesquicentennial.

せいき　生起　occurrence; incidence.
～行列 incidence matrix.

せいき　正規(性)　normality.
～の normal; normalized.　～に normally.　～行列 normal matrix.　～部分群 normal subgroup.　～(連)鎖 normal chain.　～基

normal basis. ～分布[曲線] normal distribution [curve]. ～母集団[標本] normal population [sample]. ～乱数 normal random numbers. ～環 normal ring. ～空間 normal space. ～付値 normal valuation. ～作用素 normal operator. ～連分数 normal continued fraction. ～関数 normal function. ～列 normal sequence [series]. ～族 normal family. ～確率過程 normal stochastic process. ～枠 normal frame. ～実形 normal real form. ～座標 normal coordinates. ～化 normalization. ～に平坦な normally flat. ～形 normal form; normalized form. ～系 normalized system. ～スキーム normal scheme.

せいきか 正規化 normalization.
～する normalize. ～条件 normalization condition; condition of normalization. ～(された)ベクトル normalized vector. ～(された)付値 normalized valuation. ～(された)関数 normalized function. ネーターの～定理 Noether's normalization theorem.

せいきかぐん 正規化群 normalizer.

せいきさ 正規鎖 normal chain.
～部分群 subnormal subgroup.

せいきそうきょくがた 正規双曲型(の) regularly hyperbolic.

せいきちょっこう 正規直交(の) orthonormal.
～化する orthonormalize. ～化 orthonormalization. ～基 orthonormal base. (完全)～系 (complete) orthonormal [orthonormalized] system. グラム・シュミットの～化法 Gram-Schmidt orthonormalization.

せいきひょうげん 正規表現 regular expression.

せいぎょ 制御 control.
～する control. ～関数 control function. ～変数 control variable. ～理論 control theory. 最適～ optimal control. 可容～ admissible control. 自動～ automatic control. 衝撃～ bang-bang control.

せいけい- 正型 of positive type.
～数列 sequence of positive type.

せいけい- 星型 ⇨星型(ほしがた-).

せいけいたい 星形体 star body.

せいげん 正弦 sine; sinus; sin.
～状(の) sinusoid; sinusoidal. ～法則[定理] sine rule [theorem]. ～公式 sine formula. ～曲線 sine curve. ～波 sine wave. ～対数方眼紙 logarithmic sine paper. 逆～ inverse sine. 双曲～ hyperbolic sine; sine hyperbolic. 反復～ iterated sine. 積分～ integral sine. ～積分 sine integral. ～変換 sine transformation; sine transform.

せいげん 制限 restriction; restraint.
～する restrict. 写像の～ restriction of mapping. ～直積 restricted direct product. ～ホロノミー群 restricted holonomy group. ～写像 restriction map.

せいげんかんび 制限完備な conditionally complete.

せいこう 成功 success; hit.
～する succeed; hit.

せいごう 正号 positive sign; plus sign.

せいごう 整合(的な) well-formed.
～論理式 well-formed formula.

せいごひょう 正誤表 errata; corrigenda.

せいさく 政策 policy.
最適～ optimal policy.

せいさん 生産 production.
～計画(法) production planning [scheduling]. ～管理 production control. ～指数 production index. ～年齢 production age. ～工程一覧表 (production) flow chart [sheet]. ～ライン最適編成 (production) line balancing.

せいさんかくけい 正三角形 regular triangle.

せいさんしゃ 生産者 producer.

せいし 正矢 versed sine; versine; vers.

せいし 静止 rest; quiescence; repose.
～の quiescent. ～点 rest point.

せいじ 斉次(性) homogeneity. ⇨同次(の).
～の homogeneous. ～方程式 homogeneous equation. ～座標 homogeneous coordinates. ～イデアル homogeneous ideal.

せいしき 整式 polynomial; integral expression.
有理～ rational integral expression. ～級

数 polynomial series.

せいしつ 性質 property; nature; character.
幾何学的～ geometric(al) property. 大域[局所]的～ global [local] property. ベールの～ Baire property. 普遍的～ universal property.

せいじてき 斉次的(な) homogeneous. ⇨斉次(性).

せいじてき 斉時的(な) time-homogeneous.
～マルコフ連鎖 time-homogeneous Markov chain.

せいしめんたい 正四面体 regular tetrahedron (*pl.* -rons, -dra).
～群 tetrahedral group.

せいしゃえい 正射影 orthogonal projection.
～の orthographic; orthographical. ～法 method of orthogonal projection; orthography. ～作用素 orthogonal projection.

せいじゅうにめんたい 正十二面体 regular dodecahedron (*pl.* -rons, dra).
～群 dodecahedral group.

せいじゅっかくけい 正十角形 regular decagon.

せいじゅん 正準(の) canonical.
～方程式 canonical equation. ～変換 canonical transformation. ～変数[変量] canonical variable [variate]. ～相関分析 canonical correlation analysis.

せいじょ 整除(性) divisibility.
～できる divisible; aliquot. ～できない indivisible; aliquant. ～関係 divisibility relation.

せいじょう 性状 behavio(u)r.
漸近的～ asymptotic behavio(u)r. 境界～ boundary behavio(u)r.

せいじょう 正常(な) ordinary; regular.
～解 ordinary solution. ～要素 ordinary element. ～流 regular flow.

せいじょう 星状 star; star-likeness; star-shape.
～の star; star(-)like; star-shaped; starred. ～体 star; star-like body. ～領域 star [star(-)like, star-shaped] domain. ～細分 star--refinement. ～多角形 starred polygon.

せいず 製図 drawing; drafting; protraction.
～する draw; draft. ～機械 drawing instrument; draftman's instrument. ～法 graphics.

せいすい 静水 dead water, still water.
～の hydrostatic. ～領域 dead water region.

せいすう 整数 integer; integral number.
～の integral. 正[負]の～ positive [negative] integer. ～比 ratio of integers; integral ratio. ～根 integral root. ～部分 integral part. ～基 integral basis. 有理[代数的]～ rational [algebraic] integer. 複素[ガウスの]～ complex [Gaussian] integer. p 進～ p-adic integer. ～計画法 integer programming.

せいすうろん 整数論 theory of numbers; number theory. ⇨数論.
～的な number-theoretic. 代数[解析]的～ algebraic [analytic] theory of numbers.

せいせい 生成 generation; production; creation.
～する generate; produce. ～元 generating element; generator. ～関数 generating function. ～作用素 generator; creation operator. ～(樹)木 generation tree. ～表現 generating representation. ～系 generator system; system of generators. ～された群[位相] generated group [topology]. 有限～の finitely generated. ～行列 generator matrix. ～多項式 generator polynomial.

せいせいげん 生成元 generator; generating element.

せいせいたかくけい 星正多角形 concave regular polygon.

せいせいてん 生成点 generic point.

せいせつ 正接 tangent; tan.
～の tangent; tangential. ～法則 law of tangent; tangent rule. ～曲線 tangent curve. 双曲～ hyperbolic tangent. ～対数方眼紙 logarithmic tangent paper. ～係数 tangent coefficient.

せいそく 正則(性) regularity; holomorphy.
～な regular; holomorphic(al). ～(的)に regularly; holomorphically. ～行列 regular matrix. ～曲線 regular curve. ～連分数

regular continued fraction. ～(な)埋込み[埋蔵] regular embedding. ～近似列 regular exhaustion. ～表現 regular representation. ～空間 regular space. ～事象 regular event. ～点 regular point; non-singular point. ～元 regular element. ～関数 regular function; holomorphic function. ～微分 holomorphic differential. ～写像 regular [holomorphic] mapping. ～部(分) regular [holomorphic] part. ～断面曲率 holomorphic sectional curvature. ～領域 domain of holomorphy. ～包 envelope of holomorphy. ～凸な holomorphically convex. ～ホモトープ regularly homotopic. ～化 regularization. ～局所環 regular local ring. ～錐 regular cone. ～パラメータ系 regular system of parameters. ～列 regular sequence.

せいそくほう　正則包　envelope of holomorphy.

せいそくりょういき　正則領域　domain of holomorphy.

せいぞん　生存　existence; life; survival.
～時間 life time; terminal time. ～年金 life annuity. ～者 survivor. ～率 ratio of survival. ～関数 survival function.

せいぞんほけん　生存保険　pure endowment (insurance).

せいたいしょうしき　整対称式　symmetric polynomial.

せいたかくけい　正多角形　regular polygon.

せいためんたい　正多面体　regular polyhedron (*pl.* -rons, -dra).

せいち　正値　positive-valuedness; positive-definiteness; positivity.
～の, 正定値の positive-valued; positive(-)definite. ～関数 positive-valued function. ～二次形式 positive(-)definite quadratic form.

せいちょう　生長　growth.
～曲線 growth curve; logistic curve.

せいていち　正定値(の)　positive(-)definite.

せいてき　静的な　static.

せいてき　整的に　integrally. ⇨整.

せいてん　臍点　umbilic; umbilical point; (*L.*) umbilicus (*pl.* -cuses, -lici).
～のない umbilic-free. 全～的な totally umbilical. ～曲面[超曲面] umbilical surface [hypersurface].

せいと　生徒
(高校以上の)～ student. (小中学校の)～ pupil; schoolboy[girl].

せいど　精度　precision; accuracy; exactness.
達成～ attained precision; reached precision. 要求～ prescribed precision.

せいにじゅうめんたい　正二十面体　regular icosahedron (*pl.* -rons, -dra).
～群 icosahedral group.

せいはちめんたい　正八面体　regular octahedron (*pl.* -rons, -dra).
～群 octahedral group.

せいひれい　正比例　direct proportion.
～の directly proportional. に～する be in direct proportion to.

せいぶつ-　生物　bio-.
～学 biology. ～測定学; 計量～学 biometrics; biometry. ～測定学の, 計量～学の biometric. ～検定法 bioassay.

せいぶん　成分　component.
～ごとの[に] componentwise. 接線～ tangential component. 法線～ normal component. 水平[垂直]～ horizontal [vertical] component. 直交～ orthogonal component. 既約～ irreducible component. 半単純～ semisimple component. ベキ零[単]～ nilpotent [unipotent] component. 連結～ connected component. 共変[反変]～ covariant [contravariant] component. ～ベクトル component vector. ～表示 representation by components. 対角～ diagonal element.

せいへい　整閉の　integrally closed.

せいほうぎょうれつ　正方行列　square matrix.

せいほうけい　正方形　square; quadrate.
～の square; quadrate. ～グラフ square diagram.

せいぼうけい　星芒形　asteroid; astroid.

せいほうしょうけい　正方晶系　tetragonal system.
～の tetragonal.

せいほうぶんかつひょう 正方分割表 square table.

せいみつ 精密 accuracy; precision; exactness. ⇨正[精]確.

～な accurate; precise; exact; close; detailed. ～に accurately; precisely; exactly; closely; in detail. ～科学 exact science(s). ～標本論 exact sampling theory. ～調査 close investigation.

せいめい 生命 life.

～科学 life science. ～保険 life insurance. ～表 life table; mortality table.

せいやく 制約 restriction; constraint.

～条件 (condition of) constraint. ～想定 constraint qualification. 有効～ active constraint.

せいやくそうてい 制約想定 constraint qualification.

せいやくつき 制約付き(の) constrained.

～最適化 constrained optimization.

せいりきがく 静力学 statics.

流体～ hydrostatics.

せいれき 西暦 Christian Era.

～年数 Anno Domini; A.D. ～1979年 A.D. 1979; 1979 A.D.

せいれつ 整列 well-ordering.

～順序 well-order. ～された well-ordered. ～集合 well-ordered set. 逆～集合 inversely well-ordered set.

せいろくめんたい 正六面体 regular hexahedron (*pl.* -rons, -dra); cube.

～の regular hexahedral; cubic; cubical. ～群 hexahedral group.

せかい 世界 world.

～線[点] world line [point].

セカント secant; sec. ⇨正割.

せき 積 product.

論理～ product of propositions. 内[外]～ inner [outer] product. スカラー[ベクトル]～ scalar [vector] product. 交叉[クロス]～ cross product. 括弧～ bracket product. 乗～ product. 乗～記号 product notation; Π. 無限(乗)～ infinite product. 部分～ partial product. デカルト～ Cartesian product. 合成～ composition (product); (*G.*) Faltung. 楔～ wedge product. 内[外]部～ internal [external] product. カップ～ cup product. 直～ direct product. 直～測度 product measure. テンソル～ tensor product. スラント～, 斜～ slant product. 双対～ coproduct. ～和 sum of products; product sum. ～事象 product event. ～集合[空間] product set [space]. ～計量 product metric. ～閉集合 multiplicatively closed set; multiplicative set.

せき 跡 trace; (*G.*) Spur.

被約～ reduced trace. 水平～ horizontal trace. 直立～ vertical trace. ～公式 trace formula.

せきい 赤緯 declination.

せきどう 赤道 equator.

～の equatorial. 天球～ celestial equator. ～儀 equatorial telescope. ～座標 equatorial coordinates.

せきにん 責任 responsibility; obligation; liability.

連帯～ joint responsibility [liability]; solidarity.

せきぶん 積分 integral; integration.

～の integral. ～する integrate. ～可能[可～]な integrable. ～法[学] integral calculus. 微分～法[学] differential and integral calculus; (infinitesimal) calculus. ～論 theory of integration. 不定[定]～ indefinite [definite] integral. ～変数[限界] variable [bound] of integration. ～定数 integration constant. 重(複)～ multiple integral. 二[三]重～ double [triple] integral. 累次～ repeated integral. 無限～ infinite integral. 広義[変格; 仮性]～ improper integral. 直～ direct integral. 部分～ integration by parts. 置換～ integration by substitution. 項別～ termwise integration. 実[複素]～ real [complex] integral. リーマン～ Riemann [(R-)] integral. ルベーグ～ Lebesgue [(L-)] integral. スティルチェス[ダルブー]～ Stieltjes [Darboux] integral. 抽象～ abstract integral. 中間～ intermediate integral. (曲)線[面]～ curvilinear [surface] integral. 路に沿う～ contour integral. ～路［(領)域］ path [domain] of integration. ～因子［因数］ integrating factor. ～要素 integral element. ～曲線［曲面］ integral curve [surface]. ～多様体 integral manifold. ～方程［不等］式 integral equation [inequality]. ～幾何学 integral geometry. ～二次［双一次］形式 quadratic [bilinear] integral form. ～演算 integration operation. ～変換 integral transformation; integral transform. ～作用素［演算子］ integration operator. ～公式 integral formula. ～表示［定理］ integral representation [theorem]. 正弦［対数］～ sine [logarithmic] integral (function). 楕円～ elliptic integral. 調和～ harmonic integral. 確率～ probability integral; stochastic integral. 数値～ numerical integration. 経路～ path-integral.

せきぶん- 積分 integro-.
～微分方程式 integro-differential equation.

せきぶんかのう 積分可能(性) integrability. ⇨可積(性).
～な integrable. ～関数 integrable function; summable function. ～条件 condition of integrability; integrability condition.

せきぶんろ 積分路 path of integration; integration path; integration contour.

せきりつ 積率 moment. ⇨モーメント.
～母関数 moment generating function. ～法 moment method.

せきりょく 斥力 repulsion; repulsive force.

ゼータ zeta; 【Z, ζ】.
～関数 zeta function; ζ-function. 楕円～関数 elliptic zeta function.

せだい 世代 generation.
～数 generation number.

せつ- 接 tangent; tangential.
～平面 tangent plane; tangential plane. ～超平面 tangent(ial) hyperplane. ～空間 tangent space. ～ベクトル tangent vector. ～束 tangent bundle. ～バンドル tangent bundle. ～錐 tangent cone.

せっかく 接角 adjacent angle; contiguous angle.

せっきん 接近 approach.
～する approach.

せっけい 設計 design; plan; layout.
～の品質 design requirement. 論理～ logical design. ～距離 designed distance.

せっけい 楔形 ⇨楔(くさび).

せつごう 接合 join; junction.
～積 crossed product. ～点 junction.

せっし 摂氏 Celsius; centigrade.
～四度 four degrees centigrade; 4℃.

せつじょ 切除 excision.
～する excise. ～公理 excision axiom. ～同型［形］ excision isomorphism. ～平面 cutting plane.

せっしょく 接触 osculation; contact; tangency; touch.
～する osculate; contact; touch. ～角 contact angle. ～点 point of contact [osculation]. ～円 osculating circle. ～平面 osculating plane. ～要素 contact element. ～位数 order of contact. ～構造 contact structure. n位の～ osculation of nth order. 最大～ best osculation. ～変換 contact transformation.

せっする 接する touch; contact; come in contact.

せっせん 接線 tangent; tangent(ial) line.
～の tangent; tangential. 共通～ common tangent. 二重～ double tangent. 漸近～ asymptotic tangent. ～方向 tangential direction. ～方向の tangential. ～極座標 tangential polar coordinates. ～成分 tangential component. ～曲面 tangent surface. ～標形 indicatrix of tangent. ～応力 tangential stress.

せっせん 節線 nodal line.

せっせん 截線 slit; (*G.*) Schlitz.
～のはいった slit. 円弧～ circular slit. 放射～ radial slit. 平行～ parallel slit. ～平面［領域］ slit plane [domain]. ～写像 slit mapping.

せっせん 截線 transverse line.

せっせんえい 接線影 subtangent.

せつぞく 接続 connection; connexion; continuation; prolongation.
～する connect; continue; prolong. ～可

能な prolongeable; continuable. ～幾何学 geometry of connection. 線形[アフィン] ～ linear [affine] connection. リーマン[計量; カルタン; 射影; 共形]～ Riemannian [metric; Cartan; projective; conformal] connection. ～係数[径数] coefficients [parameters] of connection. ～形式 connection form. 解析～ analytic continuation [prolongation]. 調和～ harmonic continuation [prolongation]. 直接[間接] ～ direct [indirect] continuation. 解析～可能な analytically continuable [prolongeable]. リーマン面の～ continuation of Riemann surface. ～公式 connection formula. ～定理 continuation theorem. ～行列 incidence matrix.

ぜったい 絶対 absoluteness.
～の absolute. ～(的)に absolutely. ～幾何学 absolute geometry. ～値 absolute value; modulus. ～曲率 absolute curvature. ～共変式 absolute covariant. ～誤差 absolute error. ～ノルム absolute norm. ～不等式 absolute inequality; unconditional inequality. ～不変式[量; 元] absolute invariant. ～形 absolute (form). ～微分 absolute differential. ～微分学 absolute differential calculus. ～類体 absolute class field. ～値級数 absolute series. ～収束 absolute convergence. ～収束の absolutely convergent. ～連続[可積]な absolutely continuous [integrable]. ～既約[安定]な absolutely irreducible [stable].

ぜったいち 絶対値 absolute value; modulus (*pl.* -li).

せつだん 切断 section; cut.
有理数の～ cut of rational numbers. 有理[無理]～ rational [irrational] cut. ～集合 cut set. ～面 section(al) face. 大域～ global section.

せっちゃくくうかん 接着空間 attaching space.

せってん 接点 point of contact; contact point; point of osculation; point of tangency.

せってん 節点 node; nodal point. ⇨結節点.

ゼット z; Z.
～平[球]面 *z*-plane [-sphere]. ～変換 *z*-transformation.

せっとう 切[截]頭 truncation.
～の truncate; truncated. ～体 frustum. ～錐 frustum of cone; truncate(d) cone. ～角柱 truncate(d) prism.

せつどう 摂動 perturbation.
～法 perturbation method; method of perturbation. 永年～ secular perturbation. 正則[特異]～ regular [singular] perturbation.

せっとうたい 切[截]頭体 frustum (*pl.* -tums, -ta).

せっぺん 切[截]片 segment; intercept.
y～ *y*-segment [-intercept].

せつめい 説明 explanation; exposition; interpretation; illustration; instruction.
～する explain; interpret; illustrate; instruct. ～的な explanatory; interpretative; illustrative; expository. ～変数 explanatory variable.

せまい 狭い narrow.

セミナー seminar. ⇨ゼミナール

ゼミナール seminar; (*G.*) Seminar.

セル cell.
～の cellular. ～オートマトン cellular automaton.

ゼロ zero. ⇨零(れい).
アレフ・～ aleph zero; 【$\aleph_0$】.

ぜろいちほうそく 0-1 法則 zero-one law; 0-1 law; law of naught or one. ⇨悉無律.

せろん 世論 public opinion. ⇨世[輿]論(よろん).
～調査 public opinion poll; public opinion survey.

せん 千 thousand; 1000; M; m. ⇨百万.
【米】 1000^5 quadrillion. 1000^7 sextillion; sexillion. 1000^8 septillion. 1000^9 octillion. 1000^{11} decillion. 【英】 1000^{16} octillion. 1000^{20} decillion.

せん 線 line.
実～ solid line. 点～ dotted line. 破～ broken line. 波～ wavy line. 鎖～ chained line. 折れ～; 屈折～ broken line; polygonal line. 相関折れ～ broken line of

correlation. 直～ straight line. 曲～ curved line; curve. 接[法]～ tangent [normal] (line). 母～ generating line; generator. 支持～ supporting line. 反帰～ line of regression. 力～ line of force. 基～ ground line. ～分 (line) segment. ～図式 line diagram. ～素 line element. ～野 field of lines. ～束 pencil of lines. ～叢 congruence of lines. (曲)～積分 curvilinear integral; line integral. ～密度 line(ar) density. ～状緩和 line relaxation. ～対称 axial symmetry.

ぜん- 全 total; whole; complete; all; full.
～空間 total [whole; full] space. ～微分 total differential. ～微分可能な totally differentiable. ～微分方程式[形式] total differential equation [form]. ～変動 total variation. ～曲率 total curvature. ～次数 total degree. ～順序 total order; linear order. ～順序集合 totally ordered set. ～有界な totally bounded; precompact. ～不連結な totally disconnected. ～臍面 totally umbilic surface. ～測地的な totally geodesic. ～正な totally positive. ～虚[特異]な totally imaginary [singular]. ～順列 all permutation. ～事象 whole event. ～逆像 complete inverse image. ～近傍系 complete system of neighbo(u)rhoods. ～線形群 full linear group. ～行列環 total [full] matrix ring [algebra]. ～正規な fully normal. ～相関 total correlation. ～数[部]検査 total inspection; complete count. ～透視図 total perspective. ～確率の公式 total probability law. ～商環 total ring of quotients; ring of total quotients; total quotient ring.

ぜん- 前 pre-; previous; ante-.
～コンパクト precompact. ～層 presheaf. ～概型 prescheme. ～順序 preorder. ～代数多様体 prealgebraic variety. ～ヒルベルト空間 pre-Hilbert space. ～提 premise. ～記の above- [previously] stated [mentioned]. ～項 antecedent. ～件 antecedent.

せんい 遷移 transition. ⇨推移.
～図 transition diagram. ～確率 transition probability.

せんえん 尖円 conoid.

せんが 線画 line drawing.

ぜんか 漸化 ⇨再帰.
～の, ～的な recurrent; recurring; recursive. ～(公)式 recurrence formula; recursion formula; recurring formula; recursive formula.

ぜんがいけい 前概型 prescheme.

ぜんかくりつ 全確率 total probability.
～の公式 total probability law.

ぜんきん- 漸近(的) asymptotic; asymptotical.
～的に asymptotically. ～線 asymptote; asymptotic line. ～接線 asymptotic tangent. ～的性状 asymptotic behavio(u)r. ～円[放物線] asymptotic circle [parabola]. ～錐面 asymptotic cone. ～曲線[方向] asymptotic curve [direction]. ～公式 asymptotic formula. ～表示 asymptotic representation. ～収束 asymptotic convergence; semi-convergence. ～級数[展開] asymptotic series [expansion]. ～値[路] asymptotic value [path]. ～法 asymptotic method. ～安定な asymptotically stable. ～解析 asymptotic analysis. ～正規性 asymptotic normality. ～不偏推定量 asymptotically unbiased estimator.

ぜんきんしゅうそく 漸近収束 asymptotic convergence; semi-convergence.

ぜんきんせん 漸近線 asymptote; asymptotic line.

せんけい 扇形 sector.
～の sectorial. ～領域 sectorial domain. ～グラフ sector graph.

せんけい 線形(の) linear. ⇨一次(の).
～に linearly. ～性 linearity. ～方程式 linear equation. ～代数 linear algebra. ～数学 linear mathematics. ～結合 linear combination. ～形式 linear form. ～空間 linear space. ～写像[演算] linear mapping [operation]. ～変換 linear transformation. ～作用素 linear operator. ～汎関数 linear functional. ～群 linear group. ～系 linear system. ～表現 linear represen-

tation. ～順序 linear order. ～順序集合 linearly ordered set. ～接続 linear connection. ～補間 linear interpolation. ～仮説 linear hypothesis. ～回帰 linear regression. ～相関 linear correlation. ～計画 linear programming; LP. ～モデル linear model. ～予測 linear prediction. ～回路 linear network. 区分的～な piecewise linear. ～独立[従属]な linearly independent [dependent]. ～離別[無関連]の linearly disjoint. ～化 linearization. 完備～系 complete linear system. ～同値 linear equivalence. ～符号 linear code.

せんけいぐん 線形群 linear group.
一般～; 全～ general [full] linear group. 特殊～ special linear group; unimodular group. 射影一般～ projective general linear group.

せんけいけい 線形系 linear system.
完備～ complete linear system. 特性～ characteristic linear system. 随伴～ adjoint linear system.

せんけいけいかく 線形計画(法) linear programming; LP.

せんげん 選言 disjunction.
～的な disjunctive. ～命題 disjunction proposition. ～三段論法 disjunctive syllogism.

ぜんけん 前件 antecedent.

ぜんげん 漸減 gradual decrease.

せんけんてき 先験的 (*L.*) a priori.
～確率 a priori probability.

ぜんこう 前項 antecedent.

せんこうしゃ 先行者 predecessor; antecedent.

せんさい 尖細(な) thin.
～集合 thin set.

せんざい 潜在 latency.
～(的な) latent; potential. ～価格 shadow price. ～変数 latent variable.

ぜんじしょう 全事象 whole event.

ぜんしゃ 前者 antecedent; the former.

ぜんしゃ 全射 surjection; onto(-)mapping; epimorphism.
～の surjective. ～準同型[形] surjective homomorphism; epimorphism. 標準的～ canonical surjection.

せんしゅつ 選出 choice; selection. ⇨選択.
～公理 axiom of choice. ～集合 choice set. ～定理 selection theorem.

ぜんじゅんじょ 全順序 total order; linear order.
～集合 totally ordered set.

ぜんじゅんじょ 前順序 preorder.

ぜんじゅんれつ 全順列 all permutation.

ぜんしょう 全称(の) universal.
～命題 universal proposition. ～記号 universal quantifier; ∀.

せんしょく 染色(の) chromatic.
～数 chromatic number.

せんしょくめん 線織(曲)面 ruled surface.

ぜんしん 前進 progress; advance.
～的な progressive.

ぜんしん 前進(の) forward.
～差分[方程式] forward difference [equation]. ～段階 forward step. ～消去 forward elimination. ～代入 forward substitution.

ぜんすう 全数 total number.
～検査 total [100%] inspection. ～探索 exhaustive search.

せんせきぶん 線積分 curvilinear integral; line integral.

せんそ 線素 line element.

せんそう 線叢 congruence of lines.
一次～ linear congruence.

ぜんそう 前層 presheaf (*pl.* -eaves).

せんそく 線束 pencil of lines.

センター center; centre. ⇨中心.

ぜんたい 全体 whole; totality.
～の whole; total; entire. ～集合 whole set; universal set. ～として as a whole; on the whole; totally. ～正規な fully normal.

せんたいしょう 線対称 axial symmetry; axisymmetry.

せんたく 選択 choice; selection. ⇨選出.
～する choose; select. ～公理 axiom of choice. ～集合 choice set. ～関数 choice function. ～定理 selection theorem.

せんたくし 選択肢 alternative.

ぜんたんしゃ 全単射 bijection. ⇨双射.
～の bijective.
せんだんへんけい 剪断変形 shear.
センチ centi-.
～メートル centimeter; cm.
せんてい 選定 selection.
～母数 selection parameter. ～統計量 selection statistic.
ぜんてい 前提 premise.
～の antecedent. 大[小]～ major [minor] premise.
せんてん 尖点 cusp.
～の cuspidal; cusped. ～形式 cusp form. 放物的～ parabolic cusp. 二重～ double cusp; tacnode. ～の軌跡 cusp-locus. 三～内擺線 three-cusped hypocycloid; deltoid.
せんてん 選点 collocation.
～法 collocation method.
せんど 尖度 kurtosis.
セントロイド centroid.
せんばつ 選抜 selection; choice.
～する select; choose; pick up [out]. ～試験 selective [competitive] examination.
せんぶん 線分 segment; line segment.
～の segmental; segmentary. 有向[向きづけられた]～ directed segment.
せんぶん 千分
～の一 millesimal. ～の一の millesimal; milli-. ～率 permil; permillage.
せんべつ 選別 discrimination.
～する discriminate; screen. ～能力 discriminative power. ～検査 screening inspection.
ぜんぽう 前方(へ)の forward.
ぜんめん 前面 front.
～図 elevation.
せんや 線野 field of lines; line field.
せんりゃく 戦略 strategy.
～の strategic. ～変数 strategic variable. ～集合[空間] strategy space. 純粋[混合]～ pure [mixed] strategy. 最適～ optimal strategy.

そ・ソ

そ 素(な) prime; elementary.
互いに～ relatively [mutually] prime; disjoint; coprime. ～数[因数] prime number [factor]. ～元 prime element. ～商 prime quotient. ～体 prime field. ～イデアル prime ideal. ～整域 prime integral domain. ～点 prime spot. ～論理式 prime formula. ～解 elementary solution ～事象 elementary event. ～粒子 elementary particle. ～子 element.

そ 疎(な) non-dense; discrete; sparse.
～集合 non-dense set. ～系列 discrete series. (いたるところ)～な nowhere dense. ～行列 sparse matrix.

そいんし 素因子 prime divisor; prime factor.
非孤立[孤立]～ embedded [isolated] prime divisor.

そいんすう 素因数 prime factor.
～分解 factorization into prime factors.

そう 相 phase.
～の phasic. ～空間 phase space. ～平均 phase average. ～群 phase group.

そう 層 sheaf (*pl.* sheaves); (*F.*) faisceau (*pl.* -ceaux).
～空間 sheaf space. 定数～ constant sheaf. 自明な～ trivial sheaf. 連接～ coherent sheaf. 散布的～ scattered sheaf. 解析的～ analytic sheaf. 連続写像[関数]の芽の～ sheaf of germs of continuous functions. 誘導された～ associated sheaf. 前[準]～ presheaf. 可逆～ invertible sheaf. 標準～ canonical sheaf. 豊富～ ample sheaf. 余法線～ conormal sheaf.

そう 層 stratum (*pl.* -ta, -tums).
～にする stratify. ～別; ～化 stratification. ～化抽出(法) stratified sampling.

そう 層 lamina (*pl.* -nae, -nas); lamella (*pl.* -lae, -las); layer.
～状の laminar; lamellar. ～流 lamellar stream; laminar flow. 境界～ boundary layer. 薄～ thin layer. 単一[二重]～ simple [double] layer.

そう 叢 congruence.
線～ congruence of lines. 一次線～ linear congruence. 球～ sphere congruence.

そう- 双 bi-.

ぞう 像 image.
～集合 image set. 原[逆]～ inverse image; preimage; (*G.*) Urbild. ～領域 image domain [region]. ～空間 image space. ～測度 image measure. 余～ coimage. 閉[連続]～ closed [continuous] image.

そういちじ 双一次(性) bilinearity.
～の bilinear. ～形式 bilinear form. ～関数 bilinear function.

そうか 掃過 sweep; sweeping.

そうか 層化 sheafification.

そうか 層化 stratification.
～する stratify. 二重[二段]～ sub-stratification. ～標本 stratified sample.

ぞうか 増加 increase; augmentation; growth. ⇨減少.
～の increasing; incremental. ～する increase; augment; grow. ～関数 increasing function. ～列 increasing sequence. 狭義に[単調]～の strictly [monotone] increasing. ～の状態 increasing state. ～過程 increasing process. 緩～の slowly increasing.

そうかいせき 双解析(性) bianalyticity.

〜的な bianalytic. 〜写像 bianalytic mapping.
そうがく 総額 total; total amount; totality.
そうかせい 走化性 chemotaxis.
そうかへいきん 相加平均 arithmetic mean.
そうかほう 双加法(性) biadditivity.
〜的な biadditive.
そうかん 相関 correlation.
〜(的)の correlative. 〜表 correlation table. 〜比 correlation ratio. 〜係数 correlation coefficient. 〜図 correlation diagram. 〜テンソル correlation tensor. 線形〜 linear correlation. 重〜 multiple correlation. 順位〜 rank correlation. 偏〜 partial correlation. 自己〜 autocorrelation. 無〜な uncorrelated.
そうぎょう 操業 operation; work.
〜の operating; working. 〜費用 operating cost; operating expenses.
そうぎょうれつ 双行列 bimatrix.
〜ゲーム bimatrix game.
そうきょく 双極(の) bipolar.
〜座標 bipolar coordinates; dipole coordinates. 〜子 dipole. 〜ポテンシャル bipolar potential; dipole potential.
そうきょく- 双曲 hyperbolic. ⇨双曲型(の); 双曲的.
〜幾何学 hyperbolic geometry. 〜渦[匝; 螺]線 hyperbolic spiral. 〜正[余]弦 hyperbolic sine [cosine]. 〜放物面[柱面] hyperbolic paraboloid [cylinder].
そうきょくがた 双曲型(の) hyperbolic; of hyperbolic type. ⇨双曲的.
〜変換 hyperbolic transformation. 〜空間 hyperbolic space. 〜二次曲面 hyperbolic quadrics. 〜偏微分方程式 partial differential equation of hyperbolic type; hyperbolic partial differential equation. 正規〜の regularly hyperbolic. 〜領域 hyperbolic domain; domain of hyperbolic type. 〜リーマン面 hyperbolic Riemann surface.
そうきょくし 双極子 dipole; doublet.
そうきょくせん 双曲線 hyperbola.
〜の hyperbolic; hyperbolical. 共役〜 conjugate hyperbola. 直角[等辺]〜 rectangular [equilateral] hyperbola. 共焦〜 confocal hyperbolas. 〜関数 hyperbolic function. 〜正[余]弦 hyperbolic sine [cosine].
そうきょくてき 双曲的 hyperbolic. ⇨双曲(そうきょく-); 双曲型(の).
〜変換 hyperbolic transformation. 〜点 hyperbolic point. 〜幾何学 hyperbolic geometry. 〜空間形 hyperbolic space form. 〜柱(面) hyperbolic cylinder. 〜領域 hyperbolic domain.
そうきょくめん 双曲面 hyperboloid.
〜の hyperboloidal. 〜的な hyperboloidic. 一葉[二葉]〜 hyperboloid of one sheet [two sheets]. 回転〜 hyperboloid of revolution. 共焦〜 confocal hyperboloids. 〜的位置 hyperboloidic position.
そうきょくりつ 総曲率 total curvature; integral curvature.
そうぐう 遭遇 encounter; rencounter; (*F.*) rencontre. ⇨めぐり合い.
そうけい 総計 total; total amount; totality; grand total.
〜で in total. 〜…になる amount to...; add up to....
ぞうげん 増減 increase and decrease.
そうご 相互(の) reciprocal; mutual; cross. ⇨互いに.
〜作用; 〜性 reciprocity. 〜法則 law of reciprocity. 〜エネルギー mutual energy. 〜相関 cross correlation. 〜参照 cross reference. 〜対称な mutually symmetric.
そうこう 草稿 draft; manuscript.
そうごう 総合 synthesis (*pl.* -ses).
〜的(な) synthetic. スペクトル〜 spectral synthesis. 〜幾何学 synthetic geometry.
そうごさよう 相互作用 interaction. ⇨交互作用.
そうさ 操作 operation; handling; manipulation.
〜員 operator.
そうさ 走査 sweep.
そうさん 掃散 sweeping-out; (*F.*) balayage.
〜する balayage. 〜定理 balayage theorem; sweeping-out theorem. 〜原理 balay-

age principle; sweeping-out principle. ～測度[分布] balayaged measure.

そうじ 相似 similarity; similitude.
～な similar. ～(の)中心 center of similitude [similarity]. ～の内[外]心 internal [external] center of similitude. ～法 method of similitude. ～比 ratio of similitude; similitude ratio; ratio of homothety. ～の位置 position of similarity. ～(図)形 similar figures. ～的対応 similar correspondence. ～移動 similar translation. ～変換 similar transformation; similitude; homothetic transformation; homothety. ～な行列 similar matrix. ～な表現 similar representation. ～検定 similar test.

そうじしんしゅく 相似伸縮 homothety.
～の homothetic. ～図形 homothetic figure(s).

そうしゃ 双射 bijection. ⇨全単射.

そうしゃえい 双射影 biprojection.
～的な biprojective. ～空間 biprojective space.

そうしょう 双焦(の) bifocal.
～弦 bifocal chord.

そうじょう 層状(の) lamellar; laminar; solenoidal.
～流 lamellar stream; laminar flow; solenoidal stream [flow].

そうじょうかのう 相乗可能な multiplicable.

そうじょうけん 双条件 bicondition.
～的な biconditional. ～文 biconditional.

そうじょうへいきん 相乗平均 geometric mean.

そうしんしゃ 送信者 sender.

そうせい 総正(の) totally positive.

そうせいき 双正規(な) binormal.
～性 binormality.

そうせいそく 双正則(な) biregular.
～写像 biregular mapping. ～性 biregularity.

そうせん 匝線 spiral. ⇨螺線.
逆[双曲]～ reciprocal [hyperbolic] spiral.

そうせんけい 双線形(の) bilinear.
～形式 bilinear form. ～写像 bilinear mapping. ～汎関数 bilinear functional. ～性 bilinearity.

そうたい 相対(性) relativity.
～的な relative. ～的に relatively. ～誤差 relative error. ～頻度[度数] relative frequency. ～ベクトル[テンソル] relative vector [tensor]. ～不変式[元] relative invariant. ～位相 relative topology. ～ノルム relative norm. ～境界 relative boundary. ～成分 relative component. ～次数 relative degree. ～ホモロジー代数 relative homological algebra. ～運動 relative motion. ～速度 relative velocity. ～無矛盾性 relative consistency. ～コンパクトな relatively compact. ～最短線 relatively minimal curve. ～化 relativization. ～化された定理 relativized theorem. ～論 relativity theory. ～論的な relativistic. ～危険度 relative risk.

ぞうだい 増大 growth; increase. ⇨増加.
～する grow. 関数の～ growth of function.

そうたいせいりろん 相対性理論 theory of relativity; relativity theory.
～的な relativistic. 特殊[一般]～ special [general] theory of relativity.

そうち 装置 equipment; installation; apparatus (*pl.* -; -tuses); device; unit.

そうつい 双対(性) duality.
～な dual. ～関係 duality (relation). ～原理 duality principle; principle of duality. ～定理 dual theorem; duality theorem. ～基 dual base; dual basis. ～曲線 dual curve. ～写像 dual mapping. ～表現 dual representation. ～作用素 dual operator. ～胞体 dual cell. ～空間 dual space. ～問題 dual problem. 自己～な self-dual. ～ギャップ duality gap. ～グラフ dual graph. ～変数 dual variable.

そうつい- 双対 co-.
～代数 coalgebra. ～微分 codifferential. ～積 coproduct. ～境界 coboundary. ～輪体 cocycle. ～鎖 cochain. ～根 coroot.

そうついきょうかい 双対境界 coboundary.
～輪体 coboundary. ～作用素 coboundary operator.

そうついさ 双対鎖 cochain.

～写像 cochain mapping. ～同値な cochain-equivalent. ～複体 cochain complex. 部分～複体 cochain subcomplex. 相対～複体 relative cochain complex.

そうついせいそく 双対正則(性) coregularity.
～な coregular. ～表現 coregular representation.

そうついりんたい 双対輪体 cocycle.
～群 group of cocycles. 連続～ continuous cocycle. 障害～ obstruction cocycle.

そうとう 相等 equality.

そうにゅう 挿入 insertion.
～する insert. ～整列法 sorting by insertion.

そうはん 相反(性) reciprocity. ⇨相反変換.
～な reciprocal. ～的に reciprocally. ～関係 reciprocity relation. ～系 reciprocal system. ～法則 law of reciprocity; reciprocity law. ～方程式 reciprocal equation. ～回路 reciprocal network; bilateral network.

そうはんかいろ 相反回路 bilateral network; reciprocal network.

そうはんへんかん 相反変換 correlation.
対合的～ involutive correlation.

ぞうぶん 増分 increment.
～の incremental. ～関数 increment function. ～費用 incremental cost.

そうほ 相補(的) complementary; complemented.
～束 complemented lattice. ～加群 complementary module.

そうほう 双峰(の) bimodal.
～頻度曲線 bimodal frequency curve.

そうほせい 相補性 complementarity.
～問題 complementarity problem.

そうゆうり 双有理(的) birational.
～変換[写像] birational transformation [mapping]. ～対応 birational correspondence. ～同型[形] birational isomorphism. ～同値 birational equivalence.

そうりゅう 層流 laminar [lamellar] flow [stream].

そうわ 総和 total sum; summation.
～(法) summation. ～記号 summation notation [symbol]; 【Σ】. ～規約 summation convention. ～可能な summable. チェザロ～可能な Cesàro-summable.

そえじ 添字 index (*pl.* -dices, -dexes); suffix (*pl.* -fices, -fixes). ⇨添数.
上つき～ superscript. 下つき～ subscript.

そかい 素解 elementary solution.

そく 束 pencil.
線形～ linear pencil. 直線～ pencil of lines. 平面～ pencil of planes; axial pencil. 二次曲線[曲面]～ pencil of conics [quadrics].

そく 束 bundle.
～空間 bundle space. ～写像 bundle mapping. ベクトル～ vector bundle. 接[法]～ tangent [normal] bundle. ファイバー～ fibre bundle. 主～ principal bundle. 標準～ canonical bundle. 直線～ line bundle. 空間～ space bundle. 座標[枠]～ coordinate [frame] bundle. アフィン～ affine bundle. 球面～ sphere bundle.

そく 束 lattice.
～論 lattice theory. ～論的な lattice-theoretic. 線形～ linear lattice. ベクトル～ vector lattice. 空間～ space lattice. モジュラー～ modular lattice. 分配～ distributive lattice. 可補[相補]～ complemented lattice. 完備～ complete lattice. ブール～ Boolean lattice. バナッハ～ Banach lattice. ～空間 lattice space. ～順序の lattice-ordered. ～(順序)群[集合] lattice-ordered group [set]. ～順序線形空間 lattice-ordered linear space. ～(順序)同型[形] lattice isomorphism. 部分～ sublattice.

ぞく 族 class; family.
ベール[ハーディ]～ Baire [Hardy] class. 曲線～ family [class] of curves; curve family. 関数～ class [family] of functions; function family [class]. 集合～ family of sets. 加法～ additive class. 正規～ normal family. 有向～ directed family. 部分～ subfamily; subclass. 指数型分布～ exponential family.

そくがめん 側画面 side plane.

そくじ 即時(の) instantaneous.

そくじゅんじょ 束順序(の) lattice-ordered.
～集合[群] lattice-ordered set [group]. ～線形空間 lattice-ordered linear space. ～同型[形] lattice(-ordered) isomorphism.

ぞくする 属する belong to.

ぞくせい 属性 attribute.
～の attributive. ～統計 attribute statistics.

そくせき 側跡 side trace.

そくせん 測線 measuring line.

そくち 測地(的) geodesic.
～弧 geodesic arc. ～線 geodesic (line). ～円 geodesic circle. ～半径 geodesic radius. ～座標 geodesic coordinates. ～的極座標 geodesic polar coordinates. ～的曲率[捩率] geodesic curvature [torsion]. ～流 geodesic flow. ～的変位 geodesic deviation. ～的対応 geodesic correspondence. 全～的な totally geodesic. ～的(に)完備な geodesically complete.

そくちがく 測地学 geodesy.

そくちせん 測地線 geodesic; geodesic line.
～の geodesic. 閉～ closed geodesic (line).

そくてい 測定 measurement; mensuration.
～する measure. 直接[間接]～ direct [indirect] measurement. 条件付～ conditional measurement. ～値 measured value. 体積～ stereometry. 生物～学 biometry; biometrics. 投射～ projectometry.

そくてん 測点 measuring point.

そくど 速度 velocity.
等～ constant [uniform] velocity. 絶対[相対]～ absolute [relative] velocity. 角～ angular velocity. 初[終]～ initial [final] velocity. (位)相～ phase velocity. 群～ group velocity. ～ベクトル velocity vector. ～ポテンシャル velocity potential.

そくど 測度 measure.
～論 measure theory, theory of measures. ～論的な measure-theoretic. 内[外]～ inner [outer] measure. 有限[有界]～ bounded measure. 有限[完全]加法的～ finitely [completely] additive measure. 不変～ invariant measure. 離散～ discrete measure. 確率～ probability measure. 彷徨～ random measure. 標準～ reference measure. 個数～ counting measure. 像～ image measure. 商～ quotient measure. 消滅～ killing measure. スペクトル～ spectral measure. 表現～ representing measure. ボレル～ Borel measure. ルベーグ～ Lebesgue measure. ディラック～ Dirac measure. ～空間 measure space.

そくどかせん 測度可遷(性) metrical transitivity. ⇨測度推移(性).

そくどすいい 測度推移(性) metrical transitivity.
～的な metrically transitive.

そくばく 束縛 restriction; restraint; constraint.
～する restrict; restrain; constrain; bind. ～条件 restriction [restricting] condition. ～ベクトル fixed vector. ～変数 bound variable. ～運動 constrained motion. ～力 constraining force.

そくへん 側辺 lateral edge; (*L.*) latus (*pl.* latera).

そくめん 側面 side face; lateral (face).
～の lateral. ～図 side elevation.

そくりょう 測量(法) measurement; surveying.
～問題 measurement problem. 三角～ trigonometrical measurement [surveying]; triangulation. 測鎖～ chain surveying. 六分儀～ sextant surveying. 視距～ stadia surveying. ～器械 surveying instrument. 写真～ photomeasurement.

そくりょう 側稜 lateral edge.

ソクル, ソークル socle.

そし 素子 element; cell.
論理～ logic(al) element; decision element. 二値～ binary cell.

そしきてき 組織的 systematic; organized.

そすう 素数 prime; prime number.
～分布[表] distribution [table] of prime numbers. ～定理 theorem of prime numbers; prime number theorem. 双子～ twin prime numbers.

そせい 組成 composition; formation.
～列 composition series. ～因子 composi-

tion factor.

そと 外 outside; exterior. ⇨内(うち).
～の outer; exterior. ～向き(の)法線 outward normal; outer [exterior] normal.

そとがわ 外側 outside; outer side; exterior.

そのた その他 and so on [forth]; and others; (*L.*) et alii; et al.; (*L.*) et cetera; etc.; &c.

ソフトウェア software.

そぼく 素朴な primitive.

それ 外れ deviation. ⇨偏差; 偏倚.
～る deviate.

それいかんすう 疎零関数 function with scattered zeros.

それぞれ それぞれ(に) respectively.
～の respective.

それにおうじて それに応じて accordingly.

ソレノイダル solenoidal. ⇨管状(の), 層状(の).

ソレノイド solenoid.

それゆえ accordingly, consequently, hence; therefore, thence; as a result.

そろばん 算盤 abacus (*pl.* -cuses, -ci).
～派 abacus school.

ゾーン zone.

そんえき 損益 loss and gain; profit and loss.
～分岐点 break-even.

そんざい 存在 existence.
～する exist. ～の; ～上の existential. ～している existent. ～定理 existence theorem. ～領域 existence region. ～記号[命題] existential quantifier [proposition]. ～作用素 existential quantifier.

そんしつ 損失 loss.
情報の～ loss of information. ～の歩合 rate of loss. ～関数 loss function.

た・タ

た 他 other; rest.
～の other; another. ～の点では otherwise.

た- 多 many; poly-; multi-.

たい 体 body; solid.
～の solid. 凸[卵形]～ convex body. ～心格子 body-centred lattice. 球～ solid sphere. ～調和関数 solid harmonics. ～積分 volume integral. 落～ falling body. 弾性～ elastic body. 帯電～ charged [electrified] body. 二[三; 多]～問題 two(-) [three(-); many(-)] body problem.

たい 体 field; (*G.*) Körper.
～論 field theory. ～の拡大 extension of field; field extension. 部分～ subfield. 数～ number field. 実数～ real number field; field of real numbers. 有理数～ rational number field; field of rational numbers; field of rationals. 剰余類～ residue class field. ガロア～ Galois field. 二次～ quadratic field. 可換～ commutative field. 非可換～ non-commutative field. 斜～ skew field; (*G.*) Schiefkörper; non-commutative field. 商～ quotient field. 関数～ function field. 基礎～ basic field; ground field. 係数～ field of scalars; coefficient field. 拡大～ overfield; extension field. 素～ prime field. 分岐～ ramification field. 代数的閉～ algebraically closed field. 実～ real field; (*G.*) (formal) reeller Körper. 実閉～ real closed field. 完全～ perfect field. 固定～ fixed field. 代数関数～ algebraic function field. 代数的数～, 代数～ algebraic number field. 付値～ valuation field. 不変～ invariant field; fixed field.

たい 帯 zone; strip.
～の; ～状の zonal. 球～ spherical zone; zone of sphere. ～球関数 zonal spherical function. ～球調和関数 zonal harmonics. 特性～ characteristic strip. 陪特性～ bicharacteristic strip. 境界～ boundary strip. ～状領域 strip domain [region]. 平行～(状面分) parallel strip.

だい 台 base; basis.

だい 台 frustum (*pl.* -tums, -ta).
角錐～ frustum of pyramid. 直円錐～ frustum of right (circular) cone.

だい 台 carrier; support.
関数の～ carrier [support] of function. E 上に～をもつ be supported on E. ～空間 carrier space. ～多様体 support variety.

ダイアグラム diagram.
ディンキン・～ Dynkin diagram. ヤング・～ Young diagram.

ダイアディック dyadic.
単位～ unit dyadic, idemfactor.

ダイアド dyad.

だいあん 代案 alternative.

たいいき- 大域(的) global; globally; in the large; (*G.*) im Grossen. ⇨大域的(な).
～理論 global theory. ～次元 global dimension. ～変分法 calculus of variations in the large. ～切断 global section.

たいいきてき 大域的(な) global.
～に globally; in the large; (*G.*) im Grossen. ～理論 global theory. ～性質 global property. ～解析学[微分幾何学] global analysis [differential geometry]; analysis [differential geometry] in the large. ～変分学 calculus of variations in

the large. ～計画法 global programming. ～最小解 global minimizer. ～最大解 global maximizer. ～最適化 global optimization. ～収束性 global convergence.

だいいち 第一(の) first; primary.
～に first; in the first place. ～象限 first quadrant. ～余弦公式 first cosine formula. ～積分 first Integral. ～種 first kind. ～類 first class; first category. ～可算公理 first countability axiom. ～平均値の定理 first mean value theorem. ～基本形式[量] first fundamental form [quantity]. ～変分 first variation. ～障害[碍](類) primary obstruction.

だいえん 大円 great circle.

たいおう 対応 correspondence.
～する correspond to [with]. 相似[アフィン]～ similar [affine] correspondence. 共形[射影]的～ conformal [projective] correspondence. 逆～ inverse correspondence. 一対一～ one-to-one correspondence. 一意～ univalent correspondence. 双有理～ birational correspondence. 代数的～ algebraic correspondence. 円々～ circle-to-circle correspondence; (*G.*) Kreisverwandtschaft. 測地的～ geodesic correspondence. 等長～ isometric correspondence. ～原理 correspondence principle. ～点[辺] corresponding points [sides]. ～角[頂点] corresponding angles [vertices]. ～元 corresponding element; homologous element.

たいおうさせる 対応させる attach.

たいか 退化 degeneration; degeneracy.
～した, ～の degenerate. ～した degenerated. ～し(てい)ない[非～の] non-degenerate; non (-) singular. ～する degenerate. ～性 degeneracy. ～円錐曲線 degenerate conics. ～核 degenerate kernel. ～系列 degenerate series. ～双線形形式 degenerate bilinear form. ～作用素 degeneracy operator. ～次数 nullity.

たいかく- 対角 diagonal.
～線の diagonal. ～線 diagonal; diagonal line. ～集合 diagonal set. ～行列 diagonal matrix. ～成分 diagonal element. ～(線)論法 diagonal method. ～点 diagonal point. ～三角形 diagonal triangle. ～列 diagonal sequence. ～部分和 diagonal partial sum. ～射 diagonal morphism. ～化 diagonalization. ～優位 diagonal dominance. ～和 trace.

だいがく 大学 university; college.
(総合)～ university. (分科, 単科)～ college. ～院 (post) graduate course; graduate school.

たいかくか 対角化 diagonalization.
～する diagonalize. ～可能な diagonalizable.

たいかくわ 対角和 trace.

たいきゃく 退却 retreat; withdrawal.

たいきゅうかんすう 体球(調和)関数 solid harmonic function; solid harmonics.

たいきゅうかんすう 帯球(調和)関数 zonal harmonics.

たいきゅうせい 耐久性 durability; endurance.
～のある durable; enduring.

たいきょ 退去 departure; withdrawal.

たいぐう 対偶 contraposition; contrapositive; contrapositive proposition.

たいけい 体系 system. ⇨系.
～的な systematic. 数学的～ mathematical system. 数の～ number system; system of numbers. ～的理論 systematic theory.

だいけい 台形 【英】trapezium (*pl.* -zia; -ziums). 【米】trapezoid.
～の 【米】trapezoid; trapezoidal; 【英】trapezial. ～公式[法則] trapezoidal formula [rule]. 等脚～ isosceles trapezoid. 等角～ isogonal trapezoid.

だいけい (立体の)台形 frustum (*pl.* -tums, -ta). ⇨台(だい).

だいご 第五(の) fifth.
～公準 fifth postulate. ヒルベルトの～問題 fifth problem of Hilbert.

たいごう 対合 involution.
～的な involutive. ～変換 involution (transformation). ～的自己同型[形] involutive automorphism. ～的相反変換 involutive correspondence; involutive corre-

lation. ～的接触変換 involutive contact transformation.

たいざい 滞在 stay; sojourn.
～する stay; sojourn. ～時[期]間 staying [sojourn] time.

だいさん 第三(の) third.
～の third; tertiary. ～に thirdly. ～象限 third quadrant. ～比例項 third proportional ～種 third kind. ～基本形式 third fundamental form.

だいし 第四(の) fourth.
～象限 fourth quadrant. ～比例項 fourth proportional. ～調和点 fourth harmonic point.

たいじしき 待時式 delay system.

たいしょう 対称(性) symmetry.
～(的)な symmetric; symmetrical. 中心[点]～ central [point] symmetry. 線[軸]～ axial symmetry; axisymmetry. 面～ symmetry with respect to plane. 回転～な rotationally symmetric. 球～な spherically symmetric. ～軸[面] axis [plane] of symmetry. ～点 point of symmetry. ～の位置 position of symmetry. ～配置 symmetric configuration. ～律[法則] law of symmetry; symmetric law. ～図形 symmetric figure. ～式 symmetric expression. 基本～式 elementary symmetric expression. ～多項式 symmetric polynomial. 基本～多項式 elementary symmetric polynomial. ～変換[移動] symmetric transformation; symmetry. ～行列[テンソル] symmetric matrix [tensor]. ～空間[領域] symmetric space [domain]. ～群 symmetric group. ～多元環 symmetric algebra. ～核 symmetric kernel. ～分布 symmetric distribution. 反[歪]～な anti[skew]-symmetric. p重～な p-ply symmetric. ～子 symmetrizer. ～化 symmetrization.

たいしょう 対象 object.
～変数 object variable. ～領域 object domain. 始～ initial [cofinal] object. 終～ final object. 商～ quotient object. 零～ zero object. 部分～ subobject. 群～ group object.

たいしょう 対照 contrast.

たいしょうか 対称化 symmetrization.
～する symmetrize. ～可能な symmetrizable. 中心[円]～ circular symmetrization. シュタイナー～ Steiner('s) symmetrization. ～作用素 symmetrizer.

たいしょうきごう 対象記号 individual symbol.

たいしょうすう 帯小数 mixed decimal.

たいしん 対心(的) antipodal.
～点 antipode; antipodal point.

たいすう 大数 large number.
～の強[弱]法則 strong [weak] law of large numbers.

たいすう 対数 logarithm.
～の; ～的な logarithmic. ～的に logarithmically. 常用[自然]～ common [natural] logarithm. ～の底 base of logarithm. ～表 table of logarithms. 累次[反復; 重複]～ iterated logarithm. ～目盛[方眼紙] logarithmic scale [paper]. ～尺 logarithmic scale. ～方程式[不等式] logarithmic equation [inequality]. ～的ソボレフ不等式 logarithmic Sobolev inequality. ～螺線[渦線] logarithmic spiral. ～関数[曲線] logarithmic function [curve]. ～級数 logarithmic series. ～積分 logarithmic integral. 積分～ integral logarithm. ～(的)微分 logarithmic differentiation. ～(的)導関数 logarithmic derivative. ～的留数 logarithmic residue. ～(的)特異点[分岐点] logarithmic singularity [branch point]. ～的極 logarithmic pole. ～ポテンシャル[容量] logarithmic potential [capacity]. ～(的)減衰率 logarithmic decrement. 半～的な semi-logarithmic. ～的に凸な logarithmically convex. 離散～ discrete logarithm.

だいすう 代数 algebra.
初等～ elementary algebra. 高等～ advanced algebra. 線形～ linear algebra. ブール～ Boolean algebra. リー～ Lie algebra. 分配的～ distributive algebra. 外積[テンソル]～ exterior [tensor] algebra. 部分～ subalgebra. 論理～ algebra of logic. 双対～ coalgebra. 計算機～ computer algebra. 剛～ rigid algebra. 作用素～

operator algebra.

だいすう- 代数 algebraic; algebraical.
~式 algebraic(al) expression. ~和 algebraic sum. ~方程式 algebraic equation. ~曲線[曲面] algebraic curve [surface]. ~解析 algebraical analysis. ~幾何学 algebraic geometry. ~関数 algebraic function. ~分岐点 algebraic branch point. ~特異点 algebraic singularity. ~系 algebraic system. ~複体 algebraic complex. ~群 algebraic group. ~的数体, ~体 algebraic number field. ~多様体 algebraic manifold [variety]. ~拡大 algebraic extension. ~関数体 algebraic function field. ~的重複度 algebraic multiplicity.

だいすうがく 代数学 algebra. ⇨代数(だいすう).
~の algebraic; algebraical. 線形~ linear algebra. 抽象~ abstract algebra.

だいすうがた 代数型(の) algebroid; algebroidal.
~関数 algebroid(al) function.

だいすうてき 代数的 algebraic; algebraical.
~に algebraically. ~解(法) algebraic(al) solution. ~数 algebraic number. ~整数 algebraic integer. ~元 algebraic element. ~(数)体 algebraic number field. ~構造 algebraic structure. ~対応 algebraic correspondence. ~拡大(体) algebraic extension (field). ~閉包 algebraic closure. ~整数論 algebraic theory of numbers. ~位相幾何(学) algebraic topology. ~複体 algebraic complex. ~多様体 algebraic manifold. ~特異点 algebraic singularity. ~に同値な algebraically equivalent. ~に閉じた algebraically closed. ~閉体 algebraically closed field. ~に独立[従属]な algebraically independent [dependent]. ~概型 algebraic scheme. ~重複度 algebraic multiplicity.

たいせき 体積 volume; cubic content [measure]; solid content [measure]; cubage.
内[外]~ interior [exterior] volume; inner [outer] volume. 有向~ oriented volume. ~要素 volume element. ~積分 volume integral. ~計算 cubature. ~測定 stereometry.

たいせき 対蹠(的) antipodal.
~点 antipode, antipodal point.

たいせきそくてい 体積測定 stereometry.

だいぜんてい 大前提 major premise.

だいたい 代替 alternative.

たいちょうかく 対頂角 vertically opposite angle(s); vertical angle(s).

たいてん 対点 opposite point.

たいでん 帯電 electrification.
~させる charge; electrify. ~体 charged [electrified] body.

たいとう 対等 equivalence; equivalency; equipotence; equipotency.
~な equivalent; equipotent. ~な集合 equivalent sets.

ダイナミック・プログラミング dynamic programming; DP; D.P.

だいに 第二(の) second; secondary.
~に secondly. ~象限 second quadrant. ~余弦公式 second cosine formula. ~種 second kind. ~類 second class; second category. ~可算公理 second countability axiom. ~平均値の定理 second mean value theorem. ~基本形式[量] second fundamental form [quantity]. ~変分 second variation. ~共役空間 bidual space. ~障害[碍](類) secondary obstruction.

だいにゅう 代入 substitution.
~する substitute. 逐次~ successive substitution; (successive) iteration. 後退~ backward substitution. 前進~ forward substitution. ~の原理 substitution principle.

たいひ 対比 contrast.
標準化~ normalized contrast. 処理~ treatment contrast.

だいひょう 代表 representation; representative.
~的な representative. ~値 representative. ~元 representative element. 完全~系 complete set of representative.

だいひょうほん 大標本 large sample.
~論 large sample theory; theory of large samples.

タイプ type. ⇨型(かた); 型(の)(-がた).
たいぶんすう 帯分数 mixed fraction.
たいへん 対辺 opposite side; subtense.
タイム・シェアリング time-sharing.
だいもく 題目 title; subject; topic; theme.
たいよう 太陽 sun.
～の solar. ～年 solar [tropical] year. ～暦 solar calendar.
たいよう 耐用 durability; endurance.
～年数 durable years.
だいよう 代用 substitution.
～物 substitute. ～として as substitute; alternatively.
だいようでんかほう 代用電荷法 charge simulation method.
たいら 平らな plain; flat; even.
たいりつかせつ 対立仮説 alternative hypothesis (*pl.* -ses).
たいりゃく 大略 summary; (*F.*) résumé.
たいりょう 大量の extensive.
タウ tau; 【T, τ】.
タウバー A. Tauber (1866–1942).
～の Tauberian. ～型の定理 Tauberian theorem(s).
だえん 楕円 ellipse.
～形の; ～型の; ～の elliptic; elliptical; oblong. 共焦～ confocal ellipses. 相関～ correlation ellipse. ～座標 elliptic coordinates. ～(型)変換 elliptic transformation. ～幾何学 elliptic geometry. ～関数 elliptic function. ～積分 elliptic integral. ～母数関数 elliptic modular function. ～放物面 elliptic paraboloid. ～柱[筒] elliptic cylinder. ～曲線 elliptic curve. ～(的)点 elliptic point. ～運動 elliptic motion.
だえんがた 楕円型(の) elliptic; of elliptic type.
～変換 elliptic transformation. ～二次曲面 elliptic quadrics. ～偏微分方程式 partial differential equation of elliptic type; elliptic partial differential equation. ～作用素 elliptic operator; operator of elliptic type.
たえんすい 多円錐(の) polyconic.
～投射 polyconic projection. ～図法 polyconic projection.
だえんせい 楕円(型)性 ellipticity.
だえんたいちょうわかんすう 楕円体調和関数 ellipsoidal harmonics.
回転～ spheroidal harmonics.
だえんたい[めん] 楕円体[面] ellipsoid.
～の ellipsoidal. 回転～ ellipsoid of revolution; spheroid. 扁平[長]回転～ oblate [prolate] spheroid. 慣性～ ellipsoid of inertia. ～座標 ellipsoidal coordinates. ～調和関数 ellipsoidal harmonics.
だえんてき 楕円的 elliptic. ⇨楕円型(の).
～変換 elliptic transformation. ～幾何学 elliptic geometry. ～柱(面) elliptic cylinder.
たか 多価(性) many-valuedness.
～の many-valued; multi(-)valued. ～関数 many-valued function; multi(-)valued function.
たかい 高い high.
たかいちゅうしゅつ 多回抽出 multiple sampling.
たがいに 互いに mutually; each other.
～素な (mutually) disjoint; mutually [relatively] prime; coprime. ～直交の mutually orthogonal. ～対称な mutually symmetric. ～排反な mutually exclusive.
たがいにそ 互いに素な disjoint; relatively [mutually] prime; coprime.
～集合 disjoint sets. ～数[多項式] relatively prime numbers [polynomials].
たかく 多角(の) polygonal; multangular.
～数 polygonal number. ～形 polygon. ～錐 pyramid. ～柱 prism. ～測量 traverse measurement.
たかくけい 多角形 polygon.
～の polygonal. 正～ regular polygon. 凸[凹]～ convex [concave] polygon. 平面[空間]～ planar [spatial] polygon. 内[外]接～ inscribed [circumscribed] polygon. 捻れ～ twisted [skew] polygon. 力の～ force polygon; polygon of forces. 軌道～ path-polygon. ～領域 polygonal domain [region].
たかさ 高さ height; altitude.
たかだか 高々 at most. ⇨以下.

〜可算の at most countable [enumerable].

たきょく 多極(の) multipolar.
〜座標 multipolar coordinates.

たけい 多形[型](な) polymorphic.

たげん 多元(の) hypercomplex; of several unknowns.
〜方程式 equation with several unknowns. 〜数 hypercomplex number. 〜環 hypercomplex system; algebra. 〜体 division algebra.

たげんかん 多元環 algebra; hypercomplex system.
部分〜 subalgebra. 単純[対称]〜 simple [symmetric] algebra. 商[巡回]〜 quotient [cyclic] algebra. 可換〜 commutative algebra. 群〜 group algebra. 〜(の)同型[形][準同型[形]] algebra isomorphism [homomorphism]. 〜拡大 algebra extension. 〜類 algebra class. 基本〜 basic algebra. 道〜 path algebra.

たげんたい 多元体 division algebra.

たこう 多項(の) multinomial.
〜式 multinomial. 〜展開[定理] multinomial expansion [theorem]. 〜係数 multinomial coefficient. 〜分布 multinomial distribution.

たこうしき 多項式 polynomial.
〜の polynomial. 二[三]次の〜 polynomial of degree two [three]; quadratic [cubic] polynomial. 一[多]変数の〜 polynomial of one variable [several variables]. 同次[斉次]〜 homogeneous polynomial. 既約[可約]〜 irreducible [reducible] polynomial. 原始〜 primitive polynomial. 定数〜 constant polynomial. 最小〜 minimal polynomial. 近似〜 approximation polynomial. 補間〜 interpolation polynomial. 固有[特性]〜 proper [characteristic] polynomial. 直交〜 orthogonal polynomials. 歪(わい)〜 skew polynomial. 〜環 polynomial ring. 〜表現 polynomial representation. 〜関数[近似] polynomial function [approximation]. ルジャンドルの〜 Legendre('s) polynomial. 〜級数[展開] polynomial series [expansion]. 〜述語 polynomial predicate. 基本交代〜 elementary alternating polynomial. 基本対称〜 elementary symmetric polynomial. 強〜の strongly polynomial. 交代〜 alternating polynomial. 情報〜 information polynomial. 生成〜 generator polynomial. 対称〜 symmetric polynomial. 〜時間アルゴリズム polynomial time algorithm. 分離〜 separable polynomial.

たこうしきじかん 多項式時間 polynomial time.
〜アルゴリズム polynomial time algorithm. 非決定性〜(計算可能な問題のクラス) non-deterministic polynomial time; NP.

たこうしつ 多孔質の porous.

たこうせいばいしつ 多孔性媒質 porous media.

たしか 確からしい likely; probable.
同程度に[同様に]〜 equally likely; equally probable; equiprobable.

たしかである 確かである certain, sure.

たしかに 確かに certainly.

たしかめる 確かめる make sure; ascertain; verify; confirm.

たしざん 足し算 addition.

たじゅう 多重(の) multiple. ⇨多重(たじゅう-).
〜に multiply. 〜数学的帰納法 multiple mathematical induction. 〜添数 multiple suffix. 〜積分 multiple integral. 〜共変式 multiple covariant (expression). 〜連結の multiply(-)connected. 〜推移群 multiply transitive group. 〜比較 multiple comparison.

たじゅう- 多重 multi-; pluri-; poly-.
〜指数 multi-index. 〜共線性 multicolinearity. 〜ベクトル multivector. 〜線形形式[写像] multilinear form [mapping]. 〜調和[劣調和]関数 pluriharmonic [plurisubharmonic] function.

たじゅうえんばん 多重円板 polydisc; polydisk.

ダース dozen (*pl.* -(s)); dz., doz.

たすう 多数の numerous; extensive.

だせい 楕性 inertia. ⇨慣性.
〜の inertial. 〜体 inertia field. 〜群

inertia group.

たそう 多相 multi-phase.
～抽出(法) multi-phase sampling.

ただ 唯 merely; only; simply.
～一点 only [merely] one point; unique point.

ただ ただ[無料]の[で] free(ly); gratis.

たたい 多体 many bodies.
～問題 many-body problem; problem of many(-)bodies.

ただし 但し but; however; provided [excepting] that; on condition that.

ただしい 正しい right; correct; legitimate; true.

たたみこみ 畳込み convolution; (*G.*) Faltung. ⇨合成.

ただん 多段 multistage.
～(式)の; ～階の multistage. ～抽出 multistage sampling. ～配分 multistage allocation. ～選択過程 multistage choice process.

たち 多値(の) many-valued; multi(-)valued.
～論理 many-valued [multiple-valued] logic.

たちょうわ 多調和(性) polyharmonicity.
～な polyharmonic. ～関数 polyharmonic function.

タック・ノード tacnode.

タック・ローカス tac-locus.

ダッシュ dash; 【—】.
x～ x prime; x'.

たて 縦 length; height.
～の longitudinal; lengthwise; vertical. ～波 longitudinal wave. ～振動 longitudinal vibration. ～縞 vertical stripe. ～ベクトル column vector. ～軸 (axis of) ordinate.

たてざひょう 縦座標 ordinate.

たてじく 縦軸 (axis of) ordinate.

たてせん 縦線 vertical line; ordinate.
～集合[図形] ordinate set [figure].

たてならび 縦並びの[に] tandem.

だとう 妥当(性) validity; appropriateness.
～な valid; appropriate. 一般～性 general validity.

たとえば 例えば for instance; for example; (*L.*) exempli gratia; e.g.

たば 束 bundle; bunch; cluster.

たひょうほん 多標本 *K*-sample.

ダブリュー W.
～曲面 *W*-surface. *W** 環 *W**-algebra.

ダブリュー・ケイ・ビー W.K.B.
～法 W. K. B. method, Wentzel-Kramers-Brillouin method.

タブロー (*F.*) tableau (*pl.* -leaux).
シンプレックス・～ simplex tableau.

たへんけい 多辺形 polygon. ⇨多角形.
～の polygonal; multilateral. 正～ regular polygon.

たへんすう 多変数(の) of several variables.
～関数 function of several variables. ～複素関数論 theory of functions of several complex variables.

たへんりょう 多変量(の) multivariate.
～データ multivariate data. ～解析 multivariate analysis. ～線形モデル multivariate linear model. ～正規分布 multivariate normal distribution. ～分散分析 multivariate analysis of variance; MANOVA. ～確率変数 random variable of many dimensions.

たまざん 珠算 ⇨珠算(しゅざん).

ダミー dummy.
～インデックス dummy index.

ためし 試し trial; test; check.

ためん 多面(の) multiplanar; polyhedral.
～座標 multiplanar coordinates. ～角 polyhedral angle. ～錐 polyhedral cone. ～体 polyhedron.

ためんたい 多面体 polyhedron (*pl.* -rons, -dra).
～の polyhedral. 凸[凹]～ convex [concave] polyhedron. 正～ regular polyhedron. 正～群 polyhedral group. ～図法 polyhedral projection.

たよう 多葉(性) multivalence; multivalency.
～な multivalent; many-[multi(-)] sheeted. ～関数 multivalent function.

たようたい 多様体 manifold; variety.
開～ open manifold. 部分～ submanifold; subvariety. 実[複素]～ real [complex] manifold. 可微分～ differentiable mani-

fold [variety]. 積分～ integral manifold. 位相～ topological manifold. 組合せ～ combinatorial manifold. シュタイン[リーマン]～ Stein [Riemannian] manifold. 代数～ algebraic variety. モジュラス～ variety of moduli. 境界のある～ manifold with boundary. アフィン(代数)～ affine (algebraic) variety. アーベル～ Abelian variety. 台～ support variety. 中心～ center manifold. トーリック～ toric variety. 微分～ differential manifold.

ダランベール J. R. d'Alembert (1717-1783).
～の d'Alembertian. ～演算子 d'Alembertian (operator).

たる 樽 barrel.
～型 barrel(-) shape. ～型空間 barrelled space. ～集合 barrel; barrelled set.

たん 端 end; limit.
～点 end point; extreme point; extremal point. 左[右]～ left [right] end. 上[下]～ upper [lower] limit.

たん- 単 single; simple; one; unitary; uni-; mono-. ⇨単一(の); 単純(な).
～比 simple ratio. ～利 simple interest. ～根 simple root. ～積分 simple integral. ～極 simple pole. ～格子 simple lattice. ～周期の simply periodic. ～連結の simply (-)connected. ～多項式 unitary polynomial. ～関数 simple function. ～振動 simple oscillation; simple harmonic motion; S.H.M. ～振子 simple pendulum. ～事象 simple event. ～相関 simple correlation. ～価 unit price.

たんい 単位 unit.
虚数～ imaginary unit. 抽出～ sampling unit. ～の分解[分割] partition of unity. ～分数 unit fraction. ～点 unit point. ～円 unit circle. ～円板 unit disk [disc]. ～正方形 unit square. ～球[立方体] unit sphere; unit ball [cube]. ～元 unit [identity] element. ～ベクトル unit vector. ～行列 unit [identity] matrix. ～関数 unit function. ～表現 unit representation. ～分布 unit distribution. ～質量 unit mass. ～面[体]積 unit area [volume]. ～群 unit group. ～法線 unit normal. ～胞(体) unit cell. ～イデアル unit ideal.

たんい 単位(的) unitary; unital.
～環 unitary ring. ～多元環[加群] unitary algebra [module]. ～(的)半群 monoid; unitary [unital] semigroup. ～(的)加群 unital [unitary] module.

たんいげん 単位元 identity element; unit element; neutral element.
～成分 identity component.

たんいつ 単一(の) simple. ⇨単(たん-).
～に simply. ～層 simple layer. ～(層)分布 simple (layer) distribution. ～級数 simple series. ～積分 simple integral. ～周期関数 simply periodic function. ～閉曲線 simple closed curve; Jordan curve. ～連結領域 simply(-)connected domain. ～命題 simple statement.

たんいつ 単一(性) unicity.

たんいつきばん 単一基板上の monolithic.

たんいんし 単因子 elementary divisor.
単純～ simple elementary divisor.

たんかんすう 単関数 simple function.

たんげん 単元 unit.
～群 group of units; unit group. 環の～ unit of a ring.

たんご 単語 word.
基本～ basic word(s).

たんこう 単項(の) monomial; monoidal.
～式 monomial. ～因子 monomial factor. ～表現 monomial representation. ～的変換 monoidal transformation. ～イデアル principal ideal. ～イデアル整域 principal ideal domain; PID.

たんこう 単向(の) one-way.

たんこうし 単格子 simple lattice.

たんこうしき 単項式 monomial.

たんこん 単根 simple root.

たんさく 探索 search; retrieval.
～的な exploratory. ～者 searcher. 二分～ binary [dichotomizing] search. 全数～ exhaustive search. 文献～ literature search. 情報～ information retrieval; IR. 局所～ local search. ～的データ解析 exploratory data analysis. ～方向 search

direction. 幅優先～ breadth first search. 深さ優先～ depth first search.

たんし 端子 terminal.

タンジェント tangent; tan. ⇨正接.

たんじく 短軸 minor axis.

たんしゃ 単射 injection; monomorphism.
～的な injective. ～性 injectivity. ～的分解 injective resolution. ～(的)写像 injection; injective mapping. 標準[規準]的～ canonical injection. 自然な～ natural injection. ～包絡 injective envelope; injective hull. ～加群 injective module. ～準同型[形] injective homomorphism; monomorphism. 全～ bijection.

たんしゃしょう 単斜晶(の) monoclinic.
～系 monoclinic system.

たんしゅうき 単周期(の) simply periodic.
～関数 simply periodic function.

たんしゅうごう 単集合 singleton.

たんしゅくする 短縮する curtail.

たんじゅん 単純(な) simple.
～に simply. ～化する simplify. ～化 simplification. ～平均 simple mean. ～根 simple root. ～多角形[閉曲線] simple polygon [closed curve]. ～弧 simple arc. ～拡大 simple extension. ～群[環; 多元環] simple group [ring; algebra]. ～加群 simple module. ～成分 simple component. ～指標 simple character. ～収束 simple convergence. ～事象 simple event. ～統計仮説 simple statistical hypothesis. ～任意抽出(法) simple random sampling. ～推移的な simply transitive. ～化されたゲーム simplified game. 半～な semisimple. 散在型～群 sporadic simple group. ～グラフ simple graph. ～点 non(-)singular point.

たんじょほう 短除法 short division.

たんしんどう 単振動 simple oscillation; simple harmonic motion; S.H.M.

たんすう 単数 unit; invertible element.
～群 group of units; unit group.

たんせい 単生[成]的 monogenic; (*F.*) monogène.

だんせい 弾性 elasticity.
～の elastic. ～散乱 elastic scattering. ～膜[板] elastic membrane [plate]. ～体 elastic body. ～率 elasticity modulus. ～論 theory of elasticity.

たんそうかん 単相関 simple correlation.

たんそく 単側 one side. ⇨片側(かたがわ); 両側(の)(りょうそく).
～の one-sided; unilateral. ～曲面 one-sided [unilateral] surface.

たんたい 単体 simplex (*pl.* -lexes, -lices).
～の simplicial. 順序～ ordered simplex. 有向～ oriented simplex. 辺～ boundary simplex. 曲～ continuous simplex. 特異～ singular simplex. n 次元～ n-simplex. ～写像 simplicial map. ～近似 simplicial approximation. ～分割 simplicial decomposition. ～(的)複体 simplicial complex. ～法 simplex method. ～表 simplex tableau. ～規準 simplex criterion.

たんち 探知 detection.
～する detect.

たんちょう 単調(な) monotone; monotonous; monotonic.
～に monotonously; monotonically; monotone. ～関数 monotone function. ～(数)列 monotone sequence. ～収束 monotone convergence. ～減少[増加]の monotone decreasing [increasing]. 狭義(に)～な strictly monotone. ～尤度比 monotone likelihood ratio. ～性 monotoneity; monotonicity.

たんてん 端点 end point; extremal point; extreme point.

だんどう 弾道 trajectory.
～の ballistic. ～曲線 ballistic curve. ～学 ballistics.

たんどく 単独(性) uniqueness.
～な unique. ～に uniquely. ～性の条件 uniqueness condition.

ターンパイク turnpike.
～定理 turnpike theorem.

たんぴ 単比 simple ratio.

たんぷく 短幅 shortest breadth.

たんぽうけい 単峰形(の) unimodal.
～分布 unimodal distribution.

だんめん　断面　section; cross-section.
　～の sectional.　平～ plane section.　直～ orthogonal section.　～曲率 sectional curvature.　～積 sectional area; area of cross-section.　～図 cross figure; sectional plan.

たんよう　単葉(性)　univalence; univalency.
　～な univalent; (*G.*) schlicht; simple.　～関数 univalent [(*G.*) schlicht] function.　～写像 univalent [(*G.*) schlicht] mapping.　～(性)半径 radius of univalency; univalency radius.　～領域 schlicht domain.　～型の planar; of planar character; (*G.*) schlichtartig.

たんり　単利　simple interest.
　～法 method of simple interest.

だんりょく　弾力(性)　elasticity; flexibility.
　～のある elastic; flexible.

たんれつ　単列(の)　uniserial.
　～多元環 uniserial algebra.

たんれんけつ　単連結(性)　simply-connectedness; simple connectivity.
　～の simply(-)connected.

だんわ　談話　conversation; talk.
　～会 colloquium (*pl.* -quiums, -quia); symposium (*pl.* -ums, -sia); (*It.*) conversazione (*pl.* -ni).

ち・チ

-ち 値 value.
平均～ mean value. 近似～ approximate value. 絶対～ absolute value. 極大[小]～ maximal [minimal] value. 極限～ limit(-ing) value. 集積～ accumulation value; cluster value. 特性～ characteristic value. 中間～ intermediate value. 主～ principal value. 境界[初期]～ boundary [initial] value. 最確～ most probable value. 固有～ eigenvalue. 中央～ median. 付[賦]～ valuation. 特異～ singular value.

ちいき 値域 range; range of values.
関数の～ range of function.

ちいき 地域 area; region.
～的な areal; regional. ～的に areally; regionally. ～抽出(法) areal sampling.

ちいさい 小さい[な] small; little; minor.
～部分 small part. 十分～ sufficiently small.

チェザロ E. Cesàro (1859-1906).
～総和可能な Cesàro-summable.

チェック check.
～・ビット check bit.

ちえん 遅延 delay; lag; retardation. ⇨遅れ.
～ひずみ delay distortion. ～ポテンシャル retarded potential.

ちかい 近い close; near.

ちかく 知覚 consciousness; perception.
～の perceptional; sensory. ～器官 sensorium (*pl.* -riums, -ria).

ちかづく 近づく tend to; approach.
極限に～ tend to limit; approach limit.

ちから 力 power; force.
～の平行四辺形[多角形] force parallelogram [polygon]; parallelogram [polygon] of forces. ～の関数 force function.

ちかん 置換 replacement; substitution; permutation.
～する replace; substitute; permute. ～積分(法) integration by substitution. 恒等[巡回]～ identity [cyclic] permutation. 奇[偶]～ odd [even] permutation. ～群 permutation group. ～行列 permutation matrix.

ちきゅう 地球 earth; globe.
～の terrestrial. ～の引力 terrestrial gravitation. ～儀 (terrestrial) globe. ～物理学 geophysics.

ちくごてき 逐語的 literal, (*L.*) verbatim, word-for-word.
～に literally, (*L.*) verbatim, word for word.

ちくじ 逐次(の) successive; serial; sequential.
～代入[反復] successive substitution [iteration]. ～微分 successive differentiation. ～導関数 successive derivative(s). ～近似 successive approximation. ～最小[最小点] successive minima [minimum points]. ～加速緩和法 successive over-relaxation; SOR. ～処理 serial processing. ～検査 sequential inspection. ～分析[検定] sequential analysis [test]. ～抜取り sequential sampling. ～ゲーム sequential game. ～定義可能な sequentially difinable.

ちくせき 蓄積 accumulation; storage.
～する accumulate; store.

ちけい 地形(学) topography.
～測量[図] topographical surveying [map].

ちず 地図 map; atlas.
～投影[投象](法) map projection. 縮尺～

map on a reduced scale. 等積～ equivalent map. ～製作(法) cartography.

ちずちょう 地図帳 atlas.

ちぢむ 縮む shrink. ⇨収縮.

ちぢめる 縮め(る) contract; compress; pinch. ⇨収縮.
一点に～られた空間 pinched space.

ちてき 知的(な) mental; intellectual.

ちのう 知能 intelligence.
～検査 intelligence test. 人工～ artificial intelligence. ～指数 intelligence index.

ちひょう 地表 ground; surface of earth.

ちへい 地平(の) horizontal.
～線 horizontal line; horizon. ～面 horizontal plane.

チャート chart.
フロー～ flow chart; flowchart.

チャンネル channel.

ちゅう 柱 cylinder.
～の cylindrical. ～面 cylinder (surface); cylindrical surface. 円～ circular cylinder. 楕円[二次]～(面) elliptic [quadratic] cylinder. (円)～座標 cylindrical coordinates. 直角～ right cylinder. 角～ prism. 三角～ triangular prism. 曲線～ cylindroid. ～集合 cylinder set.

ちゅう 注 remark; note.
脚～ footnote.

ちゅうい 中位(の) median.
～数; ～数の median.

ちゅういする 注意する observe; notice; warn, caution.

ちゅうおう 中央(の) middle; median; central.
～値 median. ～値不偏推定量 median-unbiased estimator. 母[標本]～値 population [sample] median.

ちゅうおうち 中央値 median.

ちゅうがいひ 中外比 extreme and mean ratio; golden ratio. ⇨黄金(の).
～分割 golden section.

ちゅうかん 中間(の) intermediate; middle.
～値の定理 intermediate value theorem. ～近似分数 intermediate convergent. ～体 intermediate field. ～積分 intermediate integral.

ちゅうかんすう 中間数 mediant.

ちゅうこう 中項 middle term; mean.

ちゅうごくしきじょうよていり 中国式剰余定理 chinese remainder theorem.

ちゅうじつ 忠実(な) faithful.
～に faithfully. ～な表現 faithful representation. ～に平坦な faithfully flat.

ちゅうしゅつ 抽出 sampling; extraction. ⇨抜取(り).
～する extract; draw. ～方式 sampling procedure. ～単位 sampling unit. ～誤差 sampling error. ～比 sampling fraction; sampling ratio. ～間隔 sampling interval. 地域～ areal sampling. 任意～ optimal sampling. 層化～ stratified sampling. 多段～ multistage sampling. 復元～ sampling with replacement. 非復元～ sampling without replacement. 確率比例～ sampling with probability proportional size. 二段～ two-stage [-step] sampling; sub-sampling. 二相～ two-phase sampling.

ちゅうしょう 抽象 abstraction; abstract.
～的な abstract. ～的に abstractly. ～する abstract. ～化 abstraction. ～性 abstractness. ～代数(学) abstract algebra. ～群 abstract group. ～空間 abstract space. ～積分 abstract integral. ～的代数多様体 abstract algebraic variety.

ちゅうじょうず 柱状図(表) histogram.

ちゅうしん 中心 center; 【英】 centre.
～の central. 円[球]の～ center of circle [sphere]. C を～とする円 circle centered at C. ～円 center circle. ～曲面 center surface. ～距離 distance between centers. 曲率～ center of curvature; curvature center. 回転[反転]の～ center of rotation [inversion]. 相似の～ center of similitude. 配景の～ center of perspectivity; perspective center. ～角[線] central angle [line]. ～点 central point; center. ～拡大 central extension. ～対称(変換) central symmetry. ～射影[投影] central projection. ～投影(図)法 central projection; perspective drawing. ～列 central series. ～差分

central difference. ～極限定理 central limit theorem. ～積率 central moment. ～傾向 central tendency. 解析的～ analytic center.

ちゅうしんか 中心化 centralization; centring; centering.
～群[環] centralizer. ～された過程 centred process.

ちゅうすう 中枢 pivot.
～の pivotal. ～要素 pivotal element.

ちゅうせん 中線 median; median line.
～定理［等式］ parallelogram theorem [identity].

ちゅうたい 柱体 cylinder.
～の cylindrical.

ちゅうだん 中断 break; interruption.
～する break; interrupt.

ちゅうてん 中点 middle point; mid(-)point; median.
～の medial. ～曲線 middle point curve. ～法則 midpoint rule [law]. ～値 midrange. ～三角形 medial triangle.

ちゅうまつひ 中末比 ⇨中外比.

ちゅうみつ 稠密(な) dense. ⇨疎(な).
～性 density; denseness. いたるところ～な everywhere dense. 自己～な dense in itself; self-dense.

ちゅうめん 柱面 cylinder (surface); cylindrical surface.
円[二次]～ circular [quadratic] cylinder (surface). 包絡～ enveloping cylinder.

ちゅうりつ 中立(の) neutral.
～元 neutral element. ～型 neutral type.

チューリング A. M. Turing (1912–1954).
決定性～機械 deterministic Turing machine.

ちょう 兆 【英】billion;【米】trillion.

ちょう- 超 hyper-; super-; ultra-; meta-.
～平面 hyperplane. ～コホモロジー hypercohomology. ～音波 supersonic wave. ～積 ultraproduct. ～数学 metamathematics. ～算術的な hyperarithmetical.

ちょういちじしゅうそく 超一次収束 superlinear convergence.

ちょうえつ 超越(性) transcendence; transcendency; transcendentality.
～の; ～的な transcendental. ～数 transcendental number. ～関数 transcendental function. ～曲線 transcendental curve. ～特異点 transcendental singularity. ～(的)元 transcendental element. ～拡大 transcendental extension. ～基底 transcendence basis. ～次数 transcendence degree; degree of transcendence.

ちょうえつちょっけい 超越直径 transfinite diameter.

ちょうえん 長円 ellipse. ⇨楕円.
～体[面] ellipsoid. ～形の elliptic; oval. ～体の ellipsoidal; ovaloidal.

ちょうおんそく 超音速(の) supersonic.
～波 supersonic wave.

ちょうおんぱ 超音波 supersonic [ultrasonic] wave.
～の supersonic; ultrasonic.

ちょうか 超過 excess.
～的な excessive. ～する exceed. ～率 coefficient of excess.

ちょうかかい 超可解(性) supersolvability.
～な supersolvable.

ちょうかく 頂角 vertical angle.

ちょうかへんどう 超過変動 overdispersion.

ちょうかんすう 超関数 distribution.
～の distributional. ～(の意味での)解 distributional solution; solution in the sense of distribution.

ちょうかんすう (佐藤)超関数 hyperfunction.

ちょうきか- 超幾何 hypergeometric.
～(微分)方程式 hypergeometric (differential) equation. ～級数 hypergeometric series. ～関数 hypergeometric function. ～多項式 hypergeometric polynomial. ～積分 hypergeometric integral. ～分布 hypergeometric distribution.

ちょうきゅう 超球 hypersphere.
～の hyperspherical; ultraspherical. ～幾何学 hypersphere geometry. ～座標 hyperspherical coordinate. 実[虚]～ real [imaginary] hypersphere. 常～ ordinary [proper] hypersphere. 点～ point hypersphere. 限界～ limiting hypersphere. ～

微分方程式 hyperspherical differential equation. ～関数 hyperspherical function. ～多項式 ultraspherical polynomial.

ちょうきょくしょてき 超局所的(な) micro-local.

ちょうきょくめん 超曲面 hypersurface.
二次～ quadratic hypersurface. 特性～ characteristic hypersurface. 座標～ coordinate hypersurface. 解［積分］～ integral hypersurface.

ちょうぐん 超群 hypergroup.

ちょうげん 超限(性) transfiniteness.
～の; ～的な transfinite. ～順序数 transfinite ordinal number. ～帰納法 transfinite induction.

ちょうげん 超弦 super(-)string.
～理論 super(-)string theory.

ちょうさ 調査 ⇨検査.

ちょうじく 長軸 major axis.

ちょうしゃほうけい 長斜方形 rhomboid.
～の rhomboid; rhomboidal.

ちょうしゅうごう 超集合 superset.

ちょうしゅうそく 超収束 overconvergence; ultraconvergence.
～する overconverge; ultraconverge. ～級数 overconvergent series; ultraconvergent series.

ちょうじょう 重畳 superposition. ⇨重ね合わせ.
～原理 principle of superposition.

ちょうすうがく 超数学 metamathematics.
～的な metamathematical.

ちょうせい 調整 adjustment; modulation; regulation.
～する adjust; modulate; regulate. 微～ fine adjustment. ～型抜取(り) sampling with adjustment.

ちょうせき 超積 ultraproduct.

ちょうだえん 超楕円(の) hyperelliptic.
～(的)な hyperelliptic. ～曲線 hyperelliptic curve. ～積分 hyperelliptic integral. ～的リーマン面 hyperelliptic Riemann surface. ～関数体 hyperelliptic function field.

ちょうたつきかん 調達期間 lead time.

ちょうてん 頂点 vertex (*pl.* -texes, -tices); summit; apex (*pl.* -xes, -pices).
～の vertical. 四～定理 four vertex theorem; theorem of four vertices. 角の～ vertex of angle. ～被覆 node cover.

ちょうねん 長年 ⇨永年(の).

ちょうふく 重複 repetition; multiplicity; reduplication; redundance. ⇨重複(じゅうふく).
～した multiple; repeated. ～組合せ［順列］ repeated combination [permutation] ～点 multiple point. ～積分 multiple integral. ～度 multiplicity. ～連結の multiply(-)connected. ～抜取り duplicate sampling.

ちょうへいめん 超平面 hyperplane.
～座標 hyperplane coordinates. 接～ tangent hyperplane. 無限遠～ hyperplane at infinity. 支持～ supporting hyperplane. ～束 pencil of hyperplanes. 回帰～ regression hyperplane. 分離～ separating hyperplane.

ちょうへんきゅう 長扁球 prolate spheroid; oblong spheroid.

ちょうほうけい 長方形 rectangle. ⇨矩形.
～の rectangular. ～行列 rectangular matrix. ～分布 rectangular distribution.

ちょうみつ 稠密 ⇨稠密(な)(ちゅうみつ).

ちょうむげん 超無限(の) ultrainfinite.
～遠点 ultrainfinite point.

ちょうやく 跳躍 leap; jump; saltus.
～関数 leap [saltus] function. ～点 leap [jump; saltus] point.

ちょうゆうかいがた 超有界型(の) (*F.*) ultra-bornologique.
～空間 (*F.*) espace ultrabornologique.

ちょうわ 調和(な) harmonic. ⇨球調和関数; 球面関数; 楕円体調和関数.
～に harmonically. ～に分かつ separate harmonically. ～化する harmonize. ～性 harmonicity. ～線束 harmonic pencil of lines. ～平均 harmonic mean. ～級数 harmonic series. ～数列 harmonic progression; H.P. ～共役点 harmonic conjugate(s); harmonically conjugate point(s). ～列点 harmonic range of points. ～関数

harmonic function; harmonics. 球面(〜)関数 spherical harmonic function; spherical harmonics; surface harmonics. 球〜関数, 体球(〜)関数 solid harmonic function; solid harmonics. 共役〜関数 harmonic conjugate; conjugate harmonic function. 〜優[劣]関数 harmonic majorant [minorant]. 〜接続 harmonic continuation [prolongation]. 〜化 harmonization. 〜積分 harmonic integral. 〜微分 harmonic differential. 〜空間 harmonic space. 〜解析 harmonic analysis. 〜形式 harmonic form. 〜測度 harmonic measure. 〜境界 harmonic boundary. 〜次元 harmonic dimension. 〜流 harmonic flow. 〜振動 harmonic oscillation. 重〜な biharmonic. 多重〜な pluriharmonic.

ちょくえんすい 直円錐 right (circular) cone.
〜台 frustum of right (circular) cone.

ちょくえんちゅう 直円柱 right (circular) cylinder.

ちょくきやく 直既約(性) indecomposability.
〜な indecomposable; directly indecomposable. 〜群 indecomposable group. 〜表現 indecomposable representation. 〜加群 indecomposable module.

ちょくきょくげん 直極限 direct limit.

ちょくさんかくちゅう 直三角柱 right triangular prism; wedge.

ちょくせき 直積 direct product; Cartesian product.
〜分解 direct product decomposition. 〜空間 (direct) product space. 〜位相 (direct) product topology. 〜測度 product measure. 〜集合 (direct) product set; Cartesian product (set).

ちょくせつ 直接(の) direct.
〜に directly. 〜法 direct method. 〜証明(法) direct proof. 〜接続 direct continuation [prolongation].

ちょくせつこう 直截口 normal section.

ちょくせん 直線 straight line.
〜の linear; rectilinear. 〜運動 linear motion. 〜座標[幾何学] line coordinates [geometry]. 〜束 line bundle; pencil of lines. 〜叢 line congruence. 半〜 half line; ray. 数〜 real line; number line. 射影〜 projective line. 無限遠〜 line at infinity.

ちょくだんめん 直断面 orthogonal section.

ちょくへいこうろくめんたい 直平行六面体 rectangular parallelepiped. ⇨直方体.

ちょくほうたい 直方体 rectangular parallelepiped; rectangular prism; cuboid; box.

ちょくりつ 直立(の) vertical; perpendicular.
〜跡 vertical trace. 〜画面 vertical plane.

ちょくりょうしめんたい 直稜四面体 orthogonal tetrahedron (*pl.* -tons, -dra).

ちょくれつ 直列 series.
〜の serial; tandem. 〜に in series; serially. 〜操作 serial operation.

ちょくろくめんたい 直六面体 rectangular parallelepiped.

ちょくわ 直和 direct sum; disjoint union.
〜集合 direct sum set. 〜因子 direct summand. 〜分解[表現] direct sum decomposition [representation]. 〜空間 direct sum space; disjoint (topological) union.

ちょさくけん 著作権 copyright; ©.

ちょしゃ 著者 author.
共〜 coauthor; joint author; collaborator.

ちょぞう 貯蔵 stock; storage; store.
〜する store.

ちょっか 直化 straightening.
〜する straighten.

ちょっかく 直角 right angle.
〜の rectangular; right-angled. 〜三角形 right(-angled) [rectangular] triangle. 〜二等辺三角形 rectangular equilateral triangle. 両〜三角形 birectangular triangle. 〜柱 right prism; right cylinder. 〜双曲線 rectangular hyperbola. 〜[直交]座標 rectangular (Cartesian) coordinates. 三〜の trirectangular.

ちょっかん 直観 intuition.
〜的な intuitive. 〜幾何(学) intuitive geometry. 〜主義 intuitionism. 〜主義的な intuitionistic. 〜主義論理(学) intuitionistic logic.

ちょっけい 直径 diameter; diam.

〜の diametral; diametric; diametrical. 〜の両端点 diametral end points. 円[球]の〜 diameter of circle [sphere]. 共役〜 conjugate diameter. 超越〜 transfinite diameter.

ちょっけい 直系 direct system.

ちょっこう 直交(の) orthogonal.

〜する intersect [cut] orthogonally; cross at right angles; be perpendicularto. 〜軸[座標] orthogonal axes [coordinates]. 〜行列 orthogonal matrix. 〜群[変換] orthogonal group [transformation]. 〜截[切]線; 〜軌道 orthogonal trajectory. 〜枠 orthogonal frame. 〜成分 orthogonal component(s). 〜分解 orthogonal decomposition. 〜曲線座標 orthogonal curvilinear coordinates. 〜補空間 orthogonal complement; orthocomplement. 〜配列表 orthogonal layout. 〜系 orthogonal system. 正規〜系 orthonormal system. 二重〜の biorthogonal. 〜関数[級数] orthogonal functions [series]. 〜多項式 orthogonal polynomials. 〜性 orthogonality. 〜化 orthogonalization. 正規〜化 orthonormalization. 〜化定理 orthogonalization theorem. 〜配列 orthogonal array.

ちらばり 散らばり dispersion.

チルダ tilde; 【〜】.

n〜 n tilde; $\tilde{n}$.

ちんじゅつ 陳述 statement; declaration.

ちんぷか 陳腐化 obsolescence; staleness.

〜損失 obsolescence loss.

つ・ツ

つい 対 pair; couple.
～にする pair; make a pair. ～をなすこと pairing. 二～ two pairs. 順序～ ordered pair. 非順序～ unordered pair. 球～ sphere pair. 胞体～ ball pair.

ついか 追加 addition; supplement; addendum (*pl.* -da).
～の additional; supplementary. ～資料 additional sample. ～された条件 additional condition.

ついせき 追跡 tracing; pursuit.
～する trace. 曲線(の)～ curve tracing. ～(曲)線 tractrix; curve of pursuit.

つうか 通過 passing; passage.
初～ first passage.

つうけい 通径(円錐曲線の) (*L.*) latus rectum (*pl.* latera recta).

つうじょう 通常(の) ordinary; proper; vulgar.
～点 ordinary point. ～曲線 ordinary curve. ～ディリクレ級数 ordinary Dirichlet series. ～特異点 ordinary singularity. ～表現 ordinary representation. ～の積分 proper integral. ～分数 vulgar fraction.

つうしん 通信 communication; correspondence.
～する communicate; correspond. ～路 (communication) channel. ～行列 communication matrix.

つうしんろ 通信路 channel.
～容量 channel capacity. 二進対称～ binary symmetric channel.

つうぶん 通分 reduction to common denominator.
～する reduce.

つうやく 通約 commensuration; reduction.
～する commensurate; reduce. ～可能な commensurable; reductive; reducible. ～不能な incommensurable.

つうやく 通訳 interpretation; interpreter.
～の interpretive. ～する interpret.

ツェノン Zenon of Elea (490?–429? B.C.).
～の背理[逆説] paradox of Zenon.

ツェルメロ E. Zermelo (1871–1953).
～の公理 Zermelo's axiom.

ツォルン M. A. Zorn (1906–1993).
～の補題 Zorn's lemma.

つぎ 次(の) next; following; second; secondary.
～に next; secondly; in the second place. ～の通り as follows.

つきがた 月形 crescent; lunar form.
～の lunar. ～領域 lunar domain; circular digon.

つぎめ 継ぎ目 joint; juncture; seam.

つづく 続く continue.
前ページより～ continued from the preceding page. 次ページに～ continued on the next page. 次号に～ to be continued in the next number.

つぶれる collapse.

つまらない trivial; insignificant.

つめこみ 詰め込み packing.
詰め込む pack.

つよい 強い strong. ⇨強(きょう-).
強く strongly. ～条件 strong(er) condition. ～意味で単調な strongly monotone.

つよさ 強さ strength; intensity.
同じ～の equally strong.

つりあい 釣合 balance; equilibrium (*pl.* -ria,

-riums); matching.
～のとれた標本 balanced sample. ～不完備ブロック計画 balanced incomplete block design; BIBD.

つるまきせん 蔓巻線 helix. ⇨螺旋.

て・テ

て 手 hand; move; trick.
右[左]～系 right [left] hand(ed) system.
～番 move.

であう 出会う meet; encounter.

てい ティ T; t.
～字形の T-shaped. ～定規 T-rule; T-square. ～・スコア T-score. ～分布 t-distribution.

てい 底 base; basis (*pl.* -ses).
～の basic. ～辺[面] base. 下[上]～ lower [upper] base. ～面積 area of base. 基～ base; basis. ～角 basic [base] angle. ～線 base line. 対数の～ base of logarithm. ～空間 base space. ～心格子 base-centred lattice.

てい- 定 fixed; constant; definite.
～刻 fixed time. ～数 constant (number). ～値 constant [fixed] value. ～点[直線] fixed point [straight line]; constant point [straight line]. ～和 constant sum. ～曲率曲面[空間] surface [space] of constant curvature. ～幅曲線 curve of constant breadth. ～傾曲線 curve of constant inclination; isocline. ～係数の with [of] constant coefficients. ～発散 definite divergence. ～発散の definitely [properly] divergent. ～積分 definite integral. ～符号 definite sign. ～符号の of definite sign; definite. ～符号形式 definite form.

ていい 低位 lower order. ⇨高位の.
～の of lower order.

ディー・エヌ dn.
～関数 dn-function; (*L.*) delta amplitudinis.

ディオファンタス Diophantus [Diophantos] (246?-330?).
～の Diophantine. ～方程式[近似] Diophantine equation [approximation].

ていか 低下 reduction; depression.
～させる reduce; depress. 階数～ reduction [depression] of order.

ていかく 定角(の) isoclinal.
～軌道 isoclinal trajectory; isocline.

ていぎ 定義 definition.
～する define. ～可能な definable. ～域 domain [region] of definition. ～方程式[関数] defining equation [function]. ～列 defining sequence. 帰納法による～ definition by induction. 帰納的～ inductive definition. 再帰的～ recursive definition.

ていぎいき 定義域 domain; domain [region] of definition.

ていけい 梯形 echelon.

ていけい 梯形 【米】trapezoid; 【英】trapezium (*pl.* -zia; -ziums). ⇨台形.

ていけいきょくせん 定傾曲線 curve of constant inclination; isocline.

ていげん 定言(の) categorical.
～命題 categorical proposition. ～三段論法 categorical syllogism.

ていさ 定差 difference. ⇨階差; 差分.

ていし 停止 stop; stoppage; halt.
～する stop; halt. ～時刻 stopping time. ～問題 halting problem; stopping problem. 最適～問題 optimal stopping problem. 最適～時刻問題 optimal stopping-time problem.

ていしきか 定式化 formulation.
～する formulate.

ていじょう 定常(な) steady; stationary.

～状態 steady [stationary] state. ～波 steady [stationary] wave; standing wave. ～流 steady flow. ～分布 steady distribution. ～点 stationary point. ～値 stationary value. ～関数[曲線; 曲面] stationary function [curve; surface]. ～過程 stationary process. ～情報源 stationary information source.

ていじょう 蹄状(の) ungual; hoof-shaped.
～体[形] ungula (*pl.* -lae).

ていすう 定数 constant; invariable.
～の constant; invariable. 比例～ proportional constant; constant of proportion [proportionality]. 絶対～ absolute constant. 普遍～ universal constant. ～値 constant value. ～項 constant term. 積分～ integration constant. 付加～ additive constant. 任意～ arbitrary constant. 誤差～ error constant. 実験～ empirical constant. 構造～ structure constant(s). ～因子 constant factor. ～係数 constant coefficient. ～多項式[関数] constant polynomial [function]. ～層 constant sheaf. ～変化(法) variation of constant(s).

ディスク disc; disk.

ディストリビューション distribution.

ていせい 訂正 correction; revision.
～する correct; revise. 誤りの～ error-correction. ～版 revised edition.

ていせいてき 定性的 qualitative.
～に qualitatively. ～性質 qualitative property.

ていち 定値
～(な) constant value; fixed value. ～[定符号]な definite. ～写像[関数] constant (-valued) mapping [function]. ～形式 definite form.

ていてん 定点 fixed point.

ていふく 定幅 constant breadth.
～曲線[曲面] curve [surface] of constant breadth.

ていへん 底辺 base.

ていめん 底面 base.

ディメンション dimension. ⇨次元.

テイラー B. Taylor (1685–1731).
～級数[展開] Taylor series [expansion].

ディラック P. A. M. Dirac (1902–1984).
～測度 Dirac measure.

ていり 定理 theorem; theor.; th.; proposition; prop.
～の theorematic; propositional. X氏の～ X's theorem; theorem of X.

ディリクレ, ディリシュレ P. G. L. Dirichlet (1805–1859).
～級数[積分] Dirichlet series [integral]. ～問題 Dirichlet problem.

ディリシュレ ⇨ディリクレ.

ていりゅう 停留 stationary. ⇨定常(な).

ていりょうかする 定量化する quantify.

ていりょうてき 定量的 quantitative.
～に quantitatively. ～性質 quantitative property.

ディレンマ dilemma.

ディンキン E. B. Dynkin (1924–2014).
～図形, ～・ダイアグラム Dynkin diagram.

デカルト R. Descartes (1596–1650).
～の Cartesian; cartesian; Descartes'. ～座標 Cartesian coordinates. ～積 cartesian product. ～の卵形 Descartes' oval.

てきおう 適応 adaptation.
～させる adapt; fit. ～制御 adaptive control. ～性 adaptibility; flexibility. ～的デザイン adaptive design.

てきごう 適合(性) conformity; compatibility; adaptation; consistence; consistency; fitting.
～する compatible. ～度 goodness of fit. ～検定 test of goodness of fit.

テキスト text.
～ブック textbook.

てきせい 適正(な) proper.
～(な)凸関数 proper convex function.

てきせつ 適切な well-posed.
非～ ill-posed.

てきちゅう 的中 hit.
～する hit; come true.

てきとう 適当な suitable; proper; adequate; appropriate; reasonable; nice.

てきよう 適用 application.

～する apply.　～できる applicable.　～できない inapplicable.

てきよう　摘要　summary; résumé; outline; abstract.

デコーダ　decoder.　⇨解読.

デコード(する) decode.

てこひ　てこ比　leverage.

てさぐり　手探り　trial and error; fumble.

～する grope; fumble.

デザルグ　G. Desargues (1593-1662).

～の Desarguesian.　非～幾何学 non-Desarguesian geometry.

デシ　deci.

～グラム decigram.　～メートル decimeter.

デシジョン·テーブル　decision table.

デジタル　digital.

～[計数型]計算機 digital computer.　～・データ digital data.　～署名 digital signature.

てじゅん　手順　procedure; process; program(me).

テスト　test.　⇨検定; 判定.

～する test.

テータ　theta; 【Θ, θ, ϑ】(θの読み方の一つ).

～関数 theta function; ϑ-function.　～級数 theta series.

データ　(*pl.*) data; (*sing.*) datum.

初期～ initial data.　多変量～ multivariate data.　不完全～ fragmentary data.　アナログ[デジタル]～ analog(-ue) [digital] data.　連続型～ continuous data.　経時的～ longitudinal data.

でたらめ　でたらめ(の)　random.

～に at random.　～振出し random start.

テック　TeX.

ラテック，レイテック LaTeX.

デデキント　J. W. R. Dedekind (1831-1916).

～の判別定理 Dedekind's discriminant theorem.

テトラディック　tetradic.

デビアンス　deviance.

てびき　手引き　manual; guidance; guide.

テフ　TeX.

ラテフ LaTeX.

テーブル　table.　⇨表(ひょう).

デシジョン・～ decision table.

テラ　tera-【10^{12}】.

デルタ　delta; 【Δ, δ】.

～関数 delta function.　クロネッカーの～ Kronecker's delta [δ].　～・アンプリトゥディニス delta amplitudinis.

デルトイド　deltoid.

てん　点　point; pt. (*pl.* pts.); dot; spot.

～の punctual.　～を打つ dot; point.　～対称 point symmetry.　～集合 point set; set of points.　～列 point sequence; sequence of points.　定[動]～ fixed [variable] point.　始[終]～ initial [terminal] point; starting [end] point.　列～ range of points.　調和列～ harmonic range of points.　二重[三重, 重複]～ double [triple, multiple] point.　通常～ ordinary point.　不動～ fixed point.　孤立～ isolated [isolate] point.　集積～ accumulation point; cluster point; limit point.　密集～ density point; point of density.　変曲～ inflection [inflexion] point; point of inflection [inflexion].　正則～ regular point; non(-)singular point.　特異～ singular point; singularity.　非特異～ non(-)singular point.　単純～ non(-)singular point.　分岐～ branch point.　単位～ unit point.　零～ zero point; zero.　無限遠～ point at infinity.　原～ origin.　対称～ symmetric point.　～質量 point mass.　～変換[関数] point transformation [function].　～推定 point estimation.　(各)～別[ごと]に pointwise.　～抽出(法) point sampling.　～線 dotted line.　～グラフ point [dot] graph.　～図表 dot diagram; scatter diagram.　～乗積 dot product.　付～スピノル dotted spinor.　得～ mark; score; point.　平均～ average mark [score].　鞍～ saddle point.　外～ exterior point.　棄却～ critical point.　端～ end point; extreme point; extremal point.　内～ interior point.

てん　天　heaven(s).

～の celestial; heavenly.　～軸 celestial axis.　～球[極] celestial sphere [pole].　～体 heavenly [celestial] body.

てんい 転位 change; inversion; transport.

てんか 転嫁 imputation.

てんか 添加 augmentation; adjunction.
～する augment; adjoin. 係数～環 augmented algebra. 代数的(な元の)～ algebraic adjunction. ～された元 adjoined element.

でんか 電荷 (electric) charge.
点～ point charge. 代用～法 charge simulation method.

てんかい 展開 expansion; development; application.
～する expand; develop. ～可能な expansible; developable; applicable. ～式 expansion; expanded expression. ～形 expanded form; extensive form. ～係数 expansion coefficient. 級数～ series expansion [development]. 級数に～する expand [develop] into series. 漸近～ asymptotic expansion [development]. テイラー[フーリエ]～ Taylor [Fourier] expansion. ～(可能)曲面 developable [applicable] surface. 余因子～ cofactor expansion.

でんかい 電界 electric field.

てんかん 転換 conversion.
～命題 converse proposition.

でんき 電気(の) electric; electrical.
～変位 electric displacement. ～回路 electric circuit [network]. ～工学 electrical engineering.

でんき- 電気 electro-.
～力学 electrodynamics. ～光学 electro-optics. 静～学 electrostatics. 動～学 electrokinetics.

てんきゅう 天球 celestial sphere.
～赤道 celestial equator. ～座標 celestial coordinates. ～儀 celestial globe.

てんきょく 天極 celestial pole.

てんきょくせん 転曲[跡]線 rolling curve; roulette.

てんじ 展示 display.
～する display.

てんじ 添字 index (*pl.* -dices, -dexes); suffix (*pl.* -fices, -fixes). ⇨添数.
三～記号 three index symbol. ～集合 index set.

でんし 電子 electron.
～の electronic. ～工学 electronics. ～計算機 electronic computer [calculator].

でんじ- 電磁 electromagnetic.
～方程式 electromagnetic equation. ～波 electromagnetic wave. ～場[界] electromagnetic field.

でんじき 電磁気 electromagnetism.

でんじきがく 電磁気学 electromagnetism; electromagnetics.

てんしゅうごう 点集合 point set; set of points.
～論 theory of point sets.

でんしん 電信 telegraph.
～の telegraphic. ～方程式 telegraph equation.

てんすう 添数 index (*pl.* -dices, -dexes); suffix (*pl.* -fices, -fixes). ⇨添字.
～集合 index set; indexing set. 共変[反変]～ covariant [contravariant] index. 無効～ dummy index; umbral index. 二[多]重～ double [multiple] suffix.

てんずひょう 点図表 dot diagram; scatter diagram.

てんせきせん 転跡線 roulette; rolling curve.

てんせん 点線 dotted line.

でんせん 伝染 infection.
～性の infectious; epidemic. ～曲線 epidemic curve.

でんせんびょう 伝染病 infectious disease; contagious disease; epidemic.
～曲線 epidemic curve. ～理論 epidemics. ～学 epidemiology.

てんそう 転送 transfer; transference; transmission; forwarding. ⇨伝送.

でんそう 伝送 transmission.
～する transmit. ～速度 transmission rate. ～容量 transmission capacity.

テンソル, テンサー tensor.
～の tensorial. 基本～ fundamental tensor. ～場 tensor field. ～計算[解析] tensor calculus. ～代数 tensor algebra. ～解析 tensor analysis; tensor calculus. ～形式 tensor(ial) form. ～束 tensor bundle.

～空間 tensor space. ～表現 tensor representation. ～積 tensor product. 計量～ metric tensor. 共変[反変]～ covariant [contravariant] tensor. 曲率[捩率]～ curvature [torsion] tensor. 相関～ correlation tensor. 対称[交代]～ symmetric [alternating] tensor. 混合～ mixed [hybrid] tensor. 連絡[縮約]～ connecting [contracted] tensor. 変形[応力]～ strain [stress] tensor. 偏差～ deflection tensor.

てんたい 天体 celestial body; heavenly body.
～の celestial. ～力学 celestial mechanics; astrodynamics. ～観測 astronomical observation.

でんたつ 伝達 transmission; communication; delivery; transfer.
～する transmit; communicate. ～関数 transfer function.

てんち 転置 transposition.
～する transpose. ～行列 transposed matrix. ～方程式 transposed equation. ～写像[作用素] transposed mapping [operator]. ～核 transposed kernel. ～表現 transposed representation.

てんちょう 天頂 zenith.
～の zenithal. ～角 zenith(al) angle. ～距離 zenith(al) distance; colatitude. ～投影 zenith projection.

てんちょく 展直 rectification.
～平面[曲面] rectifying plane [surface].

てんてい 天底 nadir.

てんとう 転倒 inversion.

でんどう 伝導 conduction.
熱～ heat conduction.

てんにゅう 転入 transgression.
～的(な) transgressive. ～写像 transgression.

でんば 電場 ⇨電界.

でんぱ 電波 electric wave.

でんぱ(ん) 伝播 propagation; dispersion.
～性の propagative; propagatory; dispersive. 誤差(の)～ propagation of errors; error propagation.

てんべつ 点別の[に] pointwise.

てんもん 天文(学) astronomy.
～の astronomical. 球面～ spherical astronomy. ～測量 celestial surveying. ～図 astronomical chart. ～三角形 astronomical triangle.

てんや 点野 field of points; point field.

てんらく 纏絡 link. ⇨まつわり.
～係数 linking coefficient.

でんりゅう 電流 electric current; electric stream.
～の強さ current intensity. ～密度 electric current density.

てんれつ 点列 sequence of points; point sequence.
～的な sequential. ～的に sequentially. 有理～ sequence of rational points. 収束～ convergent sequence of points. ～コンパクトな sequentially compact.

と・ト

ど 度 degree.
百分～ grade. 分離～ degree of separation. 写像～ mapping degree. 散布～ degree of scattering. 自由～ degree of freedom. 関連～ degree of association. 連結～ connectivity.

ど 度 time. ⇨回.

とい 問い question; inquiry.

といあわせ 問合せ enquiry, inquiry; reference.

とう 筒; 壔 cylinder. ⇨柱(ちゅう).

とう- 等 equi-; iso-; homo-; uniform.
～式 equality. ～距離の equidistant. ～濃度の equipotent. ～ポテンシャル面 equipotential surface. ～スカラー面 equiscalar surface. ～非調和列点 equianharmonic range of points. ～周 isoperimetry. ～傾螺線 isoclinal spiral. ～偏角[方位]線 isogonic line. ～分(割)の isotomic. ～線分共役点 isotomic [isometric] conjugate point. ～散布の homoscedastic. ～エントロピー流 homentropic flow. ～速運動 uniform motion. ～加速度運動 uniformly accelerated motion.

どう- 動 moving; variable.
～枠; ～標構 moving frame. ～三面角 moving trihedral. ～座標系 moving coordinate system. ～点 variable point.

とうい 等位(の) equiniveau; equipotential.
～線[面] level curve [surface]; niveau curve [surface].

どうい 同位 same order; isotopy.
～の無限小[大] infinitesimal [infinity] of the same order.

どういかく 同位角 corresponding angle(s).

とういつ 統一 unification; standardization.
～する unify; standardize. ～場理論 unified field theory.

どういつ 同一 sameness; identity; (*L.*) idem. ⇨共(きょう-).
～の same; identical; (*L.*) idem. ～直線上の collinear. ～平面上の coplanar.

どういつし 同一視 identification. ⇨等化.
～する identify. *A* を *B* と～する identify *A* with *B*.

とうえい 投影 projection. ⇨投象.
～する project. ～法 projection. ～図[面] projection chart [surface]. 正～ orthogonal projection. 斜～ oblique projection. 正軸測～法 orthogonal axonometric projection. 球面～(法) globular projection.

とうおん 等温(の) isothermal.
～媒介[助]変数 isothermal parameter. ～座標 isothermal coordinates.

とうか 等化 identification; equating.
～する identify. ～位相 identification topology. ～空間 identification space.

とうか 等価 equivalence; equivalency.
～な equivalent. ～原理 equivalence principle.

とうかく 等角(な) equiangular; isogonal; (*G.*) winkeltreu; conformal.
～に isogonally; conformally. 逆～な anti-conformal; indirectly conformal. ～同値な conformally equivalent. ～螺線 equiangular spiral. ～多角形 equiangular polygon; isogon. ～台形 isogonal trapezoid. ～共役点 isogonal conjugate point. ～変換 isogonal transformation. ～写像 conformal mapping [map]; conformal represen-

tation; conformal transformation. 逆〜写像 anti-conformal [indirectly conformal] mapping. 〜構造[不変量] conformal structure [invariant]. 〜投影[投象]図[法] conformal projection. 〜地図 conformal map.

とうかんかく 等間隔(の) equally spaced.

どうかんすう 導関数 derivative; derived function.
偏〜 partial derivative. 高階[次]〜 derivative of higher order. 対数(的)〜 logarithmic derivative.

どうき 同期 synchronization.
〜(式)の synchronous. 〜化 synchronization.

どうき 動機 motive; motivation; cause.

とうきゃく 等脚(の) isosceles.
〜台形 isosceles trapezoid. 〜三角形 isosceles triangle.

とうきゅう 等級 grade; class; order; rank.

どうきょう 同境 cobordism.
〜な cobordant. 〜類[環] cobordism class [ring]. 2を法として〜な cobordant mod 2.

とうきょり 等距離 equidistance.
〜の equidistant; equally distant; isometric. 〜(曲)線 equidistant curve. 〜超曲面 equidistant hypersurface.

とうげ 峠 pass. ⇨鞍点.
〜点 saddle point; (*F.*) col. 〜の定理 mountain pass theorem.

とうけい 統計 statistics.
〜の statistical. 〜学 statistics. 人口〜 population statistics. 動態〜 dynamic statistics. 〜力学 statistical mechanics.

とうけい 等傾(の) isoclinal.
〜(曲)線 isoclinal line; isocline. 〜螺[渦]線 isoclinal spiral.

とうけい 等径(の) isoparametric.
〜関数 isoparametric function. 〜曲線[曲面] isoparametric curve [surface].

どうけい 動径 radius (*pl.* -dii, -diuses).
〜ベクトル radius vector.

どうけい 同型[形] isomorphism.
〜な isomorphic. 〜写像 isomorphism; isomorphic mapping. 〜定理 isomorphism theorem. 〜像 isomorphic image. 順序[双対]〜 order [dual] isomorphism. 束(順序)〜 lattice isomorphism. 自己〜 automorphism. 群(の)〜 group isomorphism; isomorphism of groups. 位相〜 homeomorphism. 微分〜(写像) diffeomorphism. 局所〜な locally isomorphic. 空間的〜な spatially isomorphic. スペクトル〜な spectrally isomorphic.

とうけいがく 統計学 statistics.
〜(上)の statistical; statistic. 記述[数理]〜 descriptive [mathematical] statistics. 推測〜 stochastics. 〜者 statistician.

とうけいてき 統計的(な) statistical.
〜確率 statistical probability. 〜仮説[推論] statistical hypothesis [inference]. 〜推定 statistical estimation. 〜検定 statistical test; testing of statistical hypothesis. 〜有意性 statistical significance. 〜決定関数 statistical decision function. 〜線形モデル statistical linear model. 〜品質管理 statistical quality control.

とうけいりょう 統計量 statistic.
不偏〜 unbiased statistic. 十分〜 sufficient statistic. 補助〜 ancillary statistic. 順序〜 order statistic. 選定〜 selection statistic. 検定〜 test statistic.

とうごう 等号 sign of equality; equality sign; equal sign.

とうこうせん 等高線 contour(-) line; niveau line; level line; isopleth.

どうごはんぷく 同語反復 tautology.
〜的な tautological.

とうさ- 等差
〜数列 arithmetic progression; A.P. 〜中項 arithmetic mean. 〜級数 arithmetic series.

とうざ 当座(の) temporary; present; immediate.

どうさ 動作 action; movement(s); behavio(u)r; operation.
(テューリング機械の)〜 atomic act. 〜[検査]特性 operating characteristic.

どうざひょうけい 動座標系 moving coordinate system.

とうし 透視 perspective.
～図[画](法) perspective; perspective projection. ～面 perspective surface. 全～図 total perspective.

とうじ- 等時
～(性)の isochronous. ～曲線 tautochrone; isochrone. ～問題 problem of tautochrone [isochrone].

どうじ 同次(の) homogeneous; of the same degree [order]. ⇨斉次(せいじ).
～性 homogeneity. ～式 homogeneous expression. ～多項式 homogeneous polynomial. ～元 homogeneous element. ～部分 homogeneous part. ～座標 homogeneous coordinates. ～ホロノミー群 homogeneous holonomy group. ～の無限小[大] infinitesimal [infinity] of the same order [degree]. ～イデアル homogeneous ideal.

どうじ 同時(の) simultaneous; synchronous.
～分布 simultaneous [joint] distribution.

とうしき 等式 equality; equation.
恒～ identity; identical equation. 類～ class equation.

とうじく 等軸(の) equiaxial.
～晶系 equiaxial system; cubic system.

どうじく 同軸 ⇨共軸(きょうじく).

とうしつ 等質(性) homogeneity.
～な homogeneous. ～空間[領域] homogeneous space [domain]. ～性の検定 homogeneity test.

とうしゃ 投射 incidence; projection.
～角 angle of incidence. ～測定 projectometry.

どうしゅ 導手 conductor.
～分岐定理 conductor-ramification theorem.

どうしゅ 同種(な) isogenous; of the same kind.
～写像 isogeny.

とうしゅう 等周 isoperimetry.
～の isoperimetric. ～不等式 isoperimetric inequality. ～問題 isoperimetric problem.

とうしゅう 等終(の) residual.
～集合 residual set.

どうしゅうごう 導集合 derived set.

どうじゅんい 同順位 tied rank.

とうしょう 投象 perspective; projection. ⇨投影.
正[斜]～ orthogonal [oblique] projection. 中心～画法 (central) perspective drawing. キャビネット～(法) cabinet projection.

どうしょう 同焦(点の) confocal. ⇨共焦(の).

どうしん 同心(の) concentric.
～円 concentric circle(s). ～円環 (concentric) annulus; concentric circular ring.

とう-スカラー 等スカラー(の) equiscalar.
～面 equiscalar surface.

とうせき 等積 equivalence.
～な equivalent. ～アフィン変換 equivalent affinity. ～変換[変形] equivalent transformation [deformation]. ～な正方形で表せる quadrable. ～地図 equivalent map. ～円錐[円柱]投影(法) equivalent conic [cylindrical] projection.

どうせん 導線 generating line; generator; director curve. ⇨母線.

どうそう 同相 homeomorphism.
～な homeomorphic. ～写像 homeomorphic mapping; homeomorphism. 線形～(写像) linear homeomorphism. Aに～な homeomorphic to A. 微分～(写像) diffeomorphism.

とうそく 等速 constant velocity; uniform velocity.
～運動 uniform motion.

とうそく 等測 isometry; equimeasure.
～な isometric. ～投[射]影(法) isometric projection. ～変換 equimeasure transformation. ～包 measurable cover; (*G.*) massgleiche Hülle. ～核 measurable kernel; (*G.*) massgleicher Kern.

どうたい 導体 conductor.
～ポテンシャル conductor potential.

どうたい 動態 (population) movement.
～統計 dynamic statistics.

とうたつ- 到達
～時刻 hitting time. ～測度 hitting mea-

sure.

とうたつかのう 到達可能(性) accessibility.
～な accessible. ～境界点 accessible boundary point.

とうたつふのう 到達不能(性) inaccessibility.
～な inaccessible. ～境界点 inaccessible boundary point. ～基数 inaccessible cardinal.

とうち 等置 equivalence.
～する equate. ～法 method of equivalence.

とうち 等値 equivalence.
～な equivalent. ～な問題 equivalent problem.

とうち 等値 isoplethy.
～な isoplethic. ～(曲)線 isoplethic curve; isopleth.

どうち 同値 equivalence; equivalency; equivalent.
～な equivalent. ～でない inequivalent; not equivalent. ～法則[律] equivalence law. ～関係 equivalence relation. ～類 equivalence class. ～な距離[ノルム] equivalent distance [norm]. 位相～な topologically equivalent. ホモトピー～ homotopy equivalence. ホモトピー～な homotopy equivalent. 森田～ Morita equivalence. 森田～な Morita equivalent. ユニタリ～な unitary-equivalent. 安定～な stably equivalent. 鎖～な chain equivalent. 等角～な conformally equivalent. 安定～ stable equivalence. 線形～ linear equivalence. 双有理～ birational equivalence.

とうちゃく 到着 arrival.
～する arrive; reach. ～時刻 arrival time. 集団～ bulk arrival.

とうちょう 等長(な) isometric.
～写像 isometric mapping; isometry. ～変換 isometric transformation; isometry. ～作用素 isometric operator.

どうちょう 同調 coherence; coherency; concordance.
～的な coherent; concordant. ～的に coherently. ～的に向きづけられた coherently oriented.

とうちょうしゃ 盗聴者 tapper.

どうてい 道程 distance.

どうていど 同程度 equal grade.
～に確からしい equally likely [probable].

どうていどれんぞく 同程度連続(性) equicontinuity.
～な equicontinous.

どうてき 動的(な) dynamic.
～計画法 dynamic programming; DP.

どうてん 動点 variable point.

どうてん 同点 tie; draw; even break.
～になる tie.

どうとう 同等 equivalence; equivalency; equality.
～な equivalent; equal. ～に equivalently; equally.

どうとうのひと[もの] 同等の人[物] equal.

とうにゅう- 投入
～係数 interindustry coefficient. ～産出分析 input-output analysis.

とうのうど 等濃度 equipotency.
～の equipotent.

どうはん 同伴 association.
～の associated; associate; adjoint. ～数[元] associate number [element]. ～方程式 associated equation. ～線形写像 associated linear mapping. ～式 adjoint. ～形式[多項式] associated form [polynomial]. ～[関連]収束半径 associated convergence radii.

とうひ 等比 ratio of equality; equal [constant] ratio.
～の理 ex-aequali.

とうひ- 等比
～数列 geometric progression; G.P. ～中項 geometric mean. ～級数 geometric series.

どうひょうこう 動標構 moving frame; (*F.*) repère mobile.

とうぶん 等分 ⇨二等分; 三等分.
～する divide equally [into equal parts].

どうぶん 導分 derivation.

とうべき 等ベキ(の) idempotent. ⇨ベキ等(性).

〜元 idempotent element.

とうへん 等辺(の) equilateral.
〜三角形 equilateral triangle; regular triangle. 〜多角形 equilateral polygon. 〜双曲線 equilateral hyperbola.

とうほう 等方(性) isotropy.
〜的な isotropic. 〜群 isotropy group. 〜乱流 isotropic turbulence. 全〜的な totally isotropic. 非〜性 anisotropy. 非〜的な anisotropic.

どうぼう 同傍 same side.
〜外[内]角 exterior [interior] angle(s) on the same side.

とう-ポテンシャル 等ポテンシャル(の) equipotential.
〜線[面] equipotential line [surface].

とうめん 壔面 cylindrical surface. ⇨柱面.

どうめん 導面 directrix (*pl.* -trixes, -trices).

どうよう 同様(な) similar; analogous; of the same manner.
〜に similarly; analogously; equally; in the same manner. 〜に確からしい equally likely [probable]; equiprobable.

どうらい- 導来
〜集合 derived set. 〜群 derived group. 〜関手 derived functor. 〜列 derived series. 〜関係 induced relation. 〜圏 derived category.

どうりきがく 動力学 dynamics; kinetics.

どうるいぐん 道類群 class-group of paths.

どうるいこう 同類項 similar term; like term.

とうろく 登録 registration; entry.
〜する register; enter.

どうわく 動枠 ⇨動標構.

とおる 通る pass【along; through】.

とがった 尖った cuspidal; cusped; pointed; salient.

とがり 尖り kurtosis.

とき 時 time; moment.
〜々 sometimes; occasionally; often. 〜には起り得る eventual.

とく 解く solve; answer.
解ける solvable; soluble.

とくい 特異(な) singular.
非〜な non(-) singular. 左[右]〜な left [right] singular. 全〜な totally singular. 〜解 singular solution. 〜初期値問題 singular initial value problem. 〜部分 singular part. 〜点 singular point; singularity; irregular point. 〜積分 singular integral. 〜級数 singular series. 〜核 singular kernel. 〜積分方程式 singular integral equation. 〜単体[複体] singular simplex [complex]. 〜部分空間 singular subspace. 〜ホモロジー群 singular homology group. 〜摂動 singular perturbation. 〜性 singularity; anomaly. 〜類 ambig class. 〜イデアル ambig ideal. 〜値 singular value. 〜問題 singular problem.

とくいせい 特異性 singularity; anomaly.

とくいてん 特異点 singular point; singularity.
孤立〜 isolated singularity. 代数[対数](的)〜 algebraic [logarithmic] singularity. 通常[通性]〜 ordinary singularity. 真性〜 essential singularity. 超越〜 transcendental singularity. 直接[間接](超越)〜 direct [indirect] (transcendental) singularity. 除去可能(な)〜 removable singularity. 見掛けの〜 apparent singular point. 確定[不確定]〜 regular [irregular] singularity. 動く[かない]〜 movable [fixed] singularity. 〜集合 singularity set. 〜の合流 confluence of singularities. (収束円)周上の〜 peripheral singularity. 〜の解消 resolution of singularities.

とくいど 特異度 specificity.

どくしゃ 読者 reader.
購〜 subscriber.

とくしゅ 特殊(な) special; particular; distinguished.
〜性 speciality; particularity. 〜化 specialization; particularization. 〜曲線[曲面] special curve(s) [surface(s)]. 〜関数 special function. 〜線形[直交]群 special linear [orthogonal] group. 〜ユニタリ群 special unitary group. 〜等周問題 special isoperimetric problem. 〜相対性原理 principle of special relativity. 〜指数 speciality index; index of speciality. 〜解 particular solution. 〜計算尺 particular

slide rule. ～擬多項式 distinguished pseudo-polynomial.

とくしょう　特称(の)　existential. ⇨存在.
～記号 existential quantifier; (*G.*) Seinzeichen. ～命題 existential proposition; (*G.*) partikuläre Aussage.

とくせい　特性　(distinctive) character.
～的な characteristic. ～づけ characterization. ～関数[汎関数] characteristic function [functional]. ～方程式 characteristic equation. ～根[ベクトル] characteristic root [vector]. ～多項式 characteristic polynomial. ～曲線[帯] characteristic curve [strip]. ～線素 characteristic line element. ～多様体 characteristic manifold. ～集合[線形系] characteristic set [linear system]. ～指数 characteristic exponent. ～組成列 characteristic series. ～類 characteristic class. ～数 characteristic number; type number. ～乗数 characteristic multiplier.

どくそう　独創　originality.
～的な original.

とくちょう　特徴　characteristic.
～づける characterize. ～づけ characterization. ～関数 characteristic function.

とくてい　特定の　particular.

とくてん　得点　mark; score; point.
平均～ average mark [score].

とくべつ　特別(な)　special; particular; extra.
～号 special [extra] issue. ～品 speciality.

どくりつ　独立(性)　independence; independency.
～な independent. と～な independent of. ～に independently. ～変数 independent variable. ～事象 independent event(s). ～試行 independent trials. 一次[線形]～な linearly independent. 代数的～な algebraically independent. 関数的～な functionally independent. 線形[一次]～性 linear independence.

とけい　時計　clock.
～の針と同方向の[に] clockwise. ～の針と逆方向の[に] anticlockwise; counterclockwise.

どこにも　everywhere.
～…ない nowhere.

トーシェント　totient.
オイラーの～関数 Euler's totient function.

とじた　閉じた　closed. ⇨閉(へい-).

どすう　度数　frequency. ⇨頻度.
～分布 frequency distribution. ～分布表 frequency table. ～曲線 frequency curve. 相対～ relative frequency. 級～ class frequency. 累積～ cumulative frequency.

とちゅうぬきとり　途中抜取り　last-off sampling.

どちらでもよい　indifferent 【to】; irrespective 【of】.

とつ　凸(性)　convexity. ⇨凹(性).
～な convex. ～多角形[多面体] convex polygon [polyhedron]. ～曲線[曲面] convex curve [surface]. ～錐 convex cone. ～多面錐 convex polyhedral cone. ～体 convex body. ～関数 convex function. ～数列 convex sequence. ～集合 convex set. ～近傍 convex neighbo(u)rhood. ～(閉)包 convex hull; convex envelope; convex closure. 閉～包 closed convex hull. ～領域 convex domain. ～写像 convex mapping. ～性半径 radius of convexity; convexity radius. ～不等式 convexity inequality. ～(性)定理 convexity theorem. ～計画法 convex programming. 狭義(に)[一様(に)]～な strictly [uniformly] convex. 下に～な downwards convex. 対数的[正則]～な logarithmically [holomorphically] convex. 局所～な locally convex. 円形～な absolutely convex. 擬～な pseudo-convex. 真～関数, 適正(な)～関数 proper convex function. 真～集合 proper convex set. ～解析 convex analysis.

とっかく　凸角　salient; convex angle.
～の salient. ～点 salient point.

とっき　突起　salience; projection; protuberance.
～した salient.

とどまる　止[留]まる　stop; halt; stay; remain.

トートロジー　tautology.

とび　飛び　jump; saltus; leap.

とびこし 飛越(し) jump.
～す jump.
トポロジー topology. ⇨位相数学.
組合せ[集合]論的～ combinatorial [set--theoretic] topology.
トライアド triad.
トライセカント trisecant.
トラクトリックス tractrix.
トーラス torus (*pl.* -ri); anchor ring.
～の toric. 複素～ complex torus. ～群 torus group.
トラック track.
トラップ trap.
トラフィック traffic.
～密度 traffic intensity.
トリアディック triadic.
とりくみかた 取り組み方 approach.
とりくむ 取り組む address.
とりけし 取消(し) cancellation.
～す cancel.
とりだし 取出(し) fetch.
～す fetch.
トーリック toric.
～多様体 toric variety.
トリビアル trivial. ⇨自明(な).
～な解 trivial solution.
どりょうこう 度量衡 weights and measures.
～学 metrology.
トレース trace; (*G.*) Spur. ⇨跡(せき).
～・ノルム trace norm. ～族 trace class.
トロイド toroid.
～の toroidal.
トロコイド trochoid.
～の trochoidal. 内～ hypotrochoid. 外～ epitrochoid. ～波 trochoidal wave.
どんかく 鈍角 obtuse angle.
～の obtuse(-angled). ～三角形 obtuse triangle.

な・ナ

ない- 内 internal; interior; intra-; inner. ⇨内(うち); 外(がい-).

-ない ない not; less; un-; (-)free.
大きく[小さく]～ not greater [less] than; less [greater] than or equal to. 番号の～[無数の] numberless; unnumbered. ねじれの～ torsion-free. 臍点の～ umbilic-free. 特異点の～ singularity-free. 渦なしの; 無渦の vortex-free; irrotational. 零点の～領域 zero-free region.

ないかく 内角 internal angle; interior angle.

ないかくど 内拡度 inner [interior] content.

ないけい 内径 inside diameter; gauge.

ないざい 内在(性) immanency.
～的な intrinsic; immanent. ～的に intrinsically; immanently. ～精度 intrinsic accuracy. ～的ホモロジー intrinsic homology. ～的性質 intrinsic property.

ないしん 内心 incenter; inner center; internal center; center of incircle.

ないせい 内生(の) endogen(e)ous.
～変数 endogenous [internal] variable.

ないせき 内積 inner product.
エルミート～ Hermitian inner product. ～空間 inner product space.

ないせつ 内接 inscription.
～させる inscribe. ～円 inscribed circle; incircle. ～多角形 inscribed polygon. ～球 inscribed sphere.

ないそう 内挿(法) interpolation. ⇨補間(法).

ないそくど 内測度 inner measure.

ないたいかく 内対角 interior opposite angle(s).

ないたいせき 内体積 interior volume; inner volume.

ないてき 内的 intrinsic. ⇨内在.
～幾何学 intrinsic geometry. ～性質 intrinsic property.

ないてん 内点 interior point; inner point.

ないてんけい 内転(卵)形 inrevolvable oval.

ないはいせん 内擺線 hypocycloid.
～の hypocycloidal. 三尖点～ three-cusped hypocycloid.

ないぶ 内部 interior; inside. ⇨外部.
～の interior; inner; internal. ～に inside. ～変換 inner [interior] transformation. ～自己同型[形] inner automorphism. ～微分 inner derivation [derivative]. ～問題 interior problem. ～積 interior [internal] product. ～状態 internal state; machine configuration. ～対称性 internal symmetry. ～算法 internal law of composition.

ないぶじょうたい 内部状態 internal state; machine configuration.

ないぶん 内分 internal division.
～する divide internally. ～点 internally dividing point. ～比 internal ratio.

ないぶんさん 内分散 internal variance.

ないほう 内包 connotation; intension; comprehension; implication.
～性公理 axiom of comprehension.

ないめん 内面 inside; interior (face).

ないめんせき 内面積 inner area; interior area.

ないよう 内容 content(s).
多項式の～ content of polynomial.

ないようりょう 内容量 inner capacity.

ないよはいせん 内余擺線 hypotrochoid.
～の hypotrochoidal.

なおさら still more, much more; (*L.*) a

fortiori.

なか 中 into.
　～への写像 into-mapping.

ながさ 長さ length; linear content.
　～を求める(曲線の) rectify. 曲線の～ length of curve; curve length. 極値的～ extremal length. ～の有限な[のある]曲線 rectifiable curve.

なかやま・ただし 中山正 Nakayama Tadashi (1912-1964).

ながれ 流れ stream; flow; current.
　～の関数 stream function. 測地的～ geodesic flow. ～図 flow chart; flow sheet; flowchart.

ながれず 流れ図 flow chart; flow sheet; flowchart.

なぞ 謎 puzzle; riddle; enigma; mystery.

なぞらえ comparison; dummy treatment.

なづける 名づける call; denominate; designate.

ナット nat【natural unit】.

など and so on [forth]; and others; (*L.*) et cetera; etc.; &c.

なな- 七 seven. ⇨七(しち-).

ななめ 斜め obliqueness; obliquity; skewness.
　～の oblique; slant; skew.

ナノ nano-.
　～メートル nanometer【10^{-9} meter】.

ナブラ nabla; ∇. ⇨アトレッド.

なみ 波 wave. ⇨波(は); 波動.
　縦[横]～ longitudinal [transversal] wave. 速い[遅い]～ fast [slow] wave.

なみすう 並数 mode.

なめらか 滑らか(な) smooth.
　～に smoothly. ～さ smoothness. ～な関数 smooth function. ～な曲線[曲面] smooth curve [surface]. ～な多様体 smooth manifold. 区分的に～な piecewise smooth. ～にすること smoothing; graduation. ～さの条件 smoothness condition.

ならば if ... then.
　A～Bである if A then B; A implies B.

ならべかえ 並べかえ permutation; rearrangement. ⇨順列.
　～る permute; rearrange. ～できる permutable.

なりすまし impersonation.

ナリティ nullity.

なる 成る
　に～ become. から～ consist of; be composed of.

なんかさせる 軟化させる mollify.

なんかし 軟化子 mollifier.

なんきょく 南極 south pole.

なんじゃく 軟弱な flabby; weak.

なんもん 難問 difficult [hard] problem [question].

に・二

に　二　two.
〜個; 〜の two.　〜ダース two dozen(s).　〜次元の two-dimensional.　〜棚法 two--bin system.　〜体問題 two (-) body problem.

に-　二　bi-; di-.
〜元の binary.　〜進の binary; diadic; dyadic.

にいんし　二因子　two factors.
〜交互作用 two-factor interaction; simple interaction.

にかい　二回　⇨二度.
〜抽出[抜取] double sampling; two-step sampling.

にかい　二階　second order.
〜導関数 derivative of second order; second order derivative.　〜微分方程式 differential equation of second order; second order differential equation.

にきゅう-　二級　of second class.
〜曲線[曲面] curve [surface] of second class.

にきょく　二極　dipole.　⇨双極(の).
〜座標 dipole coordinates.

にげん　二元　two unknowns; two ways; two dimensions.
〜の binary.　〜方程式 equation with two unknowns.　〜配置法 two-way classification.　〜配置 two-way layout.　〜計画(法) two element [two-dimensional] design.

にこう-　二項　binomial; binary.
〜式 binomial.　〜方程式 binomial equation.　〜定理[係数] binomial theorem [coefficient].　〜級数[展開] binomial series [expansion].　〜分布 binomial distribution.　〜確率紙 binomial probability paper.　〜関係 binary relation.　〜演算[算法] binary operation.

にこうたいりつ　二項対立　dichotomy.

にごしん　2-5進(の)　biquinary.
〜法 biquinary notation.

にこすう　二個数(の)　dyadic.

にじ　二次(の)　quadratic; quadric; of second degree [order].
〜式 quadratic expression.　〜の項 term of second degree.　〜方程式 quadratic equation.　〜不等式 inequality of second degree.　〜曲線 quadratic curve; conics.　〜曲面 quadrics; quadratic surface; conicoid.　有心〜曲線[曲面] central conics [quadrics].　〜超曲面 quadratic [quadric] hypersurface.　〜錐面 quadratic [quadric] cone.　〜柱[筒]面 quadratic cylinder; quadric cylinder.　〜列点 range of points of second order.　〜曲線[曲面]束 pencil of quadratic curves [surfaces].　〜形式 quadratic form.　〜体 quadratic field.　〜関数 quadratic function; quadric.　〜導関数 derivative of second order.　〜微分 quadratic differential.　〜補間公式 interpolation formula of second degree.　〜計画法 quadratic programming.　〜損失関数 quadratic loss function.

にじ　二次(的)　secondary.
〜的に secondarily.　〜合成 secondary composition.

にじく　二軸(の)　biaxial.
〜球面関数 biaxial spherical surface function.

にしゃたくいつ　二者択一　alternative.

～的な alternative. ～的に alternatively. ～定理 theorem of the alternative.

にじゅう 二十 twenty; score (*pl*.-).

にじゅう 二重(の) double; duplicate; duplex; duple; two(-)fold; bi-; di-.
～に doubly. ～にする double; duplicate. ～否定 double negation [negative]. ～帰納法 double induction. ～関係 two(-)fold relation. ～分類 two way classification. ～点[要素] double point [element]. ～根号 double radical (sign). ～根 double root. ～尖点 double cusp; tacnode. ～添数 double suffix. ～数列[級数] double sequence [series]. ～積分 double integral. ～層 double layer. ～格子 double lattice. ～複体 bicomplex. ～鎖複体 double chain complex. ～システム duplex system. ～加群 bimodule. ～ベクトル bivector. ～線形の bilinear. ～直交の biorthogonal. ～連結の doubly connected. ～周期の doubly periodic. ～検査 duplication check.

にじゅうかぐん 二重加群 bimodule.

にじゅうきゅうすう 二重級数 double series.
～定理 double series theorem; (*G.*) Doppelreihensatz.

にじゅうしすう 二重指数 double exponential.
～関数型公式 double exponential formula.

にじゅうせんてん 二重尖点 double cusp; tacnode.

にじゅうふくたい 二重複体 bicomplex.

にじゅう-ベクトル 二重ベクトル bivector.

にじゅうめんたい 二十面体 icosahedron (*pl.* -rons, -dra).
～の icosahedral. 正～ regular icosahedron. (正)～群 icosahedral group.

にじゅっしんほう 二十進法 vicenary notation; base twenty system.
～の vicenary; vigesimal.

にじょう 二乗 square. ⇨平方.
～する square; raise to the second power. ～根 square root. ～比 square ratio; duplicate ratio. ～に正[反]比例する vary directly [inversely] as square. ～可積な square integrable. ～採中法 middle-square method. ～損失関数 quadratic loss function.

にしん 二進(の) binary; dyadic; diadic.
～法 binary (notation) scale; dyadic [diadic] system. ～数字 binary digit. ～整数 dyadic [diadic] integer. ～法の桁 bit. ～コード binary code. ～(符号)化された binary coded. ～化十進法 binary-coded decimal notation; BCD. ～対称通信路 binary symmetric channel.

にそう 二相 two phases.
～抽出(法) two-phase sampling.

にたいもんだい 二体問題 two(-) body problem; problem of two(-)bodies.

にだん 二段 two stages [steps].
～抽出(法) two-stage [-step] sampling; sub-sampling. ～層化 sub-stratification.

にだんかい 二段階 ⇨二段.

にだんかいほう 二段階法 two-phase method.

にち 二値(の) binary; two-valued.
～素子 binary cell. ～論理 binary [two-valued] logic.

にっていけいかく 日程計画 program; schedule; scheduling.

にてん 二点 two points.
～境界値問題 two-point boundary value problem. ～比較法 pair test.

にど 二度 twice; two times. ⇨二階.

にとうぶん 二等分 bisection.
～する bisect. ～線 bisector. 垂直～線 perpendicular bisector; midperpendicular; bisecting normal.

にとうへん 二等辺(の) isosceles.
～三角形 isosceles triangle. ～球面三角形 isosceles spherical triangle.

にないて 担い手 support; carrier.

にばい 二倍 two times; twice; twofold.
～の double; twofold; duple; duplex. ～に twice. ～にする double; duplicate. ～にすること duplication. ～精度 double precision. ～角の公式 formula for [of] double angle; double angle formula. ～比 duple ratio.

にひょうほん 二標本 two-sample.
～検定 two-sample test.

にぶ(からなる) 二部(から成る) bipartite.

～グラフ bipartite graph.

にぶん 二分 bisection; dichotomy.
～的な bisecting; dichotomizing; binary. ～可能性 bisectability. ～法 dichotomy; method of bisection. ～探索 dichotomizing search; binary search. 二重～法 double dichotomy. ～木 binary tree.

にへんりょう 二変量(の) bivariate.
～分布 bivariate distribution. ～相関 bivariate correlation.

にほうこう 二方向 two ways [directions].
～の two-way.

にめん 二面(の) dihedral; bihedral.
～角 dihedral angle. ～体群 dihedral group.

ニュー nu; 【N, ν】.

にゅうしゃ 入射(的な) injective.
～的分解 injective resolution. ～加群 injective module.

にゅうしゃかく 入射角 angle of incidence.

にゅうりょく 入力 input. ⇨出力.
～域 input area.

ニュートン I. Newton (1642-1727).
～の Newtonian; Newton's. ～・ポテンシャル Newtonian potential. ～力学 Newtonian mechanics. ～の公式 Newton's formula.

ニュートン-ほう ニュートン法 Newton's method.
減速～ damped Newton method. 非厳密～ inexact Newton method.

にりつはいはん 二律背反 antinomy.

にんい 任意(性) arbitrariness; randomness.
～な any; arbitrary; random; optional. ～に arbitrarily; at random; optionally. ～定数 arbitrary constant. ～関数 arbitrary function. ～標本 random sample. ～抽出 optional sampling.

にんげん 人間 human being; man(kind).
～の human. ～工学 human engineering; ergonomics.

にんしき 認識 recognition.
文字～ character recognition. パターン～ pattern recognition.

にんしょう 認証 certification.

ぬかす　抜かす　omit; leave out.

ぬきずり　抜刷　offprint; separate (print).

ぬきとり　抜取(り)　sampling.　⇨抽出.
　～検査 sampling inspection.　一[二]回～検査 single [double] sampling inspection.　調整型～ sampling with adjustment.　途中～ last-off sampling.

ね・ネ

ね 値 price; cost; value. ⇨価(か).

ネイピア J. Napier (1550-1617).
〜の Napierian. 〜の数 Napier's number; *e*. 〜の対数 Napierian logarithm. 〜の計算棒 Napier rod.

ねじ screw.
雌〜 female screw; nut. 雄〜 male screw; bolt.

ねじれ 捩[捻]れ torsion; twist.
〜率 torsion. 〜テンソル torsion tensor. 〜係数 torsion coefficient. 〜積 torsion product. 〜群 torsion group. 〜のない torsion(-)free. 〜のない群 torsion(-)free group. 〜四辺形 twisted [skew] quadrilateral. 〜た曲線 twisted curve. 〜の位置 twisted [skew] position. 〜元 torsion element.

ネーター E. Noether (1882-1935).
〜の; 〜的 Noetherian. 〜環 Noetherian ring. 〜加群 Noetherian module. 〜空間 Noetherian space. 〜の正規化定理 Noether's normalization theorem.

ねつ 熱 heat.
〜伝導 heat conduction. 〜源 heat source. 〜方程式 heat equation. 〜力学 thermodynamics.

ネット net. ⇨網(あみ).

ネットワーク network.
双方向性〜 bilateral network. ニューラル〜 neural network.

ねん 年 year.
暦〜 calendar [civil] year. 太陽〜 tropical [solar] year. 光〜 light-year.

ねんきん 年金 annuity; pension.
終身〜 life annuity. 〜積立 accumulation by annual instal(l)ments. 〜換算値 adjusted annual worth.

ねんせい 粘性 viscosity.
〜の viscous. 〜流体 viscous fluid. 〜解 viscosity solution.

ねんぷ 年賦 annual [yearly] instal(l)-ment; annual payment.
〜金 amount of annual [yearly] instal(l)-ments. 〜償還 annual payment.

ねんぽう 年報 annual report(s); annals.

ねんり 年利 annual interest.

の・ノ

のうど 濃度 power; potency; cardinal number.
集合の～ potency of set. 等～の; 対等な～の equipotent.

のうど 濃度 concentration.
～関数 concentration function.

のうどうてき 能動的 active.

のうりつ 能率 efficiency.
～的な efficient.

のうりつ 能率 moment. ⇨積率.
慣性～ moment of inertia. 二次～ second moment. ～行列 moment matrix.

のこり 残り difference; remainder; rest.

のぞく 除く remove; exclude. ⇨除去.
～ことができる removable. …を除いて except (for); but; exclusive of.

ノード node. ⇨節点.
～の nodal.

のべる 述べる state; say; mention; describe; express; explain.

ノーマル normal. ⇨法線; 正規(性); 標準(的).

ノモグラフ nomograph.

ノモグラム nomogram.

ノーモン gnomon.
～の gnomonic.

のりかえ 乗換 change; transfer.
～る change; transfer.

ノルム norm.
～の normic. ～を備えた normed. ～づけ可能な normable. L^p [上限]～ L^p- [sup] norm. 絶対～ absolute norm. 相対～ relative norm. 半～ seminorm; pseudo-norm. 準～ quasi-norm. 一様～ uniform norm. 被約～ reduced norm. 作用素～ operator norm. トレース・～ trace norm. スピン・～ spinorial norm. ～形式 normic form. ～線形空間 normed linear space. ～環 normed ring. ～剰余 norm--residue.

のろい 呪い curse.
次元の～ curse of dimensionality.

ノンパラメトリック(な) non-parametric.
～法 non-parametric method. ～推定 non--parametric estimation. ～検定 non-parametric test.

は・ハ

は 波 wave.
~状の wavy. ~長 wave length. ~面 wave front. 平面[球面]~ plane [spherical] wave. 表面~ surface wave. 定常~ standing [stationary; steady] wave. 正弦~ sine wave. 横[縦]~ transversal [longitudinal] wave. ~数 wave number. ~線 wavy line.

は 歯 tooth (*pl.* teeth). ⇨歯(し).

ば 場 field.
~の理論 field theory. 力の~ field of force. スカラー~ scalar field. ベクトル~ vector field. テンソル~ tensor field.

バー bar; 【—】.

ばあい 場合 case; circumstances.
有利[好都合]な~ favorable case. 例外の~ exceptional case. ~に応じて case by case. の~には in case of. この~に(は) in this case. どの~にも in any case. ~によっては under certain circumstances.

ばい 倍 times.
二~ two times; twice. 三~ three times; thrice. 一~半 one and a half times. 二~の double; twice; two(-)fold.

ばい- 倍 multiple.
~数[元] multiple.

ばい- 陪 associated; bi-.
~関数 associated function. ~微分方程式 associated differential equation. ~法線 binormal.

-ばい 倍 -fold; -tuple. ⇨倍(ばい).
二~の[に] two(-)fold. 百~(の) centuple.

パイ pi; 【Π, π, ϖ】; π【円周率】.

バイオニクス bionics.

バイオメトリ biometry.

ばいかいへんすう 媒介変数 parameter. ⇨パラメータ.
~の parametric. ~表示 parametric representation. ~方程式 parametric equation. 一意化~ uniformizing parameter; uniformizer.

ばいかく 倍角 double angle.
~(の)公式 formula for [of] double angle. ⇨二倍; 三倍.

ばいかんすう 陪関数 associated function.

はいきゃくりつ 廃却率 mortality.

はいけい 背景 background.

はいけい 配景 perspectivity.
~的な perspective. ~の位置 perspective position. ~の中心 center of perspectivity. ~的関係 perspective relation. ~写像[変換] perspective mapping [transformation].

ばいげん 倍元 multiple. ⇨倍数.

ばいすう 倍数 multiple.
~の multiple. 公~ common multiple.

バイスピノル bispinor.

はいせん 擺線 cycloid. ⇨サイクロイド.
~の cycloidal. 外~ epicycloid. 内~ hypocycloid.

はいた 排他 exclusiveness; exclusion.
~的な exclusive. ~的に exclusively. ~的論理和 exclusive or.

はいち 配置 arrangement; collocation; configuration; layout.
~する arrange; collocate. 無作為~ random arrangement. 二元~ two-way classification; two-way layout. 対称~ symmetric configuration. パスカルの~ Pascal's configuration.

はいちゅうりつ 排中律 law of excluded middle.

はいとう 配当 allocation.
最適～ optimal allocation.

ばいとくせい 陪特性(の) bicharacteristic.
～曲線[帯] bicharacteristic curve [strip].

はいはん 排反 exclusiveness.
～の exclusive; disjoint. ～事象 exclusive events; incompatible events.

はいぶん 配分 distribution; allocation; allotment.
比例～ proportional distribution. 最適～ optimal allocation.

ハイポイド hypoid.
～歯車 hypoid gear.

ばいほうせん 陪法線 binormal.
～標形 binormal indicatrix.

ハイポグラフ hypograph.

はいり 背理 paradox. ⇨逆理.
ツェノンの～ paradox of Zenon.

ばいりつ 倍率 magnifying power [rate]; magnification (ratio); scale factor.

はいりほう 背理法 (*L.*) reductio ad absurdum; reduction to absurdity; reductive absurdity; proof by contradiction.

はいれつ 配列 arrangement; array.
～する arrange; array. 直交～表 orthogonal layout. 直交～ orthogonal array.

ハウスドルフ F. Hausdorff (1868-1942).
～空間 Hausdorff space.

はかい 破壊 destruction.
～的な destructive. ～する destroy. ～検査 destructive inspection.

はかる 測[秤]る measure; weigh; estimate; survey.

はきだし 掃出し balayage; sweeping-out.
～法 sweeping-out method.

はきょく 破局 catastrophe. ⇨カタストロフィー.

ばく- 漠 vague.
～位相 vague topology. ～収束 vague convergence. ～有界[連続]な vaguely bounded [continuous].

ばくいそう 漠位相 vague topology.
～的に vaguely.

はくし 博士 doctor.
～号 doctor's degree. ～論文 thesis for doctor's degree; dissertation.

ばくしゅうそく 漠収束 vague convergence.

はくせつくうかん 迫接空間 accessible space.

ばくぜん 漠然 vagueness.
～とした vague. ～と vaguely.

はくそう 薄層 lamina (*pl.* -nae, -nas); lamella (*pl.* -lae, -las); thin layer.
～の laminar; laminal; lamellar; lamellate.

ばくはつ 爆発(解の) blow-up.

はぐるま 歯車 gear.
～の幾何学 geometry of gears. インボリュート[ハイポイド]～ involute [hypoid] gear. ～型領域 gear-shaped domain.

はこ 箱 box.
～位相 box topology. 暗～ black box.

はこひげず 箱ひげ図 box-whisker plot.

ハザード hazard.
～関数 hazard function.

はさむ (辺が角を)夾む include; contain.

はさん 破産 ruin; bankruptcy.
～させる ruin. ～確率 ruin probability.

はし 端 end; corner; edge; border; terminal; extreme.
～の extreme; extremal; terminal.

はしご 梯子 ladder. ⇨昇降演算.

はした ⇨半端, 端数.

はしゅ 把手 handle.
～体 handlebody; handle body. ～つけ操作 handle-attaching. ～つき多様体[空間] manifold [space] with handle attached.

ばしょ 場所 place; spot; position; location; situation; locus.

はすう 端数 fraction; broken number.
～の fractional; odd. ～の切上げ raising of fraction. ～の切捨て omission [discarding] of fraction.

はすう 波数 wave number.

パスカル B. Pascal (1623-1662)
～の Pascalian; Pascal's. ～の三角形[配置] Pascal's triangle [configuration].

パズル puzzle.
数学～ mathematical puzzle.

はずれち 外れ値 outlier.

はせん 破線 broken line.

はせん 波線 wavy line.

パーセント per cent; percent; percentage; %.
〜点 percentile.

はた 旗 flag.
〜多様体 flag manifold.

はたす 果たす(約束，義務などを) carry out; perform; fulfill, accomplish.

はたす 果たす(目的を) achieve.

はたす 果たす(役割を) play.

パターン pattern.
〜認識 pattern recognition. 〜解析 pattern analysis.

はち- 八 oct-; octa-; octo-; eight.
〜倍[重]の octuple; eightfold. 〜分円 octant. 〜元数 octonion. 〜角形 octagon. 〜角[辺]形の octagonal. 〜角柱 octagonal prism. 〜面体 octahedron (*pl.* -rons, -dra). 〜面の; 〜面体の octahedral. 〜面体群 octahedral group. 〜道模型 octed model. 〜進法 octenary; octal notation [scale]. 〜進(法)の octal. 〜分の三公式 three-eights rule.

はちょう 波長 wave length.

はっかく 八角 ⇨八(はち-).

パッキング packing.

パック pack.

はっけんてき 発見的(な) heuristic.
〜方法 heuristic method. 〜プログラム heuristic program.

はっさん 発散 divergence; divergency.
〜する diverge. 〜の divergent; diverging. 〜(数)列 divergent sequence. 〜級数 divergent series. スカラー場の〜 divergence of scalar field. 〜定理 divergence theorem. 〜(微分)式 divergence (differential) expression. 定[不定]〜 definite [indefinite] divergence; proper [oscillating] divergence.

ばっすい 抜粋 abstract; summary; extract; (*F.*) résumé.

はっせい 発生 appearance; genesis (*pl.* -ses); incidence; occurrence.
〜する occur. 〜的な genetic. 〜学 genetics.

バッチ batch.
〜処理 batch processing.

はっちゅう 発注 order.
〜間隔 ordering cycle [interval]. 〜点 recording point.

はってん 発展 development.
〜させる develop.

はってん 発展 evolution.
〜の evolutional; evolutionary. 〜方程式 evolution equation.

ハット hat.
x〜 x hat; $\hat{x}$.

はっぴょう 発表 presentation; publication; announcement.
〜する publish; present; announce. 未〜の unpublished.

パッフ J. F. Pfaff (1765-1825).
〜の Pfaffian. 〜形式 Pfaffian (form).

ハーディ G. H. Hardy (1877-1947).
〜族 Hardy class.

パート PERT; program evaluation and review technique.
〜・コスト PERT/COST.

はどう 波動 wave (motion); undulation.
〜の undulatory. 〜方程式[関数] wave equation [function]. 〜力学 wave mechanics. 〜光学 wave optics. 〜説 undulatory theory.

ハードウェア hardware.

バナッハ S. Banach (1892-1945).
〜空間 Banach space. 〜束 Banach lattice. 〜環 Banach algebra.

はば 幅 width; breadth.
〜広い wide; broad. 〜3フィートの 3 feet wide. 〜ひろ[ほそ]分布 platykurtic [leptokurtic] distribution. 〜優先探索 breadth first search.

パピルス Papyrus.

はぶく 省く omit; exclude.

ハミルトン W. R. Hamilton (1805-1865).
〜の; ハミルトニアン; 〜演算子; 〜関数 Hamiltonian. 〜演算子; 〜作用素 Hamiltonian operator. 〜方程式 Hamilton equation. 〜力学 Hamiltonian mechanics. 〜閉路 Hamiltonian cycle.

ハミング　R. W. Hamming (1915–1998).
～距離 Hamming distance.

ハミング-きょり　ハミング距離　Hamming distance.

はめこみ　immersion.
はめこむ immerse. はめこまれた immersed. 微分可能な～ differentiable immersion. 位相的～ topological immersion. はめこまれた部分多様体 immersed submanifold.

ハーメル　G. Hamel (1877–1954).
～基 Hamel base [basis].

はめん　波面　wave front.

はやさ　速さ　speed; velocity; rapidity.

はらう　払う　cancel; clear; reset.
分母を～ cancel denominator.

パラコンパクトな　paracompact.
～空間 paracompact space.

ばらつき　dispersion.

パラドックス　paradox. ⇨背理.

パラメータ　parameter. ⇨媒介変数.
～を定めること parametrization. ～のつけられた parametrized. ～のつけられた曲線[曲面] parametrized curve [surface]. 正則～系 regular system of parameters. ～系 system of parameters.

パラメトリックス　parametrix.

はり　針　needle.
～の実験 needle experiment.

バーリア　barrier.

はりあわせ　貼り合わせ　glueing; attaching.

パリティ　parity. ⇨奇偶(性), 偶奇(性).
～・チェック[検査] parity check. ～変換 parity transformation.

はる　張る　span.
…で張られた空間 space spanned by ….

ハール　A. Haar (1885–1933).
～測度 Haar measure.

パルス　pulse.

バレル　barrel.

はん　版　edition; ed.; edit.
普及～ popular edition. 第五～ the fifth edition. 改訂[増補]～ revised [enlarged] edition. 豪華～【特製本】 edition [édition] de luxe.

はん　半　half (*pl.* -lves).
二時～ half past [after] two; two thirty. ～時間 half an hour; a half hour. 一時間～ an hour and a half. ～～に half-and-half; in half; into halves; fifty-fifty.

はん-　半　semi-; hemi-; demi-; half-.
～円 semicircle; half-circle. ～順序 semi-order; partial order. ～群 semigroup. ～ノルム seminorm; pseudo-norm. ～単純な semisimple. ～定値[定符号]の semidefinite. ～完全な semiexact; half-exact. ～球(面) hemisphere; half-sphere. ～月 demilune; half-moon. ～直線 half-line; ray. ～平面 half-plane. ～空間 half-space; demispace; semispace. ～開[閉]区間 half-open [-closed] interval. ～極座標 semi-polar coordinates. ～対数方眼紙 semi-logarithmic paper. ～角の公式 half-angle formula.

はん-　反　anti-; contra-; re-; inverse; reciprocal.
～同値 anti-equivalence. ～対称 anti-symmetry. ～対称な anti-symmetric; skew-symmetric. ～同型[形] anti-isomorphism. ～可換な anti-commute. ～エルミート行列 anti-Hermitian matrix. ～変の contravariant. ～作用 reaction. ～比 inverse [reciprocal] ratio.

ばん　番　No.; no. (*pl.* Nos.; nos.); number; (*L.*) numero.

-ばん　～番(の)　-th.
一～ first. 二～ second. 三～ third. 四～ fourth. 二十～ twentieth. 二十一～ twenty-first. n～ n(-)th. $n+1$～ $(n+1)$(-)th.

はんあんてい　半安定(な)　semistable.
～分布 semistable distribution.

はんい　範囲　range; region; scope.
変化～ variability region. 値の～ range of values. 標本～ sample range. ～の中央 mid(-)range.

はんえん　半円　semicircle; half-circle.
～の semicircular. 上[下]～ upper [lower] semicircle. ～板 semicircular disc [disk]; semidisc; semidisk.

はんかく　半角　half(-)angle.
～(の)公式 formula for [of] half(-)angle;

half(-)angle formula.
はんかんすう 汎関数 functional.
線形～ linear functional. 加法的[特性]～ additive [characteristic] functional. 支持～ support functional. ～(微分)方程式 functional-differential equation. ～Φ作用素 functional Φ-operator.
はんかんぜん 半完全 semiexactness.
～な semiexact. ～微分 semiexact differential.
はんかんやくかのう 半簡約可能な semi-reductive.
はんききょくせん 反帰曲線 edge [line] of regression.
はんきせん 反帰線 ⇨反帰曲線.
はんきゅう 半球(面) hemi(-)sphere; half-sphere.
～の hemispherical.
はんきょくしょてき 半局所的(な) semi-local.
～環 semilocal ring.
はんくうかん 半空間 half-space; semi-space; demi-space.
支持～ supporting half-space.
はんぐん 半群 semigroup.
～環 semigroup algebra. 単位的～ monoid; unitary [unital] semigroup. 自由～ free semigroup.
はんけい 半径 radius (*pl.* -dii, -diuses).
～の radial. 曲率～ radius of curvature. 主曲率～ radius of principal curvature. 捩率～ radius of torsion. 収束～ convergence radius; radius of convergence. 有理型[形]～ radius of meromorphy. 内[外]写像～ inner [outer] mapping radius. 単葉～ univalency radius. 凸型[凸性]～ radius of convexity. 反転～ radius of inversion. 焦点～ focal radius. 測地～ geodesic radius. 慣性～ radius of inertia. ～に沿う極限(値) radial limit.
はんけい 反形 inverse figure.
～曲線 inverse curve.
はんけい 反傾(の) contragradient.
～写像 contragradient mapping. ～表現 contragradient representation.
はんげつ 半月 half-moon; demilune.
～形の semilune; semilunar; semicircular.
ハンケル H. Hankel (1839-1873).
～関数 Hankel function.
はんけん 版権 copyright; ©.
～所有 All rights reserved; Copyrighted.
はんげんし 半原始(性) semiprimitivity.
～の semiprimitive. ～環 semiprimitive ring.
はんこうしじま 半格子縞 half-plaid.
～形 half-plaid square.
はんさよう 反作用 reaction.
～の法則 law of reaction.
はんじこじゅんどうけい 反自己準同型[形] anti-endomorphism.
はんじこどうけい 反自己同型[形] anti-automorphism.
はんじもの 判じ物 puzzle.
はんしゃ 反射 reflection; reflex; reflexion.
～の reflex; reflective. ～する reflect. ～的な reflecting; reflective; reflexive. ～律[法則] reflexive law. ～原理 reflection [reflexion] principle. ～壁 reflecting [reflective] barrier. ～火[焦]線 catacaustic (curve; line).
はんしゅうき 半周期 half-period.
～的な half-periodic.
はんしゅうそく 半収束 semi-convergence. ⇨漸近収束.
はんじゅんじょ 半順序 semiorder; partial order.
～集合 semiordered set; partially ordered set.
はんじゅんそ 半準素(の) semiprimary.
～環 semiprimary ring.
はんしょう 反証 disproof.
～する; ～をあげる disprove.
はんすう 反数 opposed number; opposite number; additive inverse; reciprocal (number).
はん-スカラー 半スカラー semiscalar.
～の semiscalar. ～積 semiscalar product.
はん-スピノル 半スピノル half-spinor.
はんせいそく 半正則(性) semiregularity.
～な semiregular. ～変換 semiregular transformation. ～点 semiregular point.

はんせいち 半正(定)値(性) positive semi-definiteness.
～な positive semi-definite.

はんせんけい 半線形(性) semi-linearity.
～の semilinear. ～写像 semilinear mapping.

はんせんけい 反線形(の) reciprocal linear.
～表現 reciprocal linear representation.

はんそ 半素(の) semiprime.
～イデアル[環] semiprime ideal [ring].

はんそうせんけい 半双線形(の) sesquilinear.
～形式 sesquilinear form.

はんたい 反対 inverse; reverse; opposite; contrary.
～の inverse; reversed; opposite; contrary. ～側 opposite side. ～[逆]符号 opposite sign.

はんたいしょう 反対称 anti-symmetry.
～な anti-symmetric. ～写像 anti-symmetric mapping. ～行列 anti-symmetric matrix; skew-symmetric matrix. ～テンソル anti-symmetric tensor.

はんたいすう 半対数(的) semi-logarithmic.
～方眼紙 semi-logarithmic paper.

はんだん 判断 decision; judgement.
～表 decision table.

はんたんじゅん 半単純 semisimplicity.
～な semisimple. ～群 semisimple group. ～加群 semisimple module. ～環 semisimple ring. ～代数群 semisimple algebraic group. ～成分 semisimple component.

はんたんたい 半単体 semisimplex (*pl.* -lexes, -lices).
～の semisimplicial. ～複体 semisimplicial complex.

はんちゅう 範疇 category.
～的 categorical. ～命題 categorical proposition.

はんちょうわ 汎調和(な) panharmonic.
～級数 panharmonic series.

はんちょくせき 半直積 semidirect product.

はんちょくせん 半直線 half-line; ray.

はんちょっかんしゅぎ 半直観主義 semi-intuitionism.

はんてい 判定 decision; judgment.
～法; ～条件 criterion (*pl.* -ria; -rions); test. 収束～(法) convergence criterion [test]. 比[ベキ根]～法 ratio [radical] test. (正則性の)ヤコビ～法 Jacobian criterion (of regularity). アイゼンシュタイン既約～法 Eisenstein criterion.

はんていち 半定値(性) semi-definiteness.
～な semi-definite. 半正[負](定)値な positive [negative] semi-definite.

はんてん 反転 reflection; reflexion; inversion; reversion; reversal.
～する reflect; invert; reverse. ～原理 reflection [reflexion] principle. ～公式 inversion formula; reciprocal formula. ～幾何学 inversion geometry. ～級数 reverse [reversed] series. 級数の～ reversion of series. 時間～ time reversal. 空間～ space reflection. ～可能な invertible.

はんてん 反点 inverse point.

はんてんこうしき 反転公式 inversion formula; reciprocal formula.

バンド band.

バンド-はば バンド幅 bandwidth.

ハンドル handle. ⇨把手(はしゅ).

バンドル bundle. ⇨束(そく).
ファイバー・～ fibre bundle. フレーム・～ frame bundle. 接～ tangent bundle. 外積～ exterior power bundle.

はん-ノルム 半ノルム seminorm; pseudo-norm.

はんぱ 半端 fraction. ⇨端数(はすう).
～のない数 round number.

はんはんしゃてき 半反射的な semireflexive.

はんぴ 反比 inverse ratio; reciprocal ratio.

はんぴれい 反比例 inverse proportion; reciprocal proportion.
～の inversely proportional. に～する be in inverse proportion to.

はんぷく 反復 iteration; repetition; recurrence. ⇨累次(の).
～の iterative; repeated. ～する iterate; repeat. ～法 method of iteration; iteration method; iterative method. ～解(法) iterative solution. ～核 iterated kernel. ～対数[正弦] iterated logarithm [sine]. ～積分

repeated integral. ～重み付き最小二乗法 iteratively reweighted least squares.

はんふち 半負(定)値(性) negative semi-definiteness.
～な negative semi-definite.

はんふへん 半不変(性) semi-invariance. ～な semi-invariant. ～係数; ～式; ～量 semi-invariant.

はんぶん 半分 half (*pl.* -lves).
半分の half. ～にする halve; divide into halves. 左[右]～ left [right] half.

はんぶんすう 繁分数 complex fraction; compound fraction.

はんへいめん 半平面 half-plane.
左[右]～ left [right] half-plane. 上[下]～ upper [lower] half-plane.

はんべつ 判別 discrimination. ⇨判定.
～関数 discriminant function. デデキントの～定理 Dedekind's discriminant theorem. ～分析 discriminant analysis.

はんべつかのう 判別可能(な) distinguishable.

はんべつしき 判別式 discriminant.
基本～ fundamental discriminant. 相対～ relative discriminant.

はんぺん 反変 contravariance.
～的な contravariant. ～指標 contravariant index. ～成分 contravariant component. ～ベクトル contravariant vector. ～テンソル contravariant tensor. ～スピノル contravariant spinor. ～関手 contravariant functor. ～添数 contravariant index.

はんむげん 半無限(な) semi-infinite; half-infinite.
～区間 semi-infinite interval. ～帯(状面分) semi-infinite strip. ～計画(法) semi-infinite programming.

ばんゆう 万有(の) universal.
～引力 universal gravitation. ～曲線 universal curve. ～体 universal domain.

はんゆうげん 半有限(性) semifiniteness.
～な semifinite.

はんれい 反例 counter-example.

はんれんぞく 半連続(性) semicontinuity.
～な semicontinuous. 下[上](に)～(な)関数 lower [upper] semicontinuous function.

ひ・ヒ

ひ 比 ratio; proportion.
正～ direct ratio. 逆～; 反～ inverse [reciprocal] ratio. 公～ common ratio. 連鎖～ chain ratio. 外[内]分～ external [internal] ratio. 非調和[交叉]～ anharmonic [cross] ratio. 三角～ trigonometric ratio. ～判定法 ratio criterion [test]. ～推定量 ratio estimate. ～尺度 ratio scale. 抽出～ sampling ratio [fraction]. オッズ～ odds ratio.

ひ- 非 a-; an-; in-, im-, ir-; un-; dis-; non-. ⇨不(ふ-).
～対称 asymmetry; dissymmetry. ～調和比 anharmonic ratio. ～同値 inequivalence. ～同[斉]次な inhomogeneous; non-homogeneous. ～完全な imperfect. ～正則な irregular. ～可逆な irreversible. ～可解な unsolvable. ～分岐の unramified. ～連結な disconnected. ～線形な non(-)linear. ～孤立の non-isolated. ～可換な non-commutative. ～コンパクトな non-compact. ～コンパクト型の of non-compact type. ～ユークリッド幾何(学)[空間] non-Euclidean geometry [space]. ～アルキメデス付値[幾何学] non-Archimedian valuation [geometry]. ～デザルグ[パスカル]幾何学 non-Desarguesian [-Pascalian] geometry.

ピーアイ PI; polynomial identity.
～多元環 PI-algebra.

ビー・アイ・シー BIC; Bayesian information criterion.

ピー・アイ・ディー PID; principal ideal domain.

ひあっしゅく 非圧縮(性) incompressibility.
～の incompressible. ～流体 incompressible fluid. ～定常流 incompressible stationary flow.

ひ-アーベル 非アーベル(的な) non-abelian.

ひあんてい 非安定(な) non-stable; unstable. ⇨不安定(な).

ピー・エル・ケー-ほう P.L.K. method; Poincaré-Lighthill-Kuo method.

ひかいてんてき 非回転的(な) irrotational. ⇨無渦(の).
～ベクトル場 irrotational vector field.

ひかかい 非可解(性) unsolvability.
～な unsolvable.

ひかかん 非可換(な) non-commutative; non-abelian.
～環 non-commutative ring. ～体 non-commutative field; division ring; skew field; skewfield. ～群 non-abelian group.

ひかぎゃく 非可逆(な) irreversible.
～過程 irreversible process.

ひかく 比較 comparison.
～する compare. ～の[による] comparative. ～可能な comparable. ～可能性 comparability. ～判定(法) comparison test. ～定理 comparability theorem; comparison theorem. (と)～して in comparison with. 二点～法 pair test. 三点～法 triangle test. 多重～ multiple comparison.

ひかくだい 非拡大(の) non-expansive.
～写像 non-expansive mapping.

ひかさん 非可算(性) uncountability; unenumerability.
～の uncountable; unenumerable. ～集合 uncountable [unenumerable] set.

ひかさんすう 被加算数[量] augend.

ひかじゅつてき 非可述的 impredicative.
～数 impredicative number.
ひかんぜん 非完全(性) imperfection. ⇨不完全(性).
～な imperfect. ～体 imperfect field.
ひきこみ 引き込み retraction.
ひきざん 引き算 subtraction.
ひきすう 引数 argument.
ひきだす 引き出す derive; draw.
ひきつづく 引き続く sequential; subsequent.
ひきてい 非基底(の) nonbasic.
～変数 nonbasic variable.
ひきのばす 引き延ばす prolong; extend.
ひきもどし 引き戻し pullback; pull-back.
ひきもどす 引き戻す pull back.
ひきょうりょく 非協力(の) non-cooperative.
～ゲーム non-cooperative game.
ひく 引く subtract.
引き算 subtraction.
ひく 引く draw.
線を～ draw line.
びけいすう 微係数 differential coefficient. ⇨微分係数.
角～ angular derivative.
ひけつごうてき 非結合的(な) non-associative.
～代数 non-associative algebra.
ひけっていせい 非決定性(の) non-deterministic.
～計算 non-deterministic computation. ～多項式時間 non-deterministic polynomial time; NP.
ひけっていろんてき 非決定論的(な) non-deterministic.
ひげんし 非原始(の) imprimitive; non-primitive.
～群 imprimitive group. ～指標 non-primitive character.
ひげんすう 被減数 minuend.
ひげんみつ-ニュートン-ほう 非厳密ニュートン法 inexact Newton method.
ピコ pico-.
～メートル picometer【10^{-12} meter】.
ひこゆう 非固有(な) improper.
～回転 improper rotation.
ひこんごう 非混合(の) unmixed. ⇨純(な).
ビコンパクト bicompact.
ひさいきてき 非再帰的(な) non-recurrent; transient.
～連鎖 non-recurrent chain.
ひさんじゅつてき 非算術的 non-arithmetic.
～用法 non-arithmetic use.
ビー・シー・エイチ BCH; Bose-Chaudhuri--Hocquenghem.
～符号 BCH code.
ひしがた 菱形 rhombus (*pl.* -buses, -bi); lozenge. ⇨菱形(りょうけい).
～の rhombic; lozenge-shaped.
ひじゅう 比重 specific gravity.
ひしゅうきてき 非周期的な aperiodic.
ひじゅんかん 非循環(の) non-periodic; acyclic.
～節 non-periodic part.
ひじゅんじょつい 非順序対 unordered pair.
びしょう 微小(な) infinitesimal; minute.
～変化 infinitesimal change. ～変形 infinitesimal deformation.
ひじょうすう 被乗数 multiplicand; faciend.
ひじょうにちかい 非常に近い proximate.
ひじょうよ 非剰余 non-residue.
平方～ quadratic non-residue.
ひじょすう 被除数 dividend.
ひしん 非心 ⇨無心(の).
～母数 non-centrality parameter.
ひすいい 非推移(性) intransitivity.
～な intransitive. ～群 intransitive group.
ヒストグラム histogram.
ひずみ 歪み distortion; strain; stress; flexure; skewness. ⇨歪(わい-); 歪曲.
遅延～ delay distortion.
ひせい 非正(の) non-positive.
～の整数 non-positive integer.
ひせいじ 非斉次 ⇨非同次(性).
ひせいじょすう 非整除数 aliquant.
ひせいじょせい 非整除性 indivisibility.
ひせいすう 非整数(の) fractional; non-integral.
～位[階] fractional [non-integral] order.
ひせいそく 非正則(性) irregularity.
～な irregular. ～に irregularly. ～点 irregular point.

ひせきぶんかんすう 被積分関数 integrand.
～の有理化 rationalization of integrand.
びせきぶん(ほう) 微積分(法) ⇨微分積分法.
ひせんけい 非線形(性) non-linearity.
～の non(-)linear. ～現象 nonlinear phenomenon. ～系 nonlinear system. ～振動 nonlinear oscillation. ～方程式 nonlinear equation. ～汎関数 nonlinear functional. ～問題 nonlinear problem. ～力学 non-linear mechanics. ～計画 nonlinear programming. ～予測 nonlinear prediction.
ひそ 非素(の) non-prime.
～数 non-prime number.
ひだ 襞 fold.
写像の～ fold of mapping.
ひたいか 非退化 non-degeneration; non-degeneracy.
～の non-degenerate. ～二次曲線 non-degenerate conics [quadratic curve]. ～二次形式 non-degenerate quadratic form. ～核 non-degenerate kernel. ～臨界点 non-degenerate critical point.
ひたいしょう 非対称 asymmetry; dissymmetry; skewness.
～な asymmetric; asymmetrical; skew. ～曲線 asymmetrical curve. ～法則 asymmetric law. ～度 (degree of) skewness.
ピタゴラス Pythagoras (572–492 B.C.).
～の Pythagorean. ～の表 Pythagorean table. ～数 Pythagorean number. ～の定理 Pythagorean theorem [proposition].
ひたす 浸す immerse.
ひだり 左 left. ⇨右.
～の; ～に left. ～側(の) left; left-hand. ～回りの left-handed; anti-clockwise; sinistral. ～手系 left-hand(ed) system. ～移動 left translation. ～逆元 left inverse (element). ～連続な left continuous; continuous on the left. ～完全な left exact. ～不変な left invariant. ～正則[特異]な left regular [singular]. ～随伴の left adjoint. ～(側)極限 left-side limit. ～微分可能な left differentiable. ～下導関数 left(-hand) lower derivative. ～分解 left resolution. ～剰余類 left coset. ～*A* 加群 left *A*-module. ～整環 left order. ～イデアル left ideal. ～上りに upward(s) to the left.
ひだりまき 左巻きの sinistrorse.
ひだりまわり 左回りの[に] anti-clock-wise.
ひだんせい 非弾性 inelasticity.
～の inelastic. ～体 inelastic body. ～散乱 inelastic scattering.
ピーーち *p* 値 *p*-value.
ひちょうわひ 非調和比 anharmonic ratio; cross ratio. ⇨複比.
ひづけ 日付 date; dating.
～をかく date. ～のない undated.
ひっさん 筆算 calculation by writing.
～派 algoristic school.
ひつぜん 必然(な) necessary; inevitable; consequent.
～性 necessity.
ピッチ pitch.
～面 pitch surface. 準～ reduced pitch.
ビット bit.
ひっぱる 引っ張る strain; pull.
ひつよう 必要(性) necessity.
～な necessary. ～条件 necessary condition. ～十分条件 necessary and sufficient condition.
ひてい 否定 negation; denial.
～的な negative. ～する deny. ～文 negative sentence. ～命題 negative proposition. 二重～ double negation [negative].
ひていしき 否定式 (*L.*) modus tollens.
肯定～ modus ponendo tollens.
ひてきせつ 非適切(な) ill-posed【well-posed の反対語】.
ひと 人 human; man (*pl.* men); person.
～の human; personal. ～々 men; people.
ひどうき 非同期(の) asynchronous.
ひどうじ 非同次(性) inhomogeneity; non-homogeneity.
～の inhomogeneous; non-homogeneous. ～方程式 inhomogeneous equation. ～座標 inhomogeneous coordinates. ～格子 inhomogeneous lattice. ～鎖 non-homogeneous chain.
ひどうち 非同値(な) inequivalent.
ひとうほう 非等方(性) anisotropy.

～的な anisotropic.

ひとくい 非特異(な) non(-)singular.
～行列 non(-)singular matrix. ～点 non(-) singular point.

ひとしい 等しい equal【to】.
等しく equally. ほぼ～ nearly [approximately] equal.

ひとつ 一つ one; a unit.
～の one; single. ～にする unite; make into one. ～ずつ one by one; individually; separately. ～おきの every other; alternate. ～だけ[唯～]の unique; only one.

ひとふでがき 一筆書き
～できる unicursal. ～可能な図形 unicursal figure.

ひとやま 一山 lot.
～二百個 two hundred pieces a lot.

ひながた 雛型 model; pattern.

ひはかい 非破壊(的) non-destructive.
～検査 non-destructive inspection.

ひひょうほん 非標本(の) non-sampling.
～誤差 non-sampling error.

ひふ 非負(の) non-negative.
～の整数 non-negative integer.

ひぶ 日歩 daily interest.

ひふく 被覆 covering; cover.
～する cover. ～定理 covering theorem. 開～ open covering; open cover. 有限[可算]～ finite [countable] covering. ～変換 covering transformation. ～面 covering surface. ～空間 covering space. ～多様体 covering manifold. 普遍～群[面] universal covering group [surface]. ～(の)次数 degree of covering. ～の網目 mesh of covering. 射影～ projective cover. 頂点～ node cover.

びぶん 微分 differentiation; derivation.
～の differential. ～する differentiate. ～可能[可～]な differentiable; derivable. ～学[法] differential calculus. 項別～ termwise differentiation. 偏～ partial differentiation. 数値[図式]～ numerical [graphic] differentiation. 方向～ directed differentiation. 共変～ covariant differentiation. 交代～ alternating derivation. 外～ exterior derivative; exterior differentiation. 複素～ complex differentiation. ～係数 differential coefficient; derivative; derivate. 上[下]～係数 upper [lower] derivate. ～商 differential quotient; fluxion. 高階[次]～ higher order differentiation. 常[偏]～方程式 ordinary [partial] differential equation. ～幾何学 differential geometry. ～位相幾何学 differential topology. ～式 differential expression. ～形式 differential form. ～作用素[演算子] differential operator; differentiation operator. ～不変式[量] differential invariant. ～解析機 differential analyzer. フレシェ～ Fréchet derivative. リー～ Lie derivative. 確率～方程式 stochastic differential equation. 自動～ automatic differentiation. ～多様体 differential manifold.

びぶん 微分 differential.
二次～ quadratic differential. 第 n 階[次]～ nth differential; differential of nth order. 全～ total differential. 完全～ exact differential. 解析[調和]～ analytic [harmonic] differential. 正則[有理型]～ holomorphic [meromorphic] differential. ～形式 differential form. 劣～ subdifferential.

びぶんかのう 微分可能(な) differentiable; derivable. ⇨可微分(な).
～性 differentiability. 偏～な partially differentiable. 左[右]～な left [right] differentiable. n 回連続～な n times continuously differentiable. 全～な totally differentiable. 区分的に～ piecewise differentiable. ～な関数 differentiable function. ～多様体 differentiable manifold. ～構造 differentiable structure.

ひぶんき 非分岐
～な unramified. ～拡大 unramified extension. ～(な)被覆面 unramified covering surface.

びぶんけいすう 微分係数 differential coefficient; derivative; derivate.
偏～ partial differential coefficient; partial derivative.

びぶんしょう 微分商 differential quotient; fluxion.

びぶんせきぶんほう 微分積分法[学] differential and integral calculus; infinitesimal calculus; calculus; (*G.*) Kalkül.

びぶんどうそう 微分同相(な) diffeomorphic.
～写像 diffeomorphic mapping. ～(写像) diffeomorphism.

びぶんほうていしき 微分方程式 differential equation.
常[偏]～ ordinary [partial] differential equation. 全～ total differential equation. 連立～ system of differential equations; simultaneous differential equations. 一[高]階～ differential equation of first [higher] order; first [higher] order differential equation. 差分～ difference-differential equation. 確率～ stochastic differential equation.

ひぶんり 非分離(性) inseparability.
～的な inseparable. ～元 inseparable element. ～拡大 inseparable extension. ～的多項式 inseparable polynomial. 純～的な purely inseparable. ～指数 inseparability index. ～次数 inseparable degree.

ひぼすう 非母数(的) non-parametric.
～法 non-parametric method.

ピボット pivot. ⇨枢軸.
～演算 pivot operation; pivoting.

ひ-ホロノーム 非ホロノーム(な) non-holonomic. ⇨ホロノーム.

ひゃく 百 hundred; hecto-.
第～(の) hundredth. ～グラム hectogram(me).

ひやく 被約(の) reduced.
～表現 reduced representation. ～ノルム reduced norm. ～二次形式 reduced quadratic form. ～代数 reduced algebra. ～な環 reduced ring. ～錐 reduced cone. ～写像錐 reduced mapping cone.

ひゃくしん 百進(法) centensimal.

ひゃくばい 百倍 hundred times.
～の centuple. ～する centuplicate.

ひゃくぶん- 百分
～の五 five percent; five-hundredths. ～率 percent; percentage. ～率誤差 percentage error. 不良(～)率 percent defective. ～位数 centile; percentile. ～法 centensimal measure. ～法の centensimal. ～度 grade.

ひゃくまん 百万 million; mega-.
～の二乗[兆]【英】billion;【米】trillion. ～の三乗【英】trillion;【米】quintillon. ～の四乗[秭]【英】quadrillion. ～の五乗【英】quintillon. ～の六乗[澗]【英】sextillion; sexiilion. ～の七乗【英】septillion. ～の十乗[那由他]【英】decillion. ～分の一 micron. ～分の一の micro-.

ひゆうかい 非有界(性) unboundedness.
～な unbounded. ～関数 unbounded function. ～集合 unbounded set.

ひ-ユークリッド 非ユークリッド(の) non-Euclidean; non-euclidean.
～空間 non-Euclidean space. ～幾何学 non-Euclidean geometry. ～角 non-Euclidean angle. ～距離 non-Euclidean distance. ～計量 non-Euclidean metric.

ひょう 表 table; (*F.*) tableau (*pl.* -leaux).
～の tabular. 乗法～; 掛算九々～ multi-plication table. 対数～ logarithmic table; table of logarithms. 数～ numerical table. 関数～ function(al) table; numerical table of function. 乱数～ table of random numbers. 分散分析～ analysis of variance table. 判断～ decision table. ～差 tabular [table] difference. ～作成 tabulation. 単体～ simplex tableau.

ひよう 費用 cost; expense.
～関数 cost function. ～増加率 cost slope.

びょう 秒 second.
～速 speed per second. 毎～10 メートル speed of ten meters a second.

ひょうか 評価 evaluation; estimation; estimate.
～する evaluate; estimate; assess; price. ～式 estimate; estimating expression. 不等式～ estimation by inequality. ～ベクトル pricing vector. 粗い～ rough estimate.

ひょうけい 標形 indicatrix (*pl.* -xes, -rices).
接線～ indicatrix of tangent; tangent indic-

atrix. 従[陪]法線～ binormal indicatrix.

ひょうげん 表現 representation.
～する represent. ～的 representative; representing. ～可能な representable. ～可能性 representability. ～問題 representation problem. ～論 representation theory. 行列～ matrix representation. 正則～ regular representation. 既約[被約]～ irreducible [reduced] representation. 双対～ dual representation. 商～ quotient representation. 因子～ factor representation. モジュラー～ modular representation. 反傾～ contragradient representation. 複素～ complex representation. 部分～ subrepresentation. ～空間 representation space. ～加群 representation module. ～環 representative ring. ～関数 representing [representative] function. ～測度 representing measure. ～可能(な)事象 representable event. ～行列 representation matrix.

ひょうこう 標構 frame. ⇨枠; フレーム.
動～ moving frame; (*F.*) repère mobile. 自然～ natural frame. デュパン～ Dupin frame.

ひょうこう 標高 height; altitude; index (*pl.* -dices, -dexes).
クルルの～定理 Krull's height [altitude] theorem. ～平面(図) indexed plane.

ひょうさ 表差 tabular difference; table difference.

ひょうじ 表示 representation; expression; notation; display; listing.
～する represent. 媒介変数～ parametric representation. 級数[乗積]～ series [product] representation. 積分～ integral representation. ポアソン～ Poisson representation. 成分～ representation by components. 漸近～ asymptotic representation.

ひょうしき 標識 mark; sign; tag; indicator.

びょうしゃ 描写 description; representation. ⇨写像.

ひょうじゅん 標準 standard; level.

ひょうじゅん 標準(的) canonical; normal; normalized; standard. ⇨基準.
～形 canonical [normal; standard] form. ～座標 canonical [normal] coordinate(s). ～助変数 canonical parameter. ～因子 canonical divisor. ～基(底) canonical base [basis]. ～尺度; ～目盛 canonical [standard] scale. ～的単[全]射 canonical injection [surjection]. ～射影 canonical projection. ～近似 canonical exhaustion. ～時 standard [normal] time. ～化 standardization; normalization. ～誤[偏]差 standard error [deviation]. ～正規分布 standard normal distribution. ～複体 standard complex. ～分解 standard resolution; canonical decomposition. ～(的)写像 canonical map. ～層 canonical sheaf. ～的準同型[形](写像) canonical homomorphism.

ひょうじゅんそくど 標準測度 reference measure; standard measure.

ひょうじゅんへんさ 標準偏差 standard deviation.
標本[母集団]～ sample [population] standard deviation.

ひょうすう 標数 characteristic.
～pの of characteristic p. ～零 characteristic zero.

ひょうてき 標的 target; mark.

びょうてき 病(理(学))的 pathological.
～な例 pathological example.

ひょうてん 評点 mark(s); grade.

びょうどう 平等 uniformity; evenness; equality.
～な uniform; even.

ひょうほん 標本 sample.
～分布 sample [sampling] distribution. ～の大きさ sample size; size of sample. 大[小]～ large [small] sample. 無作為～ random sample. 確率～ random sample. 親～ master sample. ～平均[中央値; 分散; モード] sample mean [median; variance; mode]. ～特性値 sample characteristic. ～空間 sample space. ～調査 sample survey; sampling inspection. ～調査法 sampling method. ～抽出 sampling. ～設計 sample design.

ひょうめん 表面 surface; face; superficies (*pl.* -).

～の superficial. ～積 surface area. ～波 surface wave. ～張力 surface tension.

ひょうめんせき 表面積 surface area.

ひょうりょうち 秤量値 weight.

ひらいた 開いた open. ⇨開(かい-).
開いている open.

ひらく 開く extract; evolve; open; unfold.
平方[立方]に～ extract square [cubic] root. 会を～ hold a meeting.

ピラミッド pyramid. ⇨角錐.
～の pyramidal.

ひりつ 比率 ratio; proportion; scale.

ひりょうりつ 非両立(性) incompatibility.
～の incompatible.

ひりんじょう 非輪状(の) acyclic.
～複体 acyclic complex.

ヒルベルト D. Hilbert (1862-1943).
～空間 Hilbert space. ～の問題 Hilbert's problem(s).

ひれい 比例 proportion; ratio.
～の proportional. ～した proportional; proportionate. ～させる proportionate. に～する be in proportion to; be proportional to. ～式 proportional expression. ～関係 proportional relation; proportionality. ～項 proportional(s). 第四[三]～項 fourth [third] proportional. ～数 proportional numbers. ～配分 proportional distribution. ～尺 proportional scale. ～コンパス proportional compasses. ～定数 constant of proportion [proportionality]; proportional constant. ～因子 proportional factor; factor of proportionality. ～中項 mean proportional; geometric mean. ～部分 proportional part; p.p. 正～ direct proportion. 反～ inverse proportion. 逆～ reciprocal proportion; inverse proportion. 単[複]～ simple [compound] proporiton. ～抽出(法) proportionate [proportional] sampling.

ひれいちゅうこう 比例中項 mean proportional; geometric mean.

ひれいぶぶん 比例部分 proportional part; p. p.

ひれんけつ 非連結(性) disconnectedness. ⇨不連結(性).
～集合 disconnected set. 完全～な completely disconnected.

ひろさ 広さ width; breadth; extent.
広い wide; broad; large.

ひわり 日割 daily rate.

ひんしつ 品質 quality.
～標準 quality standard. ～管理 quality control. ～規格 quality specification. ～特性 quality characteristic. 平均～ average quality. 設計の～ quality of design.

ひんど 頻度 frequency. ⇨度数.
～分布 frequency distribution. ～曲線 frequency curve. ～比 frequency ratio. 累積～ cumulative frequency. 出現～ frequency of occurrence.

びんらん 便覧 catalogue; manual; handbook. ⇨要覧.

ふ・フ

ふ 負(の) negative; minus. ⇨正(の).
～に; ～の向きに negatively. ～の数 negative number. ～(の)値 negative value. ～の向き negative direction; negative orientation. ～号 minus sign; negative sign. ～(の)根 negative root. ～の無限大 negative infinity. ～領域 negative region [domain]. ～(定)値の; ～の定符号の negative definite. 半～値の negative semi-definite. ～変動 negative variation. ～の向きに漸近安定な negatively asymptotically stable. ～の二項分布 negative binomial distribution.

ふ- 不 in-; il-; im-; ir-; un-; dis-; non-. ⇨非(ひ-).
～一致 inconsistency; disagreement; non-conformity. ～等の unequal. ～等式 inequality. ～完全な incomplete; imperfect. ～安定な instabie; unstable. ～規則な; ～正則な irregular. ～連続な discontinuous. ～存在 nonexistence.

ぶ 部 part. ⇨部分.
実[虚]～ real [imaginary] part. 主要～ principal part.

ファイ phi; 【Φ; φ; ϕ】.

ぶあい 歩合 rate; ratio; percentage.

ファイバー fiber; fibre.
～バンドル[空間] fiber bundle [space].

ファイフォ FIFO (first-in first-out)【「先入れ先出し」の意】.

ファイル file.
～する file. 副～ subfile.

ファジィ(な) fuzzy.
～集合 fuzzy set.

ふあんてい 不安定(な) instable; unstable.
～解 unstable solution. ～状態 unstable state. 数値的～性 numerical instability. 完全～な completely unstable.

ふいっち 不一致 inconsistency; disagreement; nonconformity.
～な inconsistent.

フィート foot (*pl.* feet).

フィドゥーシアル fiducial.
～分布 fiducial distribution.

フィードバック feedback.
～する feedback.

フィフォ FIFO (first-in first-out)【「先入れ先出し」の意】.

フィールズ J. C. Fields (1863–1932).
～メダル Fields medal.

フィルター filter.
～基 filter base. ～づけ filtration. ～次数 filtration degree. 極大～ maximal filter; ultrafilter.

フィルタリング filtering.
確率～ stochastic filtering.

フィルトレーション filtration.

ふえき 不易 conservation.
形式[算法]～の原理 principle of conservation of form [calculation].

フェファーマン C. L. Fefferman (1949–).

ふか 付加 addition; adjunction; augmentation.
～の additional; additive. ～する add; adjoin; augment. ～定数 additive constant. ～項 additive term.

ふか 賦課 imposition.
～する impose. ～条件 imposed condition.

ふかぎゃく 不可逆(性) irreversibility.

〜な irreversible.

ふかく 俯角 angle of depression [inclination; declination].

ふかくてい 不確定(性) ambiguity; uncertainty; indeterminancy; indefiniteness.

〜な ambiguous; uncertain; indeterminate; indefinite. 〜点 ambiguous point. 〜特異点 irregular singularity.

ふかさ 深さ depth.

〜優先探索 depth first search.

ふかさん 不可算 ⇨非可算(性).

ふかのう 不可能(性) impossibility. ⇨不能.

〜な impossible.

ふかふごう 不可符号(性) non-orientability.

〜の non-orientable. 〜曲面 non-orientable surface.

ふかんぜん 不完全(性) incompleteness; imperfection. ⇨非完全(性).

〜な incomplete; imperfect. 〜ベータ[ガンマ]関数 incomplete beta [gamma] function. 〜楕円積分 incomplete elliptic integral. 〜データ fragmentary data.

ふかんび 不完備(性) incompleteness.

〜な incomplete. 〜計画 incomplete block.

ふきそく 不規則 irregularity; anomaly.

〜な irregular; anomalous.

ふきんいつぶんさん 不均一分散な heteroskedastic.

ふく 複(の) double; compound.

〜号 double sign. 〜比[比例] double ratio [proportion]. 〜格子 double lattice. 〜事象 compound event. 〜確率 compound probability.

ふく- 複 bi-; double; compound; pluri-; multiple; poly-.

ふく- 副 auxiliary; sub-.

〜軸 auxiliary [conjugate] axis. 〜基線 auxiliary ground line. 〜指数 subindex (*pl.* -dices, -dexes). 〜成分 second component. 〜逆元 adverse.

ふくかく 伏角 ⇨俯角.

ふくげん 復元 restitution; replacement.

〜の restitutive. 〜力 restitutive force. 〜する replace. 〜抽出 sampling with replacement. 非〜抽出 sampling without replacement. 符号〜 decoding.

ふくごう 複号 double sign.

ふくごう 復号 decoding.

〜する decode; decrypt. 〜化 decryption. 〜器 decoder. 最尤〜法 maximum likelihood decoding. 〜化関数 decoding function.

ふくごう 複合(の) composite; compound.

〜仮説 composite hypothesis. 〜命題 composite proposition. 〜法 composite algorithm. 〜単体法 composite simplex method. 〜事象 compound event. 〜過程 compound process. 〜文 compound statement.

ふくざつ 複雑(な) complex.

〜系 complex system. 〜性 complexity.

ふくざつど 複雑度 complexity.

ふくしき- 複式 two-way.

〜管理図 two-way control chart.

ふくしすう 副指数 subindex (*pl.* -dexes, -dices).

ふくしゃ 複写 reproduction; duplication; reprint; copy.

〜する reproduce; duplicate; copy. 〜紙 copying paper; carbon paper; tracing paper.

ふくしゃ 輻射 radiation.

ふくじゅんかん 複循環

〜節 imprimitive period.

ふくすう 複数 plural.

〜の; 〜形 plural.

ふくせい 複製 replica; reproduction.

ふくそ- 複素 complex.

〜数 complex number. 〜変数 complex variable. 〜(数値)関数 complex(-valued) function. 〜数列[級数] complex sequence [series]. 〜(数値)汎関数 complex(-valued) functional. 〜平面 complex plane. 〜球面 complex sphere. 〜ベクトル[射影]空間 complex vector [projective] space. 〜形(式) complex form. 〜表現 complex representation. 〜曲面 complex surface. 〜多様体 complex manifold. 〜微[積]分 complex differentiation [integration]. 〜

ポテンシャル complex potential. ～構造 complex structure. ～化 complexification.
ふくそすう 複素数 complex number.
共役～ conjugate complex number; complex conjugate.
ふくたい 複体 complex (*pl.* -lexes, -lices).
単体(的)～ simplicial complex. 標準～ standard complex. 鎖～ chain complex. 胞～ cell complex. 二重～ bicomplex; double complex. 部分～ subcomplex. ユークリッド～ Euclidean complex. 特異～ singular complex. 順序～ ordered complex. 代数(的)～ algebraic complex. 傾斜～ tilting complex. コスツル[コジュール, コシュール, コスール]～ Koszul complex.
ふくてん 複点 bipoint.
ふくひ 複比 double ratio; compound ratio; cross ratio; anharmonic ratio.
ふくひれい 複比例 compound ratio; compound proportion.
ふくむ 含む contain; include; imply.
ふくめいすう 複名数 compound number.
～の減法 compound subtraction. ～の除法 compound division.
ふくゆうげん 副有限(な) pro-finite.
～群 pro-finite group.
ふくり 複利 compound interest.
ふくれんけつ 複連結(性) multiple connectedness.
～領域 multiply connected domain.
ふごう 符号 sign; signature; (*L.*) signum.
正[負](の)～ positive [negative] sign; plus [minus] sign. 定～ definite sign. ～関数 signum function; sgn. ～のついた数 signed number. ～のついた測度 signed measure. ～(定)数 signature. ～検査; ～検定 sign test. ～付き順位 signed rank.
ふごう 符号 code; symbol.
～化する code; symbolize. ～化 coding; symbolization. ${}_nC_r$定マーク～ ${}_nC_r$ code. 二進～ binary code. 巡回～ cyclic code. 冗長～ redundant code. ～化十進法 coded decimal notation. ～化文字 coded character. ～化の問題 coding problem. ～化されたパターン symbolized pattern. ～復元 decoding. 誤り訂正～ error correcting code. q-進～ q-ary code. ゴレイ～ Golay code. 線形～ linear code. ハフマン～ Huffman code. ～化率 code rate; coding rate. ～化レート coding rate. ～語 code word. ～長 code length. ～理論 coding theory.
ふごう 符合 matching.
～問題 matching problem. ～数 matching number.
ふごう 負号 negative sign; minus sign.
ふごうか 符号化 encryption.
～する encrypt.
ふごうかく 不合格 disqualification; rejection.
～にする disqualify; reject. ～品 rejected goods. ～判定個数 rejection number. ～域 zone of preference for rejection.
ふごう(てい)すう 符号(定)数 signature.
二次形式の～ signature of quadratic form.
ふごうどう 不合同 incongruence.
～な incongruent; incongruous.
ふごうふくげん 符号復元 decoding.
ふごうり 不合理 contradiction; irrationality; unreasonableness; absurdity; antinomy.
～な irrational; unrcasonable; absurd.
プサイ psi; 【Ψ, ψ】.
～関数 psi function; ψ-function.
ふし 節 knot; node.
ふじんこんすう 不尽根数 surd; irrational number.
～の surd. 同類～ like surds.
ふず 付図 attached figure [chart, diagram].
ふずいする 付随する associated.
ふせいかく 不正確な inexact; inaccurate.
ふせいき 不正規(な) abnormal.
～級数 abnormal series.
ふせいごう 不整合 inconsistence; inconsistency.
～の inconsistent.
ふせいそく 不正則(性) irregularity. ⇨非正則(性).
～な irregular. ～点 irregular point.
ふせん 付箋 tag; label.
ふそく 不足 defect; deficiency; insufficiency; lack.

〜な deficient; insufficient. 〜数 deficiency; deficient number. 〜指数 defect; deficiency index (*pl.* -dexes, -dices). 〜近似 approximation in deficiency. 角〜 defect of triangle. 〜群 defect group. 〜度 degree of insufficiency. 〜和 lower sum.

ふたい 付帯(的) subsidiary; accessory; incidental.
〜条件 subsidiary [accessory; incidental] condition; side condition.

ふたご 双子 twin.
〜の twin. 〜素数 twin prime numbers.

ふたしか 不確か uncertainty; ambiguity.
〜な uncertain; ambiguous.

ふち 付値 valuation.
アルキメデス〜 Archimedean valuation. 一般〜 generalized valuation. 加法[乗法]〜 additive [multiplicative] valuation. 自明な〜 trivial valuation. 正規〜 normal valuation. 指数〜 exponential valuation. p進〜 p-adic valuation. 〜ベクトル valuation vector. 〜環 valuation ring. 擬〜 pseudo-valuation. 〜体 valuation field.

ふち 縁 border; margin. ⇨縁(へり).
〜どり bordering. 〜どった行列 bordered matrix.

ふち 負値 negative-valuedness; negative-definiteness; negativity.
〜の，負定値の negative-definite. 〜形式 negative-definite form.

ふつう 普通(の) common; ordinary.
〜公理 common axiom(s). 〜方眼紙 common section paper. 〜尺 ordinary scale. 〜目盛 ordinary [uniform] scale. 〜預金 ordinary deposit.

ふっき 復帰 return; regression.

フックス L. Fuchs (1833-1902).
〜の Fuchsian. 〜形式 Fuchsian form. 〜群 Fuchsian group. 〜型の of Fuchsian type.

ぶっしつ 物質 substance; material; matter.
〜の material; physical. 〜主義 materialism.

プッシュアウト pushout.

プッシュダウン pushdown; pd.
〜・オートマトン pushdown automaton.

ぶったい 物体 body.

ぶつり 物理(学) physics.
〜の; 〜的な physical. 〜数学 physical mathematics. 数理〜学 mathematical physics. 理論〜学 theoretical physics. 地球〜学 geophysics. 生物〜学 biophysics. 数学〜学会 physico-mathematical society.

ふてい 不定(な) indefinite; indeterminate; variable.
〜に indefinitely; indeterminately. 〜積分 indefinite integral. 〜発散 indefinite divergence. 〜符号二次形式 indefinite quadratic form. 〜元 indeterminate (element); variable (element). 〜形 indeterminate form. 〜方程式 indeterminate equation.

ふていち 負定値(の) negative-definite.

ふていめいだい 不定命題 porism.
〜の porismatic.

ふてん 付点(の) dotted.
〜スピノル dotted spinor.

ふとう 不等(の) unequal.

ふどう 浮動(する) float.
〜の floating. 〜小数点 floating point.

ふとうごう 不等号 inequality sign; sign of inequality.

ふとうしき 不等式 inequality.
絶対〜 absolute [unconditional] inequality. 条件(つき)〜 conditional inequality. 連立〜 simultaneous inequalities. 〜系 system of inequalities; inequality system. 分数[無理]〜 fractional [irrational] inequality. 三角〜 trigonometric inequality. 指数[対数]〜 exponential [logarithmic] inequality. 微分[積分]〜 differential [integral] inequality. 等周〜 isoperimetric inequality. X氏の〜 X's inequality; inequality of X. 逆向きの〜 reverse inequality. 対数的ソボレフ〜 logarithmic Sobolev inequality. 変分〜 variational inequality.

ふどうてん 不動点 fixed point.
〜定理 fixed point theorem. 〜指数 fixed point index (*pl.* -dexes, -dices).

ふとうへん 不等辺(の) scalene; of unequal sides.

～三角形 scalene; scalene triangle. ～四辺形【米】trapezium (*pl.* -zia; -s);【英】trapezoid. ～四角形の【英】trapezoidal.

ブートストラップ bootstrap.
～論法 bootstrap-argument.

ふのう 不能 impossibility; inconsistency.
～な impossible; inconsistent. ～(代数)方程式 inconsistent (algebraic) equation. 作図～な of impossible construction.

ふび 不備な deficient; imperfect; vicious.

ふひつよう 不必要 unnecessity.
～な unnecessary.

ぶぶん 部分 part; portion.
～的な partial. ～的に partially; partly. 実[虚]数～ real [imaginary] part. 整数～ integral part. 比例～ proportional part; p.p. 特異～ singular part. 円～ circular parts. 接線[法線]～ tangential [normal] part. ～積分 integration by parts. ～分数 partial fraction; parts.

ぶぶん- 部分 sub-; partial.
～バンドル subbundle. ～対象 subobject. ～理論 subtheory. ～位相空間 topological subspace. ～分数 partial fraction. ～和[積] partial sum [product]. ～分子[母] partial numerator [denominator]. ～商 partial quotient. ～収束 partial convergence. ～交絡(法) partial confounding. ～グラフ subgraph.

ぶぶんかぐん 部分加群 submodule.
A～ *A*-submodule; sub-*A*-module. 小～ small submodule. 本質～ essential submodule. 余剰～ superfluous submodule.

ぶぶんかん 部分環 subring.

ふぶんき 不分岐 ⇨非分岐(ひぶんき).

ぶぶんき 部分基 subbase; subbasis.

ぶぶんくうかん 部分空間 subspace.
トリビアルな～ trivial subspace. 真～ proper subspace. 閉～ closed subspace. 補～ complementary subspace.

ぶぶん-グラフ 部分グラフ subgraph.

ぶぶんぐん 部分群 subgroup.
不変～ invariant subgroup. 正規～ normal subgroup. 真～ proper subgroup. 巡回～ cyclic subgroup. 安定化～ stabilizer; stabilizing subgroup. 固定～ stabilizer. 自明な～ trivial subgroup.

ぶぶんけい 部分系 subsystem.

ぶぶんけん 部分圏 subcategory.
充満～ full subcategory.

ぶぶんこうし 部分格子 sublattice.

ぶぶんしゅうごう 部分集合 subset.
真～ proper subset. 閉～ closed subset. 有限～ finite subset. 共終～ cofinal subset.

ぶぶんしょう 部分商 partial quotient.

ぶぶんせき 部分積 partial product.

ぶぶんせきぶん 部分積分(法) integration by parts.

ぶぶんそく 部分束 sublattice.

ぶぶんぞく 部分族 subclass; subfamily.

ぶぶんたい 部分体 subfield.

ぶぶんたげんかん 部分多元環 subalgebra.

ぶぶんたようたい 部分多様体 submanifold.
不変～ invariant submanifold.

ぶぶんふくたい 部分複体 subcomplex (*pl.* -lexes, -lices).

ぶぶんぶんすう 部分分数 partial fraction; parts.
～分解 decomposition into partial fractions.

ぶぶんへいりょういき 部分閉領域 closed subdomain [subregion].

ぶぶんめんぶん 部分面分 subregion.

ぶぶんりょういき 部分領域 subdomain; subregion.
閉～ closed subdomain [subregion].

ぶぶんれつ 部分列 subsequence.
収束～ convergent subsequence.

ぶぶんわ 部分和 partial sum.
対角～ diagonal partial sum. 第 n～ nth partial sum.

ふへん 不変(性) invariance; permanence.
～な invariable; invariant; permanent. ～量 invariable (quantity); invariant. ～元; ～式 invariant. ～系 invariant (system). ～(微分)形式 invariant (differential) form. ～部分群 invariant subgroup. ～測度[積分] invariant measure [integral]. ～体 invariant [fixed] field. ～群 fixed group. 基本～式 basic [fundamental] invariant.

等角〜量 conformal invariant. 左[右]〜な left [right] invariant. 位相〜な topologically invariant. 〜統計量 invariant statistic. 〜推定量 invariant estimator. 〜分布 invariant distribution. 〜性原理 invariance principle. 関数関係〜の原理 principle of invariance [permanence] of functional relation (s). ホモトピー〜な homotopy invariant. ホモトピー〜量 homotopy invariant.

ふへん 不偏(な) unbiased.
〜推定量 unbiased estimator. 〜推定値 unbiased estimate. 〜回帰推定量 unbiased regression estimator. 〜分散 unbiased variance. 〜信頼域 unbiased confidence region. 平均〜な mean-unbiased.

ふへん 普遍(性) universality.
〜的な universal. 〜[万有]写像 universal mapping. 〜被覆群 universal covering group. 〜被覆面 universal covering surface. 〜被覆空間 universal covering space. 〜係数定理 universal coefficient theorem. 〜集合 universal set. 〜定数 universal constant. 〜(ファイバー)束 universal bundle. 〜公理 common axiom (s). 〜的性質 universal property.

ふへんしき 不変式 invariant.
絶対[相対]〜 absolute [relative] invariant. 基本〜 basic [fundamental] invariant. 微分[積分]〜 differential [intcgral] invariant. ベクトル〜 vector invariant.

ふへんたい 不変体 invariant field; fixed field.

ふめいすう 不名数 abstract number. ⇨無名数.

プライム prime.
〜スペクトル prime spectrum. 〜符[記]号 prime (symbol); 【′】.

フラクタル fractal.

プラス plus. ⇨正(の).
〜の plus; positive. 〜記号 plus; plus sign.

ブラック·ショールズ Black-Scholes.
〜モデル Black-Scholes model.

プラニメター planimeter.

ブランク blank. ⇨空白; 空所.

フーリエ J. B. J. Fourier (1768-1830).
〜級数[係数] Fourier series [coefficient]. 〜展開 Fourier expansion. 〜積分 Fourier integral. 〜変換 Fourier transformation; Fourier transform. 〜解析 Fourier analysis. 〜乗算作用素 Fourier multiplier.

ふりこ 振子 pendulum. ⇨振子(しんし).

プリズム prism. ⇨角柱.
〜の prismatic.

ふりだし 振出し starting point.

ふりょうじょうけん 不良条件 ill condition.
〜の場合 ill-conditioned case.

ふりょうりつ 不良率 fraction of defective; percent [proportion] defective.

ブルー BLUE; best linear unbiased estimator.

ブール G. Boole (1815-1864).
〜の Boolean. 〜代数 Boolean algebra. 〜環 Boolean ring. 〜束 Boolean lattice. 〜演算 Boolean operation (s).

ふるい 篩 sieve.
篩う sieve. 〜法 sieve method. エラトステネスの〜 sieve of Eratosthenes; Eratosthenes' sieve.

ブルバキ Nicolas Bourbaki (1938-).

プルバック pullback; pull-back.

ふるまい 振る舞い behavio(u)r. ⇨性状, 挙動.
振る舞う behave. 出力の〜 output behavio(u)r.

プレイ; プレー play.

フレシェ M. Fréchet (1878-1973).
〜微分 Fréchet derivative.

ブレジス H. Brezis ((仏) Brézis, Haïm) (1944-).

フレーム frame. ⇨枠(わく), 標構.
〜·バンドル frame bundle.

ふれんけつ 不連結(性) disconnectedness. ⇨非連結(性).
〜な disconnected. 全〜な totally disconnected.

ふれんぞく 不連続(性) discontinuity.
〜な discontinuous. 〜点 discontinuous [discontinuity] point; point of discontinuity. 〜関数 discontinuous function. 〜部分 discontinuous part. 〜群 discontinuous

group. ～体 discontinuum (*pl.* -nua). 純[真性]～な purely [properly] discontinuous.

フロー flow.
最大～ maximum flow. 多品種～ multicommodity flow.

ブローアップ blow-up.

ふろく 付録 appendix (*pl.* -dexes, -dices); supplement; addendum (*pl.* -da).

プログラム program; 【英】programme.
～をつくる program. ～言語 programming language. プログラマ programmer. プログラミング programming. ～学習 programmed instruction; PI.

プロセス process. ⇨過程; 工程.

ブローダウン blow-down.

フローチャート flow chart; flowchart.

ブロッキング blocking.

ブロック block.
～・ダイアグラム block diagram. ～・レコード block records. ～計画 block design. ～効果 block effect. 初～ initial block. ～化 blocking.

プロット plot.

ふん 分【時間・角度の単位】 minute.

ぶん 分
二～の一 one half. 三～の二 two thirds. 四～の一 quarter. 四～の三 three quarters; three fourths. 八～の三 three eighths. b～の a a over b.

ぶんいすう 分位数 quantile.
四～ quartile. 百～ centile; percentile. p～ quantile of order p.

ぶんかい 分解 resolution; decomposition; partition. ⇨因数.
～する resolve; decompose; split. ～方程式 resolvent; resolvent equation. 三次～(方程)式 cubic resolvent. 標準～ standard resolution; canonical decomposition. スペクトル～ spectral decomposition [resolution]. 単位[一]の～ resolution of identity; partition of unity. 移入～, 単射[入射]的～ injective resolution. ～定理 decomposition theorem. ～群 decomposition group. ～体 decomposition field; splitting field. 部分分数～ decomposition into partial fractions. 直和[積]～ direct sum [product] decomposition. 既約～ irreducible decomposition. 最小～体 minimal splitting field. 射影(的)～ projective resolution. シューア～ Schur decomposition. 自由～ free resolution.

ぶんかいかのう 分解可能(な) decomposable; divisible.
～性 decomposability. 無限～な infinitely divisible.

ぶんかい(ほうてい)しき 分解(方程)式 resolvent.
三次～ cubic resolvent.

ぶんかいふ(か)のう 分解不(可)能な indecomposable.
～連続体 indecomposable continuum.

ぶんかつ 分割 division; partition; decomposition. ⇨分解.
～する divide; partition. ～可能な divisible. 部分区間への～ division into subintervals. ～輪体 dividing cycle. ～問題 partition problem. ～数 number of partitions. 一[単位]の～ partition of unity. 数の～ partition of number. ～法 split plot experiment. 三角形～ triangulation. 黄金～ golden cut [section]. ジョルダン～ Jordan decomposition. 単体～ simplicial decomposition. 胞体～ cellular decomposition. ～統治法 divide and conquer method.

ぶんかつとうちほう 分割統治法 divide and conquer method.

ぶんかつひょう 分割表 contingency table.
2×2～ two-by-two contingency table. 正方～ square table.

ぶんき 分岐 branching; ramification; bifurcation.
～する branch; ramify; bifurcate. ～の, ～- branching; ramifying. ～点 branch point; ramifying point; ramification point; bifurcation point. ～過程 branching process; multiplicative process. ～被覆空間 ramified covering space. ～要素 ramified element. ～群 ramification group. ～指数

ramification exponent; ramification index. ～数 ramification number. ～体 ramification field. 導手～定理 conductor-ramification theorem. 完全～ totally ramified. 順～ tame ramification. 野性[生]～ wild ramification.

ぶんきょく 分極 polarization.

ぶんけん 文献 literature.
～表[目録] table of literatures; references; bibliography. ～探索 literature search.

ぶんさ 分叉[分差] fork.
～した forked.

ぶんさつ 分冊 fascicle, fascicule; separate volume.

ぶんさん 分散 variance; dispersion.
母～ population variance. 標本～ sample variance. 不偏～ unbiased variance. ～分析 analysis of variance. ～比 variance ratio. ～行列 variance matrix. ～関係(式) dispersion relation. ～波 dispersive wave. 過～ overdispersion.

ぶんさんしゅうごう 分散集合 scattered set.

ぶんし (分数の)分子 numerator.
部分～ partial numerator.

ぶんし 分枝 branch.
曲線の～ branch of curve. ～因子 branch divisor. ～過程 branching process; multiplicative process. ～カット branch and cut. ～限定 branch and bound.

ぶんすう 分数 fraction; fractional number; broken number.
～の fractional. 普通の～ common [vulgar] fraction. 繁～ complex fraction. 仮～ improper fraction. 真～ proper fraction. (通)常～ vulgar fraction. 既約～ irreducible fraction. 有理～ rational fraction. 部分～ partial fraction. p 進～ p-adic fraction. 連～ continued fraction. 循環連～ recurring continued fraction. ～式 fractional expression. ～関数 fractional function. ～方程[不等]式 fractional equation [inequality]. ～ベキ fractional power. ～イデアル fractional ideal. (連分数の)近似～ convergent.

ぶんせき 分析 analysis (*pl.* -ses).
～する analyze. 分散～ analysis of variance. 共分散～ analysis of covariance. 回帰～ regression analysis. 活動～ activity analysis. 判別～ discriminant analysis.

ぶんどき 分度器 protractor; graduator.

ぶんぱい 分配 distribution; partition.
～の[的な] distributive. ～する distribute; partition; allot; deal. ～性 distributivity. ～法則[律] distributive law. ～束 distributive lattice. ～的代数 distributive algebra. ～関数 partition function.

ぶんぴのり 分比の理 ⇨除比の理.

ぶんぷ 分布 distribution.
安定～ stable distribution. 母集団～ population distribution. 標本～ sample distribution. 誤差(の)～ distribution of errors. 頻度～ frequency distribution. 累積～ cumulative distribution. 確率～ probability distribution. 離散[連続]～ discrete [continuous] distribution. 二[多]項～ binomial [multinomial] distribution. 正規～ normal distribution. ガウス～ Gauss(-ian) distribution. ステューデント～ Student distribution. カイ二[自]乗～ chi-square distribution. 同時～ simultaneous [joint] distribution. 質量～ mass distribution. ～関数 distribution function. ～頻度 distribution frequency. ～曲線 distribution curve. ～によらない distribution-free. 値～ value distribution. 平衡～ equilibrium distribution. 事前～ prior distribution. 事後～ posterior distribution. 広幅[幅ひろ; 緩尖]～ platykurtic distribution. 狭幅[幅ほそ; 急尖]～ leptokurtic distribution.

ぶんぼ 分母 denominator.
公～ common denominator. ～の有理化 rationalization of denominator. 部分～ partial denominator.

ぶんみゃく 文脈 context.
～依存言語 context-sensitive language. ～自由言語 context-free language.

ぶんや 分野 field; branch; discipline.

ぶんり 分離 separation; isolation.
～する separate; distinguish; isolate. ～性

separability. ～性の separative. ～的な separative; separable. ～的に separably. ～可能な separable; distinguishable. ～不[可]能な inseparable; indistinguishable. 根の～ separation of roots. 変数～ separation of variables. ～定理 separation theorem. ～公理 separation axiom; axiom of separation. ～度 degree of separation. ～元 separable element. ～拡大 separable extension. ～閉包 separable closure. ～的多元環 separable algebra. ～生成の separably generated. ～超平面 separating hyperplane. ～位相群 separated topological group. ～次数 separable degree. ～多項式 separable polynomial. ～超越基底 separable transcendence basis.

ぶんるい 分類 classification; grouping; sorting.
～する classify; sort; label. 十進～ decimal classification. ～定理 classification theorem. ～空間 classifying space.

ぶんれつ 分裂 split.

ペー　pe【℘】
～関数 pe-function.

ペアノ　G. Peano (1858-1932).
～曲線 Peano curve.　～の公理系 Peano's axioms.

へい-　閉　closed.　⇨開(かい-).
～曲線 closed curve.　～弧 closed arc.　～曲面 closed surface.　～区間 closed interval.　～集合 closed set.　～領域 closed domain; closed region.　乗法的～集合，積～集合 multiplicative set; multiplicatively closed set.　～写像 closed mapping.　～像 closed image.　～測地線 closed geodesic.　～(微分)形式 closed (differential) form.　～被覆 closed covering.　～グラフ定理 closed graph theorem.　～凸包 closed convex hull.　局所～集合 locally closed set.

へいおう　平凹の　plano-concave.

ペイオフ　pay-off.

へいかく　平角　straight angle.

へいかつ　平滑　smoothness.
～な smooth.　～化 smoothing.　～検定 smoothness test.　～構造 smooth structure.　指数～法 exponential smoothing.

へいがめん　平画面　plan; horizontal plane.

へいきん　平均　mean; average.
～の mean; average.　～する average.　～して on an average.　算術[相加]～ arithmetic mean.　幾何[相剰]～ geometric mean.　調和～ harmonic mean.　荷重～ weighted mean.　移動～ moving average; running mean.　～値 mean; mean value; average.　～値の定理 mean value theorem.　母(集団)[標本]～ population [sample] mean.　～曲率 mean curvature.　～卵形(線) mean oval.　～収束 convergence in the mean; mean convergence.　～誤差 mean error; average error.　～二乗[平方]誤差 mean square error.　仮(の)～ working mean; temporary average.　～偏差 mean deviation; MD.　～p葉の mean p-valent.　～密度 mean [average] density.　～変化率 average rate of change.　～速度 average speed; average velocity.　～訪問回数 mean sojourn.　～点 average mark.　～法 method of averaging.　～品質 average quality.　～抜取り[検査]個数 average sample number.　～欠点数 average number of defects.　～寿命 average life span; span of human life.　～余命 expectation of life.　相～ phase average.　時間～ time average.　刈り込み～ trimmed mean.　～計算量 average-case complexity.　～絶対誤差 mean absolute error.

へいきんけんさりょう　平均検査量　expected amount of inspection.

へいきんしゅうそく　平均収束　convergence in the mean; mean convergence.

へいきんち　平均値　mean; mean value; average.
～の定理 mean value theorem.

へいきんよめい　平均余命　expectation of life.

へいけいもんだい　閉形問題　problem of enclosure.

へいこう　平行(性)　parallelism.
～な parallel.　～線 parallel lines; parallels.　～線公理 axiom of parallels.　～四辺形 parallelogram.　～六面体 parallelepiped; parallelopiped.　～体 parallelotope.　～移動 parallel displacement; (parallel) transla-

tion. ～射影[投影] parallel projection. ～座標 parallel coordinates. ～曲面 parallel surface. ～ベクトル[テンソル]場 parallel vector [tensor] field. ～定規 parallel ruler. ～角 angle of parallelism. 遠隔～性 teleparallelism; absolute parallelism. ～化可能な parallelizable.

へいこう 平衡 equilibrium (*pl.* -ums, -ria); balance.
安定～ stable equilibrium. ～した balanced. ～点 equilibrium point. ～原理 equilibrium principle. ～分布 equilibrium distribution. ～写像 balanced mapping.

へいごう 併合 merging; union.
～整列法 sorting by merging.

へいこういどう 平行移動 translation; parallel displacement.
座標軸[系]の～ translation of coordinate axes [system]. ～可能性 translatability.

へいこうせん 平行線 parallel lines; parallels.
～公理 axiom of parallels.

ベイジアン Bayesian.

へいしん 並進 translation.

ベイズ Bayes, Thomas (1702-1761).
～流の Bayesian. ～情報量基準 Bayesian information criterion; BIC.

へいたん 平坦 flatness.
～な; ～に flat; plain. ～な関数 flat function. ～な加群[部分空間] flat module [subspace]. ～点 flat point; point of flatness. ～接続 flat connection. 共形的に～な conformally flat. 射影[共円]的に～な projectively [concircularly] flat. 局所～な locally flat.

へいだんめん 平断面 plane section.

へいどう 閉道 loop; closed path. ⇨ループ.
～空間 loop space.

へいとつ 平凸(の) plano-convex.

へいほう 平方 square. ⇨二乗.
～の square; quadratic. ～数 square number. 3～メートル 3 square meters. ～根 square root. ～可積な square integrable. ～積[数] square measure [number]. ～剰余[非剰余] quadratic residue [non-residue]. ～採中法 middle-square method. ～に開く extract square root. 無～の square-free.

へいほう 閉包 closure; hull. ⇨包(ほう).
代数的～ algebraic closure. 整～ integral closure. 分離～ separable closure. 軌道～ orbit closure. ～有限な closure finite. 凸～ convex hull; convex envelope; convex closure. 射影～ projective closure.

へいほうこん 平方根 square root.
～の subduplicate. ～比 subduplicate ratio.

へいほうへんさ 平方偏差 variance; square deviation.

へいめん 平面 plane; plane [planar] surface.
～の; ～的な planar. ～幾何学 plane geometry. ～三角法 plane trigonometry. 半～ half-plane. ユークリッド～ Euclidean plane. ガウス～ Gaussian plane. 複素～ complex plane. 射影～ projective plane. 接[法]～ tangent [normal] plane. 展直～ rectifying plane. 共線～ collinear planes. ～図形 plane figure. ～曲線 plane curve. ～多角形 planar polygon. ～分 planar region. ～領域 planar domain; plane domain [region]. 点を除いた～ punctured plane. ～積 plane area. ～座標 plane coordinates. ～波 plane wave. ～束 pencil of planes; axial pencil. 切除～ cutting plane. ～グラフ plane graph. ～的グラフ planar graph.

へいめんず 平面図 plan.
標高～ indexed plan.

へいめんせき 平面積 plane area.

へいりょういき 閉領域 closed domain [region].
部分～ closed subdomain [subregion].

へいれつ 並列(の) parallel.
～計算 parallel computation.

へいろ 閉路 cycle.
ハミルトン～ Hamiltonian cycle.

へき 壁 barrier; wall.
吸収[反射]～ absorptive [reflective] barrier. 障～ barrier.

べき ベキ, べき, 冪, 巾 power.
～指数 power exponent. ～乗法 power

method. ～剰余 power(-)residue. ～関数 power function. ～級数 power series. 集合の～ power of set. ～集合 power set. 方～定理 power theorem. 分数～ fractional power. 記号～ symbolic power.

べききゅうすう ベキ級数 power series.
形式的～ formal power series. ～環 power series ring.

べきこん ベキ根 radical; power root; root.
～で解ける solvable by radicals. ～判定法 radical test. 1の～[累乗根] root of unity.

べきしすう ベキ指数 exponent (of power); power exponent.

べきたん ベキ単(性) unipotency.
～の unipotent. ～成分 unipotent component. ～行列 unipotent matrix. ～群 unipotent group.

べきとう ベキ等(性) idempotence; idempotency.
～の idempotent. ～元 idempotent element; idempotent. 原始～元 primitive idempotent. ～行列 idempotent matrix.

べきれい ベキ零(性) nilpotency.
～の nilpotent. ～元 nilpotent element. ～成分 nilpotent component. ～群 nilpotent group. ～行列 nilpotent matrix. ～イデアル nilpotent ideal. ～環 nilpotent ring. ～根基 nilpotent radical. ～元根基 nilradical.

べきわ ベキ和 power-sum.

ヘクト hecto-.
～メートル[グラム] hecto-meter[-gram]. ヘクタール hectare.

ベクトル vector.
～の vectorial. n 次元～ n-vector. 行～ row vector. 列～ column vector. ～量 vector quantity. 速度～ velocity vector. 単位～ unit vector. 零～ null [zero] vector. 特性～ characteristic vector. 付値～ valuation vector. 基底～ basic [basis] vector. 数～ number vector. 幾何～ geometrical vector. n 項～ n-tuple vector. 縦[横]～ column [row] vector. 位置～ position vector. 動径～ radius vector. 接[法]～ tangent [normal] vector. 固有～ eigenvector. ～解析 vector analysis. ～場 vector field. ～流 vector flux. ～・ポテンシャル vector potential. ～関数 vector function. ～空間 vector space. 共[反]変～ covariant [contravariant] vector. 巡回～ cyclic vector. ～積 vector product. ～三重積 vector triple product. ～束 vector bundle. ～不変式 vector invariant. ～方程式 vectorial equation. ～値の vector-valued. ～値関数 vector-valued function.

ページ[頁] page; p. (*pl.* pp.).

へそ 臍 umbilicus (*pl.* -ci, -cuses). ⇨臍点(せいてん).
～の umbilic; umbilical.

ベゾフ O. V. Besov (1933-　　).
～空間 Besov space.

ベータ beta; 【B, β】.
～関数 beta function; B-function.

べつ 別(の) another.
～証明 another proof.

べつずり 別刷 offprint; separate (print).

ヘッセ L. O. Hesse (1811-1874).
～の Hessian. ～形式 Hessian (form). ～行列 Hessian matrix.

ベッセル F. W. Bessel (1784-1846).
～関数 Bessel function. 変形～関数 modified Bessel function.

ペーパー paper.

へらす 減らす decrease; reduce; lessen.

へり 縁 border. ⇨縁(ふち).
～を付ける border. ～集合 border set. ～を付けた行列 bordered matrix.

ヘリコイド helicoid. ⇨螺旋.

へる 減る decrease; reduce; lessen.

ベール R. Baire (1874-1932).
～(関数)族 Baire class. ～空間 Baire space.

ベルソル versor.

ベルヌーイ D. Bernoulli (1700-1782), Jak. B. (1654-1705), Joh. B. (1667-1748).
～試行 Bernoulli trial.

ヘルマンダー L. V. Hörmander (1931-2012).

へん 辺 side; (*L.*) latus (*pl.* latera); edge; boundary; member.

隣～ adjacent side. 始[終]～ initial [terminal] side. 対応～ corresponding sides. ～準同型[形] edge homomorphism. ～単体 boundary simplex. 左[右]～ left [right]-hand member [side].

へん-　偏　partial.
～微分 partial differentiation. ～微分可能な partially differentiable. ～導関数 partial derivative. ～微分方程式 partial differential equation. ～相関 partial correlation. ～微分係数 partial differential coefficient; partial derivative.

ベン　J. Venn (1834–1923).
～(の)図(式) Venn diagram.

へんい　変位　displacement.
垂直～ perpendicular displacement. 法線(方向の)～ normal displacement. 仮想～ virtual displacement. 測地的～ geodesic deviation.

へんい　偏倚　deviation; deviate. ⇨偏差.

へんいき　変域　domain; variability domain.

へんいすう　変位数　quantile.

へんえん　辺縁　margin. ⇨周辺.
～の marginal. ～度数 marginal frequency.

へんえん　扁円　ellipse.

へんか　変化　change; variation.
～する vary. ～できる variable. 微小～ infinitesimal change. …に正[逆]比例して～する vary directly [inversely] as …. 定数～(法) variation of constant(s). ～範囲 variability region. ～成分[点] variable component [point].

へんかく　偏角　argument; amplitude.
～の原理 argument principle.

へんかく　変格(の)　improper.
～直交行列 improper orthogonal matrix. ～[広義]積分 improper integral.

へんかん　変換　transformation; transform; change.
～する change; transform. 変数～ change of variable(s). 座標～ change of coordinate(s). 恒等～ identity [identical] transformation. 合同～ congruent transformation. 相似～ similar transformation; similitude; homothety. アフィン～ affine transformation, affinity. 射影[共形]～ projective [conformal] transformation. 一次[線形]～ linear transformation. メービウス～ Möbius transformation. 双有理～ birational transformation. 対称～ symmetric transformation. 直交[ユニタリ]～ orthogonal [unitary] transformation. 主軸～ principal axis transformation. 積分～ integral transformation. 無限小～ infinitesimal transformation. 内部～ inner transformation. 接触～ contact transformation. フーリエ[ラプラス]～ Fourier [Laplace] transformation; Fourier [Laplace] transform. 逆～ inverse transformation; inverse transform. 正[余]弦～ sine [cosine] transform. ～公式 transformation formula. ～行列 transformation matrix. ～群 transformation group. ～母数 transformation parameter. ～子 transformation quantity; transformator. 基底～ base change.

へんかんすう　変関数　argument function.

へんきゅう　扁球　oblate spheroid.
～の oblate.

へんきょく　変曲　inflection; inflexion.
～点 inflection [inflexion] point; point of inflection [inflexion].

へんきょく　偏極　polarization.
～させる polarize. ～半平面[多様体] polarized half-plane [variety].

へんけい　変形　deformation; modification; strain; transformation.
～させる deform; modify; strain. ～可能な deformable. 無限小～ infinitesimal deformation. 構造の～ deformation of structures. ～率 deformation ratio. ～ベッセル関数 modified Bessel function. ～テンソル strain tensor. 剪断～ shear. 基本～ elementary transformation. 射影的～ projective deformation. ～理論 deformation theory.

へんこう　変更　change; alteration; modification.

へんこう　偏光　polarization; polarized light.

〜させる polarize. 〜計 polarimeter. 〜子 polarizer.

へんさ 偏差 deviation; deflection.
〜点 deviation point. 平均〜 mean deviation. 標準〜 standard deviation. 四分位〜 quartile deviation. 〜値 T-score. 〜テンソル deflection tensor.

へんさい 返済 pay-back; repayment.
〜期間 pay-back period.

へんしゅう 編集 editing; compilation.
〜する edit; compile. 〜上の editorial.

へんじる 変じる vary; change. ⇨変わる.

へんしんきょり 辺心距離 apothem.

へんすう 変数 variable; argument.
独立〜 independent variable. 従属〜 dependent variable. 媒介[助]〜 parameter. 実[複素]〜 real [complex] variable. 束縛〜 bound variable. 確率〜 random variable. 人為〜 artificial variable. 緩和[余裕]〜 slack variable. 命題〜 proposition variable. 一[多]〜の関数 function of one [several] variable[s]. 〜変換 change of variable(s). 〜分離 separation of variables. 質的〜 qualitative variable. 主〜 primal variable. 説明〜 explanatory variable. 双対〜 dual variable. 量的〜 quantitative variable.

へんそうかん 偏相関 partial correlation.
〜係数 partial correlation coefficient.

ペンタグラム pentagram.

へんたんたい 辺単体 boundary simplex (*pl.* -lexes, -lices).

へんちょう 変調 modulation.
〜速度 modulation rate. 周波数〜 frequency modulation.

へんちょう 扁長(な) prolate.
〜回転楕円体 prolate spheroid.

へんてん 変点 variable point.

へんとう 返答 return; answer; reply; response.

へんどう 変動 variation; fluctuation.
全〜 total variation. 正[負]〜 positive [negative] variation. 有界〜の of bounded variation. 〜係数 coefficient of variation.

へんどうかんすう 偏導関数 partial derivative.

へんびぶん 偏微分 partial differentiation.
〜する partially differentiate. 〜係数 partial differential coefficient; partial derivative. 〜可能な partially differentiable.

へんぶん 変分 variation.
〜の variational. 〜学[法] calculus of variations. 〜問題 variational problem. 〜導関数 variational derivative. 第一[二]〜 first [second] variation. 〜原理 variational principle. 〜方程式 variational equation. 〜不等式 variational inequality.

へんぺい 扁平 flatness.
〜な oblate; flat. 〜回転楕円体 oblate spheroid.

へんりょう 変量 variable; variate.
正規化〜 normalized variate. 多〜 multi-variate.

へんりょう- 変量 random.
〜効果 random effect. 〜模型 random effect model.

へんりょうけい 偏菱形 rhomboid.
〜の rhomboid; rhomboidal.

ほ・ホ

ほ-　補　complementary.
～角 supplement; supplementary angle. ～数[元] complementary number [element]; complement. ～集合 complementary set; complement. ～母数 complementary modulus. ～系列 complementary series. ～空間 complementary space; complement. ～加群 complementary module. ～グラフ complement graph.

ぼ-　母-　generating; population.
～関数 generating function. 確率[積率]～関数 probability [moment] generating function. ～線 generator; generatrix (*pl.* -trices); generating line. ～点 generatrix (*pl.* -trices). ～面 generator; generatrix (*pl.* -trices). ～空間 generating space. ～集団 population. ～平均 population mean. ～分散 population variance. ～相関 population correlation. ～数 modulus; parameter.

ポアソン　S. D. Poisson (1781-1840).
～分布 Poisson distribution. ～積分[表示] Poisson integral [representation].

ほい　補遺　supplement; appendix (*pl.* -dices, -dixes); addendum (*pl.* -da).

ほう　包　hull; closure; envelope; cover.
凸～ convex hull; convex envelope; convex closure. 閉凸～ closed convex hull. ～核 hull-kernel. 等測～ closure with equal measure; measurable cover; isometric closure. 下～ lower envelope. 正則～ envelope of holomorphy.

ほう　法　modulus (*pl.* -li); mod.; divisor.
nを～として modulo n (=to modulus n).

ほう-　法　normal.
～線 normal. ～平面 normal plane. ～ベクトル normal vector. ～曲率 normal curvature. ～束 normal bundle. ～錐 normal cone.

-ほう　法　method. ⇨方法; 論法.
直[間]接～ direct [indirect] method. 消去[代入]～ method of elimination [substitution]. 帰納～ induction; inductive method. 微積分～ differential and integral calculus. 変分～ calculus of variations. 緩和～ relaxation method. 勾配～ gradient method. 最急降下～ method of steepest descent; steepest descent method. 単体～ simplex method. 反復～ iterative method. 分割統治～ divide and conquer method.

ぼう　棒　bar; stick.
～グラフ bar [rectangular] graph. ～構成 bar-construction.

ほうい　方位　direction; orientation; azimuth.
～の directional; azimuthal. ～角 azimuth(al) (angle).

ほうい　包囲　enclosure; inclosure.
～する enclose; surround.

ほうかい　崩壊　collapse.
～する collapse.

ぼうがい　妨害　obstruction; interference; disturbance.
～物 obstracle; obstructive. ～検出 interference detection.

ほうかく　方格　square. ⇨方陣.
グレコラテン～ Greco-Latin square. 格子～(法) lattice square (method).

ほうがん　包含　inclusion.
～する include. ～関係 relation of inclusion; inclusion relation. ～写像 inclusion

mapping.

ほうがんし 方眼紙 section [graph] paper. 対数～ logarithmic paper. 半対数～ semi-logarithmic paper. 確率(～)紙 probability paper.

ほうきゅうかんすう 方球(調和)関数 ⇨縞球(調和)関数.

ほうけい 方形 square.
～に squarely.

ほうこう 方向 direction; sense. ⇨向き; 時計.
～余弦 direction cosine. ～比 direction ratio. ～因子 direction factor. 主～ principal direction. 特性～ characteristic direction. 正の～ positive direction [sense]. 負の～ negative direction [sense]. 降下～ descent direction. 探索～ search direction.

ほうごう 包合 involution.
～系 involution system.

ほうこうづけ 方向づけ orientation. ⇨向きづけ.
～する orient(ate).

ほうごうていり 縫合定理 sewing theorem.

ほうこうりょう 方向量 vector. ⇨ベクトル.

ほうこく 報告 report; bulletin.
～する report.

ほうさく 方策 measure; policy.

ほうしき 方式 method; system; way; procedure; mode; form; formula (*pl.* -las, -lae).
抽出～ sampling procedure.

ほうしゃ 放射 radiation.
～状の radial. ～状に radially. ～截線 radial slit.

ほうしゃせん 放射線 ray; radial ray; radiating line.

ほうじょ 包除 inclusion and exclusion.
～原理 principle of inclusion and exclusion.

ほうじん 方陣 square. ⇨方格.
ラテン[オイラー]～ Latin [Euler] square. 魔～ magic square.

ぼうしん 傍心 excenter.

ぼうせつ 傍接 escription.
～させる escribe. ～円 escribed circle.

ほうせん 法線 normal.
～の normal. 単位～ unit normal. 主～ principal normal. 従[陪]～ binormal. 余～ conormal. 内[外]向き～ inward [outward] normal. ～成分 normal component. ～導関数 normal derivative. ～応力 normal stress. ～的に平坦な normally flat.

ほうせんえい 法線影 subnormal.

ほうそく 法則 law; rule; principle.
基本～ fundamental law. 計算～ law of calculation. 交換[結合, 推移]～ commutative [associative, transitive] law. 反射～ reflexive law. 吸収～ absorptive [absorption] law. 指数～ law of exponents. 正弦[余弦]～ sine [cosine] law; sine [cosine] rule. 大数の～ law of large numbers. ～収束 convergence in law. 慣性～ law of inertia. 反作用の～ law of reaction. 梯[台]形～ trapezoidal rule. シンプソンの～ Simpson's rule.

ほうたい 胞体 cell.
～の cellular. 単位～ unit cell. 凸～ convex cell. 双対～ dual cell. ～分割 cellular decomposition. ～近似 cellular approximation. ～写像 cellular mapping.

ほうたいつい 胞体対 ball pair.

ぼうちょう 膨張 expansion; dilatation; dilation. ～する expand; dilate. ～写像 expanding mapping; inflation. ～変換 dilatation. ～率 expansion coefficient [rate].

ほうていしき 方程式 equation.
～的な equational. 一次[線形]～ linear equation. 二次[三次; 四次]～ quadratic [cubic; quartic] equation. n次～ equation of degree n. 代数[超越]～ algebraic [transcendental] equation. 無理～ irrational equation. 三角[指数; 対数]～ trigonometric [exponential; logarithmic] equation. 連立～ simultaneous equations. ～系 system of equations, equation system. 不定～ indeterminate equation. ディオファンタス～ Diophantine [Diophantus] equation. 固有～ proper equation. 永年～ secular equation. 微分[積分]～ differential [integral] equation. 微(分)積分～ integro-differential equation. 差分～

difference equation. 変分～ variational equation. 同伴～ associated equation. 転置～ transposed equation. 随伴～ adjoint equation. 補助～ adjoint equation; subsidiary equation. 定義～ defining equation. 特性～ characteristic equation. 規[標]準～ canonical equation; normal equation. 構造～ structure equation. 運動(の)～ equation of motion. 波動[熱; 電信]～ wave [heat; telegraph] equation. ロジスティック～ logistic equation. 発展～ evolution equation. 連続の～ equation of continuity. ～的な特徴づけ equational characterization. 捕食者-被食者～ predator-prey equation.

ほうでん 放電 (electric) discharge.

ほうふ 豊富な ample; abundant.
～因子 ample divisor. (非常に)～系 (very) ample system. 豊富層 ample sheaf.

ほうふくたい 胞複体 cell complex.
幾何学的～ geometric cell complex.

ほうぶつ- 放物; 抛物 parabolic. ⇨放物型(の); 放物的(な).
～幾何学 parabolic geometry. ～点 parabolic point. ～柱[筒] parabolic cylinder. ～柱関数[座標] parabolic cylinder function [coordinates].

ほうぶつがた 放物型(の) parabolic; of parabolic type.
～変換 parabolic transformation. ～二次曲面 parabolic quadrics. ～偏微分方程式 partial differential equation of parabolic type; parabolic partial differential equation. ～の面 surface of parabolic type; parabolic surface.

ほうぶつせん 放物線; 抛物線 parabola.
～の parabolic; parabolical. 共焦～ confocal parabolas. 漸近～ asymptotic parabola. ～座標 parabolic coordinates.

ほうぶつてき 放物的(な) parabolic; parabolical. ⇨放物型(の).
～変換 parabolic transformation. ～領域 parabolic domain [region]. ～尖点 parabolic cusp.

ほうぶつめん 放物面 paraboloid.
～の paraboloidal. 楕円[双曲]～ elliptic [hyperbolic] paraboloid. 回転～ paraboloid of revolution. 回転楕円～ elliptic Paraboloid of revolution. ～座標 paraboloidal coordinates.

ほうべき 方ベキ power.
～定理 power theorem.

ほうほう 方法 method; manner; procedure; process.
～論 methodology. 代数[解析]的～ algebraic(al) [analytic(al)] method. 幾何学的～ geometric(al) method. 帰納的～ inductive method. 構成的～ constructive method.

ほうらく 包絡(する) envelop.
～線 envelope. ～面 envelope; enveloping surface. ～錐[柱](面) enveloping cone [cylinder]. ～多元環 enveloping algebra. ～環 (universal) enveloping algebra. 単射[移入]～ injective envelope. ～検出力関数 envelope power function.

ほうりゃく 方略 strategy. ⇨戦略.

ほうわ 飽和 saturation.
～する be saturated. ～した saturated. ～モデル saturated model. 近似の～ saturation of approximation.

ほがい 補外(法) extrapolation. ⇨補間(法).
～する extrapolate. ～公式 extrapolation formula.

ほかく 補角 supplement; supplementary angle.
～の supplementary.

ほかく·さいほかく 捕獲·再捕獲 capture-recapture.
～法 capture-recapture method.

ほかん 補完 supplementation; imputation.
～する supplement.

ほかん 補間(法) interpolation.
～的な interpolatory. ～する interpolate. ～公式 interpolation formula. ～多項式 interpolation polynomial. 線形～ linear interpolation. 多項式～ polynomial interpolation. 逆～ inverse interpolation. 図式～ graphic interpolation. ラグランジュ～ Lagrange's interpolation. ～空間 interpo-

lation space.

ぼかんすう 母関数 generating function.
指数(的)～ exponential generating function. 確率[積率]～ probability [moment] generating function. 重み～ weight enumerator.

ほくうかん 補空間 complementary space; complement.
直交～ orthogonal complement; orthocomplement.

ほ-グラフ 補グラフ complement graph.

ほけい 保型(の) automorphic.
～関数 automorphic function. ～形式 automorphic form. ～因子 factor of automorphy.

ほけん 保険 insurance.
～数学 insurance mathematics; calculus of contingency. ～率 insurance rate; premium rate. 生命～ life insurance.

ほげん 補元 complement; complementary element.

ほげん 補弦 complementary [supplementary] chord.

ほけんすうりし 保険数理士 actuary.

ほこ 補弧 supplementary arc; complementary arc.
～の supplementary.

ほし 星 star.
～印 star; asterisk; ＊.

ほじ 保持 conservation; maintenance.

ほしがた 星形[型] star-shape; pentagram; ☆.

ほしがた- 星型 star; star(-)like; star-shaped.
～集合 star (set). ～領域 star(-)like [star--shaped] domain. ～体 star body. ～細分 star-refinement. ～関数[写像] star-like function [mapping]. ～半径 radius of starlikeness. ～有限性 star-finite property. ～収束 star convergence.

ほじゅう 補充 supplement; complement.
～の supplementary; complementary. ～する supplement; complement; augment. ～法則 complementary law. ～次数 complementary degree.

ほしゅうごう 補集合 complement; complementary set. ⇨余集合.

ぼしゅうだん 母集団 population.
～分布 population distribution. ～特性値 population characteristic. ～平均[分散] population mean [variance].

ほじょ 補助(の) auxiliary; subsidiary; ancillary; adjoint.
～方程式 adjoint equation; subsidiary equation. ～線 adjoint [auxiliary] line. ～円 auxiliary circle. ～統計量 ancillary statistic. ～定理 lemma.

ほしょう 保証 guarantee; assurance; security; certification.
～する guarantee; assure; secure; certificate. ～水準 security level.

ほしょくしゃ-ひしょくしゃ 捕食者-被食者 predator-prey.
～方程式 predator-prey equation.

ほしょくどうぶつ 捕食動物 predator.

ほじょていり 補助定理 lemma (*pl.* -mas, -mata).
基本～ fundamental lemma. シュワルツの～ Schwarz(') lemma.

ほすう 補数 complementary number; complement.

ぼすう 母数 modulus (*pl.* -li).
～の modular. 周期～ periodicity modulus. 補～ complementary modulus. ～群 modular group. ～関数 modular function.

ぼすう 母数 parameter; population parameter. ⇨パラメータ.
～の parametric. ～によらない parameter--free; non-parametric. ～関数 paramertic fnction. ～空間 parameter space.

ほせい 補正 correction.
連続～ continuity correction.

ぼせん 母線 generator; generating line; generatrix (*pl.* -trices).
～群[族] family of generators. ～系 system of generators; generator system.

ほそい 細い thin; narrow.

ほそく 補足 complement; supplement.
～の complementary; supplementary.

ほそく 保測 measure preservation.
～変換 measure preserving transformation;

equimeasure transformation.

ほぞん 保存 preservation; conservation.
～的な conservative. ～する preserve. ～関数 preserving function. ～則 conservation law. 領域～の定理 theorem of domain preservation. ～力 conservative force.

ほだい 補題 lemma (*pl.* -mas, -mata). ⇨補助定理.

ほっきょく 北極 north pole.

ボックス box. ⇨箱.
ブラック・～ black box.

ポテンシャル potential.
～の potential. ～論 potential theory. ～[位置]エネルギー potential energy. 対数～ logarithmic potential. ニュートン～ Newtonian potential. 双極～ bipolar [dipole] potential. 導体～ conductor potential. 遅延～ retarded potential. スカラー[ベクトル]～ scalar [vector] potential. 速度～ velocity potential. 複素～ complex potential.

ホドグラフ hodograph.
～法 hodograph method. ～面 hodograph plane. ～変換 hodograph transformation.

ほとんど 殆んど almost.
～いたるところ almost everywhere; a.e.; (*F.*) presque partout; p.p. ～すべての almost all; a.a. ～確実に almost surely; a.s.; almost certainly. ～効果的に almost effectively.

ほぼ nearly; almost; about; approximately.
～等しい nearly [approximately] equal.

ほぼく 補木 co-tree.

ほぼすう 補母数 complementary modulus.

ホモトピー homotopy.
～同値 homotopy equivalence. ～同値な homotopy equivalent. ～群 homotopy group. ～型 homotopy type. ～類 homotopy class. ～完全系列 homotopy exact sequence. ～作用素 homotopy operator. ～球面 homotopy sphere. ～不変量 homotopy invariant. 制限～ restricted homotopy. ～不変な homotopy invariant.

ホモトープ(な) homotop; homotopic.
A に関して～ homotopic relative to *A*. ゼロに～ homotopic to 0.

ホモローグ(な) homologous; (*F.*) homologue; (*G.*) homolog.
ゼロに～ homologous to 0.

ホモロジー homology.
～の; ～的な homological. ～代数(学) homological algebra. ～群 homology group. ～類 homology class. ～完全系列 homology exact sequence. ～写像 homological mapping. ～次元 homological dimension. 巡回～ cyclic homology.

ポリアディック polyadic.

ポリガンマ polygamma.
～関数 polygamma function.

ポリシー policy. ⇨方策.

ポリトロープ(な) polytropic.
～方程式 polytropic equation.

ぼりょう 母量 generatrix (*pl.* -trices).

ボルノロジー bornology.

ボルノロジック (*F.*) bornologique.

ボレル É. F. E. J. Borel (1871–1956).
～集合 Borel set. ～可測な Borel(-) measurable.

ホロ- horo-.
～球面 horosphere. ～サイクル horocycle.

ホロサイクル horocycle.

ホロノミー holonomy.
～群 holonomy group. 同次～群 homogeneous holonomy group. 制限～群 restricted holonomy group.

ホロノーム(な) holonomic.
～空間 holonomic space. ～系 holonomic system. 非～系 non-holonomic system.

ほんしつてき 本質的(な) essential.
～に essentially. ～完備類 essentially complete class. ～拡大 essential extension. ～上限 essential supremum. ～に有界な essentially bounded. 本質部分加群 essential submodule.

ほんやく 翻訳 translation.
～する translate.

ほんらい 本来の proper; original.
～類 proper class.

ほんろん 本論 main discourse [subject].

ま・マ

まい- 毎 every; each.
〜回 every [each] time. 〜秒五糎 five centimeters a second; 5 cm per sec; 5 cm/sec.
まいきょ 枚挙 enumeration. ⇨列挙.
〜する enumerate. 〜(可能)定理 enumeration theorem.
マイクロ micro-.
〜メーター micrometer. 〜束 microbundle.
まいぞう 埋蔵 embedding, imbedding. ⇨埋込(み).
〜する embed, imbed. 〜定理 embedding theorem. 正則(な)〜 regular embedding. ユークリッド空間に〜する embed into Euclidean space.
マイナス minus. ⇨負(の).
〜の minus.
まいぼつひよう 埋没費用 sunk cost.
まえしょり 前処理 precondition.
まえばらい 前払 advance (payment); prepayment.
まえむきけんきゅう 前向き研究 prospective study.
まきすう 巻き数 winding number. ⇨回転数.
まく 幕 curtain; screen.
マーク mark.
マクシミン maximin; max-min.
〜値 maximin [max-min] value. 〜原理 maximin [max-min] principle.
マクローリン C. Maclaurin (1698-1746).
〜級数[展開] Maclaurin series [expansion].
マーケット market. ⇨市場.
〜・リサーチ market research.
ましぶん 増分 ⇨増分(ぞうぶん).
まじわり 交わり intersection; meet.
交わる intersect; meet. 〜の数 intersection number. 〜の定理 intersection theorem. 〜分割 cross-partition.
ます 増す increase; augment.
マスター master.
〜方程式 master equation. 〜・サンプル master sample.
または or; either ... or.
まち 待(ち)
〜行列 queue. 〜時間 queuing (time). 〜合せ queuing. 〜合せ(行列)理論 queuing theory.
まちがいなく 間違いなく certainly.
まちぎょうれつ 待ち行列 queue.
集団〜 bulk queue. 連続〜 tandem queue.
まつい 末位 last location [place].
〜の数字 last significant digit; L.S.D.
まっこう 末項 last term.
まっしょう 抹消 deletion; cancellation.
〜する delete; cancel.
まったん 末端 terminal; end.
〜の terminal.
マッチング matching.
完全〜 perfect matching. 極大〜 maximal matching. 最大〜 maximum matching.
まつわり linking; link.
まつわる link. 〜数 linking number [coefficient]. 自己〜数 self-linking coefficient. 〜複体 linked complex.
マトリックス matrix (*pl.* -trices, -trixes). ⇨行列.
マトロイド matroid.
〜束 matroid lattice.

マニュアル manual. ⇨便覧.
まばらな sparse.
まほう 魔法 magic.
～の magic. ～[方]陣 magic square.
まるい 丸い round; circular; circled; globular.
マルコフ A. A. Markov (1856–1922).
～過程[連鎖] Markov process [chain]. 隠れ～モデル hidden Markov model; HMM.
マルチンゲール martingale.
優[劣]～ super [sub] martingale. 半～ semi-martingale.
まるめ 丸め rounding.
～る round; round-off [-up]; round-down. ～ること rounding. ～誤差 rounding error; round-off error. ～(られ)た数 round number.
まん 万 ten thousands.
まんぞく 満足 satisfaction.
～させる satisfy. ～させうる satisfiable.

み・ミ

みおとし 見落し missing; oversight.

みかいけつ 未解決の unsolved; open; unsettled; pending.
～問題 unsolved [open] problem [question].

みかけ 見掛けの apparent; formal; dummy.
～特異点 apparent singular point [singularity]. ～次数 apparent [formal] degree.

みぎ 右 right. ⇨左.
～の; ～に right. ～側(の) right; right-hand. ～回りの right-handed; clockwise; dextrorse. ～手系 right-hand(ed) system. ～移動 right translation. ～逆元 right inverse. ～連続な right continuous; continuous on the right. ～完全な right-exact. ～正則[特異]な right regular [singular]. ～随伴の right adjoint. ～不変な right invariant. ～(側)極限 right-side limit. ～微分可能な right differentiable. ～下導関数 right(-hand) lower derivat(iv)e. ～分解 right resolution. ～剰余類 right coset. ～*A* 加群 right *A*-module. ～整環 right order. ～イデアル right ideal. ～上りに upward(s) to the right.

みきはず 幹葉図 stem and leaf plot.

みぎまき 右巻きの dextrorse.

みぎまわり 右回りの[に] clockwise.

みこまれる 見込まれる prospective.

みだし 見出し title; caption; table of contents; index (*pl.* -dices, -dexes).

みたす み[満]たす satisfy.
～こと satisfaction. 条件[方程式]を～ satisfy condition [equation].

みだれ 乱れ turbulence; disturbance.

みち 道; 路 path; contour.
道の幾何学 geometry of paths. 道の空間 path space. シュトルツの路 Stolz path. 積分の路 path of integration. 路に沿う積分 contour integral. 逆の～ inverse [reverse] path. 道多元環 path algebra.

みち 未知(の) unknown.
～量 unknown; unknown quantity. ～数 unknown. ～関数 unknown function.

みちのり 道程 distance.

みつ 密(な) dense.
～点 point of density; density point. ～行列 dense matrix.

みつぐみ 三つ組 triple; triplet; triplex; triad.

みっしゅうてん 密集点 density point; point of density.
～定理 density theorem.

みっつ 三つの three; ternary. ⇨三つ組.

みつど 密度 density.
線[面; 体]～ line(ar) [surface; volume] density. 確率[分布]～ probability [distribution] density. ～関数 density function. 平均～ mean [average] density. 一様～ uniform density. ～定理 density theorem. 充填～ packing density.

みつば 三葉 clover-leaf.
～結び目 clover-leaf knot.

みてい 未定(の) undetermined; indeterminate.
～係数(法) (method of) undetermined [indeterminate] coefficient(s).

ミート meet.

みとりず 見取図 (rough) sketch.

ミドレンジ midrange.

みなみ 南 south.
～の south; southern. ～半球 south(ern)

hemisphere.

ミニマックス minimax.

～値 minimax value. ～解 minimax solution. ～検定 minimax test. ～原理 minimax principle.

みね 峰, 峯 peak.

みほん 見本 sample; specimen.

～関数 sample function. ～過程 sample process. ～点 sample point.

みまん 未満 below; under; less than.

みゃくたい 脈体 nerve.

みゃくどう 脈動 pulsation.

～する pulsate.

ミュー mu; 【M, μ】.

みよ 見よ see; (*L.*) vide. ⇨参照.

前を～ see before; (*L.*) vide ante. 下を～ see below; (*L.*) vide infra.

ミリ milli-.

～メートル[グラム] millimeter [-gram]; mm. [mg.].

む・ム

む 無 nothing; nought; naught; nullity.

むいみ 無意味な meaningless; empty; insignificant; of no sense.

むえん 無縁(の) irrelevant.
～イデアル irrelevant ideal.

むえん 無縁 insignificance.
～な insignificant; extraneous. ～根 extraneous [insignificant] root. ～成分 insignificant component.

むか 無渦(の) irrotational; vortex-free; lamellar. ⇨渦(うず).
～運動 irrotational motion; vortex-free [lamellar] motion.

むかい 向い(の) opposite.
～側 opposite side.

むかんけい 無関係な unrelated; irrelevant【to】; independent【of】; extraneous.

むき 向き sense; direction; orientation.
正[負]の～ positive [negative] sense; positive [negative] orientation. 逆[反対]の～ opposite sense. ～づけ orientation. ～づけ可能な orientable. ～づけ不(可)能な non-orientable. ～を保存する[逆にする]写像 sense-preserving [-reversing] mapping; orientation preserving [reversing] mapping.

むきおく 無記憶の memoryless.

むきおくせい 無記憶性 lack of memory.

むきじゅん 無規準(の) free.
～ゲーム free game.

むきづけ 向きづけ orientation.
～る orient. ～可能[不(可)能]な orientable [non-orientable]. ～られた oriented. ～られた線分 oriented (line) segment. ～られた単体[多様体] oriented simplex [manifold]. ～られた座標近傍系 oriented system of coordinate neighbo(u)rhoods.

むきょく 無極(性) apolarity.
～の apolar. ～定理 apolarity theorem.

むげん 無限 infinity; (*L.*) infinitum. ⇨有限(の).
～の infinite; endless. ～に[まで] infinitely; to infinity; (*L.*) ad infinitum. ～公理 axiom of infinity. ～小数[級数] infinite decimal [series]. ～(乗)積[連分数] infinite product [continued fraction]. ～(数)列 infinite sequence. ～和 infinite sum. ～積分 infinite integral. ～区間 infinite interval. ～集合 infinite set. ～(次)行列 infinite matrix. ～群[体] infinite group [field]. ～分割性 infinite divisibility. ～母集団 infinite population. ～位数の of infinite order. ～次元の of infinite dimension; infinite(-)dimensional. 可算～の denumerably [countably; enumerably] infinite. ～に多くの infinitely many. ～多価の infinitely many-valued. ～回微分可能な infinitely differentiable. ～連結の infinitely connected; of infinite connectivity. ～分解可能な infinitely divisible. $x_1, x_2, \ldots$ (～まで) x_1, x_2 ad inf. [ad infinitum].

むげんえん 無限遠 infinity.
～点 point at infinity. ～直線[空間] line [space] at infinity.

むげんしょう 無限小 infinitesimal.
～の infinitesimal. ～の位数 order of infinitesimal. ～算法 infinitesimal calculus. ～運動[変化] infinitesimal motion [change]. ～変換 infinitesimal transfor-

mation. ～変形 infinitesimal deformation. ～回転 infinitesimal rotation. ～等長変換 infinitesimal isometry. 高[同]位の～ infinitesimal of higher [the same] order.

むげんだい 無限大 infinity; 【∞】; (*L.*) infinitum.
～の位数 order of infinity. 高[同]位の～ infinity of higher [the same] order. 正[負]の～ positive [negative] infinity.

むこう 無向(の) undirected.
～グラフ undirected graph.

むこうてんすう 無効添数 dummy index; umbral index (*pl.* -dices).

むさくい 無作為(な) random.
～に at random. ～化 randomization. ～化された randomized. ～標本[抽出] random sample [sampling]. ～配置 random arrangement.

むし 無視できる[してもよい] negligible; ignorable; disregardable.
～量 negligible quantity. ～ほど小さい negligibly small.

むじゅん 矛盾 contradiction; inconsistency; absurdity; antinomy; discrepancy.
～する contradict; be contradictory; be inconsistent [incompatible]. ～した contradictory; inconsistent; incompatible; paradoxical. ～律 law [principle] of contradiction. ～命題 contradiction; contradictory proposition.

むじゅんなくていぎされる(た) 矛盾なく定義される(た) well-defined.

むじゅんのない 矛盾のない compatible.

むじょうけん 無条件(の) unconditional; absolute.
～に unconditionally; absolutely. ～収束 unconditional [absolute] convergence.

むじょうほう 無情報(の) noninformative.
～事前分布 noninformative prior distribution.

むしん 無心(の) non(-)central. ⇨有心(の).
～二次曲線 non-central conics; conics without center. ～二次曲面 non-central quadrics. ～(性) non-centrality.

むすう 無数の infinite; infinitely many; countless; unnumbered.

むすび 結び join; cup(-operation).
結ぶ join; tie.

むすびいと 結び糸 knot. ⇨結び目.

むすびつける 結びつける attach.

むすびめ 結び目 knot; node.
～群 knot group. ～射影 knot projection. 交代～ alternating knot. 三葉～ clover-leaf knot. ～型 knot type. ～理論 knot theory.

むすぶ 結ぶ knot; bind.
結ばれている[いない] knotted [unknotted]; bound [unbound].

むそうかん 無相関の uncorrelated; non-correlated.

むだがない irredundant; non(-)redundant.

むていぎ 無定義(の) undefined.
～概念 undefined concept; primitive notion. ～用[術]語 undefined term.

むてん 無点(の) undotted.
～スピノル undotted spinor.

むふごう 無符号(の) non-signed.
～数 non-signed number.

むへいほう 無平方の square-free.

むむじゅん 無矛盾(性) non(-)contradiction; consistency.
～な non(-)contradictory; consistent. ～律 law of non(-)contradiction. 相対～性 relative consistence. ～な仮定 consistent assumption(s). ～の要請 consistency requirement.

むめいすう 無名数 absolute number; abstract number.

むり 無理(の) irrational. ⇨有理(の).
～量 irrational quantity; surd. ～数 irrational; irrational number. ～式 irrational expression. ～方程[不等]式 irrational equation [inequality]. ～関数 irrational function. ～切断 irrational cut. ～性 irrationality.

むりすう 無理数 irrational number.

むりょう 無料(の) free (of charge); gratuitous; gratis.
～で free (of charge); cost-free; gratis.

むりんじょう 無輪状(の) acyclic.

め・メ

め 芽 germ.
解析関数[集合]の～ germ of analytic function [set].
めいぎしゃくど 名義尺度 nominal scale.
めいじ 明示(的な) explicit.
～公式 explicit formula.
めいじ 名辞 term.
大[小]～ major [minor] term. 中[媒]～ middle term.
めいすう 名数 denominate [denominated] number; concrete number.
めいすう 命数(法) numeration.
めいだい 命題 proposition; statement.
～の propositional. 肯定[否定]～ affirmative [negative] proposition. 全称[特称]～ universal [existential] proposition. 範疇～ categorical proposition. 複合～ composite proposition. 合成～ compound proposition [statement]. 逆[転換]～ converse proposition. 合接[連合]～ conjunction proposition; conjunctive proposition. 様相～ modal proposition. 論理～ logical proposition. ～(計)算 calculus of propositions; propositional calculus. ～論理(学) propositional logic; propositional calculus. ～変項[定項] propositional variable [constant]. ～関数 propositional function. ～値 propositional value. ～結合記号 propositional connective.
めいはく 明白な clear; evident; obvious; explicit; apparent.
めいはくにする 明白にする explain.
めいめい 命名 denomination; naming; designation.
～する denominate; name; designate.
めいもくじょう 名目上の nominal.
めいりょう 明瞭に clearly.
めいれい 命令 order; command; direction; instruction.
～する order.
めいろ 迷路 maze; labyrinth.
メガ mega-.
～サイクル[ワット] megacycle [megawatt].
めぐりあい めぐり合い encounter; rencounter; (*F.*) rencontre.
～の確率 probability of rencounter.
メジアン median.
メタ meta-.
～アーベル群 meta-abelian [metabelian] group. ～理論 meta-theory.
メッセージ message.
めのこざん 目の子算 rough calculation [estimation].
メビウス A. F. Möbius (1790–1868).
～変換 Möbius transformation. ～の帯 Möbius' band. ～関数 Möbius function.
めもり 目盛 scale; graduation.
～る graduate. 普通～ ordinary [uniform] scale. 関[対]数～ functional [logarithmic] scale.
めん 面 surface; face; front.
～の facial. 曲～ (curved) surface. 錐[柱]～ conical [cylindrical] surface. 回転～ surface of revolution. 可展[展開(可能)曲]～ developable surface; applicable surface. 線織～ ruled surface. 極小～ minimal surface. 包絡～ enveloping surface; envelope. エルゴード～ ergodic surface. 等位～ level [niveau] surface. ピッチ～

pitch surface. 開いた[閉じた]～ open [closed] surface. 単側～ one-sided [unilateral] surface. 両側～ two-sided [bilateral] surface. リーマン[被覆]～ Riemann [covering] surface. ～角 face angle; facial angle. ～作用素 face operator. ～心格子 face-centred lattice. ～素 surface element. ～積分 surface integral. ～片 surface patch. ～密度 surface density. 波～ wave front.

めんせき 面積 area; superficial content [measure].
～の areal; superficial. 平[曲; 表]～ plane [surface] area. 底[側]～ area of basc [side (surface)]; basic [lateral] area. 内[外]～ inner [outer] area. 混合～ mixed area. ～グラフ area graph. ～座標 area coordinates. ～要素 area(l) element; surface element. ～定理 area theorem. ～汎関数 areal functional. ～確定の of definite area; measurable.

めんせきけい 面積計 planimeter.
アムスラーの～ Amsler's integrator.

めんせきぶん 面積分 surface integral.

めんそ 面素 surface element; area (l) element.

メンバーシップ membership.
～関数 membership function.

めんぶん 面分 region; domain; area.
(平) ～ plane region [domain]; planar region [domain]. 開[閉]～ open [closed] region.

も・モ

もうしたて 申し立て claim.
(事実として)～る claim.
もぎじっけん 模擬実験 simulation.
もくてき 目的 aim; object; objective; purpose; target. ⇨目標.
～式[関数] objective expression [function]. ～言語 target language.
もくひょう 目標 target; object; aim. ⇨目的.
～(物); target; object. ～変数 target variable. ～関数 objective function; criterion. ～精度 aimed at precision.
もくろく 目録 table; list; catalog(ue).
出版～ catalogue of publications.
もけい 模型 model; pattern.
線形～ linear model. 決定[確率]論的～ deterministic [stochastic] model. 配分～ allocation model. 学習～ learning model.
モザイク mosaic.
～片 piecc of mosaic; tessera (*pl.* -rae).
もじ 文字 letter; character.
～の; ～で表した literal. ～通りに literally. 大[頭]～ capital letter. ～式 literal expression. ～方程式[係数] literal equation [coefficient]. ～認識 character recognition. ギリシア～ Greek letter. 楔形～ cuneiform (character).
モジュラー modular.
～群 modular group. ～指標 modular character. ～束 modular lattice. ～関数 modular function. ～形式 modular form. ～表現 modular representation. ～曲線 modular curve.
モジュラス modulus (*pl.* -li); mod.
対数的～ logarithmic modulus.
モジュール module.
もちあげ (曲線, ベクトル場の)持上げ lift.
モデュレーション modulation.
～空間 modulation space.
モデル model. ⇨模型.
確率(的)～ stochastic model. ～理論 model theory. 可算～ enumerable model. ～化 model(l)ing. 位相(的)～ topological model. 隠れマルコフ～ hidden Markov model; HMM.
モード mode.
標本～ sample mode.
もどる 戻る return; come back; recur.
モニック(な) monic.
～多項式 monic polynomial.
モノイダル monoidal.
～変換 monoidal transformation.
ものさし 物差 rule; scale.
モノジーン(な) monogenic; (*F.*) monogène.
モノドロミー monodromy.
～群 monodromy group. ～行列 monodromy matrix.
モノリシック monolithic.
モーメント moment. ⇨積率.
～問題 moment problem.
もりた・きいち 森田紀一 Morita Kiichi (1915–1995).
森田同値 Morita equivalence. 森田同値な Morita equivalent.
もんだい 問題 problem; question.
幾何学的～ geometric(al) problem. 作図～ problem of construction. 四色～ four-colo(u)r problem. 固有値～ eigenvalue problem. 境界値～ boundary value problem. フェルマーの～ Fermat's problem.

係数～ coefficient problem. モーメント～ moment problem. 未解決(の)～ unsolved [open] problem [question]. 解けない[解決不能な]～ unsolvable problem. 練習～ exercise. 雑～ miscellaneous problems. 主～ primal problem. 巡回セールスマン～ traveling salesman problem. 双対～ dual problem. 相補性～ complementarity problem. 特異～ singular problem.

モンテカルロ Monte Carlo.
～法 Monte Carlo method.

や・ヤ

や 矢 arrow; sagitta.
～線 arrow; (*G.*) Speer. ～線図 arrow diagram.

や 野 field.
線～ field of lines; line field. 点～ field of points.

やく 約 about; nearly; (*L.*) circa; ca.

やく- 約 divisor; reduced.
～イデアル divisor ideal. ～錐 reduced cone. ～写像錐 reduced mapping cone. ～結 reduced join. ～懸垂 reduced suspension. ～鎖 division chain.

やくげん 約元 divisor. ⇨約数.

やくさ 約鎖 division chain.
～律 division chain condition.

やくすう 約数 measure; divisor; submultiple; aliquot.
公～ common measure [divisor]. ～関数 divisor function. ～問題 (*G.*) Teilerproblem.

やくぶん 約分 abbreviation; reduction.
～する abbreviate; reduce.

ヤコビ C. G. J. Jacobi (1804-1851).
～の; ～行列式 Jacobian. ～行列 Jacobian matrix. ～多様体 Jacobian variety. (正則性の)～判定法 Jacobian criterion (of regularity).

ヤコビアン Jacobian.

やせい 野性[生](的な) wild.
～的空間 wild space. ～分岐 wild ramification.

やせん 矢線 arrow; (*G.*) Speer; directed (line) segment.
～図 arrow diagram.

ヤード yard.
～ポンド法 yard-pound system.

ヤング A. Young (1873-1940).
～図形, ～・ダイアグラム Young diagram.

ゆ・ユ

ユー U; u.
〜型分布 U-shaped distribution. 〜管理図 u-chart.

ゆいいつ 唯一(の) unique. ⇨単独(性).
〜に uniquely. 〜の解 unique solution. 〜性 uniqueness; unicity.

ゆう- 優 major; dominant; super- ⇨劣(れつ-).
〜弧 mojor arc. 〜級数[関数] majorant (series) [(function)]. 〜ゲーム dominant game. 〜マルチンゲール supermartingale. 〜調和な superharmonic.

ゆうい 有意(性) significance.
〜な significant. 〜性 significance. 統計的〜性 statistical significance. 〜性(の)検定 significance test. 〜水準 significance level. 〜状態 significant condition. 〜な差異 significant difference [distinction].

ゆうい 有意(な) purposive; non-random.
〜標本 purposive sample. 〜選出 purposive selection.

ゆうい 優位 dominance.
対角〜 diagonal dominance.

ゆうえつ 優越 dominance; dominancy; domination; superiority.
〜の dominant. 〜する dominate. 〜原理 domination [dominance] principle. 〜関係 dominance relation.

ゆうかい 有界(な) bounded.
〜(的)に boundedly. 非〜な unbounded. 上[下]に〜な bounded (from) above [below]; bounded upwards [downwards]. 全〜な totally bounded. 一様(に)〜な uniformly bounded. 〜集合 bounded set. 〜領域 bounded domain. 〜数列[関数] bounded sequence [function]. 〜作用素 bounded operator. 〜変動の of bounded variation. 〜収束の boundedly convergent. 〜収束定理 theorem of bounded convergence. 〜的完備な boundedly complete. 〜性 boundedness. 〜(横)座標 abscissa of boundedness. 〜量化子 bounded quantifier.

ゆうかいがた 有界型(の) of bounded type; (*G.*) beschränktartig; (*F.*) bornologique.
〜関数 function of bounded type; (*G.*) beschränktartige Funktion. 〜空間 (*F.*) espace bornologique.

ゆうかいへんどう 有界変動 bounded variation.
〜(の)関数 function of bounded variation.

ゆうかく 優角 superior angle; major angle.

ゆうかく 有角(の) oblique.
〜透視図 oblique perspective; angular perspective.

ゆうかんすう 優関数 majorant; majorant function; superior function.
調和〜 harmonic majorant.

ゆうぎ 遊戯 play; game.

ゆうきゅうすう 優級数 majorant; majorant series; dominant series.
〜判定法 majorant test; M-test.

ゆうけってい 優決定 overdetermination.
〜の overdetermined. 〜系 overdetermined system.

ゆうげん 有限(の) finite. ⇨無限.
〜(的)に finitely. 〜性 finiteness. 〜小数 finite [terminating] decimal. 〜点[直線] finite point [line]. 〜集合 finite set. 〜順序数 finite ordinal number. 〜区間 finite

interval. ～列 finite sequence. ～級数 finite series. ～和[積] finite sum [product]. ～連分数 finite continued fraction. ～被覆 finite covering. ～交叉性 finite intersection property. ～群[(次)拡大] finite group [extension]. ～部分 finite part. ～値関数 finite-valued function. ～母集団 finite population. ～オートマトン finite automaton. ～数学 finite mathematics. ～次元の finite(-)dimensional. ～型の of finite type. ～生成の finitely generated. ～加法族 finitely additive class [family]. ～加法的測度 finitely additive measure. ～(性)条件 finiteness condition. 局所～な locally finite. ～グラフ finite graph. ～差分 finite difference. ～要素法 finite element method.

ゆうげんしょうすう 有限小数 finite decimal; terminating decimal; decimal of [with] finite figures.

ゆうこ 優弧 superior arc; major arc.

ゆうこう 有向(の) directed; oriented.
～線分 directed [oriented] segment. ～接線 directed tangent. ～曲線 oriented [directed] curve. ～グラフ directed graph; digraph; quiver. ～弧[円] oriented arc [circle]. ～平面[曲面] oriented plane [surface]. ～単体 oriented simplex. ～距離[体積] oriented distance [volume]. ～集合 directed set. ～族 directed family. (部分)～点族 (partial) directed family of points. ～木 directed tree.

ゆうこう 有効(な) efficient; effective; significant; valid.
～性 efficiency; effectiveness; significance; validity. ～点 efficient point. ～統計[推定]量 efficient statistic [estimate]. ～距離[射程] effective range. ～次数[種数] effective degree [genus]. ～数字 significant figure [digit]. ～因子 effective divisor. ～制約 active constraint.

ゆうごう 融合 amalgamation; fusion.
～する amalgamate. ～積 amalgamated product.

ゆうじょうほう 有情報(の) informative.
～事前分布 informative prior distribution.

ゆうしん 有心(の) central. ⇨無心(の).
～二次[円錐]曲線 central conics; conics with center. ～二次曲面 central quadrics. ～(性) centrality.

ゆうすう 友数 amicable number. ⇨親和数.

ゆうせん 優先 priority; preference.
～法 priority method. ～権 priority; preference.

ゆうそうてき 遊走的(な) wandering.
非～な non-wandering.

ゆうちょうわ 優調和(な) superharmonic.
～関数 superharmonic function. ～変換 superharmonic transformation. ～性 superharmonicity.

ゆうど 尤度 likelihood.
～比 likelihood ratio. ～方程式 likelihood equation. ～関数 likelihood function. 最大～ maximum likelihood. ～比検定 likelihood ratio test.

ゆうどう 誘導 induction; derivation.
～する induce; derive. ～された induced. 磁気～ magnetic induction. ～方程式 induced equation; induction equation. ～準同型[形]写像 induced homomorphism. ～表現 induced representation. ～指標 induced character. ～(された)関係[位相] induced relation [topology]. ～されたカルタン接続 induced Cartan connection. ～単位 derived unit. ～計量 induced metric.

ゆうびん 郵便 mail; post.
～調査 mail survey.

ゆうり 有理(の) rational. ⇨無理(の).
～的な; ～数の rational. ～的に rationally. ～数 rational number. ～式[分数式] rational expression [fractional expression]. ～整数 rational integer. ～演算 rational operation. ～根 rational root. ～点[要素] rational point [element]. ～曲線[曲面] rational curve [surface]. ～関数 rational function. ～整関数 rational integral function; polynomial (function). ～切断 rational cut. ～作用 rational action. ～整数環 ring of rational integers. ～関数体 field

of rational functions. ～方程式 rational equation. ～表現 rational representation. ～的微分方程式 rational differential equation. ～同値な rationally equivalent. ～化 rationalization. 分母の～化 rationalization of denominator.

ゆうりけい 有理型[形] meromorphy.
～の meromorphic. ～関数 meromorphic function. ～半径 radius of meromorphy; meromorphy radius. ～曲線 meromorphic curve. ～写像 meromorphic mapping.

ゆうりすう 有理数 rational number.
～体 rational number field; field of rational numbers; field of rationals.

ゆうりせいかんすう 有理斉関数 quantic.
～の quantical.

ゆえに 故に therefore; hence; accordingly; consequently; so that.

ユー・エフ・ディー UFD; unique factorization domain.

ゆがみ 歪み distortion; skewness; strain. ⇨歪曲.

ユークリッド Euclid; Eukleides (330?-275? B.C.).
～の; ～的な Euclidean; euclidean. ～幾何学 Euclidean geometry. 非～幾何学 non-Euclidean geometry. ～空間 Euclidean space. ～の互除法 Euclidean algorithm. ～の公理 Euclidean axiom(s). ～(的)距離[計量] Euclidean distance [metric]. ～整域 Euclidean domain. ～多面体 Euclidean polyhedron. ～座標系 Euclidean [rectilinear] coordinate system. ～接続 Euclidean connection. ～単体複体 Euclidean [geometrical; rectilinear] (simplicial) complex. ～胞複体 Euclidean [geometric] cell complex. 局所～的な locally Euclidean.

ユーザー user.

ゆすう 輸数 deficient number.

ゆそう 輸送 transportation; transshipment.
～計画 planning of transportation. ～問題 transportation [transshipment] problem. ～方程式 transport equation.

ユニタリ unitary.
～性 unitarity. ～行列[群] unitary matrix [group]. ～・シンプレクティック群 unitary symplectic group. ～変換 unitary transformation. ～表現 unitary representation. ～空間 unitary space.

ユニモジュラー(な) unimodular.
～群 unimodular group. ～行列 unimodular matrix. ～対等 unimodular equivalence. 完全～(な) completely unimodular.

ゆび 指 digit; finger.
～の digital. ～文字 finger alphabet.

ユプシロン upsilon; 【Υ, υ】.

ゆみがた 弓形 ⇨弓形(きゅうけい).

ゆらい 由来 origin; source; cause.
～する originate in; result from.

ゆらぎ 揺ぎ fluctuation. ⇨揺動.

ゆるい 緩[弛]い loose; slack; relaxed.

ゆるす 許す admit; permit; allow.

よ・ヨ

よ- 余 co-; complementary.
～域 codomain. ～因子 cofactor. ～行列 comatrix. ～弦 cosine. ～接 cotangent. ～割 cosecant. ～角 complementary angle. ～法線 conormal. ～対数 cologarithm. ～生成素 cogenerator. ～集合 complementary set; complement. ～事象 complementary event. ～関数 complementary function.

よ- 四 ⇨四(し-).

よいき 余域 codomain.

よいんし 余因子 cofactor.
～行列 cofactor matrix. ～展開 cofactor expansion.

よいんすう 余因数 cofactor.

よう 葉 sheet.
一[単]～双曲面 hyperboloid of one sheet. 二[複]～双曲面 hyperboloid of two sheets. *p*[多]～の面 *p* [many; multi]-sheeted surface. *p*～関数 *p*-valent function. 平均～数 mean number of sheets.

よういにする 容易にする facilitate.

よういん 要因 factor.
～の factorial. ～分析(法) factorial experiment. ～単位 factorial unit. ～配置計画[実験] factorial design.

ようかんすう 陽関数 explicit function.

ようきが 用器画 instrumental [mechanical] drawing.

ようきゅう 要求 demand; requirement; claim.
～する demand; require; claim.

ようけい 葉形 folium (*pl.* -lia).

ようけん 要件 requirement.

ようご 用語 term; terminology.
～集 glossary.

ようし 要旨 substance; gist; argument; point; summary; résumé; outline.

ようじょうきょくせん 葉状曲線 folium (*pl.* -lia).
デカルトの～ folium of Descartes.

ようじょうにする 葉状にする foliate.

ようせい 要請 request; claim; demand; requirement; postulate.

ようせき 容積 volume; (solid) content; capacity; cubic(al) content; cubage; solidity; bulk; dimensions.

ようせん 葉線 folium (*pl.* -lia). ⇨葉状曲線.

ようそ 要素 element; constituent.
集合の～ element of set. 弧(長)～ arc element. 面積[体積]～ surface [volume] element. 関数～ function element. 境界～ boundary element. 行列～ matrix element. 逆～ inverse element. 積分～ integral element. 極～ polar element. 三角形の～ elements of triangle. 有限～法 finite element method.

ようそう 様相 mood; mode; model.
～の modal. ～論理 modal logic. ～命題 modal proposition.

ようそう 葉層 foliation.
単純～ simple foliation. 解析～ analytic foliation. ～構造 foliation; foliated structure. ～多様体 (*F.*) variété feuilletée.

ようてき 陽的(な) explicit.
～に explicitly.

ようどう 揺動 fluctuation.
～散逸定理 fluctuation-dissipation theorem.

ようにんできる 容認できる acceptable.

ようやく 要約 summary; résumé; outline; abstract.

～する summarize; outline; abstract. 5数～ five number summary.

ようらん 要覧 survey; outline; catalog(ue); handbook.

ようりょう 容量 capacity.
外[内]～ outer [inner] capacity. ニュートン～ Newtonian capacity. 対数～ logarithmic capacity. 解析(的)～ analytic capacity. ～関数 capacity function. 記憶～ capacity of memory; memory size; storage capacity. 通信路～ channel capacity.

ようりょう 容量 cubic measure.

よかく 余角 complementary angle.

よかく 余核 cokernel.

よかつ 余割 cosecant; cosec.

よかんすう 余関数 complementary function.

よき 予期 expectation.
～する expect.

よぎょうれつ 余行列 comatrix.

よく 翼 wing; airfoil.
薄～ thin wing.

よくせい 抑制 inhibition; control; suppression.
～する inhibit; control; suppress.

よけい 余形 complementary figure.

よげん 余弦 cosine; cos.
～法則[定理] cosine rule [theorem]. 第一[第二]～公式 first [second] cosine formula. ～円 cosine circle. ～曲線 cosine curve. 逆～ inverse cosine. 双曲～ hyperbolic cosine. 積分～ integral cosine. ～変換 cosine transformation; cosine transform.

よこ 横 width.
～の horizontal; lateral; transversal; row. ～波 transversal [transverse] wave. ～振動 transversal vibration. ～ベクトル row vector. ～軸 (axis of) abscissa; transverse (axis).

よこぎる 横切る cross; traverse; intersect.

よこざひょう 横座標 abscissa (*pl.* -sas, -sae).
収束[正則]～ abscissa of convergence [regularity]; convergence [regularity] abscissa. 有界～ abscissa of boundedness.

よこじく 横軸 (axis of) abscissa; transverse (axis).

よこせん 横線 bar.
～工程表 bar chart.

よこなみ 横波 ⇨横波(おうは).

よさ 良さ goodness.
当てはまりの～ goodness of fit.

よし 余矢 coversed sine; coversine; covers; cvs.

よじ 四次(の) quartic; biquadratic.
～方程式 quartic [biquadratic] equation; equation of degree 4 [four]. ～曲線 quartic. ～関数 quartic function; quartic.

よじげん 余次元 codimension.

よじしょう 余事象 complementary event.

よしゅうごう 余集合 complement; complementary set. ⇨補集合.

よじょう 余剰(な) superfluous.
～部分加群 superfluous submodule.

よすう 余数 complement; complementary number.

よせいせいそ 余生成素 cogenerator.

よせざん 寄せ算 addition. ⇨加法.

よせつ 余接 cotangent; cot.
～バンドル cotangent bundle.

よそう 予想 conjecture.
フェルマ[ビーベルバッハ]の～ Fermat's [Bieberbach's] conjecture.

よぞう 余像 coimage.

よそく 予測 forecast; forecasting; prediction.
～する forecast. 要求～ demand forecasting. ～の問題 problem of prediction. ～値 predictor. 線形～ linear prediction. 最良線形～値 optimal linear predictor.

よそくし 予測子 predictor.
～修正子法 predictor-corrector method.

よたいすう 余対数 cologarithm.

よつぐみ 四つ組 quadruple.

よって accordingly; consequently; hence.

よていひょう 予定表 schedule; program(me).

よはいせん 余擺線 trochoid. ⇨トロコイド.

よび 予備(の) preliminary; preparatory.
～試験 preliminary examination. ～考察 preliminary [preparatory] consideration. ～調査 pilot survey; pretest. ～定理 prepa-

ration theorem; (*G.*) Vorbereitungssatz.

よぶん 余分 excess; extra; surplus; redundance.
～な excessive; extra; surplus; superfluous; redundant. ～のない表現 irredundant representation.

よほうせん 余法線 conormal.
～層 conormal sheaf.

よむ 読む read.
読み出す read out.

よゆう 余裕 allowance.
～のある slack. ～変数 slack variable.

よろん 世[輿]論 public opinion. ⇨世論(せろん).
～調査 survey of public opinion.

よわい 弱い weak. ⇨弱(じゃく-).
弱く weakly. 弱める weaken. ～位相 weak topology. ～条件 weak(er) condition. ～解 weak solution.

よん 四 ⇨四(し).

よんか 四科【中世大学の算術・幾何・天文・音楽】 quadrivium.

よんこう 四項(の) quadrinomial.
～式 quadrinomial.

よんじょうべき 四乗ベキ biquadrate; biquadratic.
～の biquadratic.

よんしょく 四色 four colo(u)rs.
～問題 four colo(u)r problem.

よんばい 四倍 four times.
～の quadruple; quadruplicate. ～に quadruply. ～にする quadruplicate. ～数 quadruple.

ら・ラ

ライフォ LIFO (last-in first-out)【「後入れ先出し」の意】.

らくすう 絡数 linkage coefficient; linking number.

らくたい 落体 falling body.

ラグランジュ J. L. Lagrange (1736–1813).
～の Lagrangian. ～の括弧式 Lagrange's [Lagrangian] bracket. ～乗数 Lagrange multiplier. ～(の)関数 Lagrangian function.

ラジアン radian; rad.

らしせん 螺獅線 conchoid. ⇨コンコイド.

らせん 螺線 spiral.
～形[状]の spiral; spiro-. アルキメデス～ Archimedes [-dean] spiral. 代数～ algebraic spiral. 双曲～ hyperbolic spiral. 対数～ logarithmic spiral. 等傾～ isoclinal spiral. コルニュの～ Cornu's spiral; clothoid. ～運動 spiral motion.

らせん 螺旋 helix.
～の helicoidal. ～状の helicoid; helicoidal. 常[一般]～ ordinary [general] helix. ～(曲)面 helicoid; helicoidal surface. ～体 helicoid; screw. ～運動 helicoidal motion; screw motion.

らっか 落下 fall; drop.
自由～ free fall.

ラックス P. D. Lax (1926–).

ラッセル B. Russell (1872–1970).
～の逆理 paradox of Russell.

ラテン Latin.
～(人)の Latin. ～方格[方陣] Latin square. グレコ～方格[方陣] Greco-Latin square.

ラプラシアン Laplacian.

ラプラス P. S. Laplace (1749–1827).
～演算子 Laplace operator; Laplacian. ～変換 Laplace transformation; Laplace transform.

ラベル label. ～をつける label.

ラムダ lambda;【Λ, λ】.
～関数 lambda function; λ-function. ～行列 lambda matrix.

らんかい 乱塊 randomized block.
～法 method of randomized blocks; randomized block method.

ランク rank. ⇨順位; 階数.
ランキング ranking.

らんけい 卵形 oval; ovaloid; ovoid.
～の oval; ovaloid; ovoid; ovoidal. 平均～ mean oval.

らんけいせん 卵形線 oval; convex curve.
～の oval.

らんけいたい 卵形体 ovaloid; ovoid; convex body.
～の ovaloid; ovoid; ovoidal.

らんけいめん 卵形面 ovaloid.

らんすう 乱数 random number(s).
～表 table of random numbers. ～さい(ころ) random die. 擬似～ pseudo-random numbers.

ランダウ E. G. H. Landau (1877–1938).
～の記号 Landau's symbol; O, o.

ランダム random. ⇨無作為(な).
～な random. ～に at random. ～化 randomization.

らんぽ 乱歩 random walk.

らんりゅう 乱流 turbulent flow; turbulence.
一様～ homogeneous turbulence. 等方～ isotropic turbulence.

り・リ

り 利 interest; gain.
～子[息] interest. ～率 rate of interest. 単[複]～ simple [compound] interest. 元～ principal and interest. 年～ annual interest. ～益 gain(s); profits.
リー M. S. Lie (1842-1899).
～群 Lie group. ～環[代数] Lie algebra.
リアルタイム(の) real-time.
りえき 利益 gain(s); profits.
～をえる gain.
リオンス，ジャック＝ルイ J-L. Lions (1928-2001).
リオンス，ピエール＝ルイ P-L. Lions (1956-).
りきがく 力学 dynamics; mechanics.
～(上)の dynamic; dynamical; mechanical. 静～ statics. 古典[ニュートン]～ classical [Newtonian] mechanics. 流体～ hydrodynamics; fluid dynamics; hydromechanics; fluid mechanics. 電気～ electrodynamics. 熱～ thermodynamics. 天体～ astrodynamics; celestial mechanics. 剛体の～ dynamics of rigid bodies. 量子～ quantum mechanics. 統計～ statistical mechanics. 図式～ graphical mechanics. 非線形～ nonlinear mechanics. ～系 dynamical system.
りきせき 力積 impulse.
りきせん 力線 line of force.
りきりつ 力率 power factor; moment.
りさん 離散(的) discrete.
～化 discretization. ～(的)量 discrete quantity. ～データ discrete data. ～分布 discrete distribution. ～(型)確率変数 discrete random variable. ～集合 discrete set. ～空間 discrete space. ～位相 discrete topology. ～付値 discrete valuation. ～測度[スペクトル] discrete measure [spectrum]. ～一様性 discrete uniformity. ～最適化 discrete optimization. ～数学 discrete mathematics. ～対数 discrete logarithm.
りさんか 離散化 discretization.
～誤差 discretization error.
りし 利子 interest. ⇨利; 利息.
りしん 離心
～の eccentric. ～円[角] eccentric circle(s) [angle]. ～率 eccentricity. ～近点離角 eccentric anomaly.
リスク risk.
～関数 risk function.
リスト list.
りせつ 離接 disjunction. ⇨選言.
～的な disjunctive.
りそう 理想 ideal.
～的な ideal. ～点 ideal point. ～境界 ideal boundary.
りそく 利息 interest. ⇨利.
～算 calculation of interest.
リチュウス lituus【螺線の一種】.
りつ 率 rate; modulus (*pl.* -li); proportion; index (*pl.* -dices, -dexes).
円周～ ratio of the circumference of a circle to its diameter; circular ratio; π. 利～ rate of interest. 連続～ modulus of continuity. 呼損～ loss probability. 屈折～ refraction index. 収束～ rate of convergence.
りつ 律 law; principle. ⇨法則.
交換～ law of commutativity; commutative

law. 結合～ law of associativity; associative law. 分配～ law of distributivity; distributive law. 悉無～ zero-one law; 0-1 law; law of naught or one.

りっしょう 立証 proof; verification.
～する prove; verify; establish; show.

りったい 立体 solid (body).
～の solid; three-dimensional; stereo-. ～幾何学 solid geometry. ～数 solid number. ～図形 solid figure. ～角 solid angle. ～性 solidity. ～で (*L.*) in solido.

りったい- 立体 stereo-.
～画法 stereography. ～画法の stereographic. ～射影 stereographic projection.

りったいきゅうせき 立体求積(法) cubature.

リッチ G. Ricci-Curbastro (1853-1925).
～曲率 Ricci curvature. ～・フロー Ricci flow. ～・テンソル Ricci tensor.

リットル liter;【英】litre.
ミリ～ milliliter.

りっぽう 立方 cube. ⇨三乗.
～の cubic. ～根 cube root; cubic root. ～数 cubic number. ～格子 cubic lattice. 完全～ perfect cube. ～体 cube. ～倍積 duplication of cube. ～に開く extract cubic root. ～晶系 cubic system.

りっぽうこん 立方根 cube root; cubic root.
～の subtriplicate. ～比 subtriplicate ratio.

りっぽうたい 立方体 cube; regular hexahedron (*pl.* -rons, -dra).
～の cubic; cubical. 単位～ unit cube.

りつめんず 立面図 elevation; vertical plane.

りていひょう 里程標 milestone.

りとく 利得 profit; gain; pay-off.
～関数 pay-off function. ～行列 pay-off matrix.

リニア linear. ⇨線形(の).

リマソン (*F.*) limaçon. ⇨蝸牛形.

りまわり 利回り yielding interest.

リーマン G. F. B. Riemann (1826-1866).
～の Riemannian; Riemann's. ～計量 Riemannian metric. ～幾何学 Riemannian geometry. ～空間 Riemannian space. 対称～空間 Riemannian symmetric space. ～曲率 Riemannian curvature. ～積 Riemannian product. ～接続 Riemannian connection. ～多様体 Riemannian manifold. ～積分 Riemann integral. ～可積な Riemann [(R) -] integrable. ～球(面) Riemann sphere. ～面 Riemann surface.

りゃっきほう 略記法 abridged notation.

りゆう 理由 reason; cause; grounds.
…の～で by reason of; on account of.

りゅうぎ 流儀 style; way; fashion.

りゅうこうせい 流行性(の) epidemic.
～疾病 epidemic (disease).

りゅうし 粒子 particle.
微～ corpusc(u)le.

りゅうしゅつ 流出 outflow; efflux; exit; escape.
～境界 exit boundary.

りゅうすう 留数 residue; residuum (*pl.* -dua).
～定理 residue theorem. ～解析[計算] residue calculus. 対数的～ logarithmic residue.

りゅうせん 流線 stream line; flow line.
～形(物体) stream-line body.

りゅうたい 流体 fluid.
～の fluidal; liquid; fluid. ～力学 fluid dynamics; hydro-dynamics; fluid mechanics; hydromechanics. ～静力学 hydrostatics. 圧縮(性)[粘性]～ compressible [viscous] fluid. 完全～ perfect fluid.

りゅうつう 流通 circulation; currency.
～座標 current coordinate(s).

りゅうつうかんり 流通管理 distribution management.
物的～ physical distribution management.

りゅうどう 流動 flux; fluxion.
～法 method of fluxion.

りゅうにゅう 流入 inflow; influx; afflux; entrance.
～境界 entrance boundary.

りゅうりつ 流率 fluxion.
～法 method of fluxion.

りゅうりょう 流量 flux.
～計 flow meter.

りょう 量 quantity; quantum (*pl.* -ta); amount; magnitude.

〜を限定する quantify. 幾何学的〜 geometric(al) quantity. 既知[未知]〜 known [unknown] quantity. ベクトル[スカラー]〜 vector [scalar] quantity. 離散[連続]〜 discrete [continuous] quantity.

りょう 稜 edge.
側〜 lateral edge. 〜角 dihedral angle.

りょう- 両 both; bi-; ambi-; amphi-.
〜辺 both sides. 〜直角三角形 birectangular triangle. 〜側加群 bimodule. 〜側曲面 two-sided [bilateral] surface. 〜連続な bicontinuous. 〜解析的な biholomorphic. 〜生の amphibious.

りよう 利用 utilization; use.
〜率 utilization factor [rate]. 〜可能な available.

りょういき 領域 domain; region.
〜の regional. 平面〜 plane domain; planar domain. 空間〜 space domain; spatial domain. 閉〜 closed domain [region]. 有界[非有界]〜 bounded [unbounded] domain. 正[負]〜 positive [negative] region. 単[複]連結〜 simply [multiply] (-) connected domain. 円〜 circular domain [region]. 角[円環]〜 angular [annular] domain. 凸[星型]〜 convex [starlike; star-shaped] domain. 単葉〜 schlicht domain. 截線〜 slit domain. 基本〜 fundamental region. 正則〜 domain of holomorphy. ジョルダン〜 Jordan domain. ラインハルト〜 Reinhardt domain. 依存[影響]〜 domain of dependence [influence]. 〜保存の定理 theorem of domain preservation. 実行可能〜, 許容〜 feasible region. 信頼〜法 trust region method.

りょうか 量化 quantification.
〜する quantify.

りょうかいせきてき 両解析的 biholomorphic.
〜写像 biholomorphic mapping.

りょうかきごう 量化記号 quantifier.

りょうかく 稜角 dihedral angle.

りょうかし 量化子 quantifier.
有界〜 bounded quantifier.

りょうがわ 両側(の) two-sided. ⇨両側(の)(りょうそく).
〜イデアル two-sided ideal. 〜検定 two-sided test.

りょうきゃくき 両脚器 compasses.

りょうけい 菱形 lozenge; rhombus (*pl.* -buses, -bi). ⇨菱形(ひしがた).

りょうし 量子 quantum (*pl.* -ta).
〜化する quantize. 〜化 quantization. 〜力学 quantum mechanics. 〜情報理論 quantum information theory.

りょうそく 両側(の) two-sided; double; both; bilateral; bi-. ⇨単側(たんそく).
〜イデアル two-sided ideal. 〜曲面 two-sided surface; bilateral surface. 〜検定 two-sided test. 〜剰余類 double coset. 〜加群 bimodule. 〜安定な stable in both directions.

りょうてき 量的(な) quantitative.
〜変数 quantitative variable.

りょうとうろんぽう 両刀論法 dilemma.

りょうめんたい 菱面体 rhombohedron (*pl.* -rons, -dra).
〜の rhombohedral. 〜晶系 rhombohedral system.

りょうりつ 両立(性) compatibility.
〜する compatible【with】. 〜しない incompatible. 〜条件 compatibility condition.

りょうれんぞく 両連続(性) bicontinuity.
〜な bicontinuous. 〜写像 bicontinuous mapping.

りりつ 利率 rate of interest.

りろん 理論 theory.
〜(上)の; 〜的な theoretic; theoretical. 〜的に theoretically. ガロア〜 Galois theory. 情報〜 information theory. 〜物理学 theoretical physics. 〜式 theoretical formula [expression] 〜的分布 theoretic distribution. 最適化〜 optimization theory. 符号〜 coding theory. 結び目〜 knot theory.

りんかい 臨界(の) critical.
〜点 critical point. 〜値 critical value. 〜角 critical angle. 〜行列式 critical determinant. 〜格子 critical lattice.

りんかく 輪郭 outline; contour.

りんかん　輪環　cyclic permutation.
～の順 cyclic order.
りんかんぐん　輪環群　torus group.
りんかんたい　輪環体　torus (*pl.* -ri). ⇨円環体; トーラス.
りんかんめん　輪環面　torus (*pl.* -ri); anchor ring. ⇨円環面; トーラス.
リンク　link; linkage.
りんじょう　輪状(の)　ring-shaped; circular.
りんせつ　隣接　adjacency.
～行列 adjacency matrix.
りんせつ　隣接(の)　neighbo(u)ring; adjacent; adjoining; contiguous.
～部分 adjacent part. 隣(接)辺 adjoining [adjacent] side. ～角 adjacent angle. ～関数 contiguous function.
りんたい　輪体　cycle.
基本～ fundamental cycle. 境界～ boundary cycle; boundary. 分割～ dividing cycle. 双対～ cocycle.
りんてんきょくせん　輪転曲線　roulette; rolling curve.
りんぺん　隣辺　adjacent [adjoining] side.

る・ル

るい　類　class; category.
同値～ equivalence class.　剰余～ residue class.　特性～ characteristic class.　ホモトピー[ホモロジー]～ homotopy [homology] class.　～数 class number.　～関数 class function.　～体 class field.　～体論 class field theory.　～構造 class formation.　第一[二]～集合 set of first [second] category.　～等式 class equation.

るいか　累加　cumulation.
～平均 progressive mean.

るいじ　類似　analogy; similarity.
～な analogous; similar; like.

るいじ　累次(の)　repeated; iterated.　⇨反復.
～級数 repeated series.　～極限 repeated [iterated] limit.　～積分 repeated [iterated] integral.　～対数 iterated logarithm.

るいじ-　類似
～中線 symmedian.　～重心 symmedian point.　～角柱 symmedian prism.

るいじかくちゅう　類似角柱　symmedian prism.

るいじょう　累乗　repeated multiplication; power.　⇨ベキ.
～指数 (power) exponent.

るいじょう-　累乗　radical.
～根 radical root.　～(根)数 radical number.

るいじょうこん　累乗根　radical root; root.
1の～[ベキ根] root of unity.

るいすう　類数　class number.

るいせき　累積　cumulation; accumulation.
～的な cumulative.　～度数 cumulative frequency.　～分布 cumulative distribution.　～度数[分布]曲線図 cumulative frequency [distribution] curve; ogive.　～分布関数 cumulative distribution function; C.D.F.　～誤差 accumulated error.

るいたい　類体　class field.
～論 class field theory.

るいひ　類比　analogy.

るいべつ　類別　classification; sorting.
～する classify; sort.　粗い～ rough classification.

ルジャンドル　A. M. Legendre (1752-1833).
～変換 Legendre('s) transformation.

ルーチン　routine.

ルート　root.　⇨根(こん).
～系 root system.

ルビ　ruby (英); agate (米).

ループ　loop.　⇨自閉線; 閉道.
～空間 loop space.　～定理 loop theorem.

ルベーグ　H. L. Lebesgue (1875-1941).
～測度 Lebesgue measure.　～積分 Lebesgue integral.　～可積な Lebesgue [(L)-] integrable [summable].　～数 Lebesgue number.

ルール　rule.

ルーレット　roulette.

れ・レ

れい 零 zero; null; naught; nought.
～の zero; null; nil. ～でない non-zero; non-vanishing. ～になる vanish; become zero. ～にする make zero; annihilate. ～元 zero element. ～ベクトル[行列] zero vector [matrix]. ～因子 zero divisor; nil(-) factor. ～点 zero (point). ～鎖 zero chain. ～相関 zero correlation. ～対象 zero object. ～系 null system. ～列 null sequence. ～関数 null function. ～境界 null boundary. ～複体 null complex. ～再帰の null recurrent. ～和 zero-sum. ～または一の法則 zero-one law; 0-1 law; law of naught or one. 標数～ characteristic zero. ～空間 null space.

れい 例 example; ex.; instance; illustration; case.
～をあげる give example. ～をあげれば for instance; for example. 反～ counter--example.

れいか 零化 annihilation.
～する annihilate. ～子 annihilation operator. ～元 annihilator. ～イデアル annihilator; annihilator ideal. ～域 annihilator. ～群 annihilator. ～集合 annihilator set.

れいがい 例外 exception; anomaly.
～の exceptional. ～的に exceptionally. ～曲線 exceptional curve. …は～として except (for) …; with the exception of…. ～群 exceptional group.

れいき 励起 excitation.
～する excite.

れいだい 例題 exercise; example.

れいてん 零点 zero; zero point.
k位の～ zero point of k th order [order k].
～のない領域 zero-free region.

れいりつ 捩率 torsion. ⇨捩[捻]れ.
測地(的)～ geodesic torsion. ～半径 radius of torsion. ～線 line of torsion. ～形式 torsion form. ～テンソル torsion tensor.

れいわ 零和 zero-sum.
～ゲーム zero-sum game. ～二人ゲーム zero-sum two person game. 非～ゲーム non-zero-sum game.

レオノーム(な) rheonomic; rheonomous.

レゾルベント resolvent.
～方程式 resolvent equation. ～作用素 resolvent operator. ～集合 resolvent set.

れつ 列 column.
～ベクトル column vector. ～有限行列 column-finite matrix. ～基本変形 elementary column operation.

れつ 列 sequence; progression; series.
～の sequential. (点)～的に sequentially. 数～ sequence (of numbers); progression. 点～ sequence of points; point sequence. 関数～ sequence of functions. 有限[無限] ～ finite [infinite] sequence. 集合[事象] ～ sequence of sets [events]. 収束[発散] ～ convergent [divergent] sequence. 基本[コーシー]～ fundamental [Cauchy] sequence. 減少[増加]～ decreasing [increasing] sequence. 収縮～ shrinking sequence. 最小化[最大化]～ minimizing [maximizing] sequence. 許容～ admissible sequence. ～極限 sequential limit. 正規～ normal sequence; normal series. 組成～ composition series. 中心～ central series. 部分～ subsequence. ～的完備な sequentially complete. 正則～ regular

sequence.
れつ- 劣 sub-; inferior; minor. ⇨優(ゆう).
～弧 minor arc. ～マルチンゲール submartingale.
れっかく 劣角 minor angle; inferior angle.
れつかほう 劣加法(性) subadditivity.
～的な subadditive. ～(的)汎関数 subadditive functional.
れつかんすう 劣関数 minor function; minorant.
れっきょ 列挙 enumeration. ⇨枚挙.
～する enumerate; list. ～解 enumerative solution.
れつけってい 劣決定 underdetermination.
～の underdetermined. ～系 underdetermined system.
れっこ 劣弧 minor arc; inferior arc.
れつこうばい 劣勾配 subgradient.
れっし 列氏 Réaumur.
～十八度 18°R.
れつせいき 劣正規(な) subnormal.
～作用素 subnormal operator.
れつちょうわ 劣調和(性) subharmonicity.
～な subharmonic. ～関数 subharmonic function. ～拡大 subharmonic extension. 多重～な plurisubharmonic.
れってん 列点 range of points.
調和～ harmonic range of points.
れつびぶん 劣微分 subdifferential.
れつふへん 劣不変 subinvariance.
～な subinvariant. ～測度 subinvariant measure.
れつ-モジュラー 劣モジュラー submodular.
レトラクト retract.
変位～ deformation retract. 絶対近傍～ absolute neighbo(u)rhood retract.
レプリカ replica.
レベル level. ⇨水準.
レポート report.
レムニスケート lemniscate. ⇨ renzyukei.
れんかん 連環[関] link; linkage.
～比率[指数] link relatives. ～率 linkage value.
れんかん 連関 coupling; linkage.
れんかん 連関 ⇨関連.
～収束半径 associated convergence radii.
れんけい 連携 coalition.
れんけい 連係 linkage. ⇨連関.
れんけつ 連結 link.
～関数 link function.
れんけつ 連結 connection; connexion; connectedness.
～な connected; connective. ～する connect. ～性 connectedness; connectivity. ～度 connectivity. ～写像 connecting mapping. ～集合 connected set. ～成分 connected component. 局所～の locally connected. 弧状～の arcwise connected. 単[複]～の simply- [multiply-] connected. n重～の n-ply connected. 無限～の infinitely connected; of infinite connectivity. 実数の～性 connectedness of real numbers. 鎖状～の well-chained. ～グラフ connected graph. ～準同型[形](写像) connecting homomorphism.
れんごう 連合 conjunction; association.
～命題 conjunction [conjunctive] proposition. 観念～ association of ideas. ～準線[準円] associated director line [circle]; joint directrix [director circle].
れんごう 連合 union.
線[面]素～ union of line [surface] elements.
れんさ 連鎖 chain.
～比 chain ratio. ～条件 chain condition. ～律 chain rule. 解析～ analytic chain. マルコフ～ Markov chain. 出生死亡～ birth and death chain. 吸収～ absorbing chain. ～式 sorites. ～論法 sorites.
レンジ range. ⇨範囲.
れんしゅう 練習 exercise; ex.; training: practice.
～問題 exercise; ex.
れんじゅけい 連珠形 lemniscate.
れんじょう 連乗 continued multiplication.
～積 continued product; factorial.
れんじょほう 連除法 division algorithm.
レンズ lens (*pl.* -ses).
～空間 lens space.
れんせつ 連接 coherence; coherency.

〜した coherent. 〜層 coherent sheaf.

れんぞく 連続 continuity.

〜な continuous. 左[右]〜な left [right] continuous; continuous on the left [right]. 半〜な semicontinuous. 不〜な discontinuous. 〜的に continuously. 〜性 continuity. 〜写像 continuous mapping. 〜関数 continuous function. 〜像 continuous image. 〜弧[曲線] continuous arc [curve]. 〜群 continuous group. 〜幾何(学) continuous geometry. 〜分布 continuous distribution. 〜スペクトル continuous spectrum. 〜流 continuous flow. 〜度[率] continuity modulus; modulus of continuity. 一様〜な uniformly continuous. 絶対〜な absolutely continuous. 各個〜な separately continuous. 同程度〜な equicontinuous. 〜的微分可能な continuously differentiable. 〜性(の)公理 axiom(s) of continuity. 〜定理 continuity theorem. 〜体 continuum (*pl.* -nua). 〜修正, 〜補正 continuity correction.

れんぞくがた 連続型(の) of continuous type; continuous.

〜確率変数 continuous random variable. 〜確率過程 stochastic process of continuous type. 〜データ continuous data.

れんぞくせい 連続性 continuity. ⇨連続.

〜性(の)公理 axiom(s) of continuity. 〜原理 continuity principle. 一様〜 uniform continuity. 同程度〜 equicontinuity.

れんぞくたい 連続体 continuum (*pl.* -nua).

〜濃度の記号 notation [symbol] of power of continuum; aleph; ℵ. 〜仮説 continuum hypothesis.

れんぞくど 連続度 continuity modulus; modulus of continuity.

れんぴ 連比 continued ratio.

れんぴれい 連比例 continued proportion.

れんぶんすう 連分数 continued fraction.

有[無]限〜 finite [infinite] continued fraction. 正則[正規]〜 regular [normal] continued fraction. 循環〜 recurring continued fraction. 〜展開 expansion into continued fraction.

れんらく 連絡(する) connecting.

〜テンソル connecting tensor.

れんりつ 連立(の) simultaneous.

〜方程式 simultaneous [system of] equations. 〜一次方程式 simultaneous [system of] linear equations. 〜不等式 simultaneous [system of] inequalities. 〜一次合同式 simultaneous [system of] congruences of first degree.

ろ・ロ

ロー rho; 【P, ρ】.
ろうとめん 漏斗面 funnel.
ログ log; logarithmic.
～概型 logarithmic scheme. ～構造 logarithmic structure.
ろく- 六 hexa-; sex-.
～次曲線 sextic curve.
ろくじゅう 六十 sixty.
～の sexagesimal. ～進法 sexagesimal system.
ろくせんけい 六線形 hexagram.
ロクソドローム loxodrome. ⇨斜航(的).
～の loxodromic.
ろくてんずけい 六点図形 quadrangular set of six points.
ろくばい 六倍 sextuple; six times.
～の sixfold; sextuple.
ろくぶんえん 六分円 sextant.
ろくぶんぎ 六分儀 sextant.
～測量 sextant surveying.
ろくぼうせいけい 六芒星形 hexagram.
ろくめんたい 六面体 hexahedron (*pl.* -rons, -dra).
～の hexahedral. 正～ regular hexahedron; cube. (正)～群 hexahedral group.
ロジスティック logistic.
～曲線 logistic curve. ～方程式 logistic equation.
ろっかくけい 六角形 hexagon; sexangle.
～の hexagonal; sexangular. 正～ regular hexagon. ～織 hexagonal web.
ろっかくすう 六角数 hexagonal number.
ろっかくちゅう 六角柱 hexagonal prism.
ロット lot. ⇨仕切.
～許容不良率 lot tolerance percent defective; LTPD.
ろっぺんけい 六辺形 hexagon.
～の hexagonal. 等角～ equiangular hexagon.
ろっぽうしょうけい 六方晶系 hexagonal system.
ローマ Rome.
～人[の] Roman. ～字 Roman letter. ～数字 Roman cypher; Roman figure [numeral]. ～字体 Roman type.
ろん 論 theory. ⇨論(-ろん).
誤差～ theory of errors. 微分方程式～ theory of differential equations. 関数～ function theory; theory of functions. 整数～ number theory; theory of numbers. 測度～ measure theory; theory of measures. 虚数乗法～ theory of complex multiplication.
-ろん 論 -logy; theory of. ⇨論(ろん).
方法～ methodology.
ろんぎ 論議 discussion; argument.
ろんしょう 論証 proof; demonstration; argument.
～する prove; demonstrate. ～できない undemonstrable; indemonstrable; unprovable. ～され(てい)ない undemonstrated. ～演算 calculus of logic. ～幾何 proof geometry.
ロンスキ H. J. M. Wronski (1778–1853).
～の行列式 Wronskian.
ロンスキアン Wronskian.
ろんせつ 論説 essay; article.
ろんだい 論題 subject; topic; theme.
ろんぶん 論文 paper; treatise; thesis (*pl.* -ses); article; memoir.

共著～ joint paper; collaborated paper. 卒業～ thesis. 学位～ thesis; dissertation. 博士～ doctoral thesis.

ろんぽう 論法 (method of) argument; reasoning; logic.

ろんり 論理 logic.

～の; ～的 logical. ～学 logic. 記号～(学) symbolic logic. 命題[述語]～ propositional [predicate] logic. 二値～ binary [two-valued] logic. 多値～ many-valued logic. 様相～ modal logic. ～主義 logicism. ～的な公理 logical axiom. ～命題 logical proposition. ～演算子 logical operator. ～記号 logical symbol. ～関数 logical function. ～和[積] logical sum [product]. ～回路 logical circuit. ～機械 logical machine. ～設計 logical design. 一階述語～ first order predicate logic.

ろんりしき 論理式 formula; logical formula; well-formed formula; logical expression. ⇨(論理)式.

ろんりせき 論理積 logical product; product of propositions; and.

ろんりそし 論理素子 logic(al) element; decision element.

ろんりわ 論理和 logical sum; sum of propositions; or.

排他的～ exclusive or. 非排他的～ inclusive or; and/or.

わ・ワ

わ 和 sum.
総～ total sum. 有限[無限]～ finite [infinite] sum. 二乗[平方]～ sum of squares; square-sum. 積[ベキ]～ sum of products [powers]; product- [power-] sum. 部分～ partial sum. 二重～ double sum. 三角～ trigonometric sum. 直～ direct sum. 論理～ logical sum. 順序～ ordered sum. 状態～ sum over states. ～事象 sum event. ～集合 sum of sets; union. 級数の～ sum of series.

わ 輪 ring; link; loop.

わい- 歪 skew(-); anti-.
～度 skewness. ～エルミート形式[行列] skew-Hermitian form [matrix]. ～対称な skew-symmetric; anti-symmetric. ～疾走線 skew cissoid. ～多項式 skew polynomial.

ワイエルシュトラス K. Weierstrass (1815–1897).
～の Weierstrassian; Weierstrass'.

わいきょく 歪曲 distortion.
～する distort. ～不等式 distortion inequality. ～定理 distortion theorem.

わいど 歪度 skewness.

わかつ 分かつ divide.
調和に～ divide harmonically.

わかりやすい 分かり易い intelligible; lucid; simple; clear; plain.

わかりやすく 分かり易く clearly; lucidly; simply; intelligibly; plainly.

わきだし 湧出し source.
～なしの source-free; solenoidal.

わく 枠 frame. ⇨標構.
直交～ orthogonal frame. 射影～ projective frame. アフィン～ affine frame. 動～ moving frame. フルネー～ Frenet frame. 正規～ normal frame. ～族 family of frames.

わけまえ 分け前 share; portion; quota.

わざわい 災い curse.

わさん 和算 mathematics developed in Japan; Japanese mathematics.

わざん 和算 ⇨和算(わさん).

わしゅうごう 和集合 union; union [sum] of sets.

わな 罠 trap.

わぶん 和分 summation; sum.
～する sum. ～記号 summation notation.

わりあい 割合 rate; ratio; proportion; percentage.

わりあて 割当 assignment; allotment; distribution; allocation; quota.
～る allot; allocate; prorate. ～の問題 assignment problem; problem of allocations. 最適～ optimal allocation. ～法 quota method.

わりきれる 割り切れる (exactly) divisible; contain.
割り切れない indivisible.

わりこみ 割り込(み) interruption; intrusion.
～む interrupt; intrude; break in.

わりざん 割算 division. ⇨除法.

わりつけ 割り付け layout.

わりびき 割引(き) discount; reduction.
～く discount; reduce. ～高 amount of discount. ～率 discount rate.

わる 割る divide.
4を2で～ divide four by two.

わんきょく 彎曲 bend(ing).
～した curved; bent; tortuous.

III

付　録

1．記号・式の英語での読み方

2．数の呼称

3．各種の字体

1. 記号・式の英語での読み方

( )	parentheses
{ }	braces
[ ]	brackets
$\frac{1}{2}$	one half; one over two
$\frac{3}{2}$	three halves; three over two
$\frac{1}{3}$	one third; one over three
$\frac{2}{3}$	two thirds; two over three
$\frac{1}{4}$	one fourth; one over four; a quarter
$\frac{3}{4}$	three fourths; three over four; three quarters
$\frac{1}{21}$	one twenty oneth; one over twenty one
$\frac{2}{21}$	two twenty oneths; two over twenty one
$\frac{3}{22}$	three twenty seconds; three over twenty two
$2\frac{2}{3}$	two and two thirds
12.345	twelve point three four five
$0.12\dot{3}\dot{4}$	zero point one two thirty four recurring [repeating]
$\therefore$	therefore
$\because$	because
X	capital X
$\forall x$	for all x; for any x
$\exists x$	for some x; there exists an x such that
$p \wedge q$	p and q; conjunction of p and q
$p \vee q$	p or q; disjunction of p and q

$\sim p$; $\neg p$; $\neg p$	not p; negation of p
$p \rightarrow q$; $p \Rightarrow q$	p implies q; if p then q
$p \leftrightarrow q$; $p \rightleftarrows q$; $p \Leftrightarrow q$	p is equivalent to q; p if and only if [iff] q
x_0	x (sub) nought [naught; zero]
a'	a prime
a''	a double prime; a second
a'''	a triple prime; a third
$\bar{a}$	a bar
$\tilde{a}$	a tilde
a^*	a star [asterisk]
$\hat{a}$	a hat [roof; wedge]
$\pm a$	plus or minus a
$\|a\|$	absolute value of a
$2a$	two (times) a; twice a
a^2	a square [squared]
a^3	a cube [cubed]
a^4	a to the four [fourth (power)]
a^n	a to the n [nth (power)]
a^{-n}	a to the minus n [nth (power)]
$a^{1/2}$	a to the one-half (power)
$a^{-1/2}$	a to the minus one-half (power)
$\sqrt{a}$	square root of a
$\sqrt[3]{a}$	cube [cubic] root of a
$\sqrt[n]{a}$	nth root of a
$\frac{2}{3}[\{a+b(c-d)\}+e]$	two thirds open bracket open brace a plus b open parenthesis c minus d close parenthesis close brace plus e close bracket
$a=b$	a equals [is equal to] b
$a \neq b$; $a \neq b$; $a \neq b$	a is not equal to b
$a=0$	a equals [is equal to] zero; a vanishes
$a \fallingdotseq b$; $a \approx b$	a is approximately [nearly] equal to b
$a \equiv b$	a is identical to [with] [identically equal to] b
$a > b$	a is greater [larger] than b

$a<b$	a is less [smaller] than b
$a \geqq b$; $a \geqslant b$; $a \geq b$	a is greater than or equal to [not less than] b
$a \leqq b$; $a \leqslant b$; $a \leq b$	a is less than or equal to [not greater than] b
$a+b$	a plus b
$a-b$	a minus b
$a \times b$	a multiplied by b; a times b; a cross b
$a \cdot b$	a times b
$a \div b$	a divided by b
$\frac{a}{b}$; a/b	a divided by b; a over b; a by b
$a : b$	ratio of a to b
$a+b=c$	a plus b equals [is equal to] c
$(a+b)c$	a plus b in parentheses times c
$\frac{a-b}{c+d}$	a minus b over c plus d
$(a+b)^2=a^2+2ab+b^2$	a plus b square [squared] equals [is equal to] a square [squared] plus two [twice] a b plus b square [squared]
$a^2-b^2=(a+b)(a-b)$	a square [squared] minus b square [squared] equals [is equal to] a plus b times a minus b
$\frac{-b \pm \sqrt{b^2-4ac}}{2a}$	minus b plus or minus square root of b square [squared] minus four a c over two a
$a : b = c : d$	the ratio of a to b equals [is equal to] the ratio of c to d; a is to b as c is to d
$n!$	n factorial; factorial (of) n
${}_nP_r$	permutation of n things (taken) r at a time
${}_nC_r$	combination of n things (taken) r at a time
$a \equiv b \pmod{n}$	a is congruent to b modulo n
$a \mid b$	a is a divisor of b; b is divisible by a
$\left(\frac{a}{p}\right)$	Legendre's symbol of a with respect to p
$\left(\frac{m}{n}\right)$	Jacobi's symbol of m with respect to n
$\angle \mathrm{A}$	angle (at) A
$\angle \mathrm{AOB}$	angle A O B

$\overline{\mathrm{AB}}$	segment A B
$\overset{\frown}{\mathrm{AB}}$	arc A B
$\triangle \mathrm{ABC}$	triangle A B C
▱ABCD	parallelogram A B C D
⊲OAB	sector O A B
$\mathrm{AB} \parallel \mathrm{CD}$	A B is parallel to C D
$\mathrm{AB} \perp \mathrm{CD}$	A B is perpendicular to C D
$\triangle \mathrm{ABC} \equiv \triangle \mathrm{DEF}$; $\triangle \mathrm{ABC} \cong \triangle \mathrm{DEF}$	triangle A B C is congruent to triangle D E F
$\triangle \mathrm{ABC} \backsim \triangle \mathrm{DEF}$	triangle A B C is similar to triangle D E F
$\Sigma \barwedge \Sigma'$	sigma is in projective relation with sigma prime
$\Sigma \doublebarwedge \Sigma'$	sigma is in perspective relation with sigma prime
$\overrightarrow{\mathrm{AB}}$	vector A B
$\begin{pmatrix} a_{11} & \cdots & a_{1n} \\ \cdots & \cdots & \cdots \\ a_{m1} & \cdots & a_{mn} \end{pmatrix}$	m times n matrix a one one to a m n
(a_{ij})	matrix a i j
$\det A$; $\lvert A \rvert$	determinant of (matrix) A
$\det(\mathfrak{x}, \mathfrak{y}, \mathfrak{z})$	determinant with the column vectors German x y z
A^{-1}	inverse (matrix) of A; inverse of matrix A
${}^{t}A$; A^{t}; A^{T}	transposed matrix of A; transpose of (matrix) A
$\mathrm{adj}\, A$	adjoint of (matrix) A
$\mathrm{tr}\, A$	trace of (matrix) A
$A \otimes B$	Kronecker product of A B
$M \otimes N$	tensor product of M N
$\{x \mid \mathrm{C}\}$; $\{x : \mathrm{C}\}$	set of x [all x's] satisfying (the condition) C
$x \in E$	x is an element of [belongs to] E
$E \ni x$	E contains x
$A \subset B$	A is contained [included] in B; A is a subset of B
$A \supset B$	A contains [includes] B
$A \subsetneqq B$	A is properly contained [included] in B; A is a proper subset of B
$A \cup B$; $A \cup B$; $A \cup B$	A union B; A cup B; union of A and B
$A \cap B$; $A \cap B$; $A \cap B$	A intersection B; A cap B; intersection of A and B

$A \triangle B$	symmetric difference of A and B
E^c; $\complement E$	complement of E
E°; E^i; Int E	interior of E
$\overline{E}$; Cl E	closure of E
E'	derived set of E
$\overline{\mathrm{co}}\, E$; $\mathscr{H}(E)$	closed convex hull of E
∂D; bdry D	boundary of D
$\bigcup_{k=1}^{n} E_k$	union of E_k k running from one to n
$\bigcap_{k=1}^{n} E_k$	intersection of E_k k running from one to n
$H \dotplus K$; $H \oplus K$	direct sum of H and K
$\varnothing$; $\emptyset$	empty [void] set
dim Ω	dimension of capital omega
$[x]$	the greatest integer not greater than [exceeding] x; x in brackets; integral part of x
sgn x	sign [signum] (of) x
sin x	sine (of) x
csc x	c s c (of) x; cosecant (of) x
arcsin x; $\sin^{-1}x$	arcsine [inverse sine] (of) x
e^x	e to the x
exp x	exponential of x
sinh x	sine hyperbolic [hyperbolic sine] (of) x
log x	log (of) x; logarithm (of) x
$\log_a x$	log (of) x to the base a
ln x; log nat x	natural logarithm of x
log com x; $\log_{10} x$	common logarithm of x
$\log^+ x$	plus-logarithm of x
cis θ	cis [c i s] (of) theta
sn u	s n (of) u
$\cos^2\theta+\sin^2\theta=1$	the square of the cosine of θ plus the square of the sine of θ equals unity
$\bar{z}$	conjugate of z

$\Re z$; $\mathrm{Re}\, z$	real part of z
$\Im z$; $\mathrm{Im}\, z$	imaginary part of z
$\arg z$	argument of z
x_n	x sub n
$\{x_n\}$	sequence x sub n
$x_1,\ x_2,\ \cdots$	x (sub) one x (sub) two and so on; x_1, x_2 ad infinitum [to infinity]
$\max(x_1,\ \cdots,\ x_n)$	maximum (value) of the sequence x (sub) one to x (sub) n
$\min(x_1,\ \cdots,\ x_n)$	minimum (value) of the sequence x (sub) one to x (sub) n
$\lim\limits_{n\to\infty} x_n$	limit of x_n as n tends to infinity
$\overline{\lim\limits_{n\to\infty}}\, x_n$; $\limsup\limits_{n\to\infty} x_n$	upper limit [superior limit] of x_n as n tends to infinity
$\underline{\lim\limits_{n\to\infty}}\, x_n$; $\liminf\limits_{n\to\infty} x_n$	lower limit [inferior limit] of x_n as n tends to infinity
$x_n \to \alpha\ (n\to\infty)$	x_n tends to [converges to; approaches] alpha as n tends to infinity
$x_n \to -\infty\ (n\to\infty)$	x_n tends to minus infinity as n tends to infinity
$\lim\limits_{n\to\infty} \frac{1}{n}(x_1+\cdots+x_n)$	limit as n tends to infinity of one over n times the sum of the x's from sub one to sub n
$\sum\limits_{k=1}^{n} x_k$	sum of x_k k equals [running from] one to n
$\sum\limits_{n=1}^{\infty} x_n$	sum of (all) x_n n equals [running from] one to infinity
$\prod\limits_{k=1}^{n} x_k$	product of x_k k equals [running from] one to n
$\prod\limits_{n=1}^{\infty} x_n$	product of (all) x_n n equals [running from] one to infinity
$f(x)$	(function) f of x
$f : X \to Y$	function f mapping X into Y; mapping f from X into Y
$f : x \mapsto y$	function f mapping x to y
$f\|E$	(function; mapping) f restricted [contracted] on E
f^{-1}	inverse (function) of f
f^{+}	positive part of f
f^{-}	negative part of f
$\mathrm{supp}\, f$	support of f
$\mathrm{Im}\, f$; $\mathrm{im}\, f$	image of f

Coim f; coim f	coimage of f
Ker f; ker f	kernel of f
Coker f; coker f	cokernel of f
$f \circ g$	f composition g
$f * g$	f convolution g
$f \sim g$	f is equivalent to g
$\mathrm{Hom}_A(M,\ N)$	module of A-homomorphisms from M to N
$\max_E f(x)$	maximum (value) on E of f of x
$\min_E f(x)$	minimum (value) on E of f of x
$\sup_E f(x)$	supremum on E of f of x
$\inf_E f(x)$	infimum on E of f of x
$\mathrm{ess}\sup_E \lvert f(x) \rvert$	essential supremum of E of the absolute value of f of x
$\mathrm{vrai}\max f(x)$	vrai maximum of f of x
$x \to a$	x tends to [approaches] a
$\lim_{x \to a} f(x)$	limit of f of x as x tends to [approaches] a
$f(x) \to b\ (x \to a)$	f of x tends to [approaches; converges to] b as x tends to [approaches] a
$\lim_{x \to \infty} \frac{f(x)}{g(x)}$	limit of f of x over g of x as x tends to infinity
$\varinjlim A_\lambda$	inductive limit of A sub lambda
$\varprojlim A_\lambda$	projective limit of A sub lambda
$\lVert x \rVert$	norm of x
$dx \wedge dy$	d x wedge d y
Δx	delta x; increment of x
$\frac{dy}{dx}$	d y over [by] d x
$\frac{d^2y}{dx^2}$	d two y over [by] d x square
$\frac{\partial z}{\partial x}$	round d z over round d x
$\frac{\partial^2 z}{\partial x \partial y}$	round d two z over round d x times round d y

$p_0(x)y''+p_1(x)y'+p_2(x)y=\varphi(x)$	p sub zero of x times y double prime plus p sub one of x times y prime plus p sub two of x times y equals phi of x
$F'(x)$, $DF(x)$	derivative [derivate] of F at x
$\mathrm{grad}\,f$	gradient of f
$\mathrm{div}\,f$	divergence of f
$\mathrm{rot}\,u$	rotation of u
$\mathrm{curl}\,u$	curl of u
$\nabla\cdot f$	del [nabla] dot f
$\nabla\times u$	del [nabla] cross u
Δf	Laplacian of f
$\Box f$	d'Alembertian of f
$\Delta f\equiv\nabla^2 f$	Laplacian of f is identical with del [nabla] squared of f
$\int y\,dx$	integral of y with respect to x; integral y d x
$\int f(x)dx$	(indefinite) integral of f of x
$\int_a^b f(x)dx$	integral from a to b of f of x; integral of f with respect to x from a to b
$(\mathrm{P})\int_a^b f(x)dx$; $\mathrm{p.v.}\int_a^b f(x)dx$	principal value of the integral from a to b of f of x
$\iint z\,dxdy$	double integral of z with respect to x (and) y; integral z d x times d y
$\iint f(x,\ y)dxdy$	double integral of f of x y
$\iint_E f(x,\ y)dxdy$	double integral over E of f of x y
$\int_E f\,d\mu$	integral over E of f with respect to μ
$\|f\|_p$	L^p norm of f
$\int_C f(z)dz$	(contour) integral along C of z
$\mathrm{Res}(c;\ f)$; $\mathrm{res}_c\, f(z)$	residue at c of f (of z)
$f\perp g$	(function) f is orthogonal to g
$\mathrm{Pr}\,E$; $\mathrm{Pr}(E)$; $P(E)$	probability of E
$\mathrm{Pr}(E\vert F)$; $P(E\vert F)$	probability of E under the condition [when] F

$E(X)$	expectation of X
$\mathrm{Var}\,X$; $\mathrm{Var}(X)$	variance of X

2. 数の呼称

10^n	n [米・仏系]	n [英・独系]
one, unity	0	0
ten	1	1
hundred	2	2
thousand	3	3
million	6	6
milliard	9	9
billion	9	12
trillion	12	18
quadrillion	15	24
quintillion	18	30
sextillion	21	36
septillion	24	42
octillion	27	48
nonillion	30	54
decillion	33	60
undecillion	36	66
duodecillion	39	72
tredecillion	42	78
quantuordecillion	45	84
quindecillion	48	90
sexdecillion	51	96
septendecillion	54	102
octodecillion	57	108
novemdecillion	60	114
vignitillion	63	120

3. 各種の字体

1. アルファベット

1	2	3	4	5	6	1	2	3	4	5	6
A a	*A a*	***A a***	A	A	A	N n	*N n*	***N n***	N	N	N
B b	*B b*	***B b***	B	B	B	O o	*O o*	***O o***	O	O	O
C c	*C c*	***C c***	C	C	C	P p	*P p*	***P p***	P	P	P
D d	*D d*	***D d***	D	D	D	Q q	*Q q*	***Q q***	Q	Q	Q
E e	*E e*	***E e***	E	E	E	R r	*R r*	***R r***	R	R	R
F f	*F f*	***F f***	F	F	F	S s	*S s*	***S s***	S	S	S
G g	*G g, g*	***G g***	G	G	G	T t	*T t*	***T t***	T	T	T
H h	*H h*	***H h***	H	H	H	U u	*U u*	***U u***	U	U	U
I i	*I i*	***I i***	I	I	I	V v	*V v*	***V v***	V	V	V
J j	*J j*	***J j***	J	J	J	W w	*W w*	***W w***	W	W	W
K k	*K k, k*	***K k***	K	K	K	X x	*X x, x*	***X x***	X	X	X
L l	*L l*	***L l***	L	L	L	Y y	*Y y, y*	***Y y***	Y	Y	Y
M m	*M m*	***M m***	M	M	M	Z z	*Z z*	***Z z***	Z	Z	Z

1：センチュリー・オールド，2：センチュリー・オールド(イタリック)，
3：センチュリー・ボールド(イタリック)，4：ピラネシ，
5：プリンス・スクリプト，6：タイプ・フェース．

2. ギリシア文字

文字	発音	文字	発音	文字	発音
Α α	ǽlfə	*Κ* κ	kǽpə	*Τ* τ	tau, tɔ:
Β β	bí:tə, béitə	*Λ* λ	lǽmdə	*Υ* υ	jú:psilən,
Γ γ	gǽmə	*Μ* μ	mju:, mu:		ju:psáilən
Δ δ	délta	*Ν* ν	nju:, nu:	*Φ* φ, ϕ	fai
Ε ε, ϵ	épsilən/-lan,	*Ξ* ξ	gzai, ksai	*Χ* χ	kai
	epsáilən	*Ο* ο	óumikrən,	*Ψ* ψ	*p*sai, psi:/-sai
Ζ ζ	zí:tə		oumái-	*Ω* ω	óumigə,
Η η	í:tə, éitə	*Π* π, ϖ	pai		oumégə/-mí:-
Θ θ, ϑ	θí:tə, θéitə	*Ρ* ρ	rou		
Ι ι	íoutə, aióutə	*Σ* σ, ς	sígmə		

3. ドイツ文字

活字体		筆記体		ラテン文字		発音
𝔄	𝔞	*A*	*a*	A	a	a:
𝔅	𝔟	*B*	*b*	B	b	be:
ℭ	𝔠	*C*	*c*	C	c	tse:
𝔇	𝔡	*D*	*d*	D	d	de:
𝔈	𝔢	*E*	*e*	E	e	e:
𝔉	𝔣	*F*	*f*	F	f	ɛf
𝔊	𝔤	*G*	*g*	G	g	ge:
ℌ	𝔥	*H*	*h*	H	h	ha:
ℑ	𝔦	*I*	*i*	I	i	i:
𝔍	𝔧	*J*	*j*	J	j	jɔt
𝔎	𝔨	*K*	*k*	K	k	ka:
𝔏	𝔩	*L*	*l*	L	l	ɛl
𝔐	𝔪	*M*	*m*	M	m	ɛm
𝔑	𝔫	*N*	*n*	N	n	ɛn
𝔒	𝔬	*O*	*o*	O	o	o:
𝔓	𝔭	*P*	*p*	P	p	pe:
𝔔	𝔮	*Q*	*q*	Q	q	ku:
ℜ	𝔯	*R*	*r*	R	r	ɛr
𝔖	ſ, 𝔰	*S*	*ſ, s*	S	s	ɛs
𝔗	𝔱	*T*	*t*	T	t	te:
𝔘	𝔲	*U*	*u*	U	u	u:
𝔙	𝔳	*V*	*v*	V	v	faʊ
𝔚	𝔴	*W*	*w*	W	w	ve:
𝔛	𝔵	*X*	*x*	X	x	iks
𝔜	𝔶	*Y*	*y*	Y	y	ˈypsi·lɔn
ℨ	𝔷	*Z*	*z*	Z	z	tsɛt
𝔄̈	𝔞̈	*Ä*	*ä*	Ä	ä	ˈa:ˈʊmlaʊt
𝔒̈	𝔬̈	*Ö*	*ö*	Ö	ö	ˈo:ˈʊmlaʊt
𝔘̈	𝔲̈	*Ü*	*ü*	Ü	ü	ˈu:ˈʊmlaʊt
	𝔠𝔥		*ch*		ch	ˈtse:ˈha:
	𝔠𝔨		*ck*		ck	ˈtse:ˈka:
	ß		*ß*		ß	ˈɛsˈtsɛt
	𝔱𝔷		*tz*		tz	ˈte:ˈtsɛt

4. ロシア文字

活字体		筆記体		発音	対応するラテン文字
А	а	*А*	*а*	á	a
Б	б	*Б*	*б*	bɛ́	b
В	в	*В*	*в*	vɛ́	v
Г	г	*Г*	*г*	gɛ́	g
Д	д	*Д*	*д*	dɛ́	d
Е	е	*Е*	*е*	jɛ́	e
Ё	ё	*Ё*	*ё*	jó	ë
Ж	ж	*Ж*	*ж*	ʒɛ́	ž
З	з	*З*	*з*	zɛ́	z
И	и	*И*	*и*	í	i
Й	й	*Й*	*й*	í krátkəjə	ĭ
К	к	*К*	*к*	ká	k
Л	л	*Л*	*л*	ɛ́l	l
М	м	*М*	*м*	ɛ́m	m
Н	н	*Н*	*н*	ɛ́n	n
О	о	*О*	*о*	ó	o
П	п	*П*	*п*	pɛ́	p
Р	р	*Р*	*р*	ɛ́r	r
С	с	*С*	*с*	ɛ́s	s
Т	т	*Т*	*т*	tɛ́	t
У	у	*У*	*у*	ú	u
Ф	ф	*Ф*	*ф*	ɛ́f	f
Х	х	*Х*	*х*	xá	h
Ц	ц	*Ц*	*ц*	tsɛ́	c
Ч	ч	*Ч*	*ч*	tʃʼɛ́	č
Ш	ш	*Ш*	*ш*	ʃá	š
Щ	щ	*Щ*	*щ*	ʃʼtʃʼá	šč
Ъ	ъ		*ъ*	tʼvʼórdɨj znák	”
Ы	ы		*ы*	ɨ	y
Ь	ь		*ь*	mʼáxkʼij znák	’
Э	э	*Э*	*э*	ɛ́	è
Ю	ю	*Ю*	*ю*	jú	yu
Я	я	*Я*	*я*	já	ya

数学 英和·和英辞典 増補版

検印省略

NDC 410.3

© 1979, 2016

1979 年 7 月 15 日　初版 1 刷発行
2011 年 9 月 15 日　初版76刷発行
2016 年 2 月 25 日　増補版 1 刷発行

編著者　小　松　勇　作
東京理科大学
数学教育研究所

発行者　南　條　光　章
東京都文京区小日向 4 丁目 6 番19号

印刷者　加　藤　純　男
東京都千代田区三崎町 2 丁目15番 6 号

発行所　東京都文京区小日向 4 丁目 6 番19号
電話東京 3947 局 2511 番(代表)
〒112-0006／振替 00110-2-57035 番
URL http://www.kyoritsu-pub.co.jp/
共立出版株式会社

印刷・加藤文明社／製本・ブロケード　Printed in Japan

一般社団法人
自然科学書協会
会員

ISBN 978-4-320-11150-9

JCOPY ＜出版者著作権管理機構委託出版物＞
本書の無断複製は著作権法上での例外を除き禁じられています．複製される場合は，そのつど事前に，出版者著作権管理機構（ＴＥＬ：03-3513-6969，ＦＡＸ：03-3513-6979，e-mail：info@jcopy.or.jp）の許諾を得てください．

学習・実力養成の好パートナー!!

▼“やさしく学べる”▼ シリーズ

やさしく学べる基礎数学 線形代数・微分積分
石村園子著…………A5判・246頁・本体2000円(税別)

やさしく学べる線形代数
石村園子著…………A5判・224頁・本体2000円(税別)

やさしく学べる微分積分
石村園子著…………A5判・230頁・本体2000円(税別)

やさしく学べるラプラス変換・フーリエ解析 増補版
石村園子著…………A5判・268頁・本体2100円(税別)

やさしく学べる微分方程式
石村園子著…………A5判・228頁・本体2000円(税別)

やさしく学べる統計学
石村園子著…………A5判・230頁・本体2000円(税別)

やさしく学べる離散数学
石村園子著…………A5判・230頁・本体2000円(税別)

100ページの文章術
—わかりやすい文章の書き方のすべてがここに—
酒井聡樹著…………A5判・100頁・本体1000円(税別)

これから論文を書く若者のために
【究極の大改訂版】
酒井聡樹著…………A5判・326頁・本体2700円(税別)

これから研究を始める高校生と指導教員のために
酒井聡樹著…………A5判・350頁・本体2600円(税別)

これからレポート・卒論を書く若者のために
酒井聡樹著…………A5判・242頁・本体1800円(税別)

これから学会発表する若者のために
—ポスターと口頭のプレゼン技術—
酒井聡樹著…………B5判・182頁・本体2700円(税別)

理数系のための技術英語練習帳
—さらなる上達を目指して—
金谷健一著…………B5判・254頁・本体2700円(税別)

これなら使える100例
技術英語論文の書き方
英語論文作成研究会編…A5判・232頁・本体2700円(税別)

詳解演習シリーズ

詳解 線形代数演習
鈴木七緒・安岡善則他共編‥本体2500円

詳解 微積分演習 I
福田安蔵・安岡善則他共編‥本体2200円

詳解 微積分演習 II
鈴木七緒・黒崎千代子他共編 本体2000円

詳解 微分方程式演習
福田安蔵・安岡善則他共編‥本体2500円

詳解 物理学演習 上
後藤憲一・山本邦夫他共編‥本体2500円

詳解 物理学演習 下
後藤憲一・西山敏之他共編‥本体2400円

詳解 物理/応用数学演習
後藤憲一・山本邦夫他共編‥本体3400円

詳解 力学演習
後藤憲一・神吉　健他共編‥本体2500円

詳解 電磁気学演習
後藤憲一・山崎修一郎共編‥本体2700円

詳解 理論/応用量子力学演習
後藤憲一・西山敏之他共編‥本体4200円

詳解 構造力学演習
彦坂　熙・崎山　毅他共著‥本体3700円

詳解 測量演習
佐藤俊朗編…………本体2500円

詳解 建築構造力学演習
蜂巣　進・林　貞夫著……本体3400円

詳解 機械工学演習
酒井俊道編…………本体2900円

詳解 材料力学演習 上
斉藤　渥・平井憲雄共著…本体3600円

詳解 材料力学演習 下
斉藤　渥・平井憲雄共著…本体3400円

詳解 制御工学演習
明石　一・今井弘之共著…本体4000円

詳解 流体工学演習
吉野章男・菊山功嗣他共著‥本体2900円

詳解 電気回路演習 上
大下眞二郎著…………本体3500円

詳解 電気回路演習 下
大下眞二郎著…………本体3500円

■各冊：A5判・176～454頁(税別価格)

http://www.kyoritsu-pub.co.jp/　**共立出版**　※価格は変更される場合がございます。